PERGAMON INTERNATIONAL LIBRARY
of Science, Technology, Engineering and Social Studies

The 1000-volume original paperback library in aid of education, industrial training and the enjoyment of leisure
Publisher: Robert Maxwell, M.C.

INTRODUCTION TO PHYSICS

Other Titles of Interest

BUCHDAHL:	Twenty Lectures on Thermodynamics
BUSCH & SCHADE:	Lectures on Solid State Physics
CLARK & STEPHENSON:	The Historical Supernovae
ELTON & MESSEL:	Time and Man
GINZBURG:	Problems in Undergraduate Physics Vol 1: Mechanics Vol 2: Electricity and Magnetism Vol 3: Optics Vol 4: Molecular Physics, Thermo-dynamics Atomic and Nuclear Physics
HALL:	Radiation and Life
HEY:	The Radio Universe, 2nd Edition
LANDAU:	General Physics: Mechanics and Molecular Physics
MEADOWS:	Stellar Evolution, 2nd Edition

This volume has been edited under the general supervision of
D. ter Haar

P. FRAUENFELDER and P. HUBER

INTRODUCTION TO PHYSICS

MECHANICS, HYDRODYNAMICS
THERMODYNAMICS

Translated by

F. S. LEVIN and J. L. WEIL

PERGAMON PRESS
OXFORD · NEW YORK · TORONTO
SYDNEY · PARIS · FRANKFURT

U.K.	Pergamon Press Ltd., Headington Hill Hall, Oxford OX3 0BW, England
U.S.A.	Pergamon Press Inc., Maxwell House, Fairview Park, Elmsford, New York 10523, U.S.A.
CANADA	Pergamon of Canada Ltd., 75 The East Mall, Toronto, Ontario, Canada
AUSTRALIA	Pergamon Press (Aust.) Pty. Ltd., P.O. Box 544, Potts Point, N.S.W. 2011, Australia
FRANCE	Pergamon Press SARL, 24 rue des Ecoles, 75240 Paris, Cedex 06, France
FEDERAL REPUBLIC OF GERMANY	Pergamon Press GmbH, 6424 Kronberg-Taunus, Pferdstrasse 1, Federal Republic of Germany

First English edition 1966
Reprinted 1978

Library of Congress Catalog Card No. 66-16367

This is a translation from the German of Vol. 1 of *Einfuhrung in die Physik* (2nd revised edition), published in Basel, 1963, by Ernst Reinhardt Verlag AG. The problems at the end of this volume are taken from *Problems in Undergraduate Physics, Volumes I and IV*, © 1965, Pergamon Press Ltd.

Printed in Great Britain by Bookmag, Henderson Road, Inverness

ISBN 0 08 013521 8 flexicover

CONTENTS

TRANSLATORS' PREFACE

IN PREPARING the English edition of this book, we have incorporated a number of changes, including those corrections and changes communicated to us by the authors. Section 28 of the German text, on the use of measuring instruments, has been omitted, and the section on heat transfer originally found in Chapter 11 has been placed at the end of Chapter 8. The order of Sections 79-4, 79-5, and 79-6 has been changed. Additional references and also more recently measured values of various physical quantities have been inserted both into the text and the tables. The mass scale based on $C^{12} = 12{\cdot}00000$ AMU and the absolute temperature scale based on the triple point of water have been introduced in accordance with recent international adoption. Discussions of the Miller indices and the fourteen space lattices have been added to Chapter 9 on Crystal Structure, since these concepts are much used in Britain and the U.S.A. Some other slight changes in the textual material have been made in an effort to make the presentation in English both simple and lucid. However, no major departures from the original text have been made.

In contrast to the German edition, numerous problems have been included in the present volume. These were selected from Volumes I and IV of the collection of problems compiled by Strelkov and Yakolev, and Ginzburg, Levin, Rabinovich, and Sivukhin respectively. The complete volumes of problems are being published in English translation by Pergamon Press. The addition of these problems makes the present text eminently well suited to the modern physics curricula now being developed in many places.

F. S. LEVIN
J. L. WEIL

Motto: Precise definitions form the framework of every quantitative science. They are the invisible guests at all objective discussions.

Daniel Bernoulli (1700–1782)

[*By courtesy of Miss Sophie Bernoulli, Basle*]

CHAPTER 1

INTRODUCTION

1. General Survey

An examination of the present state of scientific knowledge leads one to conclude that physical phenomena are extremely complex. The exact sciences introduce simplifications by using certain *basic principles*, which are postulated to be valid without exception. Such principles are characterized by their universality; their range of validity is independent of the nature of the phenomena to which they are applied. These principles have been verified by a vast number of experiments. Well-known examples are Newton's three laws of motion and the principle of the conservation of energy.

In addition to these basic principles, there are in all fields of physics certain *empirical laws*, whose formulations contain phenomenological constants, e.g., Hooke's law of elasticity and the law of refraction. The phenomenological constants can be related to one another by means of the basic principles. The following requirements restrict the applicability of empirical laws to processes occurring in our environment: (1) an understanding of the significance of the phenomenological constants and of the units in which they are expressed, (2) a knowledge of their range of validity, and (3) a knowledge of the basis on which the laws were formulated. (Often, this basis may change with time.)

In both the verbal and the mathematical formulations of a law, it is necessary that words and symbols have a clear, rigorous meaning. This unique meaning is obtained by means of definitions. Hence a definition is the precise formulation of a scientific concept. Quantities which can vary in space or time (most quantities can) must be defined in differential form.

To correlate data obtained in the laboratory with the results of applications of the basic principles and empirical laws to real processes, we must consider the theory of errors (see Section 7).

2. The Three Standards of Measurement

Space, *time*, and *matter* (mass) are the independent elements of mechanics. The units of space, time, and mass are then the independent basic units of measurement and thus have a fundamental significance. From them, we can derive all the other units of mechanics by using the basic principles and empirical laws.

Physical phenomena can be understood either on a qualitative level, by careful description without the use of mathematics, or on a quantitative level, by means of numerical verification of the relationships in question.

To measure a quantity means to compare it with a standard unit of that quantity. The fundamental units are established by means of definitions; that is, they are completely arbitrary. However, the definitions should be formulated so that the units are as exact as possible, can be easily reproduced, and lead to simple numerical values for the most commonly occurring measurements.

2-1 The Standard of Length

In the classical physics of Newton, the concept of space is understood to be intuitively obvious. It exists independently of time and of any objects embedded in it. A further postulate of classical physics, and one which is broadly supported by experience, is that the geometry of space is Euclidean.†

In Euclidean geometry, the measurement of space is based on a unit of length. A definition of the standard of length determines the units of area and volume as well as the unit of length.

DEFINITION. The standard of length is the *meter*. Redefined in 1960 as the length equal to 1,650,763.73 wavelength *in vacuo* of the radiation corresponding to the transition between the levels $2p_{10}$ and $5d_5$ of the isotope $^{86}_{36}Kr$. This definition abrogated the platinum-iridium meter bar as the standerd of length. One meter is equal to 39.37 inches.

2-2 The Standard of Time

In classical physics, time is postulated to be independent both of space and of any matter occupying it and is considered to be flowing uniformly. Hence any absolutely uniform, periodically recurring process can be used to determine the unit of time. The standard of time is related to the rotation of the earth about its axis, a motion known to be sufficiently regular for this purpose.‡

† The special and general theories of relativity relax these assumptions. In the special theory of relativity, spacelike and timelike quantities are interrelated. The general theory of relativity postulates that matter causes space to become distorted and gives rise to departures from Euclidean geometry.

‡ Through modern experimental techniques, periodic systems with much greater uniformity than the rotation of the earth can be utilized as clocks. In all probability, atomic clocks will eventually be used to establish the unit of time. For a brief discussion of the pertinent methods and also of the three fundamental units, see *The Fundamental Constants of Physics*, by E. R. Cohen, K. M. Crowe, and J. W. M. Dumond, Interscience Publishers, Inc., New York, 1957.

DEFINITION. The common standard of time is the *mean solar day*, which is the average length of time between two successive passages of the sun across a meridian at any arbitrary location.

The earth completes its elliptical orbit about the sun in 365.24 days. As seen from the sun, it regresses about 1 degree per day, and the sun, as seen by an observer on the earth, advances by about 1 degree per day. Therefore the true solar day is somewhat longer than the time it takes the earth to make one rotation. Since the speed of the earth in its elliptical path about the sun is different at different times of the year, the apparent daily motion of the sun varies in time. To obtain a constant mean solar day, one replaces the real sun by a fictitious, mathematical sun which is postulated to move with a constant speed in the equatorial plane (not in the plane of the ecliptic).

The standard of time is not subdivided decimally in classical physics (although such subdivisions are common in atomic and nuclear physics):

$$\begin{aligned} 1 \text{ mean solar day} &= 24 \text{ hours (hr)} \\ &= 24 \times 60 = 1440 \text{ minutes (min)} \\ &= 24 \times 60 \times 60 = 86{,}400 \text{ seconds (sec)}. \end{aligned}$$

DEFINITION. The *second* is 1/86,400 part of a mean solar day.†

2-3 The Standard of Mass

Matter occurs in many varied forms in the space–time structure of our world. Chemistry classifies these forms as follows:

(a) *Elements*, which are forms of matter that cannot be further decomposed by chemical means. Elements consist of atoms, of which there are 103 different species known as of 1964.

(b) *Compounds*, which are forms of matter that consist of chemically bound combinations of elements. The simplest example of a compound, and the basic building block for all compounds, is called a molecule. Most substances fall into the category of compounds.

(c) *Mixtures*, which are forms of matter that are composed of chemically unbound combinations of elements or compounds.

It is empirically known that all these kinds of matter have a number of common properties, which we will consider later. In place of the concept of matter, physics employs the concept of mass, since the latter has been

† A slightly different unit of time is used by astronomers for precision work. This unit is a *second*, and it is defined as 1/31,556,925.9747 of the tropical year for 1900. The two seconds are so nearly equivalent that the accumulated time difference of the two systems between 1900 and 1960 is about 30 seconds. For our purposes it makes no difference which unit is used. For a more complete discussion see G. M. Clemence, in *Revs. Modern Phys.* **29**, 2 (1957). Studies are in progress for redefining the *second* in terms of atomic physics standards.

shown to be independent of the kind of matter, of the position of matter in space, and of time. The choice of a standard of mass is dictated only by the requirement that it be chemically stable.

DEFINITION. The standard of mass is called the standard kilogram. It is the mass of a cylinder of platinum-iridium, which is kept at Sèvres, near Paris.

The rigorous definitions of the standards of length, time, and mass have been given great attention. Maintenance of the standards of length and of mass is the permanent responsibility of the International Bureau of Weights and Measures; the astronomical observatories of the world determine the standard of time.

In the *gravitational* system of measurement, the standard kilogram is used to define the unit of force. The mass of the standard kilogram is attracted by the mass of the earth with a force that depends on geographic latitude. The unit of force is defined as that force with which the earth attracts the mass of the standard kilogram at sea level and a latitude of 45°. This force is called a *kilogram force* or kgf. We will use this unit of force throughout our discussion of statics.

3. SUBUNITS

Most units are subdivided decimally. The notation in Table 1-1 is used to designate the decimal multiples.

TABLE 1-1

Factor	Prefix	Abbreviation	Example
10	deca-	D	decalumen
100	hecto-	H	hectoliter
1000	kilo-	k	kilometer
10^6	mega-	M	megawatt
10^{-1}	deci-	d	deciliter
10^{-2}	centi-	c	centimeter
10^{-3}	milli-	m	millivolt
10^{-6}	micro-	μ	microampere
10^{-9}	nano-	n	nanosecond
10^{-12}	pico-	p	picofarad

In addition to these, certain units in common use have been given special names:

$$
\begin{aligned}
1\,\mu &= 1 \text{ micron} &&= 10^{-6}\text{ m} = 10^{-3}\text{ mm},\\
1\,\text{m}\mu &= 1 \text{ millimicron} &&= 10^{-6}\text{ mm},\\
1\,\text{Å} &= 1 \text{ angstrom} &&= 10^{-8}\text{ cm},\\
1\,\text{f} &= 1 \text{ fermi} &&= 10^{-13}\text{ cm},\\
1\,\text{b} &= 1 \text{ barn} &&= 10^{-24}\text{ cm}^2.
\end{aligned}
$$

4. Quantities, Magnitudes, Units, and Dimensions

Algebra deals exclusively with *pure numbers*, such as a, b, x, y, etc. The symbol a represents an arbitrary number, it is an expression signifying the addition of a definite number of ones, or units. Physics, on the other hand, is concerned with *quantities*. Examples of physical quantities are: mass, m; time, t; temperature, T; velocity, $\boldsymbol{v}$; displacement, $\boldsymbol{s}$; force, $\boldsymbol{F}$; stress, σ; torque, $\boldsymbol{M}$.

Definition. A physical quantity is the product of a pure number (magnitude) and the unit of that quantity. The magnitude specifies the number of units present in the quantity.

Equations between magnitudes are called scalar equations and obey algebraic relations. The actual physical quantities themselves also obey equations, but these are not necessarily algebraic.

A quantity can be measured in different units of the same kind; e.g., time in seconds, minutes, hours, or days; and length in centimeters, meters, or kilometers. Relations exist between different units of the same kind, e.g.,

$$
\begin{aligned}
1\ \text{hr} &= 3600\ \text{sec},\\
1\ \text{km} &= 10^3\ \text{m} = 10^5\ \text{cm},\\
1\ \text{km/hr} &= 1000\ \text{m}/3600\ \text{sec} = 0.278\ \text{m/sec}.
\end{aligned}
$$

If we wish to avoid equations between units, then we must agree at the outset which units will characterize the fundamental quantities of length, time, and mass. In this way, we clearly define a system of measurement. There are three systems of measurement currently used in physics:

(1) The centimeter-gram-second system (cgs), in which the unit of length is the centimeter (cm), the unit of mass is the gram mass (gm), and the unit of time is the second (sec). (Units for the remaining physical quantities can then be expressed in terms of these three fundamental units.)

(2) The Giorgi system (mks), in which the unit of length is the meter (m), the unit of mass is the kilogram-mass (kgm), and the unit of time is the second (sec).

(3) The British system (fps), in which the unit of length is the foot (ft), the unit of mass is the pound-mass (lbm), and the unit of time is the second (sec). This system is related to, but different from, the British gravitational system used almost exclusively for engineering work.

Another system of measurement that is widely used today is the gravitational system, but it employs peculiar units. We shall discuss it when Newton's second law of motion is introduced.

Systems of measurement are somewhat arbitrary and are used merely to determine units of measurement. Equations relating physical quantities apply, however, for any arbitrary choice of units. Thus the units of calculated quantities may often occur in unfamiliar forms. As an example, consider density. The defining equation for density, ϱ, is $\varrho = m/V$, where m is the mass and V is the volume occupied by the mass. Choosing the gram as the unit of mass and the millimeter as the unit of length, we find, for the case of iron, $\varrho = 7.88 \times 10^{-3}$ gm/mm³. By employing the equation 1 mm³ = 10^{-3} cm³, we can express the preceding value for the density in the more common form of the cgs system:

$$\varrho = 7.88 \times 10^{-3}\ \text{gm}/10^{-3}\ \text{cm}^3 = 7.88\ \text{gm/cm}^3.$$

In mechanics, each physical quantity can be uniquely expressed in terms of the three fundamental quantities: length, time, and mass. The manner in which a quantity depends on the fundamental quantities is known as its dimension.

DEFINITION. The *dimension* of a physical quantity is the expression that specifies the power with which each of the fundamental quantities (length L, mass M, and time T) appears in the unit of that particular quantity; i.e. the dimension of x, denoted by $[x]$, is given by the equation $[x] = [L^a M^b T^c]$, where a, b, and c are positive or negative integers or zero.

The concept of dimension is more general than the concept of unit. Symbols for dimensions, as noted above, are placed in angular brackets. The expression for the unit can be obtained from the formula for the dimension, since we can replace the symbols L, M, and T by the units in the particular system of measurement. For example, in the case of a constant velocity, which is defined by the equation $v = s/t$, it follows that the dimension of v is $[v] = [LT^{-1}]$, so that the unit of velocity in the cgs system is 1 cm · sec^{-1} and in the mks system it is 1 m · sec^{-1}.

In addition, the formula for the dimension indicates how the units change when we transform to another set of arbitrary units by means of the equations satisfied by the fundamental units. As an example, let us consider acceleration, with dimension $[a] = [LT^{-2}]$. The unit is 1 cm · sec^{-2} in the cgs system. If we choose 1 km as the unit of length and 1 hr as the unit of time, the unit of acceleration becomes

$$1\ \text{km/hr}^2 = 10^5\ \text{cm}/3600^2\ \text{sec}^2 = 7.716 \times 10^{-3}\ \text{cm/sec}^2.$$

Finally, the formula for the dimension supplies information about the measuring process. The dimension of velocity, $[v] = [LT^{-1}]$, indicates that v can be determined from the ratio of a length to a time.

Two physical quantities having the same dimension (and therefore also the same units in a given measuring system) are not necessarily equivalent; it can only be said that their units transform in the same way when the fundamental units are changed. Thus work is the product of a force and a distance, whereas the product of a force and the length of a lever represents a torque.

Both quantities are products of force and length, having the dimension $[ML^2T^{-2}]$, but they are entirely different in their character: work is a scalar and torque is a vector (see Section 9-2).

Frequently we use the units of measurement or the dimension to test a formula which was obtained by a lengthy calculation. Only terms with the same dimension can occur in a series, since only quantities of the same kind can be added. The complete expression must have the same dimension as the quantity to be computed.

EXAMPLE. The height y of the center of mass S of a symmetrical T-shaped object of uniform mass (Fig. 1-1) is calculated from the formula

$$y = a^2b + cd^2 + 2abd/2(ab + cd).$$

This formula is dimensionally correct, for the numerator is a sum of terms of dimension $[L^3]$, the denominator is correspondingly of dimension $[L^2]$, and the quotient has the dimension of length. Numerical factors cannot be checked by means of dimensional arguments.

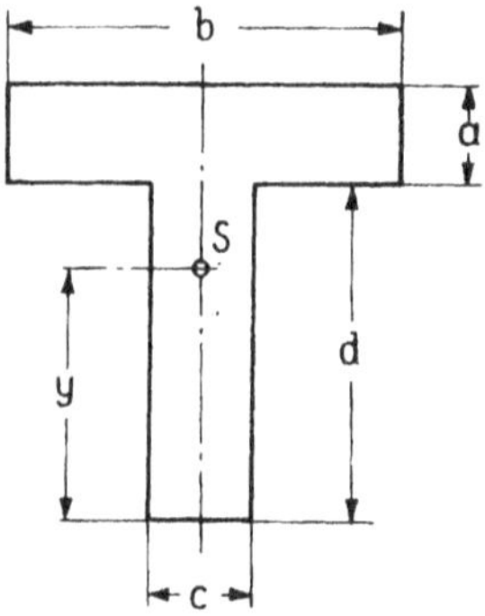

FIG. 1-1. Center of mass of a T-shaped object.

5. SCALARS AND VECTORS

The various kinds of quantities that occur in physics are distinguished by the number of statements necessary to define them unambiguously. For our purposes, it suffices to mention two kinds, scalars and vectors.

DEFINITION. Scalars are quantities that are completely defined by means of a magnitude and a unit.

Examples of scalars are mass (gm), time (sec), volume (cm^3), and temperature (°C).

DEFINITION. Vectors are quantities that are uniquely determined by means of a magnitude, a unit, and a direction in space.

Examples of vectors are position vector (***r***), velocity (***v***), force (***F***), and torque (***M***). The straight line along which a vector acts is called the line of action, or orientation, of the vector.

To clearly distinguish scalars from vectors, scalars will be printed in italics, and vectors in boldface. When only the magnitude of a vector is considered, the symbol will be placed between vertical bars or written in italics. For example the symbol $\boldsymbol{F}$ represents a force with a given magnitude and direction; F or $|\boldsymbol{F}|$ represents the magnitude of the force.

Scalars, which are determined by a magnitude and a unit, obey the laws of elementary algebra. The situation is more complicated in the case of vectors, since the direction of a vector must be considered, as well as its magnitude and units. Thus, a vector equation $\boldsymbol{A} = \boldsymbol{B}$ contains three statements.

(1) The magnitude of $\boldsymbol{A}$ must be equal to the magnitude of $\boldsymbol{B}$: $A = B$.

(2) The units of $\boldsymbol{A}$ must be equal to the units of $\boldsymbol{B}$.

(3) The direction of $\boldsymbol{A}$ must be the same as the direction of $\boldsymbol{B}$.

6. Vector Algebra

6-1 Graphical Representation and Measurement of Vectors

The simplest vector is the displacement vector $\boldsymbol{s}$, which connects a point A in space to a point B in space. It has a magnitude of s meters and is directed from A as origin to B as endpoint. It is represented by an arrow whose length is proportional to the magnitude of the vector. The direction of the vector is fixed by the orientation and the sense of the line representing it. The sense is indicated by the arrow at the end of the line. By a suitable choice of scale it is possible to represent any arbitrary vector by a displacement vector. For example, to represent the velocity $\boldsymbol{v} = 300$ m/sec, we choose a unit of 100 m/sec and draw, as in Fig. 1-2, an arrow in the prescribed direction with a length of three units.

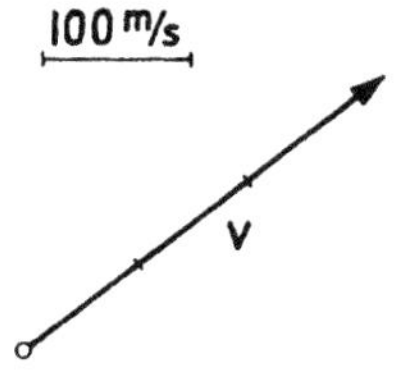

Fig. 1-2. A velocity vector.

6-2 Multiplication of a Vector by a Scalar

Let $\boldsymbol{c} = \alpha \boldsymbol{b}$, where α is a scalar quantity. The equality sign requires that the orientation of $\boldsymbol{c}$ be the same as that of $\alpha \boldsymbol{b}$. Multiplication of a vector by a scalar changes only its magnitude, never its direction. If the scalar is a pure number, the unit is unaltered, but if the scalar has dimensions, the unit is changed.

We obtain a vector of magnitude b in a prescribed direction when the magnitude b is multiplied by a unit vector in the given direction. Unit vectors, which always have unit magnitude, will be here characterized by the subscript 0. A vector of magnitude b lying along the tangent to a curve, for example, will be expressed as $\boldsymbol{b} = b\boldsymbol{t}_0$, where $\boldsymbol{t}_0$ is a unit vector in the direction of the tangent. However, should the vector $\boldsymbol{b}_1$ lie along the normal to a curve, it will be expressed as $\boldsymbol{b}_1 = b\boldsymbol{n}_0$, where $\boldsymbol{n}_0$ is a unit vector normal to the curve.

6-3 Graphical Addition of Two Vectors Acting at the Same Point

In Fig. 1-3 the two vectors $\boldsymbol{A}$ and $\boldsymbol{B}$ act at the point P. Their sum is given by $\boldsymbol{A} + \boldsymbol{B} = \boldsymbol{R}$, where $\boldsymbol{R}$ is the resultant vector and $\boldsymbol{A}$ and $\boldsymbol{B}$ are the components of $\boldsymbol{R}$. This equation states (1) that $\boldsymbol{A}$ and $\boldsymbol{B}$ are the same kind of quantity, since only similar kinds of quantities may be added together; and (2) that the resultant is also a vector which singly produces the same effect as the two components acting together. We can construct $\boldsymbol{R}$ by the following process. The vector $\boldsymbol{B}$ is displaced so that it begins at the endpoint of $\boldsymbol{A}$. The endpoint of $\boldsymbol{B}$ is now represented by the point Q. The vector $\boldsymbol{R}$ which effects the displacement from P to Q in one step is the resultant. Hence, we obtain the following theorem.

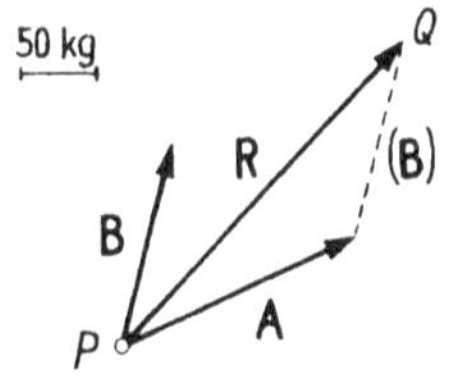

FIG. 1-3. Addition of two vectors acting at the same point. $A = 175$ kg, $B = 135$ kg, $R = 280$ kg.

ADDITION THEOREM. The resultant of two vectors acting at the same point extends from the origin of the first vector to the endpoint of the second, displaced, vector.

Note. For the purpose of constructing diagrams, parentheses will always be used to distinguish the displaced vector $(\boldsymbol{B})$ from the vector $\boldsymbol{B}$. Incorrect results are often obtained in statics because the displaced vectors are drawn in place of the original vectors.

The commutative law of algebra also holds for vector addition. If one first constructs $\boldsymbol{B}$ with its given magnitude and direction, and then adds $(\boldsymbol{A})$ to it, it is clear on geometrical grounds that the commutative law is obeyed:

$$\boldsymbol{R} = \boldsymbol{A} + \boldsymbol{B} = \boldsymbol{B} + \boldsymbol{A}.$$

The sum of two vectors is therefore independent of their order. We readily see that this theorem can be extended to includeaddition of an arbitrary number of component vectors (see Fig. 1-4).

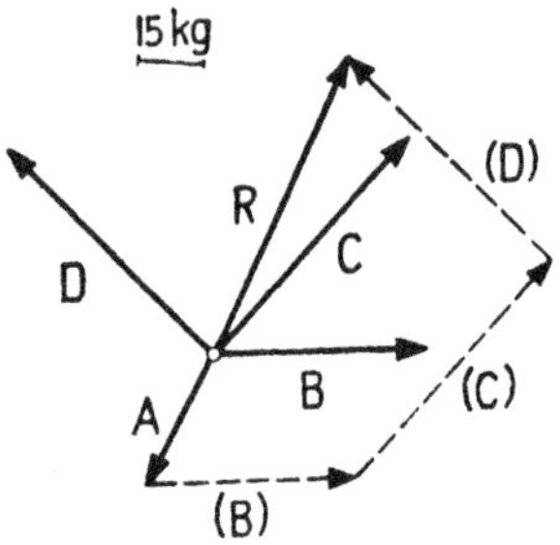

FIG. 1-4. Addition of several vectors acting at the same point. $A = 30$ kg, $B = 45$ kg, $C = 60$ kg, $D = 60$ kg, $R = 66$ kg.

6-4 *Subtraction of Two Vectors Acting at the Same Point*

Subtraction of a vector is addition with the sense of the arrow reversed. Subtraction is thus reduced to addition (see Fig. 1-5).

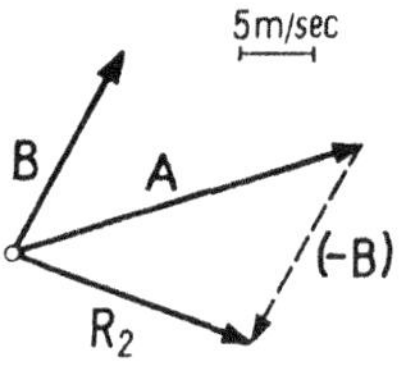

FIG. 1-5

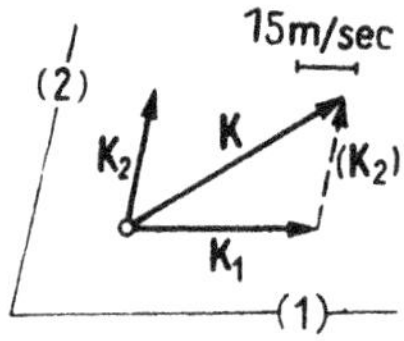

FIG. 1-6

FIG. 1-5. Subtraction of two vectors acting at the same point. $A = 25$ m/sec, $B = 15$ m/sec, $R = 17.3$ m/sec.

FIG. 1-6. Reduction of a vector to components in a plane. $K = 60$ m/sec, $K_1 = 40$ m/sec, $K_2 = 30$ m/sec.

6-5 *Decomposition of a Vector into Two Components Lying in Prescribed Directions*

Any vector $\boldsymbol{K}$ can be interpreted as the resultant of two components $\boldsymbol{K}_1$ and $\boldsymbol{K}_2$. If the directions of the components are given, then $\boldsymbol{K}$ in turn can be separated into these components.

In Fig. 1-6, let the specified component directions be labeled by (1) and (2). From the addition theorem it follows that lines in the directions of (1) and (2) must be drawn through the origin and endpoint of $\boldsymbol{K}$, respectively. We extend these auxiliary lines until they intersect. Then the component $\boldsymbol{K}_1$ must run from the origin of $\boldsymbol{K}$ to the point of intersection, while $(\boldsymbol{K}_2)$ extends from the point of intersection to the endpoint of $\boldsymbol{K}$ in such a way that $\boldsymbol{K}$ is the sum $\boldsymbol{K}_1 + \boldsymbol{K}_2$. It is to be understood that $\boldsymbol{K}_2$ also acts at the origin of $\boldsymbol{K}$.

The decomposition of a vector into more than two components in specified directions in a plane is possible in infinitely many ways, since the problem is not uniquely defined. In Fig. 1-7 we see two possible methods of decomposing a vector K into three components with prescribed directions.

While the plane decomposition of a vector into more than two components with given orientations is ambiguous, the spatial decomposition of a vector into three noncoplanar components with specified directions is again unique. The two-dimensional character of a plane and the three-dimensional character of space are reflected in this fact. However, the purely graphical methods used up to now are no longer suitable for spatial systems of vectors; henceforth, figures drawn in perspective will be necessary. Here, the analytical treatment of vector quantities has important advantages.

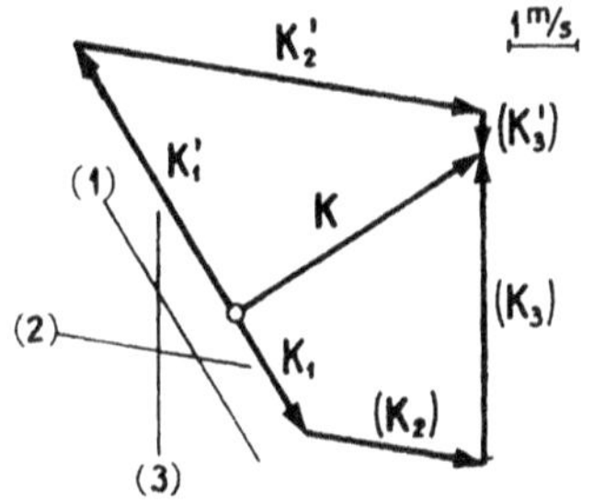

FIG. 1-7. Two possible decompositions of a vector into three components. $K_1 = 2.00$ m/sec, $K_2 = 2.70$ m/sec, $K_3 = 4.40$ m/sec, $K_1' = 4.50$ m/sec, $K_2' = 6.20$ m/sec, $K_3' = 0.60$ m/sec, $K = 4.3$ m/sec.

6-6 Analytical Representation of Vectors in a Cartesian Coordinate System

Systems of coplanar vectors. A system of n similar vectors acts at the point P (Fig. 1-8). Let us choose a cartesian coordinate system with the origin at P and decompose each vector A_i into its rectangular components parallel to the coordinate axes. This can be done analytically as soon as we know the angle α_i that the vector A_i makes with the positive x-axis. Thus for all quadrants, we have

$$A_{ix} = A_i \cos \alpha_i, \quad A_{iy} = A_i \sin \alpha_i .$$

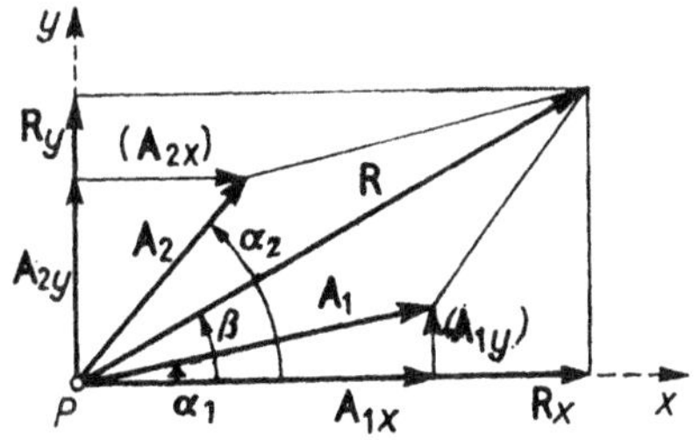

FIG. 1-8. Cartesian components of vectors in a plane.

Since all x-components have the same orientation (irrespective of sign), and the same is true for all y-components, the components along each of the axes may be added as scalars, yielding

$$\sum_i A_{ix} = R_x = \sum_i A_i \cos \alpha_i, \quad \sum_i A_{iy} = R_y = \sum_i A_i \sin \alpha_i.$$

Thus R_x and R_y are the components of the resultant with respect to this coordinate system. Consequently

$$R = \sqrt{(R_x^2 + R_y^2)} \quad \text{and} \quad \tan \beta = R_y/R_x,$$

where β is the angle that $\boldsymbol{R}$ makes with the positive x-axis. Since all vectors act at the origin, the resultant also acts at this point.

Spatial systems of vectors. Let n spatially distributed vectors $\boldsymbol{A}_i$ act at a point P. Let us choose a cartesian coordinate system with origin at P and determine the three angles α_i, β_i, and γ_i that each vector $\boldsymbol{A}_i$ makes with the positive coordinate axes. Each vector can then be uniquely decomposed into three components A_{ix}, A_{iy}, and A_{iz}. This decomposition results from the orthogonal projection of the vector $\boldsymbol{A}_i$ on the three coordinate axes:

$$A_{ix} = A_i \cos \alpha_i, \quad A_{iy} = A_i \cos \beta_i, \quad A_{iz} = A_i \cos \gamma_i.$$

Furthermore, since $A_{ix}^2 + A_{iy}^2 + A_{iz}^2 = A_i^2$, it follows that

$$\cos^2 \alpha_i + \cos^2 \beta_i + \cos^2 \gamma_i = 1,$$

which means that only two of the three directions cosines are independent; the third can be computed from the first two. This implies that only one unique decomposition of the vector is possible (Fig. 1-9), since the specification of two angles and a length completely defines a vector.

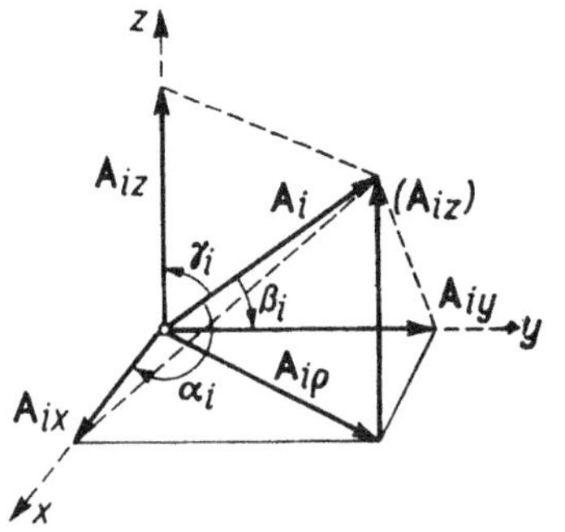

FIG. 1-9. Cartesian components of a vector in space.

The components of the resultant are found by scalar addition of the components along the three coordinate axes:

$$R_x = \sum_i A_{ix} = \sum_i A_i \cos \alpha_i, \qquad R_y = \sum_i A_{iy} = \sum_i A_i \cos \beta_i,$$
$$R_z = \sum_i A_{iz} = \sum_i A_i \cos \gamma_i.$$

From this we find the magnitude R of the resultant, and the angles φ_i that $\boldsymbol{R}$ makes with the coordinate axes:

$$R^2 = R_x^2 + R_y^2 + R_z^2$$

and

$$\cos\varphi_1 = R_x/R, \quad \cos\varphi_2 = R_y/R, \quad \cos\varphi_3 = R_z/R.$$

6-7 The Nature of Vectors

Vectors are quantities that are uniquely defined by their three components with respect to an arbitrary cartesian coordinate system and by specification of their units. The component representation of a vector is an expression relating the vector to the three unit vectors $\boldsymbol{e}_x$, $\boldsymbol{e}_y$, and $\boldsymbol{e}_z$ that span the cartesian coordinate system:

$$\boldsymbol{A} = A_x\boldsymbol{e}_x + A_y\boldsymbol{e}_y + A_z\boldsymbol{e}_z = \boldsymbol{A}_x + \boldsymbol{A}_y + \boldsymbol{A}_z.$$

We call the scalar quantities A_x, A_y, and A_z the magnitudes of the components and the vector quantities $\boldsymbol{A}_x$, $\boldsymbol{A}_y$, and $\boldsymbol{A}_z$ the vector components.

A vector is a quantity that is independent of the choice of a particular coordinate system. Therein lies the importance of the vector concept. On the other hand, the three vector components $\boldsymbol{A}_x$, $\boldsymbol{A}_y$, and $\boldsymbol{A}_z$ refer to the description in terms of a particular coordinate system; these components are dependent on the choice of coordinates. In a new coordinate system with unit vectors $\boldsymbol{e}'_x$, $\boldsymbol{e}'_y$, and $\boldsymbol{e}'_z$, the vector has the new components A'_x, A'_y, and A'_z, and is represented by the vector equation

$$\boldsymbol{A} = A'_x\boldsymbol{e}'_x + A'_y\boldsymbol{e}'_y + A'_z\boldsymbol{e}'_z.$$

The linear combination of the magnitude of the components and the corresponding unit vectors can be transformed from one coordinate system to another and always represents the same vector. This invariance property of a vector under coordinate transformations is the heart (or perhaps some totally different phrase) of the vector concept.

6-8 Multiplication of Vectors

Let $\boldsymbol{A}$ and $\boldsymbol{B}$ be two arbitrary vectors. Both quantities are invariant under coordinate transformations. If a product of two vectors is to be defined, it must also be invariant under a transformation of coordinates. Only then can it claim to be of physical significance. The refined description of a vector, using tensor algebra, shows that there exist only two possibilities for which the product of two vectors is an invariant. They are (a) the *scalar* product, $\boldsymbol{A}\cdot\boldsymbol{B}$, and (b) the *vector* product, $\boldsymbol{A}\times\boldsymbol{B}$. The scalar product is a scalar quantity, while in three-dimensional space, the vector product yields a quantity that can be interpreted as a vector. It is, however, a vector having a completely different character from the type discussed above, since it is associated with a particular direction of rotation. It is no longer invariant under a transformation that converts a right-handed coordinate system to a left-handed one. Such a transformation is called an inversion. Vectors that are invariant under inversions are called *polar vectors*; those that are not invariant, like the vector product, are called *axial vectors*.

The scalar product of two vectors. Let $\boldsymbol{A}$ and $\boldsymbol{B}$ be two vectors acting at the same point and intersecting at an angle φ (Fig. 1-10). The angle φ is that

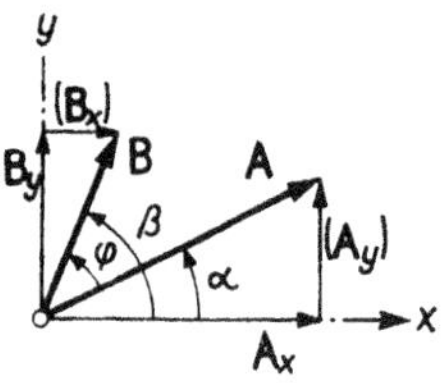

FIG. 1-10. The scalar product.

through which $\boldsymbol{A}$ must be rotated (in the positive sense) to make it parallel to $\boldsymbol{B}$.

DEFINITION. The *scalar product* of two vectors is the product of their magnitudes multiplied by the cosine of the angle between them:

$$\boldsymbol{A} \cdot \boldsymbol{B} = AB \cos \varphi .$$

This product is a scalar.

The properties of the scalar product are:

(a) The scalar product of a vector with itself is the same as the square of its magnitude:

$$\boldsymbol{A} \cdot \boldsymbol{A} = A^2 \cos 0° = A^2 .$$

(b) The scalar product of two unit vectors in a cartesian coordinate system is either 1 or 0:

$$\boldsymbol{e}_i \cdot \boldsymbol{e}_k = \begin{cases} 1, & \text{for} \quad i = k \\ 0, & \text{for} \quad i \neq k \end{cases} \qquad (i, k = x, y, z).$$

(c) In a cartesian coordinate system, the scalar product equals the sum of the products of the magnitudes of those components which have the same index.

To prove (c) let us consider two vectors $\boldsymbol{A}$ and $\boldsymbol{B}$ inclined at an angle φ (Fig. 1-10). The angles of the vectors measured from the positive x-axis in a plane rectangular coordinate system are α and β. Thus $\varphi = \beta - \alpha$ and

$$\begin{aligned} \boldsymbol{A} \cdot \boldsymbol{B} &= AB \cos (\beta - \alpha) \\ &= AB[\cos \beta \cos \alpha + \sin \beta \sin \alpha] \\ &= (A \cos \alpha)\,(B \cos \beta) + (A \sin \alpha)\,(B \sin \beta); \\ \boldsymbol{A} \cdot \boldsymbol{B} &= A_x B_x + A_y B_y . \end{aligned}$$

For three dimensions we have

$$\boldsymbol{A} \cdot \boldsymbol{B} = (A_x \boldsymbol{e}_x + A_y \boldsymbol{e}_y + A_z \boldsymbol{e}_z) \cdot (B_x \boldsymbol{e}_x + B_y \boldsymbol{e}_y + B_z \boldsymbol{e}_z),$$

and applying the product rule for unit vectors in a cartesian coordinate system, we obtain

$$\boldsymbol{A} \cdot \boldsymbol{B} = A_x B_x + A_y B_y + A_z B_z .$$

(d) A scalar product can be zero only if one of the magnitudes is zero or if the vectors are perpendicular ($\cos 90° = 0$).

(e) The two factors in a scalar product are interchangeable, i.e., it obeys the commutative law:

$$\boldsymbol{A} \cdot \boldsymbol{B} = \boldsymbol{B} \cdot \boldsymbol{A},$$

since $\cos \varphi = \cos (360° - \varphi)$.

(f) The scalar product obeys the distributive law:

$$\boldsymbol{A} \cdot (\boldsymbol{B} + \boldsymbol{C}) = \boldsymbol{A} \cdot \boldsymbol{B} + \boldsymbol{A} \cdot \boldsymbol{C}.$$

The proof follows by inspection of the above product in the component form.

The vector product of two vectors. The vector product $\boldsymbol{A} \times \boldsymbol{B}$ is a vector which is perpendicular to the plane determined by the vectors $\boldsymbol{A}$ and $\boldsymbol{B}$ (Fig. 1-11). It has a magnitude $AB \sin \varphi$. Its sense is that of the forward motion of a right-hand screw as $\boldsymbol{A}$ turns toward $\boldsymbol{B}$ (right-hand rule,† see Section 9). If $\boldsymbol{n}_0$ is a unit vector whose direction is defined by the right-hand rule, then $\boldsymbol{A} \times \boldsymbol{B} = AB \sin \varphi \boldsymbol{n}_0$.

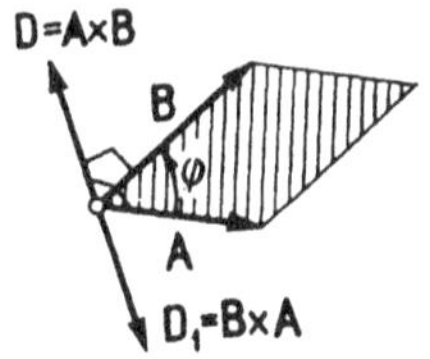

FIG. 1-11. The vector product and its interpretation as an area.

Properties of the vector product are:

(a) A vector product is zero if either of the two vectors is zero, or if the two vectors are either parallel or antiparallel.

(b) Interchanging the order of the factors in a vector product causes the product to change sign; the vector assumes the opposite sense: $\boldsymbol{A} \times \boldsymbol{B} = -\boldsymbol{B} \times \boldsymbol{A}$. To prove this recall that $\sin \varphi = -\sin(2\pi - \varphi)$. The commutative law is not valid for the vector product.

† If one turns the x-axis toward the y-axis, then the positive z-axis indicates the forward motion of a right-hand screw.

(c) The vector product of a unit vector with itself is zero. In a right-handed cartesian system we have†

$$e_x \times e_y = e_z, \quad e_y \times e_z = e_x, \quad e_z \times e_x = e_y.$$

The sign is preserved under cyclic permutation of the indices x, y, z (Fig. 1-12).

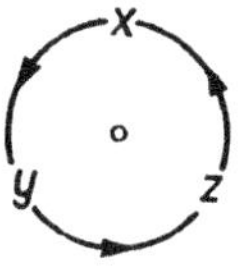

FIG. 1-12. Cyclic permutations.

(d) The vector product can be represented using components. A rectangular system of coordinates is fixed by the three unit vectors e_x, e_y, and e_z. Multiplying the vectors A and B in their component representations gives

$$\begin{aligned} A \times B &= (A_x e_x + A_y e_y + A_z e_z) \times (B_x e_x + B_y e_y + B_z e_z) \\ &= A_x B_x (e_x \times e_x) + A_y B_y (e_y \times e_y) + A_z B_z (e_z \times e_z) \\ &\quad + (A_x B_y - A_y B_x)(e_x \times e_y) + (A_y B_z - A_z B_y)(e_y \times e_z) \\ &\quad + (A_z B_x - A_x B_z)(e_z \times e_x). \end{aligned}$$

The first three terms are zero, and using the rules for the vector product of unit vectors, we obtain

$$A \times B = (A_y B_z - A_z B_y)\, e_x + (A_z B_x - A_x B_z)\, e_y + (A_x B_y - A_y B_x)\, e_z.$$

The cyclic permutation of indices should be noted.

An elegant condensation of this result is the representation of the vector product as a determinant. We write the three unit vectors as the first row of a third-rank determinant. For the second and third rows, we take the three components of each of the vectors entering the vector product. The component representation is thus obtained by expanding the determinant in terms of the minors of the first row:

$$\begin{aligned} A \times B &= \begin{vmatrix} e_x & e_y & e_z \\ A_x & A_y & A_z \\ B_x & B_y & B_z \end{vmatrix} \\ &= (A_y B_z - A_z B_y)\, e_x + (A_z B_x - A_x B_z)\, e_y + (A_x B_y - A_y B_x)\, e_z. \end{aligned}$$

The area of a parallelogram. Two displacement vectors A and B determine a parallelogram whose area is $AB \sin \varphi$. This suggests representing the area of the parallelogram by the vector product of the vectors generating it. The vector $A \times B$ representing the area is perpendicular to the plane of the

† Unless otherwise stated, we shall assume all coordinate systems to be right-handed.

parallelogram. This agrees with the general features of a surface since the direction of the vector area is different for the orderings $\boldsymbol{A}$, $\boldsymbol{B}$ and $\boldsymbol{B}$, $\boldsymbol{A}$. $\boldsymbol{A} \times \boldsymbol{B}$ represents the upper surface of the parallelogram and $\boldsymbol{B} \times \boldsymbol{A}$ the lower surface.

The volume of a parallelepiped. Three noncoplanar displacement vectors $\boldsymbol{A}$, $\boldsymbol{B}$ and $\boldsymbol{C}$ determine a parallelepiped. If the parallelogram $\boldsymbol{A} \times \boldsymbol{B}$ is considered as the base of the parallelepiped, the vector area of the parallelogram is perpendicular to the base. The projection of $\boldsymbol{C}$ onto the vector $\boldsymbol{A} \times \boldsymbol{B}$ is the height of the parallelepiped, so that the volume can be represented as the scalar product of the surface vector and $\boldsymbol{C}$. The ordering of the factors in this scalar product is arbitrary,

$$(\boldsymbol{A} \times \boldsymbol{B}) \cdot \boldsymbol{C} = \boldsymbol{C} \cdot (\boldsymbol{A} \times \boldsymbol{B}),$$

but the noncommutative character of the vector product does affect the result:

$$(\boldsymbol{A} \times \boldsymbol{B}) \cdot \boldsymbol{C} = -(\boldsymbol{B} \times \boldsymbol{A}) \cdot \boldsymbol{C}.$$

A cyclic interchange of the vectors preserves the sign of the volume:

$$\boldsymbol{A} \cdot (\boldsymbol{B} \times \boldsymbol{C}) = \boldsymbol{B} \cdot (\boldsymbol{C} \times \boldsymbol{A}) = \boldsymbol{C} \cdot (\boldsymbol{A} \times \boldsymbol{B}).$$

These formulas all represent the same parallelepiped: in the first expression $\boldsymbol{B} \times \boldsymbol{C}$ is chosen as the base, in the second, $\boldsymbol{C} \times \boldsymbol{A}$, and in the third, $\boldsymbol{A} \times \boldsymbol{B}$. The magnitude of the volume is obtained from the third-rank determinant formed from the components of the three vectors, as follows readily from he component representation of the vector product:

$$V = \begin{vmatrix} A_x & A_y & A_z \\ B_x & B_y & B_z \\ C_x & C_y & C_z \end{vmatrix}.$$

The triple vector product. In physics we often find a twofold vector product. That such a quantity represents a vector is shown in the following derivation. Using the determinantal form, we have

$$\boldsymbol{A} \times (\boldsymbol{B} \times \boldsymbol{C}) = \begin{vmatrix} \boldsymbol{e}_x & \boldsymbol{e}_y & \boldsymbol{e}_z \\ A_x & A_y & A_z \\ B_yC_z - B_zC_y & B_zC_x - B_xC_z & B_xC_y - B_yC_x \end{vmatrix}.$$

Writing out the x-component of the vector product, we obtain

$$\begin{aligned}[A_y(B_xC_y - B_yC_x) - A_z(B_zC_x - B_xC_z)]\, \boldsymbol{e}_x \\ = [B_x(A_yC_y + A_zC_z) - C_x(A_yB_y + A_zB_z)]\, \boldsymbol{e}_x.\end{aligned}$$

Adding and subtracting $A_xB_xC_x$, we obtain,

$$[B_x(A_xC_x + A_yC_y + A_zC_z) - C_x(A_xB_x + A_yB_y + A_zB_z)]\, \boldsymbol{e}_x.$$

An analogous computation puts the y and z components in exactly the same form:

$$[B_y(A_xC_x + A_yC_y + A_zC_z) - C_y(A_xB_x + A_yB_y + A_zB_z)]\, e_y$$

and

$$[B_z(A_xC_x + A_yC_y + A_zC_z) - C_z(A_xB_x + A_yB_y + A_zB_z)]\, e_z.$$

The terms in the parentheses are clearly the scalar products $\boldsymbol{A}\cdot\boldsymbol{C}$ and $\boldsymbol{A}\cdot\boldsymbol{B}$, so that the result can be expressed as

$$A \times (B \times C) = B(A \cdot C) - C(A \cdot B),$$

or, noting the symmetry of the scalar product, as

$$A \times (B \times C) = B(C \cdot A) - C(B \cdot A).$$

The formulas are invariant under cyclic permutations of the three vectors. Since the result is the difference of two vectors, it also is a vector.

7. Theory of Errors

A consideration of the *errors* associated with a measured quantity leads to the following important results.

(1) It shows which of the several factors in a measurement makes the main contribution to the total error, i.e., it indicates how the method of measurement can be improved in order to reduce the error.

(2) It enables us to decide whether a result is consistent with an established law, within the limits of error. If every result of a particular measurement differs by more than the limits of error from the result predicted by the law, then a significant discovery has been made.

7-1 The Concept of Error

All physical quantities are determined by measurement; i.e. by comparing them with the units of measurement established by means of definitions. The true value of a quantity cannot be known from a measurement; indeed, there is no physical meaning in asking about the true value of a measured quantity. Even if the limits of sensitivity of the measuring instrument and of the sense organs are reached in carrying out a measurement, the results of several independent measurements will differ from each other by a small amount. Two measurements are considered to be independent if they have not been influenced in any way by either the observer or the measuring instrument. Any choice may be made among the various independent results of a measurement, since all of these results are equally valid. However, the following points should be noted in this connection: (1) Errors arising from mistakes are to be excluded. (2) Measured quantities should not contain systematic errors. Systematic errors can be detected by the presence of a bias in the measured values of a known quantity. This means that some undetermined effects are present in the measuring process, which must be ascertained and either eliminated or taken into account.

Measured values fluctuate about some *mean value.* The cause of this fluctuation lies in the imperfections of the measuring apparatus and the sense organs that are employed. A prerequisite for a suitable method of measurement is that the *random* errors be small compared with the mean value of the measured quantity.

POSTULATE. If n independent measurements of a quantity are made, with results l_i, the *most probable value* of the measured quantity is equal to the arithmetic *mean* $\bar{l}$ of all measured values l_i:†

$$\bar{l} = (1/n)\sum_{i=1}^{n} l_i = (1/n)(l_1 + l_2 + \cdots + l_n).$$

Physics is concerned only with these most probable values.

DEFINITION. The *deviation* Δ_i of a single measured quantity is the difference between the measured value and the mean value:

$$\Delta_i = l_i - \bar{l}.$$

Deviations have the same dimension as the measured quantity.

EXAMPLE. The temperature of a room is measured with six different thermometers and the following results are found:

t_i	Δ_i	Δ_i^2
36.14	−0.027	729×10^{-6}
36.21	+0.043	1849×10^{-6}
36.18	+0.013	169×10^{-6}
36.22	+0.053	2809×10^{-6}
36.11	−0.057	3249×10^{-6}
36.14	−0.027	729×10^{-6}
$\bar{t} = 36.167$°C	$\sum \Delta_i = -0.002$°C	$\sum \Delta_i^2 = 9534 \times 10^{-6}$ (°C)^2

REMARK. For the purpose of calculating the deviations, the mean value is computed to one more significant figure than the single values. However the final result may be presented with more significant figures than the individual measurements *only* if the number of measurements is sufficiently large; i.e., n additional significant figures may be included in the result if at least 10^n individual measurements have been made.

The sum of the deviations of a series of measurements must be zero:

$$\sum_{i=1}^{n} \Delta_i = \sum_{i=1}^{n} (l_i - \bar{l}) = \left(\sum_{i=1}^{n} l_i\right) - n\bar{l} = 0,$$

This is not true in the preceding example because the mean value has been rounded off.

† The mean value is designated by a bar over the symbol representing the quantity.

Suppose that one additional measurement of the temperature were to be made. The expected deviation of this measurement from the previously determined mean can be estimated using a quantity called the *standard deviation*, σ:

$$\sigma = \sqrt{[\sum_i \Delta_i^2/(n-1)]},$$

where n is the number of measurements. If the measurements have a normal distribution, there is a 68 per cent probability that the new measurement lies between $\bar{t} - \sigma$ and $\bar{t} + \sigma$.

Now suppose that a whole new series of n measurements of the temperature were to be made. The expected deviation of the new mean $\bar{t}'$ from the old mean $\bar{t}$, can be estimated using a quantity called the *standard error of the mean*, $\bar{\sigma}$:

$$\bar{\sigma} = \sigma/\sqrt{n}.$$

If the measurements have a normal distribution, there is a 68 per cent probability that $\bar{t}'$ lies in the interval, $\bar{t} \pm \bar{\sigma}$.

For the series of measurements above,

$$\bar{\sigma} = \sqrt{(9534 \times 10^{-6}/30)} = 17.8 \times 10^{-3}\,^{\circ}\mathrm{C}.$$

The result of this series of measurements is presented in the form $t = 36.17 \pm 0.02^{\circ}\mathrm{C}$.

It should be noted that although the standard error of the mean, $\bar{\sigma}$, can be calculated for a series of independent measurements, it is an indeterminate quantity for a single measurement. It is often impossible to measure a quantity more than once, in which case the error can only be estimated (see Section 7-2).

DEFINITION. The *relative error* r of a quantity is the ratio of the error to the mean value. For the relative error of a single measurement,

$$r_i = \Delta_i/\bar{l},$$

and for the relative error of the mean value,

$$r = \bar{\sigma}/\bar{l},$$

The errors dealt with above are random observational errors made in measuring the magnitude of a quantity. In addition to these errors, we must always consider the calibration errors of the measuring instruments (see Section 7-3).

7-2 Estimation of the Error in a Single Measurement

A single measurement has no assignable limits of error. However, the result of the measurement must have a certain degree of reliability assigned to it. To do this, we estimate the precision with which the reading was made, the fluctuations of the pointer of a measuring instrument, etc. Considering the total of these factors as the deviation, we divide it by the measured value and obtain the relative error. Since these errors are not known exactly, they should not be specified to more than two significant figures.

7-3 Errors of Derived Quantities; Propagation of Error

Most of the results of measurements are used in functional relationships to determine new quantities. Such indirectly determined quantities are called *derived quantities*. We shall show how the error of the derived quantity can be calculated from the errors of the component quantities.

Sums and differences. When a physical quantity X is obtained from the *sums* and/or *differences* of independent observations on two or more physical quantities,

$$X = a - b + c \ldots,$$

the standard error of the derived quantity is

$$\bar{\sigma}_x = \sqrt{(\bar{\sigma}_a^2 + \bar{\sigma}_b^2 + \bar{\sigma}_c^2 + \ldots)},$$

where $\bar{\sigma}_a$, $\bar{\sigma}_b$, ... are the standard errors of the mean values of the measured quantities. The same expression can be applied to the standard deviations or to the estimated errors of single measurements.

EXAMPLES.

(a) $(50 \pm 3) + (55 \pm 4) = 105 \pm 5.$

(b) $(200 \pm 3) - (204 \pm 4) = -4 \pm 5.$

In example (b), $|r_x| = 1$, from which we see the danger of determining a quantity from the difference of two numbers nearly equal in value. The result is then not very useful, even though the single measurements can be very accurate. When applying so-called *difference methods*, the careful experimenter will ensure that the differences are not small. Commonly occurring cases where this state of affairs invalidates the measurement are:

(a) difference weighing (a heavy container with light contents),

(b) weighing the mass of a gas in a heavy vessel,

(c) density determination of a solid using a pyknometer,

(d) all simultaneous current and voltage measurements if the total current is of the same order of magnitude as the current in the voltmeter.

Products and quotients. If a physical quantity X is obtained by *multiplication* or *division* of results of independent observations on two or more physical quantities a, b, c, ..., the relative error r_x in the resulting value of X depends on the relative errors r_a, r_b, r_c, ... in the measurements of a, b, c, ... and is given by

$$r_x = \sqrt{(r_a^2 + r_b^2 + r_c^2 + \ldots)}.$$

The relative errors may be those determined from the standard error of the mean, standard deviation, or estimated error. If, for instance, $r_a, r_b, r_c, \ldots$ are relative errors, then the standard error in X is

$$\bar{\sigma}_x = X\sqrt{(r_a^2 + r_b^2 + r_c^2 + \ldots)}.$$

EXAMPLES.

$$(100 \pm 3)(6 \pm 4) = 600 \pm 600\sqrt{[(0.03)^2 + (0.66)^2]} = 600 \pm 400.$$

$$(100 \pm 3)(100 \pm 4) = 10^4 \pm 10^4\sqrt{[(0.03)^2 + (0.04)^2]} = 10^4 \pm 500.$$

$$(100 \pm 40)/(10 \pm 3) = 10 \pm 10\sqrt{[(0.4)^2 + (0.3)^2]} = 10 \pm 5.$$

Powers and roots. The error that results from raising a quantity a to some power n, $X = a^n$, is a special case of the rule for products. For example,

$$X = a^3 = (a) \times (a) \times (a), \quad r_x = \sqrt{(r_a^2 + r_a^2 + r_a^2)} = r_a\sqrt{3}.$$

In general, if $X = a^n$, then $r_x = r_a\sqrt{n}$.

If we solve the last two expressions for a and r_a, we have

$$a = X^{1/n}, \quad r_a = r_x\sqrt{1/n}.$$

Thus we have the result that the relative error in the nth root of a quantity is $\sqrt{1/n}$ times smaller than the relative error in the quantity.

Calibration errors. Every physical quantity is the product of a magnitude and a unit. By the laws of the theory of errors, the square of the relative error of this product is equal to the sum of the squares of the relative error of the magnitude (measuring error) and the relative error of the unit of measurement (*calibration error* of the measuring instrument). Since no measuring instrument, however accurate, is free from error, it follows that the relative error of any measurement is a combination of the relative measuring error and the relative calibration error:

$$r_{\text{total}} = \sqrt{(r_M^2 + r_C^2)}.$$

The calibration error of all instruments necessarily enters into the measurement. The calibration error for the length, time, and mass standards is usually less than 0.5 per cent. The best laboratory types of electrical measuring instruments have a calibration error between 0.5 and 1 per cent.

CHAPTER 2

STATICS OF RIGID BODIES

8. Newton's Laws

The subjects of statics and dynamics are based on Newton's three laws and the hypothesis that force is a vector quantity. At the beginning of our study of statics, these principles will be stated without extensive commentary. The entire subject of mechanics is derived from them, and a detailed discussion will be postponed until we take up the study of dynamics.

(1) Newton's first law (the law of inertia). Every body (every mass) remains in a state of rest or of uniform rectilinear motion (so-called *inertial motion*) until some external agency causes it to change this state.

(2) Newton's second law. The causes of change of motion are *forces*. Every force acting on a body produees an *acceleration* which is proportional to the force and is in the same direction as the force. The proportionality factor between a force and the acceleration it produces on a body is called the *mass* of the body:

$$F = ma.$$

(3) Newton's third law (action and reaction). Whenever one body exerts a force on a second body, the second exerts an equal and oppositely directed force on the first. Action equals reaction.

(4) Vector addition of forces. Forces add as vectors.

9. Fundamental Concepts of the Statics of Rigid Bodies

The subject of *statics* consists of the study of the forces acting on bodies which remain at rest or move with uniform rectilinear motion. A state of uniform rectilinear motion is equivalent to one of rest, since in both cases the state of motion does not change. This equivalence is based on Newton's first law.

We shall use the term *particle* to mean a body of finite, nonzero mass whose dimensions are extremely small compared with other lengths involved in a problem. A particle can be considered as a geometric point, and is also called a *point mass*.

The statics of rigid bodies deals with *systems* of particles. As a consequence, Newton's third law plays an especially important role in this subject. In contrast to a particle or a point mass whose motion can be one of *translation* only, an extended system of particles can also *rotate* about an axis. The statics of a rigid body deals with both translations and rotations, whereas the statics of a particle is concerned only with translations. This will be recognized from the definition of *equilibrium*, the basic concept of statics.

DEFINITION. A body (or a system of particles) is (a) in *translational equilibrium* if it remains in a state of rest or of uniform rectilinear motion, and (b) is in *rotational equilibrium* if it remains in a state of rest or of uniform rotation.

Five concepts are necessary in treating the statics of rigid bodies: (1) the concept of force, (2) the concept of the moment of a force with respect to an axis, (3) the concept of a rigid body, (4) the concept of equilibrium and (5) Newton's third law.

9-1 The Concept of Force

Our muscular sensations give us an idea of the concept of force, but they are obviously not suited for a precise description or measurement of forces. A better basis for understanding this concept lies in the fact that a force deforms a solid body on which it acts. In certain cases there is a simple relation between a force and the deformation produced by it. For example, when one end of a coil spring is fixed and a force $\boldsymbol{F}$ is exerted on the other end, the elongation $\boldsymbol{x}$ of the spring is directly proportional to the force, provided that the spring is not prestressed and the elongation does not exceed the elastic limit for proportionality. That is,

$$\boldsymbol{F} = k\boldsymbol{x}.$$

The proportionality factor k is called the *force constant*, or the *spring constant*. By the use of a spring it is possible to recognize and to compare forces. Anything which is capable of extending a spring is called a force.

This introduction to the concept of force has the advantage of being generally understandable. However, it is not wholly satisfactory from a physical point of view, since it is based on a phenomenological law. We shall show in Section 21-2 that another and more satisfactory definition of force can be given, namely, the cause of a change of motion.

Force is a vector quantity, and hence three quantities must be given to specify it uniquely: its magnitude, direction, and unit. As units of force in statics, we will use the pound force (lbf), the kilogram force (kgf), and the gram force (gf), defined in Section 2-3. To avoid possible mistakes, it should be pointed out again that the pound mass (lbm), kilogram mass (kgm), and gram mass (gm) are *scalar* quantities, while the pound force, kilogram force, and gram force are *vectors* (see Fig. 2-1). When speaking of a force, we must always state *on* what body it acts and *by* what body it is exerted.

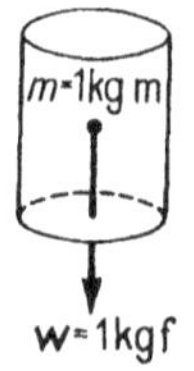

FIG. 2-1. Kilogram mass and kilogram force.

An especially important force quantity in statics is the *weight* of an object.

DEFINITION. The *weight* w of a body is defined as the force with which the body is attracted by the earth.

Because of the polar flattening of the earth, the earth's attraction for a given body is smaller at the equator than at the poles. As a result, the weight of a body varies from point to point on the earth's surface.

9-2 The Moment of a Force, or the Torque, with Respect to an Axis

A force acting on a body is not in itself sufficient to set the body in rotation about a fixed axis. The line of action of the force must not intersect the axis if the force is to affect the rotational motion.

Consider a plane E, designated in what follows as the *plane of rotation*, perpendicular to the axis of rotation D. For the present, we shall consider only forces parallel to this plane. The *moment arm*, or *lever arm*, h, of a force parallel to the plane of rotation (Fig. 2-2) is defined as the perpendicular distance from the line of action of the force to the axis of rotation D. The *moment* of the force, or the *torque*, M, is now defined.

DEFINITION. The *moment*, M, of a force which is parallel to the plane of rotation of an axis is defined as the product of the force F and the moment arm h:

$$M = Fh.$$

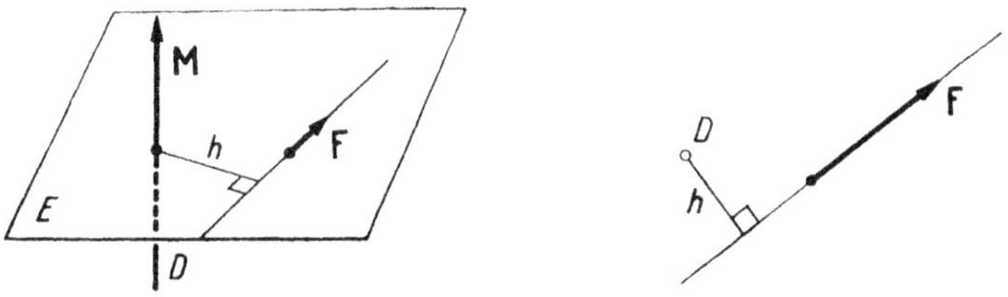

FIG. 2-2. The moment of a force in the plane of rotation.

According to this definition, the moment of a force about an axis is not changed when the force is displaced along its line of action, since this does not change the moment arm h.

The definition given above does not assign a direction to the moment of a force. However, by adopting an arbitrary convention, a direction can be assigned to a rotation, and thus a moment can be represented by a vector. This arbitrary convention is provided by the *right-hand screw* rule, or *right-hand* rule, for short. A rotating right-hand screw (Fig. 2-3) has both a direction of rotation and a direction of advance. The connection between these motions, the right-hand screw rule, is used in assigning a direction to a moment vector.

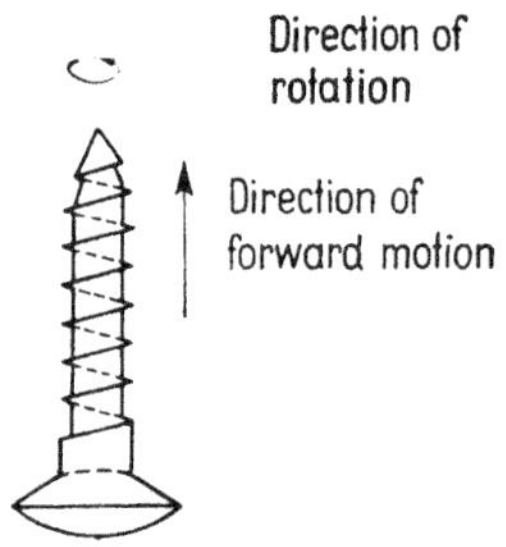

FIG. 2-3

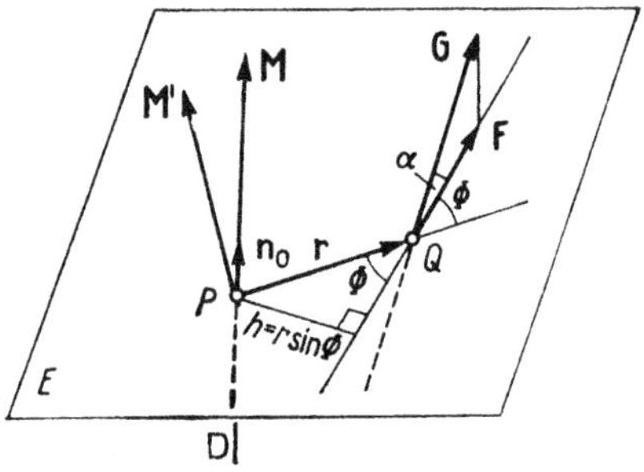

FIG. 2-4

FIG. 2-3. A right-hand screw.

FIG. 2-4. The moment of an arbitrary force.

The force F in Fig. 2-2 has a certain sense, or direction of rotation with respect to the axis. If a right-hand screw is turned in this direction, its direction of advance is taken as the direction of the moment vector or torque vector. Perhaps a more convenient way to remember the relationship between a force and its moment with respect to an axis is to think of one's right hand as grasped around the axis with the fingers pointing in the direction of the force. The thumb then points in the direction of the moment of the force. The moment vector is perpendicular to the plane of F and h, and hence is parallel to the axis of rotation D. Its magnitude is Fh, and its unit is one pound-force · foot (1 lbf · ft), one kilogram-force · meter (1 kgf · m), or one gram-force · centimeter (1 gf · cm).

A moment can also be produced by a force which is not parallel to the plane of rotation. In Fig. 2-4, the force G makes an angle α with the plane of rotation E. *Only* the component F lying in the rotation plane, of magnitude $G \cos \alpha$, has a moment about the axis D. The component perpendicular to the plane does have a moment about an axis perpendicular to D, but its only effect is to give rise to reaction forces in the bearings which keep the direction of the axis fixed.

The moment of a force about a fixed axis can be represented mathematically by the vector product

$$\boldsymbol{M} = \boldsymbol{r} \times \boldsymbol{F},$$

where $\boldsymbol{F}$ is the force component in the plane of rotation E, applied at point Q, and $\boldsymbol{r}$ is the radius vector from the axis of rotation to point Q (Fig. 2-4). According to the definition of a vector product,

$$\boldsymbol{M} = \boldsymbol{r} \times \boldsymbol{F} = hF\boldsymbol{n}_0,$$

since

$$h = r \sin \Phi .$$

The vector $\boldsymbol{n}_0$ is a unit vector along the axis of rotation. The vector $\boldsymbol{M}$ is then perpendicular to the plane of $\boldsymbol{r}$ and $\boldsymbol{F}$ and has a direction given by the right-hand screw rule. If only the point P is fixed, then the moment of $\boldsymbol{G}$ is given by

$$\boldsymbol{M}' = \boldsymbol{r} \times \boldsymbol{G},$$

where $\boldsymbol{M}'$ is perpendicular to the plane determined by $\boldsymbol{r}$ and $\boldsymbol{G}$.

If a number of forces act on a body simultaneously, their moment vectors can be added. The resultant of these vectors is the resultant moment and is equivalent to the effect of the individual moments (*law of moments*). Since the moment vectors of any arbitrary set of forces in a plane are all parallel to one another, their magnitudes can be added algebraically.

9-3 The Concept of a Rigid Body

In a system consisting of many particles, the particles will usually be displaced relative to each other when the system is acted upon by an external force. The body then experiences a change in shape. There are, however, many instances in which the relative displacement is extremely small. The concept of a *rigid body*, which is of the greatest importance in statics, is based on this fact. A rigid body, however, represents a limiting case which is never actually encountered in practice.

DEFINITION. A *rigid body* is one which does not change in shape when acted on by external forces.

The significance of the rigid-body concept lies in the fact that since the shape of such a body does not change, the line of action of an applied force remains fixed and the moment of the force does not change with time if the body is stationary. This makes possible a simple treatment of problems in static equilibrium, since the conditions for rotational equilibrium are then exactly like those for translational equilibrium. Because the lines of action of all forces acting on a rigid body do not change, the forces can be displaced at will along their lines of action.

9-4 The Concept of Equilibrium; Necessary and Sufficient Conditions

The definition of equilibrium requires that two conditions shall be satisfied. (1) The initial translational motion shall not change. (2) The initial rotational motion shall not change.

The necessary and sufficient *conditions of equilibrium* therefore consist of two requirements. (1) The resultant of all forces acting on the system must be zero. (2) The vector sum of all torques acting on the system, about any arbitrary axis, must vanish.

The first condition ensures a state of inertial translational motion (rest or uniform rectilinear motion). The second ensures that the rotational motion shall not change.

The analytic expressions of the equilibrium conditions are

$$\boldsymbol{R} = \sum_i \boldsymbol{F}_i = 0,$$

that is, the three following translational conditions must be satisfied:

$$R_x = \sum_i F_{xi} = 0,$$

$$R_y = \sum_i F_{yi} = 0,$$

$$R_z = \sum_i F_{zi} = 0.$$

The forces F_{xi}, F_{yi}, F_{zi} are the components of the force $\boldsymbol{F}_i$ in any arbitrary coordinate system:

$$\boldsymbol{M} = \sum_i \boldsymbol{M}_i = 0,$$

which is equivalent to the three rotational conditions

$$M_x = \sum_i (y_i F_{zi} - z_i F_{yi}) = 0,$$

$$M_y = \sum_i (z_i F_{xi} - x_i F_{zi}) = 0,$$

$$M_z = \sum_i (x_i F_{yi} - y_i F_{xi}) = 0.$$

In these equations, x_i, y_i, and z_i are the coordinates of the point of application of the force $\boldsymbol{F}_i$, expressed in the same coordinate system as the components of $\boldsymbol{F}_i$.

Thus, for the general case of a *spatial* or three-dimensional system of forces, there are *six* equations of equilibrium. For the special case of a system of *coplanar* forces, the number of equations is reduced to *three.*

The equations of equilibrium can be solved graphically or analytically. The graphical method is suitable only for systems of coplanar forces, but for such systems it is often more convenient than the analytical method. For a system of coplanar forces, the graphical conditions of equilibrium can be stated as follows:

(1) The force polygon of all the external forces must be closed.

(2) The algebraic sum of the moments of all forces must vanish for an axis through any arbitrary point. This point is customarily chosen in such a way that the smallest number of unknown quantities appears in the moment equation.

9-5 Newton's Third Law

Newton's third law is the most important principle in statics. It states that whenever one body exerts a force F_1 (*action*) on a second body, the second body exerts on the first a force F_2 (*reaction*) as in Fig. 2-5. The action and reaction have equal magnitudes, opposite directions, and the same line of action.

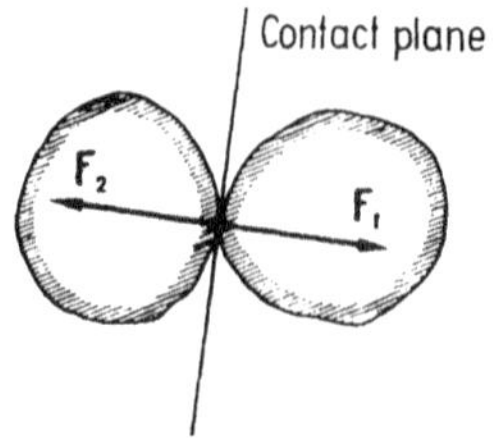

FIG. 2-5. Action and reaction.

In the absence of friction, the action and reaction forces occurring when two bodies make contact are perpendicular to the common plane of contact of the two bodies. The contact plane is the common tangential plane at the point of contact. The contact plane, and hence the line of action of an action–reaction pair of forces, is uniquely defined only for the special case of point contact. When contact is made along a line or over a surface the location of the line of action must be computed from the conditions of equilibrium and is not known in advance.

10. CALCULATION OF THE COMPLETE SYSTEM OF FORCES

We now see that the entire problem of any investigation of equilibrium has been reduced to that of developing a method of finding all of the forces that act on a rigid body in equilibrium. For systems composed of several rigid bodies or "partial systems", the usual procedure is to consider the equilibrium of these partial systems one at a time.

10-1 Separation Into Partial Systems

When considering the equilibrium of a part of a complex system, we start by *isolating* that part from the rest of system. We first imagine the earth to be removed and replaced by the force it exerts on the system, namely, the weight of the system. The weight acts at the "center of gravity" of the part of the system considered. (The center of gravity is a uniquely determined point whose position can be calculated by the methods to be discussed in Section 16. For the present, we consider its position to be known.) The line of action of the weight is vertically downward.

Next, we remove, one after the other, all bodies which make contact with the part of the system considered, and replace them with the forces which they formerly exerted at their points of contact.

When this process has been completed, the part of the system which we have selected is completely isolated, and we are now ready to formulate the equilibrium conditions for all the forces acting on it.

10-2 The Combination Principle

Suppose we have a system made up of a number of rigid bodies connected by pin joints. It is often useful to consider the system as a whole as a single rigid body. For this purpose, we assume that the possibility of motion at the joints has been eliminated. This has the advantage that the unknown forces at the joints drop out of the calculation because the action and reaction forces at a joint annul each other. The action and reaction both become a part of the internal force system of the body, while our equations involve only the external forces acting on the system. When some of the unknown external forces have been computed by this means, it then becomes an easy matter to compute the unknown contact forces acting at the joints by considering the equilibrium of various parts of the system.

11. Resultant of Forces which Do Not Act at the Same Point

11-1 Graphical Calculation of the Resultant of Several Coplanar Forces

When two forces acting on a rigid body do not have the same point of application, we can imagine the forces displaced along their lines of action to the point where these lines intersect. The forces can then be added as in Fig. 1-3. For a rigid body, this displacement of a force along its line of action is always permissible. We see that the line of action of the resultant of the forces passes through the point of intersection of the lines of action of its components.

To avoid confusion in the diagram of the force system by showing both the original and the displaced forces, an auxiliary diagram, called the *polar diagram* is constructed. The figure constructed in the polar diagram is called the *force polygon*. From this diagram, the correct magnitude and direction of the resultant force can be found. The line of action of the resultant in the main diagram can then be found by a parallel displacement. The line of action of a force will be indicated by the corresponding lower-case letter enclosed in parantheses. If the force polygon closes on itself, implying that the resultant force is zero, then the system of forces is in translational equilibrium.

EXAMPLE. Figure 2-6(a) shows a body acted on by three coplanar forces $\boldsymbol{F}_1$, $\boldsymbol{F}_2$, and its weight $\boldsymbol{w}$. In Fig. 2-6(b), we select a point P as the pole.

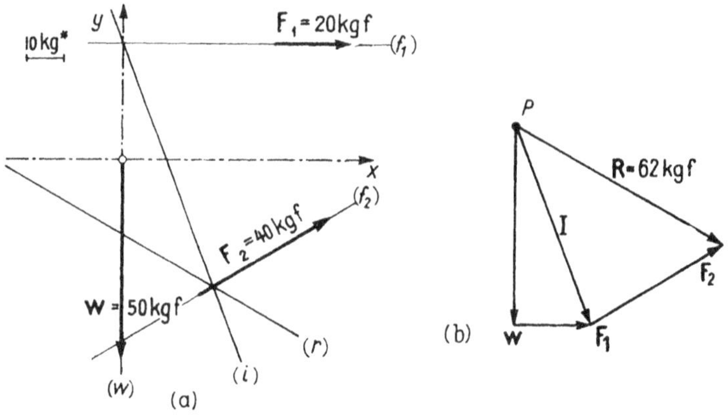

FIG. 2-6. Addition of three coplanar forces that do not act at the same point.

The forces $\boldsymbol{w}$ and $\boldsymbol{F}_1$ are first combined at P. This provides the intermediate resultant $\boldsymbol{I}$ in magnitude and direction. The line of action (i) of $\boldsymbol{I}$ is parallel to $\boldsymbol{I}$ and passes through the point of intersection of lines (w) and (f_1).

To the intermediate resultant $\boldsymbol{I}$, we now add the force $\boldsymbol{F}_2$ in the polar diagram, thus obtaining the resultant $\boldsymbol{R}$ of all three forces. The line of action (r) of the resultant passes through the point of intersection of the lines (i) and (f_2), and the problem is solved.

12. Systems of Parallel Forces Acting on a Rigid Body

It is possible to decompose any set of forces into their components parallel to the three coordinate axes. Thus, in determining the resultant of several nonparallel forces, we can first combine the parallel components, as discussed below. Then we combine these intermediate resultants vectorially to give the complete resultant.

12-1 Analytical Calculation of the Resultant of two Parallel Forces

Parallel forces in the same direction (Fig. 2-7). We consider force components parallel to the y-axis,

$$\sum F_{yi} = F_1 + F_2 = R.$$

Let us now select an axis through the point O on the (as yet) unknown line of action of the resultant. The moment of the resultant about this axis is then zero, and hence

$$\sum M = O = F_1 a_1 + F_2 a_2,$$

where a_1 and a_2 are the moment arms of $\boldsymbol{F}_1$ and $\boldsymbol{F}_2$ about O. From this it follows that

$$-a_1/a_2 = F_2/F_1.$$

If $F_1 > F_2$, then $a_2 > a_1$. From the relative sign difference of a_1 and a_2, we conclude that $\boldsymbol{F}_1$ and $\boldsymbol{F}_2$ are on opposite sides of O. We conclude, therefore, that the resultant of two parallel forces in the same direction lies between the forces, nearer the larger force, and its line of action divides the line joining the component forces internally in inverse ratio to the magnitude of the forces.

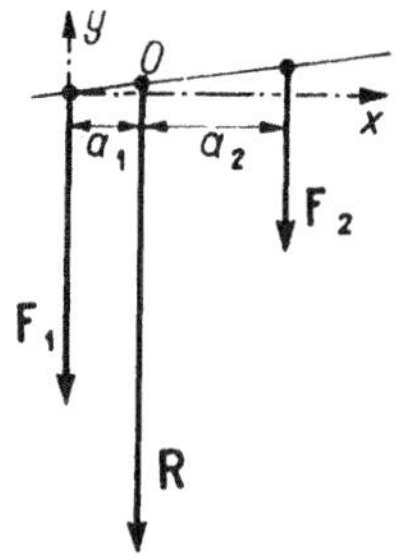

FIG. 2-7

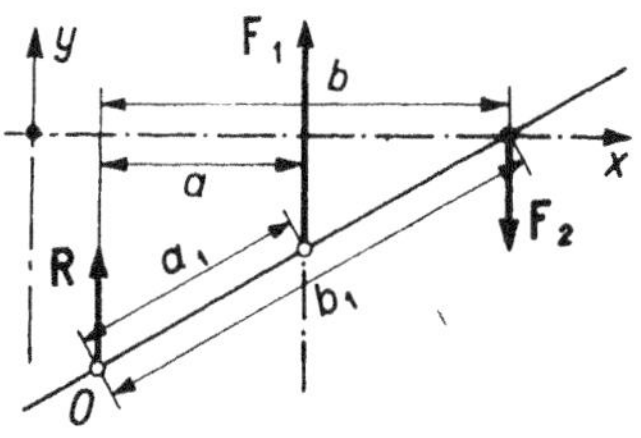

FIG. 2-8

FIG. 2-7. Parallel forces in the same direction.

FIG. 2-8. Parallel forces in opposite directions.

Parallel, oppositely directed forces (Fig. 2-8). We now consider force components parallel to the y-axis, acting in opposite directions,

$$\sum_i F_{yi} = F_1 - F_2 = R.$$

With respect to an axis through point O,

$$\sum M = O = F_1 a - F_2 b,$$

$$a/b = a_1/b_1 = F_2/F_1.$$

If $F_1 > F_2$, then $b_1 > a_1$. Since a_1 and b_1 have the same sign, $\boldsymbol{F}_1$ and $\boldsymbol{F}_2$ are on the same side of O. Hence the resultant of two parallel, oppositely directed forces lies outside both forces, on the side of the larger force. Its magnitude equals the difference between the forces. Its line of action divides the line joining the points of application of the forces (or any other line) externally in inverse proportion to the magnitudes of the forces.

12-2 Graphical Determination of the Resultant of two Parallel Forces

Parallel forces in the same direction (Fig. 2-9). Let us add to the force $\boldsymbol{F}_1$ the auxiliary force $\boldsymbol{H}_1$, which acts along the line joining the points of application of the forces $\boldsymbol{F}_1$ and $\boldsymbol{F}_2$. The intermediate resultant of $\boldsymbol{F}_1$ and $\boldsymbol{H}_1$ is the force ***I***. Next, let us add to the force $\boldsymbol{F}_2$ the auxiliary force $\boldsymbol{H}_2$, having the same magnitude as $\boldsymbol{H}_1$ but oppositely directed. The intermediate resultant of $\boldsymbol{H}_2$ and $\boldsymbol{F}_2$ is the force ***II***. Since $\boldsymbol{H}_1$ and $\boldsymbol{H}_2$ are in the same line and have the same magnitude, they cancel each other and so do their moments. The resultant of ***I*** and ***II*** passes through the point of intersection of the lines of action (1) and (2) and represents the desired resultant of $\boldsymbol{F}_1$ and $\boldsymbol{F}_2$. Its magnitude is given by

$$R = F_1 + F_2 .$$

Parallel, oppositely directed forces (Fig. 2-10). Addition of the auxiliary forces $\boldsymbol{H}_1$ and $\boldsymbol{H}_2$, as above, leads to the desired result:

$$R = F_1 - F_2 .$$

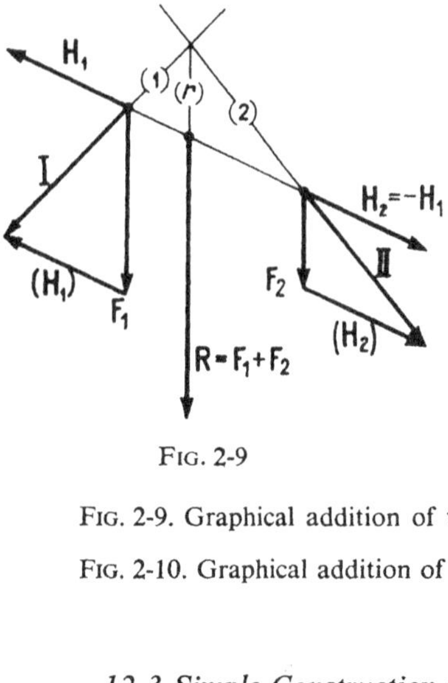

FIG. 2-9

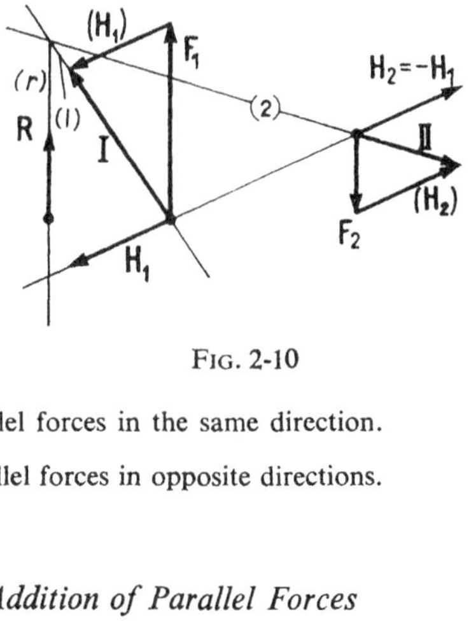

FIG. 2-10

FIG. 2-9. Graphical addition of two parallel forces in the same direction.

FIG. 2-10. Graphical addition of two parallel forces in opposite directions.

12-3 Simple Construction for the Addition of Parallel Forces

We now describe a simple construction which can always be applied to obtain the line of action of the resultant of two parallel forces. It is particularly useful because it can also be applied to the inverse process of resolving a force into two parallel components.

For two parallel forces ***A*** and ***B*** in the same direction, we have found that the line of action of the resultant divides a line joining the forces in the proportion $|a_1|/|a_2| = B/A$. This division of the line joining the points of application of the forces is made by the following geometrical construction.

We connect the points of application of the forces A and B, as in Fig. 2-11 (a), displace A to the point of application of B, displace B reversed by 180° to the point of application of A, and connect the tips of the displaced forces. This connecting line intersects the line joining the points of application in the desired point P, through which the line of action of the resultant R must pass.

We will prove this. Since the triangles x_1x_2P and y_1y_2P are similar (Fig. 2-11), it follows immediately that $|a_1|/|a_2| = B/A$.

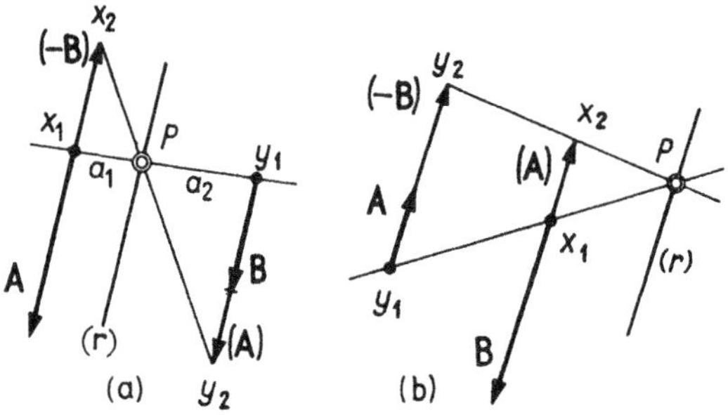

FIG. 2-11. Simple construction for the addition of two parallel forces.

The external division point for two parallel, oppositely directed forces can be found by exactly the same procedure, as shown in Fig. 2-11 (b).

The point P is called the *center of force*, abbreviated CF. The center of force of a system of two parallel forces is the point of intersection of the line of action of their resultant with the line joining the points of application of the forces. If the point of application of the resultant is at the center of force, the moment about any axis perpendicular to the plane of the forces is the same as the moment of the component forces.

Because of its importance, we shall indicate the center of force in following diagrams by two concentric circles.

12-4 Parallel, Oppositely Directed Forces of Equal Magnitude (the Couple)

The analytical method gives for the resultant (Fig. 2-12):

$$\sum_i F_{yi} = 0 \quad (F_1 = F_2 = F).$$

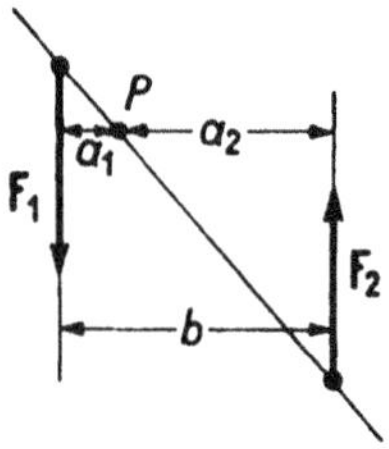

FIG. 2-12. A couple.

With respect to an arbitrary point P,

$$\sum M = F_1 a_1 + F_2 a_2 = F(a_1 + a_2) = Fb.$$

Hence *the resultant of a couple is zero*. Its moment about *any* point is equal to the magnitude of *either* force multiplied by the perpendicular distance b between the lines of action of the forces. The only property of a couple is its moment, i.e., a couple represents a pure moment.

12-5 Resolution of a Force into two Parallel Components having Specified Lines of Action

We wish to find the forces $\boldsymbol{A}$ and $\boldsymbol{B}$ whose lines of action are (a) and (b) and whose resultant is $\boldsymbol{R}$ (Fig. 2-13). First we construct an arbitrary line (p) which cuts across (a), (b), and (r). Let the intersections of (p) with (a), (b), and (r) be the points of application of the forces $\boldsymbol{A}$, $\boldsymbol{B}$, and $\boldsymbol{R}$. Then $\boldsymbol{R}$ is displaced to the point of application of $\boldsymbol{B}$ and a line is constructed which connects the tip of ($\boldsymbol{R}$) with the point of application of $\boldsymbol{A}$. Then ($\boldsymbol{B}$) is the vector from the point of application of $\boldsymbol{R}$ to the intersection of the connecting line with (r). We determine $\boldsymbol{A}$ by subtracting $\boldsymbol{B}$ from $\boldsymbol{R}$.

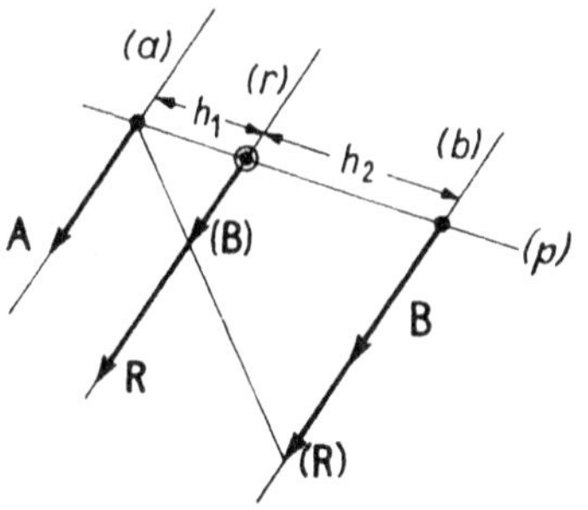

FIG. 2-13. Resolution of a force into two parallel components.

From the similar triangles in Fig. 2-13, we see that

$$\frac{R}{B} = \frac{A + B}{B} = \frac{h_1 + h_2}{h_1}.$$

This is equivalent to the moment condition used in Section 12-1, since from the relation

$$1 + A/B = 1 + h_2/h_1,$$

it follows that $A/B = h_2/h_1$.

12-6 A Useful Relation

If, in a system of two parallel forces, the forces are rotated about their point of application, keeping them parallel to each other, the resultant of the forces also rotates about the point of intersection of its line of action with the line joining the points of application of the forces, i.e., the center of force. Thus, in Fig. 2-14, $\boldsymbol{F}_1$ and $\boldsymbol{F}_2$ are two forces whose points of appli-

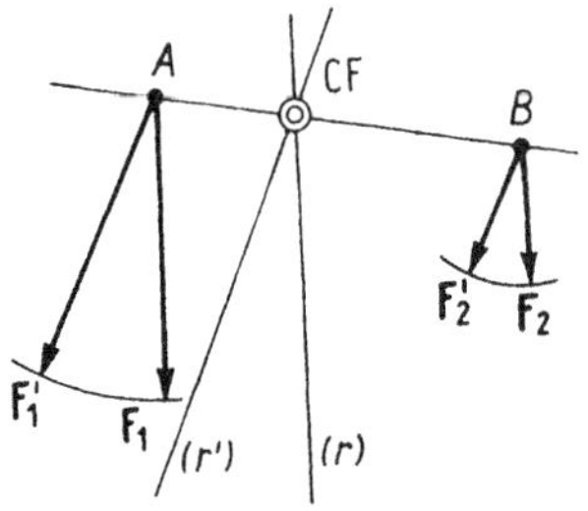

FIG. 2-14. Invariance of center of force under rotation.

cation are A and B. The line of action of the resultant divides the line segment AB (internally or externally) in inverse ratio to the magnitudes of the forces. The segment AB is kept fixed in our rotation process, and the magnitudes of the forces remain constant. It follows that the division point CF also remains fixed. This can be stated in the following brief form.

The center of force of parallel forces remains fixed when the forces are rotated about their points of application, keeping them parallel to one another.

13. Elementary Examples in Plane Statics

In this section, we give some detailed examples showing how the principles of statics are applied to a number of specific problems. In particular we will use some of the methods of combining forces given above.

13-1 Brief Summary of the Graphical Conditions of Equilibrium for a System of Coplanar Forces

When a number of known forces act on a system, they should first be combined to give their resultant. If the system is in equilibrium, the resultant is zero and the force polygon must close on itself (condition for translational equilibrium). The condition for rotational equilibrium namely, that the resultant moment about any axis be zero, differs in form, depending on the number of forces which act on the system.

(1) *Two forces.* The forces must have the same line of action, equal magnitude, and opposite direction.

(2) *Three forces.* The three lines of action must intersect at a common point. The resultant of any two of the forces must be equal in magnitude, opposite in direction, and have the same line of action as the third force.

(3) *Four forces.* The four forces can be combined in two pairs. The two resultants of these pairs must have the same line of action, equal magnitude, and opposite direction.

13-2 The Weightless, Perfectly Flexible Cord

A *perfectly flexible* cord is a linear system which can exert only tensile forces (i.e., pulls) in the direction of its axis.

In Fig. 2-15 a weightless, perfectly flexible cord is attached to a wall at one end and a tensile force $\boldsymbol{H}$ is exerted on it at the other end. The forces on the cord are: (1) the force $\boldsymbol{H}$, of given magnitude, exerted by the hand on the cord (its line of action is in the direction of the axis of the cord at its right end); and (2) the force $\boldsymbol{W}$ exerted by the wall on the cord (its magnitude is unknown and its line of action is that of the axis of the cord at its left end).

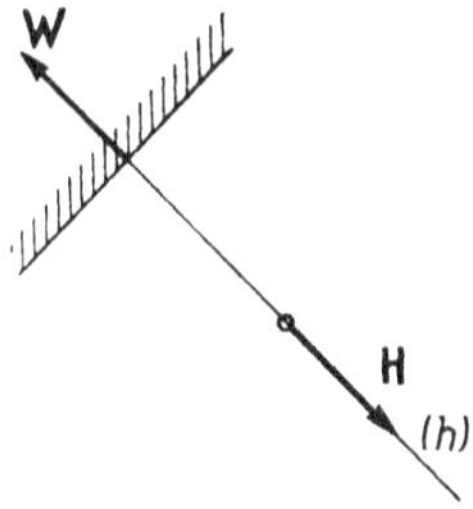

FIG. 2-15. Equilibrium of a weightless cord.

Two forces act on the cord. Equilibrium is possible only if the axes of the cord, at its right and left ends, lie on the same line and if the forces $\boldsymbol{W}$ and $\boldsymbol{H}$ have equal magnitudes. Therefore a weightless flexible cord in equilibrium always lies on a straight line. The surroundings of the cord exert on its ends oppositely directed forces of equal magnitude, whose lines of action coincide with the axis of the cord.

13-3 Some Sample Problems

A ladder of weight $\boldsymbol{w}$ leans against a wall. There are no friction forces at either end of the ladder. The ladder is kept in equilibrium by a cord, as shown in Fig. 2-16. *Four forces* act on the ladder:

(1) Its weight $\boldsymbol{w}$: magnitude known, line of action vertical.

(2) The force $\boldsymbol{B}$ exerted by the ground: magnitude unknown, line of action vertical.

(3) The force $\boldsymbol{A}$ exerted by the wall: magnitude unknown, line of action perpendicular to the wall.

(4) The force *C* exerted by the cord: magnitude unknown, line of action along the axis of the cord.

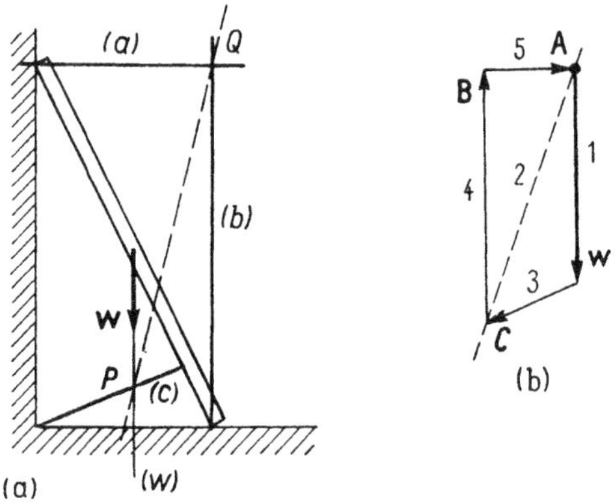

FIG. 2-16. Ladder in equilibrium.

Equilibrium is possible only if the lines of action of two pairs of the forces coincide, and if the force polygon closes on itself. To determine the common line of action we first combine *C* and *w*. Their resultant passes through point *P* in Fig. 2-16 (a). We then combine *A* and *B*. Their resultant passes through point *Q*. The line *PQ* is therefore the common line of action of the partial resultants.

In constructing the force polygon, it is convenient to begin with the known force *w* and to add to it its partner *C*. Since the resultant of these forces must have the direction *PQ*, the magnitude of *C* is determined. The force polygon can now be completed by the addition of the forces *B* and *A*. Since they must be put in the diagram with their prescribed directions, their magnitudes are completely determined. The numbers by the lines in the polar diagram give the order in which the lines are constructed.

A uniform rod of weight $w = 50$ lbf is pivoted on a bearing at one end and rests at its other end on a horizontal plane (Fig. 2-17). *Three forces* act on the rod:

(1) Its weight *w*: magnitude known, line of action vertical.

(2) The force *A* exerted on the rod by the ground: magnitude unknown, line of action vertical.

(3) The force *B* exerted on the rod by the bearings: both magnitude and line of action unknown.

The force exerted by the bearing on the rod must pass through the center-line of the bearing, because both cylindrical surfaces have a common tangent plane which is perpendicular to the radius at the point of contact.

Three forces act on the system, and for equilibrium the three lines of action must intersect at a common point. That is, the line of action of the

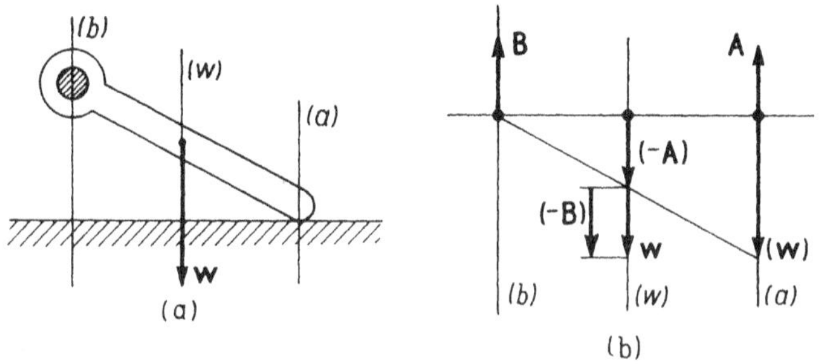

FIG. 2-17. Equilibrium of a rod on a bearing. $w = 50$ lbf, $A = 25$ lbf, $B = 25$ lbf.

bearing reaction must be parallel to the line of action of the resultant of the other two forces. The forces A and B can therefore be found by the method explained in Section 12-5.

A cart (Fig. 2-18) rests on an inclined plane as a result of frictional forces on its locked wheels. *Four forces* act on the cart:

(1) Its weight w: both magnitude and line of action known.

(2) The normal force A_1 exerted on the left wheel: magnitude unknown, line of action perpendicular to the inclined plane.

(3) The normal force A_2 exerted on the right wheel: magnitude unknown, line of action perpendicular to the inclined plane.

(4) The frictional forces on both wheels, which can be combined to a resultant R: magnitude unknown, line of action lies along the surface of the inclined plane.

Let us combine the forces w and A_2, and R and A_1, to form two resultants. The rest of the solution is clear from Fig. 2-18.

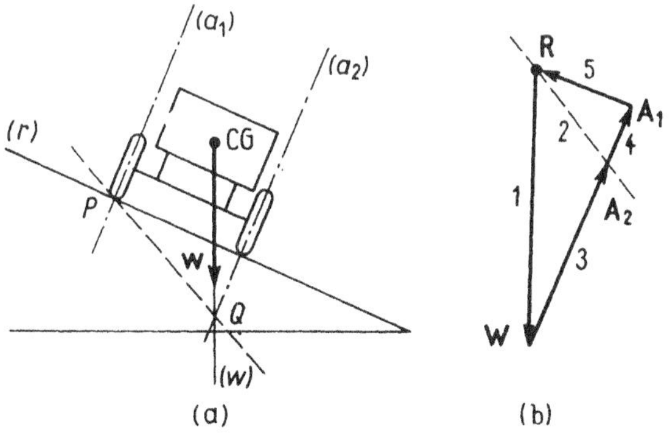

FIG. 2-18. Cart in equilibrium on an inclined plane.

13-4 The Cylindrical Pulley in Contact with a Frictionless Cord

A cylindrical pulley can rotate without friction on a fixed cylindrical axle (Fig. 2-19). It supports a weightless cord, which can slide over it without friction. One end of the cord is attached to a wall, and a tensile force $\boldsymbol{T}$ is exerted on the other end. The weight of the pulley is $\boldsymbol{w}$.

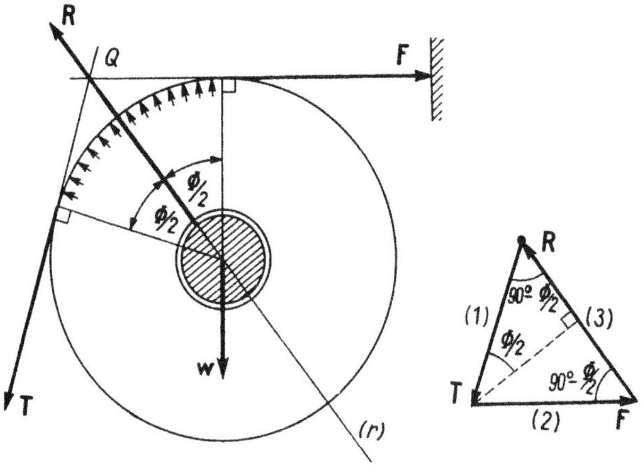

FIG. 2-19. Equilibrium for the frictionless cord.

There are two systems to consider, the pulley and the cord. One of the forces on the pulley is the reaction at the axle, which is unknown in both magnitude and direction. Hence, we first consider the equilibrium of the cord. It is acted on by *three forces*:

(1) The force $\boldsymbol{T}$: magnitude known, line of action lies along the axis of the cord.

(2) The force $\boldsymbol{F}$ exerted by the wall: magnitude unknown, line of action is the axis of the cord.

(3) The force $\boldsymbol{R}$ exerted by the pulley: both magnitude and line of action unknown.

The force $\boldsymbol{R}$ is the resultant of an infinite number of infinitesimal forces, all perpendicular to the tangents to the circle at the points of contact of the cord. The line of action of each of these forces therefore passes through the center of the axle, and they can be combined into a resultant, whose line of action will also pass through the center of the axle. Equilibrium is possible only if the three lines of action intersect at a common point, namely, point Q at the intersection of the lines of action of $\boldsymbol{T}$ and $\boldsymbol{F}$. The polar diagram can therefore be constructed as in Fig. 2-19, and all forces acting on the cord are known. Since the line of action (r) passes through the center of the circle to which the lines of action of $\boldsymbol{T}$ and $\boldsymbol{F}$ are tangent, it bisects the angle between the two tangents, and the force polygon is always an isosceles triangle.

Therefore, the forces exerted on a cord passing over a frictionless pulley are the same at left and right, whatever the magnitude of the angle Φ, and the reaction of the pulley passes through the center of the axle:

$$R = 2T \sin(\Phi/2), \quad F = T.$$

The statements above are *not* true if there is friction between the cord and the pulley.

To investigate the equilibrium of the pulley, *three forces* must be considered (Fig. 2-20):

(1) The weight $\boldsymbol{w}$ of the pulley: both magnitude and line of action known.

(2) The force $\boldsymbol{C}$ exerted on the pulley by the cord: this force is the reaction to the force $\boldsymbol{R}$, and it therefore has the same magnitude and line of action as $\boldsymbol{R}$, and is oppositely directed.

(3) The reaction $\boldsymbol{A}$ of the axle on the pulley: both the magnitude and direction of its line of action are unknown. It passes through the centerline of the pulley and axle.

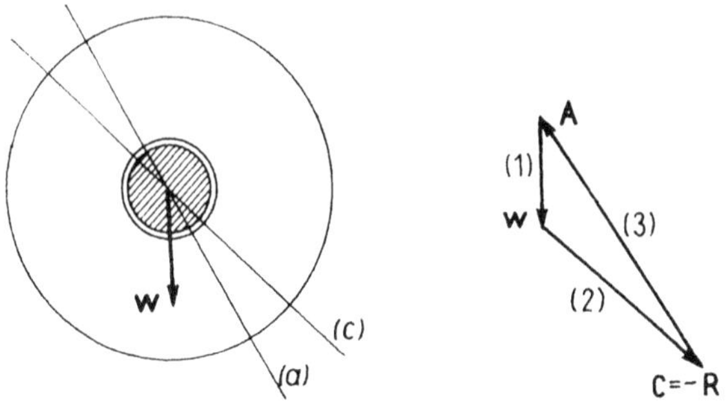

FIG. 2-20. Equilibrium for the pulley.

Since two of the forces are known, we first combine these into a resultant, which must then be counteracted by the force $\boldsymbol{A}$. This is possible only if the line of action of $\boldsymbol{A}$ is the same as that of the resultant of $\boldsymbol{C}$ and $\boldsymbol{w}$ (Fig. 2-20). The point of application of the force $\boldsymbol{A}$ on the axle is therefore determined.

13-5 Frictionless Inclined Plane; Finite Area of Contact

A block of weight $\boldsymbol{w}$ rests on a frictionless inclined plane (angle of inclination α) and is prevented from sliding by a cord (Fig. 2-21). *Three forces* act on the block:

(1) Its weight, $\boldsymbol{w}$: both magnitude and line of action known.

(2) The force *C* exerted on the block by the cord: magnitude unknown, line of action lies along the cord.

(3) The force *A* exerted on the block by the plane: both magnitude and line of action unknown. It is, however, perpendicular to the surface of the plane and is the resultant of an infinite number of parallel forces which are distributed nonuniformly along the lower surface of the block.

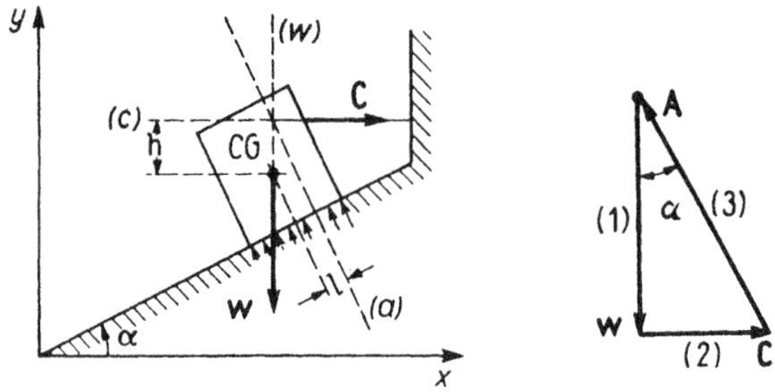

FIG. 2-21. Block on an inclined plane, no friction.

Since three forces act on the block, equilibrium is possible only if the line of action of *A* passes through the point of intersection of the lines of action of *w* and *C*. A measurement on the polar diagram gives the magnitude of *A*. Note that *A* is greater than *w*, since the force exerted by the cord has a component in the direction of the normal to the surface.

The line of action of *A* has been located graphically by using the condition for rotational equilibrium under the action of three forces. Let us now illustrate the analytical method of calculating *A* and the position of its line of action. Selecting *x*- and *y*-axes in the horizontal and vertical directions, respectively, we obtain the equations

$$\sum F_x = 0 = C + A_x, \quad \sum F_y = 0 = w + A_y,$$

from which it follows that

$$A_x = -C, \quad A_y = -w, \quad A = \sqrt{(C^2 + w^2)}.$$

Therefore,

$$A_x/A_y = \tan\alpha = C/w,$$

$$C = w\tan\alpha,$$

$$A = w\sqrt{(1 + \tan^2\alpha)} = w/\cos\alpha.$$

(Compare with the polar diagram in Fig. 2-21.)

To determine the distance *l* in Fig. 2-21, we take moments about an axis through the center of gravity, *CG*:

$$\sum M = 0 = -Ch + Al,$$

$$l = \frac{Ch}{A} = \frac{w\tan\alpha\, h\cos\alpha}{w} = h\sin\alpha.$$

13-6 The Wedge and the Screw: Statically Indeterminate Problems

It often happens that an attempt is made to solve equilibrium problems by the methods of the statics of rigid bodies, when the problem actually belongs in the field of elasticity. The simplest of these problems is that of a beam supported at three points, or of a table with four legs.

All of these problems are characterized by the fact that one of the contacts between the system and its surroundings can be removed without destroying the equilibrium. If we remove the central support from a beam resting on three supports, the beam (considered as a quasi-rigid-system) remains in equilibrium. In the same way, a table with four legs usually remains in equilibrium if one leg is removed. The forces which act in such *statically indeterminate* problems are actually determined by the elastic deformations of the system, and cannot be computed by the methods of the statics of rigid bodies.

Quite similar statements can be made regarding the wedge, the screw, etc. Apart from the fact that at the surface of contact there act an infinite number of forces whose lines of action are unknown, there is the further complication that the forces vary from point to point as a result of local elastic (or even plastic) properties of the system. These problems, therefore, do not lie in the realm of the statics of rigid bodies. The same criterion of statically indeterminate problems applies here also, namely, that a portion of the contact surfaces can be removed without appreciably disturbing the equilibrium.

14. The Law of Static Friction

14-1 Experimental Facts

When a body rests on a flat, level surface, it is in equilibrium. Furthermore, it remains in equilibrium if we exert a small force on it, in a horizontal or an inclined direction. Not until the applied force $\boldsymbol{P}$ (see Fig. 2-22) exceeds a certain limiting value does the body begin to accelerate. This experimental fact can be reconciled with Newton's second law of motion

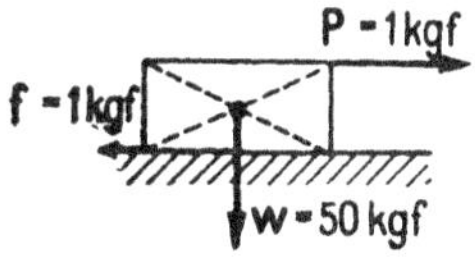

FIG. 2-22. Friction force.

only if a *friction* force acts between the body and the surface. These static friction forces have the following properties:

(1) The frictional force on a body is zero if no forces having a component parallel to the surface act on the body.

(2) If a small external force $\boldsymbol{P}$, parallel to the surface, is exerted on the body, the surface exerts on the body a friction force which counteracts the effect of $\boldsymbol{P}$. The friction force is equal in magnitude to $\boldsymbol{P}$, opposite in direction, and its line of action is in the plane of contact between the body and the surface.

(3) As the force $\boldsymbol{P}$ is increased, the friction force increases also until it reaches a certain limiting value which it cannot exceed. This limiting value is called the *fully developed* friction force, f_{max}.

(4) The friction force obeys Newton's third law. When the surface exerts a friction force on the body, the body exerts a friction force on the surface.

An empirical law governing the fully developed friction force was discovered by Coulomb. It agrees reasonably well with experiment. Coulomb's law of friction states that

$$f_{max} = \mu_0 N.$$

The factor μ_0 is called the *coefficient of static friction*, and N is the *normal force* which the body exerts perpendicularly against the surface. The empirical constant μ_0 is independent of the normal force, and depends only on the properties of both of the surfaces in contact. For example,

$$\mu_{0\,\text{iron-iron}} = 0.15 \text{ to } 0.20,$$

$$\mu_{0\,\text{leather-metal}} = 0.4 \text{ to } 0.6.$$

If the force $\boldsymbol{P}$ exceeds the magnitude of the fully developed friction force, then according to Newton's second law, the body accelerates. The limiting case of no motion is that for which $\boldsymbol{P} = f_{max}$.

The law of static friction therefore states that for equilibrium,

$$f \leqq \mu_0 N.$$

The inequality sign holds for a state of rest, the equal sign for the limiting case in which motion impends.

The law of static friction is verified only roughly by experiment. The four preceding statements are therefore simplifications and correspond only approximately to the actual facts. Experimental proof of the law is possible only for fully developed friction. The condition $f < \mu_0 N$ only removes the conflict with Newton's second law, but cannot be directly verified experimentally. Hence, points (2) and (3) must be considered as postulates.

In applying the law of friction to the statics of rigid bodies, it is convenient to combine the friction force $\boldsymbol{f}$ with the reaction force $\boldsymbol{A}$ of the surface to form the total reaction $\boldsymbol{T}_0$. When one body makes frictional contact with another, the total reaction makes an angle with the surface normal at the point of contact. In equilibrium, the total reaction balances the resultant of all other forces.

14-2 Friction Angle and Friction Cone

When a body rests on an inclined plane with friction, it is acted on by three forces. If the angle of inclination of the plane is small enough so that the body does not slide, the friction force is not fully developed (Fig. 2-23). The forces on the body are:

(1) Its weight w: both magnitude and line of action known.

(2) The friction force f: magnitude unknown, line of action along the surface of the plane and directed upward.

(3) The reaction of the plane, A: magnitude and the location of its line of action unknown. Its direction is perpendicular to the surface of the plane.

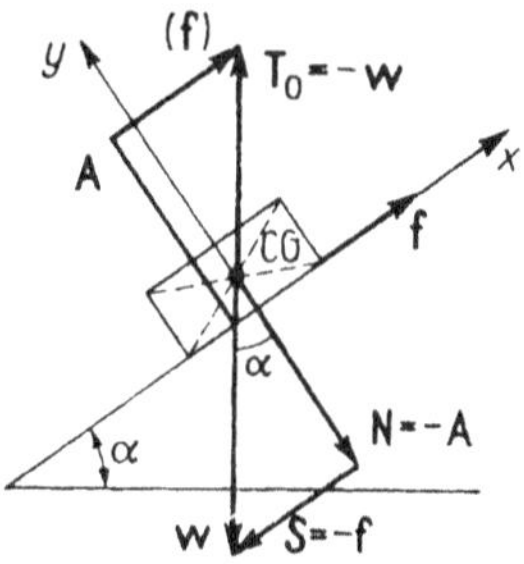

FIG. 2-23. Block in equilibrium on an inclined plane with friction.

To satisfy the condition for rotational equilibrium under three forces, the line of action of A must pass through the point of intersection of the lines of action of w and f. This also follows from the fact that the lines of action of w and T_0 both pass through the center of gravity. The following conditions must be satisfied for translational equilibrium:

$$\sum F_x = 0 = f - w \sin \alpha, \quad f = w \sin \alpha = |S|;$$
$$\sum F_y = 0 = A - w \cos \alpha, \quad A = w \cos \alpha = |N|.$$

Thus we see that

$$T_0 = \sqrt{(f^2 + A^2)} = w.$$

If the angle of inclination of the plane is increased to its limiting value without slipping, then at this limiting angle α_0

$$f_{\max} = w \sin \alpha_0 = \mu_0 w \cos \alpha_0,$$
$$\tan \alpha_0 = \mu_0 .$$

The angle α_0, the limiting angle of inclination for which the block will not slip, is called the *angle of repose* (or sometimes the *angle of friction*). The tangent of this angle is equal to the coefficient of static friction between the given surfaces, and its measurement provides a convenient method for determining this coefficient.

When $\alpha = \alpha_0$, the total reaction makes an angle α_0 with the normal to the surface. If $\alpha < \alpha_0$, the total reaction makes a smaller angle with the normal. If a force $\boldsymbol{P}$ acts on the body, parallel to the plane and directed downward, then f increases with $\boldsymbol{P}$ until it reaches the value $f_{\max}$. The total reaction then makes an angle α_0 with the normal to the surface, inclined toward the right (Fig. 2-24). If an upward force $\boldsymbol{P}'$, parallel to the surface, is

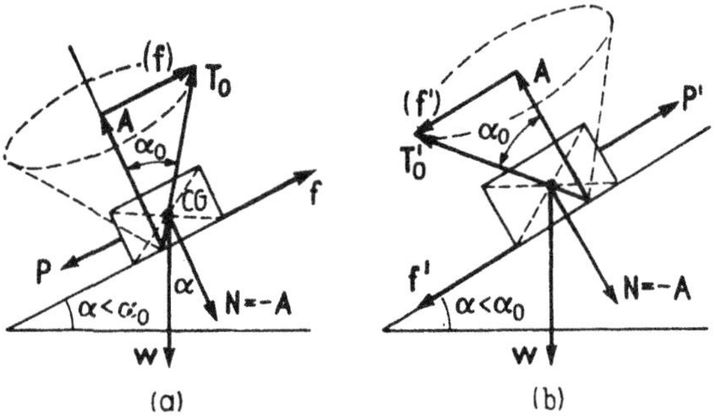

FIG. 2-24. The cone of friction for the case of static friction.

exerted on the body, the friction force first decreases to zero as $\boldsymbol{P}'$ is increased, and then increases in the opposite direction until it is fully developed. The limiting case in which upward sliding impends is then reached. The total reaction then makes an angle α_0 with the normal, inclined now toward the left. Both cases are independent of the angle of inclination α of the plane if $\alpha < \alpha_0$. In equilibrium, the total reaction lies within a cone of half-angle α_0, whose axis is normal to the surface on which the body slides. In the limiting case of impending motion, it lies on the surface of this cone. This cone, whose apex is always at the point of intersection of the lines of action of the friction force and the reaction of the surface, is called the *cone of friction*. The concept of the cone of friction is often helpful in the discussion of problems of equilibrium when frictional forces are present.

15. EXAMPLES USING THE LAW OF STATIC FRICTION

A block on a plane inclined at an angle α with the horizontal is pressed against the plane by a horizontal force $\boldsymbol{P}$. Let the line of action of $\boldsymbol{P}$ pass through the center of gravity of the block. We wish to compute the magnitude of $\boldsymbol{P}$ for the limiting cases of impending motion, up and down the plane (Fig. 2-25).

Downward motion imminent:

I. $\sum F_x = 0 = \mu_0 A - w \sin\alpha + P\cos\alpha$:

$$P\cos\alpha = w\sin\alpha - \mu_0 A. \tag{15-1}$$

II. $\sum F_y = 0 = -w \cos \alpha - P \sin \alpha + A$:

$$A = w \cos \alpha + P \sin \alpha. \tag{15-2}$$

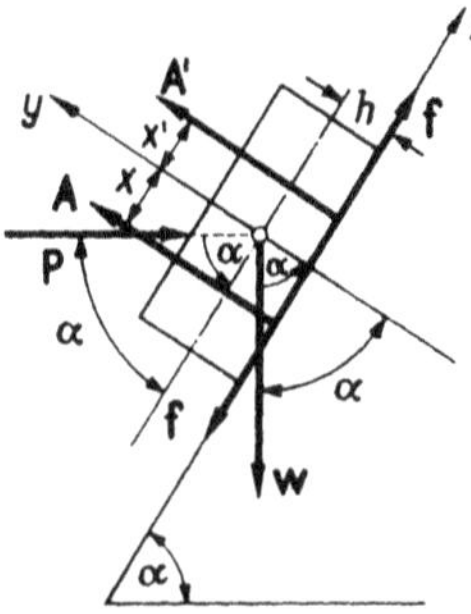

FIG. 2-25. Block pressed against an inclined plane.

It follows from Eqns. (15-1) and (15-2) that

$$P = \frac{w(\sin \alpha - \mu_0 \cos \alpha)}{\cos \alpha + \mu_0 \sin \alpha},$$

$$A = \frac{w}{\cos \alpha + \mu_0 \sin \alpha}.$$

III. With respect to the center of gravity, the moment condition is

$$\sum M = 0 = \mu_0 A h - A x, \quad x = \mu_0 h.$$

Upward motion impends:

I. $\sum F_x = 0 = -\mu_0 A' - w \sin \alpha + P \cos \alpha.$

II. $\sum F_y = 0 = -w \cos \alpha - P \sin \alpha + A'$:

$$A' = w \cos \alpha + P \sin \alpha,$$

$$P \cos \alpha = w \sin \alpha + \mu_0(w \cos \alpha + P \sin \alpha),$$

$$P = \frac{w(\sin \alpha + \mu_0 \cos \alpha)}{\cos \alpha - \mu_0 \sin \alpha},$$

$$A' = \frac{w}{\cos \alpha - \mu_0 \sin \alpha}.$$

III. With respect to the center of gravity,

$$\sum M = 0 = -\mu_0 A' h + A' x', \quad x' = \mu_0 h.$$

The line of action of the reaction force lies to the right of the center of gravity when upward motion impends, and at the same distance to the left of the center of gravity when downward motion impends.

A block having two projections on its lower face rests on a horizontal surface (Fig. 2-26). It is drawn toward the right by a force ***P***, whose line of

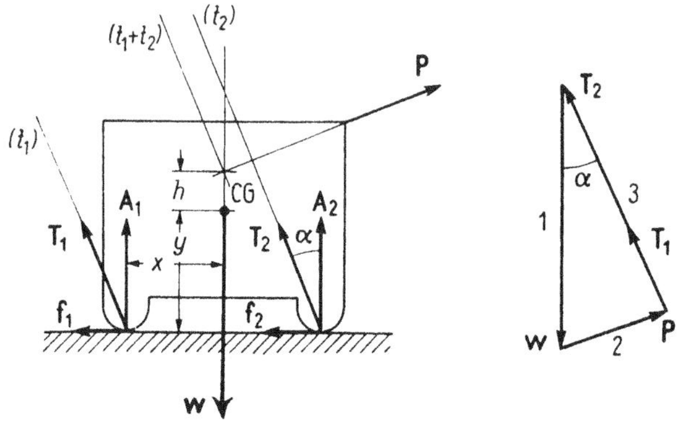

FIG. 2-26. Block with two points of contact, in equilibrium on a horizontal plane.

action is given. We wish to investigate the limiting case of equilibrium and to compute the force $\boldsymbol{P}$. The block is acted on by *four forces*:

(1) The weight $\boldsymbol{w}$.

(2) $\boldsymbol{P}$, having a prescribed line of action.

(3) The total reaction $\boldsymbol{T}_1$ with line of action (t_1).

(4) The total reaction $\boldsymbol{T}_2$ with line of action (t_2).

Since $f_1/A_1 = f_2/A_2 = \mu_0$, $\boldsymbol{T}_1$ and $\boldsymbol{T}_2$ are parallel. The resultant of $\boldsymbol{w}$ and $\boldsymbol{P}$ must have the same line of action as the resultant of $\boldsymbol{T}_1$ and $\boldsymbol{T}_2$. This resultant is parallel to $\boldsymbol{T}_1$ and $\boldsymbol{T}_2$, and must pass through the point of intersection of the lines of action of $\boldsymbol{w}$ and $\boldsymbol{P}$. After the force $\boldsymbol{P}$ is determined, the components $\boldsymbol{T}_1$ and $\boldsymbol{T}_2$ can be determined by the method described in Section 12-5.

Let us investigate the limiting case in which a body remains at rest on an inclined plane, acted on only by its own weight (Fig. 2-27). The body can slide only toward the left, and this will happen when the plane tangent to the cone of friction is vertical. The inclined plane must then make an angle α_0 with the horizontal. Since the vector $\boldsymbol{w}$ determines the line of action of $\boldsymbol{T}_0$, which is on the surface of the cone in this case, the apex of the cone lies at

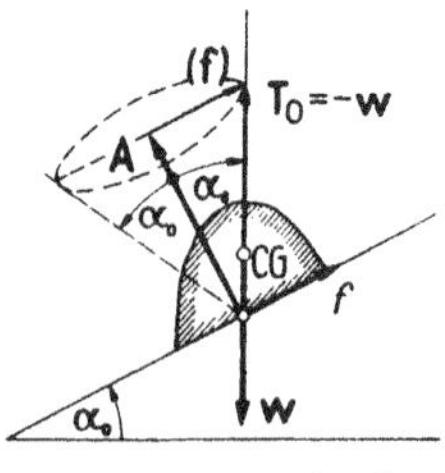

FIG. 2-27. Sliding on an inclined plane.

the point of intersection of the line of action of w with the inclined plane. The forces A and f_{max} can therefore be constructed.

A rectangular plate rests on its edge on an inclined plane. We wish to find the magnitude of the maximum horizontal force P that can be exerted against one corner, as in Fig. 2-28 (a), without causing the plate to slide up

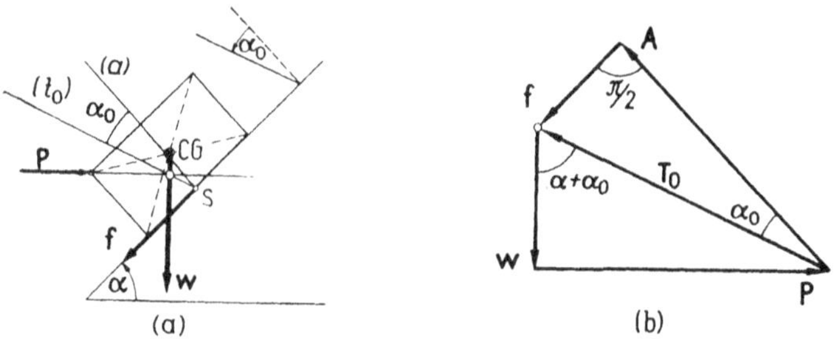

FIG. 2-28. Sliding plate on an inclined plane.

the plane. The magnitude of the angle of repose, α_0, is given. *Three forces* act on the plate:

(1) The force w, which is given.

(2) The force P.

(3) The force $T_0 = f_{max} + A$.

The resultant of P and w must lie on the line of action of T_0. In addition, the three lines of action must intersect at a common point, the point of intersection of the lines of action of w and P. The apex S of the cone of friction is therefore determined. The magnitudes of the unknown forces can be found from the polar diagram, since we know w and the angles between w and T_0, w and P, T_0 and A, and A and f.

16. Theory of the Center of Gravity

16-1 The Concept of the Center of Gravity

Imagine a system divided into a number of small particles of mass Δm_i and weight Δw_i. The weights Δw_i of the individual particles of a system form a set of parallel forces, which has a center of force about which the resultant w rotates when all the individual forces Δw_i are rotated about their points of application, remaining parallel to each other.

DEFINITION. The *center of gravity* of a body is defined as the center of force of the set of forces consisting of the weights of all of its particles.

The significance of the center of gravity can be made clear from the following imaginary experiment. The orientation of a body relative to the earth can be changed either by rotating the body relative to a fixed earth, or by rotating the earth about the fixed body. Both processes lead to the same result, but the second is simpler to consider, since then the weights of all the particles of a body rotate about fixed points of application; i.e., the positions of the particles. As shown in Section 12-6, this does not affect the position of the center of gravity. The line of action of the total weight rotates about the center of gravity and always coincides with a plumb line extending from the body to the earth (Fig. 2-29). The center of gravity of a body, which will be abbreviated as CG, is therefore characterized by the fact that the line of action of the weight passes through this point for all orientations of the body with respect to the earth.

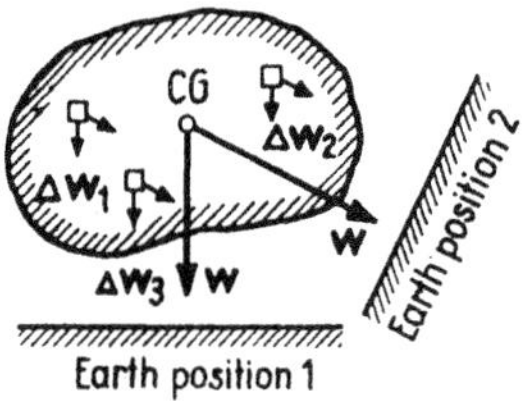

FIG. 2-29. Diagram of the center of gravity.

16-2 The Specific Weight of a Homogeneous Body

It is found convenient to divide all bodies qualitatively into two classes. (a) A body is said to be *homogeneous* if the smallest volume elements which are considered are found to be alike; for example, gases, liquids, and solutions. (b) A body is said to be *inhomogeneous* if the smallest parts into which it is considered to be subdivided are not all alike; for example, wood, stone, and suspensions. These descriptions are far from being strict definitions of the two classes. Whether a given body is considered as homogeneous or inhomogeneous depends on the size of the smallest volume element we choose to consider. If a body can be considered homogeneous for volume elements of 1 cm^3, it will be homogeneous for any volume elements larger than this, but it will not necessarily be homogeneous for elements as small as 1 mm^3. A strict formulation enables us to speak at best only of "quasi-homogeneous" bodies.

If a body is homogeneous, we can define a property of the body called its *specific weight*.

DEFINITION. The specific weight $\boldsymbol{\varrho}$ of a body is defined as the ratio of its total weight $\boldsymbol{w}$ to its volume V:

$$\boldsymbol{\varrho} = \boldsymbol{w}/V.$$

Since the specific weight $\boldsymbol{\varrho}$ is the ratio of the vector $\boldsymbol{w}$ to the scalar V, it is also a vector. Its line of action is always vertical, like that of the weight vector. The unit of specific weight in the mks system is 1 kgf/m³ and is 1 lbf/ft³ in the fps system.

16-3 Center of Gravity of Homogeneous Bodies

Since all volume elements of a homogeneous body are alike, the weight of a volume element ΔV can always be found when the specific weight ϱ is given:

$$\Delta w = \varrho \Delta V.$$

From the discussion in Section 12-1, the center of force of the weights w_1 and w_2 of two volume elements (see Fig. 2-30) is found from the equation

$$\frac{a_1}{a_2} = \frac{\Delta w_2}{\Delta w_1} = \frac{\varrho \Delta V_2}{\varrho \Delta V_1} = \frac{\Delta V_2}{\Delta V_1}.$$

The position of the center of force is independent of the specific weight and depends only on the geometrical distribution of volume elements. This result also holds for the center of gravity of the entire body. The center of gravity of a homogeneous body does not depend on the material of which it is composed; its position depends only on the geometrical distribution of the volume elements of the body.

We will now consider a number of methods for determining the center of gravity of a homogeneous body.

Calculation of the position of the center of gravity of homogeneous bodies by considerations of symmetry. Consider the following types of symmetry: (a) orthogonal axes or planes of symmetry, (b) axes or planes of inclined symmetry, and (c) a center of symmetry.

If a homogeneous rigid body has an axis of symmetry, the center of gravity lies on this axis.

To prove this, we let a (Fig. 2-31) be an axis of inclined symmetry. A typical line of symmetry is s. Then, according to the definition of inclined

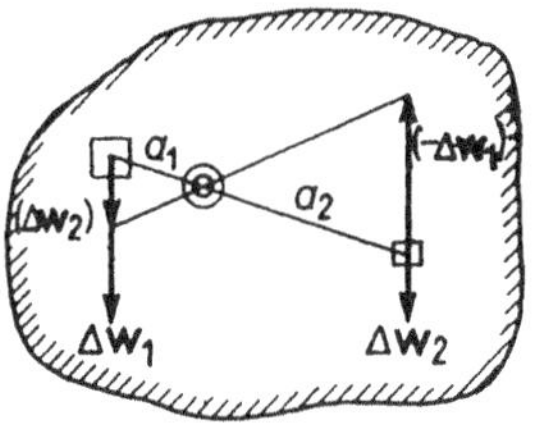

FIG. 2-30

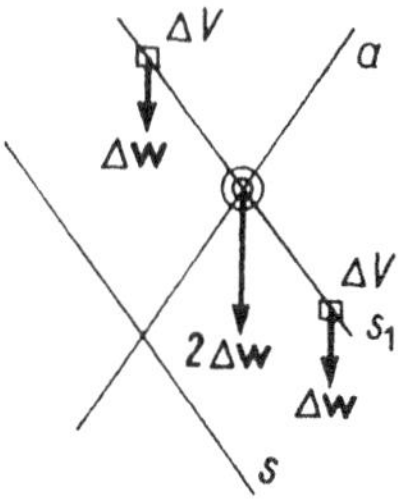

FIG. 2-31

FIG. 2-30. The center of gravity of a homogeneous body is independent of the material of the body.

FIG. 2-31. Determination of the center of gravity when the body has a symmetry axis.

symmetry, for every volume element ΔV at the left of the axis there is an equal volume element ΔV at the right of the axis. Both are at the same distance from the axis of symmetry, and lie on a line s_1 parallel to s. Since the weights of these volume elements are equal, their center of force lies on the axis of symmetry. But since all such partial resultants have their points of application on a, the center of gravity must also lie on the symmetry axis. Its position along this axis, however, is still undetermined.

It follows from these considerations that if a homogeneous body has two axes of symmetry, the center of gravity lies at their point of intersection. If a homogeneous body has several axes of symmetry, they must all intersect at the center of gravity. If a homogeneus body has a center of symmetry, this point is also the center of gravity.

Some important special cases. Center of gravity of plane figures.

(a) *Homogeneous thin rod.* By symmetry, the center of gravity lies at the midpoint of the rod.

(b) *Triangle* (Fig. 2-32). A triangle has three axes of inclined symmetry, namely, the three medians. The center of gravity lies at their point of intersection.

(c) *Parallelogram* (Fig. 2-33). A parallelogram has four axes of inclined symmetry, the two diagonals and the two lines joining the midpoints of opposite sides. The CG is at the point of intersection of these axes.

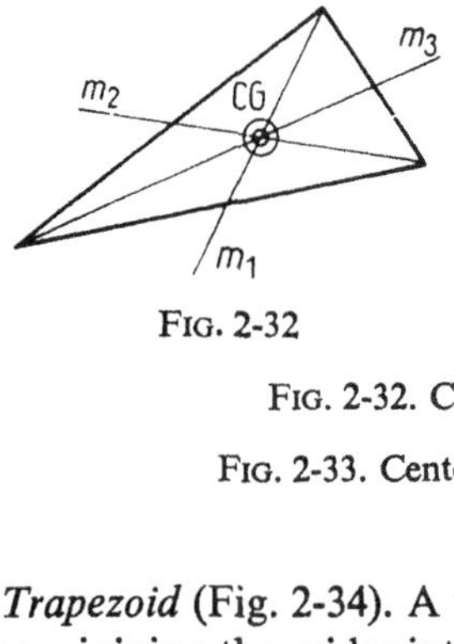

FIG. 2-32

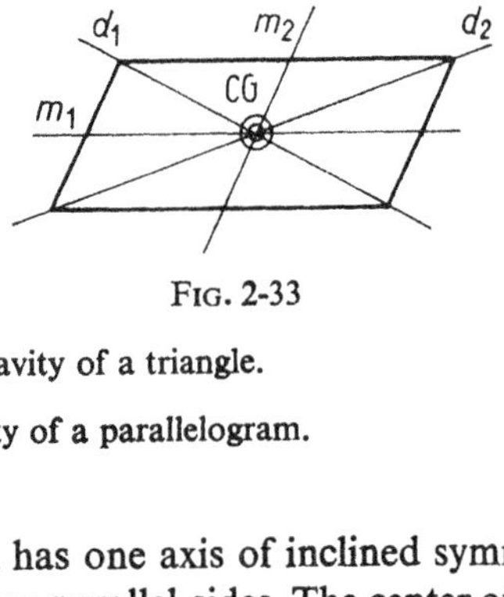

FIG. 2-33

FIG. 2-32. Center of gravity of a triangle.

FIG. 2-33. Center of gravity of a parallelogram.

(d) *Trapezoid* (Fig. 2-34). A trapezoid has one axis of inclined symmetry, the line s joining the midpoints of its two parallel sides. The center of gravity must lie on this axis, and also on the line joining the centers of gravity C_1 and C_2 of the component triangles.

(e) *Circle.* The center of gravity is at the center, the center of symmetry.

(f) *Ellipse.* An ellipse has two orthogonal axes of symmetry and an infinite number of inclined symmetry axes (the principal axes and any arbitrary diameter). The center of gravity is at the center, the point of intersection of all these axes of symmetry.

(g) *Composite figures.* These can be subdivided into areas whose centers of gravity are known. Weights are assigned to these centers in proportion to the corresponding areas, and the center of force of this system of parallel forces is computed.

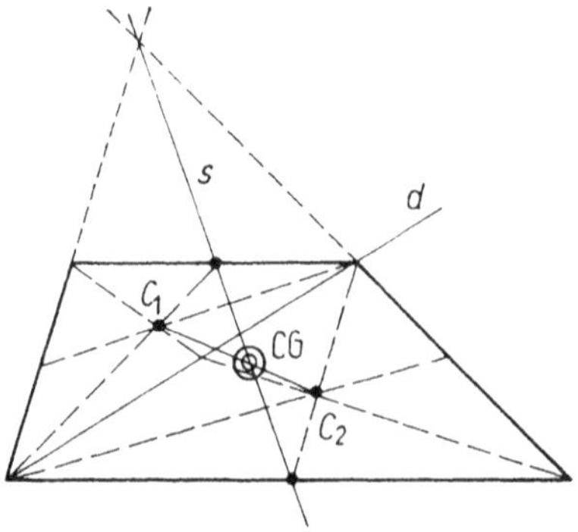

FIG. 2-34. Center of gravity of a trapezoid.

It may appear superfluous to consider the center of gravity of plane figures. However, the theories of the bending of a beam and the torsion of a rod of arbitrary cross-sectional shape involve a knowledge of the position of the center of gravity of the cross section of the beam or the rod.

Center of gravity of a body. (a) *Parallelepiped.* A parallelepiped has three planes of symmetry which intersect at the center of gravity.

(b) *Cylinder* (orthogonal or inclined). A cylinder has a plane of symmetry parallel to its end faces and at one-half its height (orthogonal or inclined), and an infinite number of planes of symmetry which all intersect along the axis of the cylinder. The center of gravity therefore lies on the axis at one-half the height.

(c) *Triangular pyramid* (Fig. 2-35). Consider any triangular face to be placed horizontally, and divide the pyramid into an infinite number of infinitesimally thin, horizontal, triangular plates. The partial weights of these plates can be considered to act at the centers of gravity of the triangles. These centers of gravity lie on an axis connecting the apex of the pyramid with the center of gravity of the base. By analogy with the concept of the *center* of gravity, this line can be called the *axis* of gravity. The center of gravity of the pyramid therefore lies on this axis. The pyramid has four faces, any one of which can be considered the base. Hence there are four such axes of gravity, which intersect at the center of gravity.

To determine the vertical height of the center of gravity construct a plane as in Fig. 2-35 (a) through any edge such as AS and the axis of gravity S_1S. This plane intersects the pyramid in a triangle ADS, as in Fig. 2-35 (b). Point S_1, the center of gravity of the lower face, divides the

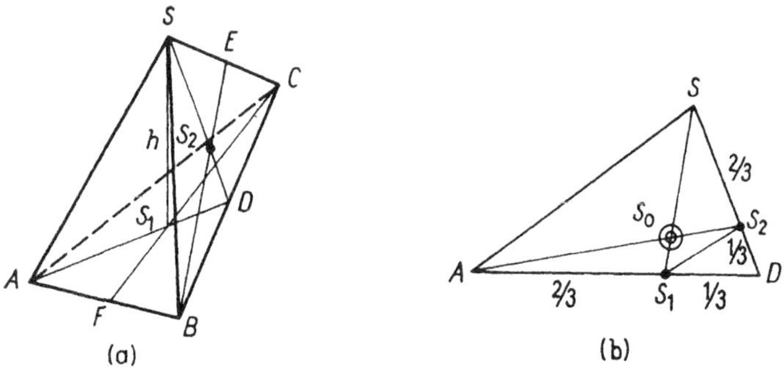

FIG. 2-35. Center of gravity of a triangular pyramid.

side AD in the ratio 2 : 1. The same is true of the point S_2 and the side SD. It follows from plane geometry that S_1S_2 is parallel to AS, and hence, ASS_0 is similar to $S_2S_1S_0$, so that

$$S_1S_2/AS = \tfrac{1}{3} = S_0S_1/S_0S.$$

That is, the center of gravity S_0 of a triangular pyramid divides every axis of gravity in the ratio 3 : 1 as measured from the apex. It is therefore at a height $h/4$ above the base, where h is the height of the apex.

(d) *Arbitrary cones and pyramids.* When these objects are subdivided into triangular pyramids by dividing their bases into triangles, the center of gravity of each of these pyramids is at one-quarter of the height above the base. The combined center of gravity therefore lies in this horizontal plane. On the other hand, if these bodies are subdivided into thin plates by sections parallel to the base, we find that the center of gravity of each lies on the line joining the apex with the center of gravity of the bottom face (the axis of gravity). The center of gravity therefore lies on the axis of gravity at a point one-quarter of the height above the base.

Calculation of the center of gravity of homogeneous plates by the law of moments. Any arbitrary figure bounded by straight lines can always be subdivided into portions whose centers of gravity are known. Weights proportional to the areas and acting at the centers of gravity of each part can then be assigned to each portion. If we choose an appropriate coordinate system, the law of moments can be applied about an axis through any arbitrary point. Two useful points are the origin of the coordinate system and the desired center of gravity itself.

The coordinates of the center of gravity can be calculated from the moment law by considering the weight first to be parallel to the x-axis, then parallel to the y-axis. If the figure has an axis of symmetry, the CG must lie on this axis, and it is convenient to select it as one of the coordinate axes.

EXAMPLE. We wish to find the coordinate of the center of gravity of the T-shaped plate in Fig. 2-36. The CG lies on the symmetry axis of the figure, which we select as the y-axis.

The x-coordinate of the center of gravity, $\bar{x}$, is therefore zero. To calculate the y-coordinate, we have from the moment equation, about an axis through the origin,

$$-cd\frac{d}{2} - ab\left(d + \frac{a}{2}\right) = -(ab + dc)\,\bar{y},$$

$$\bar{y} = \frac{a^2b + 2abd + cd^2}{2(ab + cd)}$$

With respect to an axis through the CG,

$$-ab\left(d + \frac{a}{2} - \bar{y}\right) + cd\left(\bar{y} - \frac{d}{2}\right) = 0,$$

$$2abd + a^2b + cd^2 = 2\bar{y}(ab + cd),$$

$$\bar{y} = \frac{a^2b + 2abd + cd^2}{2(ab + cd)}.$$

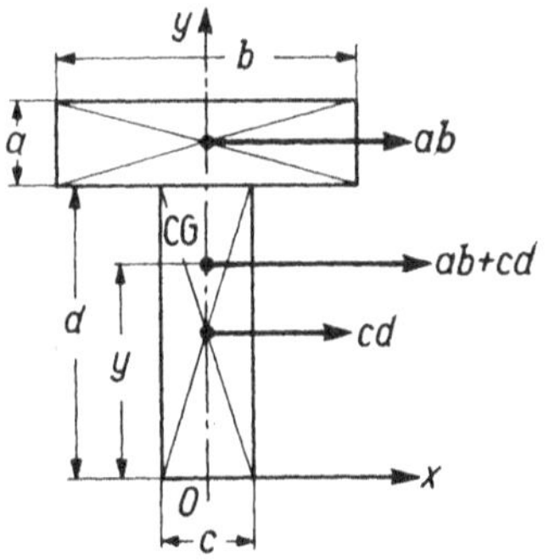

FIG. 2-36

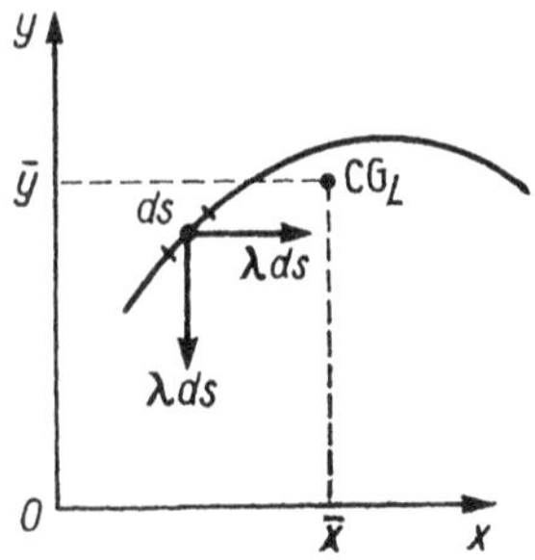

FIG. 2-37

FIG. 2-36. Center of gravity of a T-shape.

FIG. 2-37. Center of gravity of a curve.

16-4 General Formulas for Computing the Coordinates of the Center of Gravity

Center of gravity of a curve (Fig. 2-37). Let $y = f(x)$ be the equation of a curve in the coordinate system x, y. For the calculation of its center of gravity, we consider that mass is distributed uniformly along the curve. Let λ be the weight per unit length. Let us divide the curve into elements of length ds, assign to each element a weight $\lambda\, ds$, and set up the condition that the sum of the moments of the weights of these elements equals the moment of the total weight with respect to an axis through the origin. For vertical forces, the moment law gives

$$\int_L x\lambda\, ds = \bar{x}L\lambda,$$

and hence

$$\bar{x} = \frac{1}{L}\int_L x\,ds, \tag{16-1}$$

where L is the length of the curve and $\bar{x}$ is the abscissa of its center of gravity, CG_L.

For horizontal forces, the moment law gives

$$\int_L y\lambda\,ds = \bar{y}\lambda L, \quad \bar{y} = \frac{1}{L}\int_L y\,ds,$$

where $\bar{y}$ is the ordinate of the center of gravity, CG_L.

Center of gravity of a surface (Fig. 2-38). An arbitrary plane figure lies in the plane of a two-dimensional coordinate system. We assume that mass is distributed uniformly over the surface, the weight per unit area being σ.

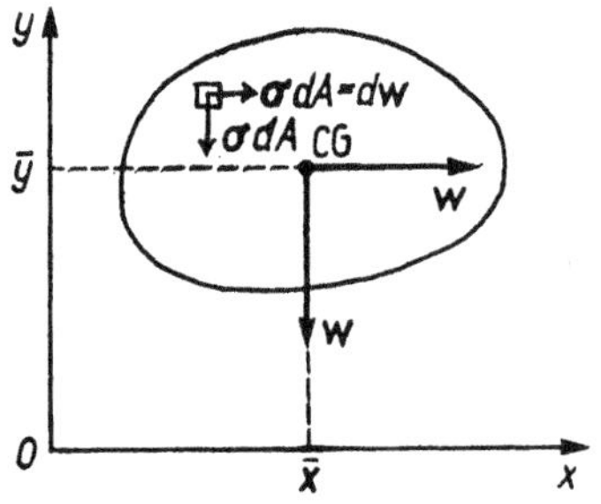

FIG. 2-38. Center of gravity of a plane surface.

Let us divide the surface into elements of area dA, assign a weight $\sigma\,dA$ to each, and set up the moment law with respect to the origin. For vertical forces, we have

$$\int_A x\sigma\,dA = \bar{x}\int_A \sigma\,dA = \bar{x}\sigma A,$$

$$\bar{x} = \frac{\int_A x\,dA}{\int dA} = \frac{1}{A}\int_A x\,dA. \tag{16-2}$$

For horizontal forces, we have

$$\int_A y\sigma\,dA = \bar{y}\int_A \sigma\,dA = \bar{y}\sigma A,$$

$$\bar{y} = \frac{1}{A}\int_A y\,dA.$$

16-5 The Theorems of Pappus-Guldin

We now use the above formulas in the derivation of two relations called the *theorems of Pappus*. The general formulas for the coordinates of the centers of gravity of curves and of surfaces lead to an interesting relation between the surface area of a figure of rotation and the center of gravity of its generating curve on the one hand, and on the other to a relation between the volume of a figure of revolution and the center of gravity of its generating surface.

First theorem of Pappus. Let $y = f(x)$ be the equation of the generating curve of a figure of rotation (Fig. 2-39). If this curve rotates about the

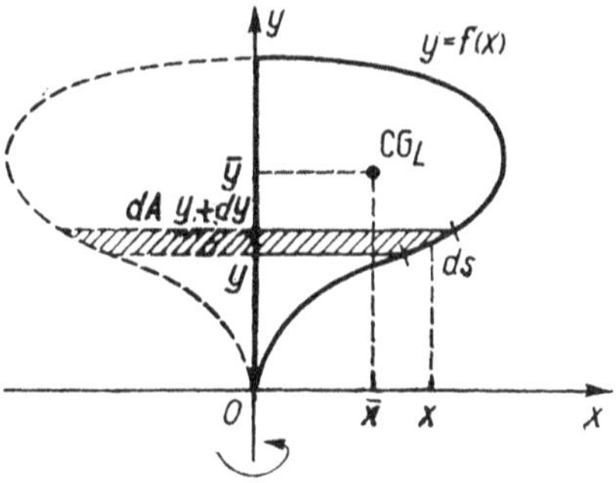

FIG. 2-39. The first theorem of Pappus.

y-axis, a line element ds of the curve generates on the figure of revolution an element of surface area dA given by

$$dA = 2\pi x \, ds.$$

The total surface area A is therefore

$$A = 2\pi \int_L x \, ds,$$

and by comparison with Eqn. (16-1), we see that

$$A = 2\pi \bar{x} L,$$

which is the first theorem of Pappus. The product $2\pi\bar{x}$ is the length of the path covered by the center of gravity CG_L of the generating line.

FIRST THEOREM OF PAPPUS. The surface area of a figure of rotation is the product of the length of the generating curve and the length of path described by the center of gravity of the generating curve in one revolution about the rotation axis.

This theorem can be applied to find the area of a surface of revolution when the center of gravity of the generating curve is known. Conversely, it can be used to find the center of gravity of the generating curve of a figure of revolution if the surface area and the length of the generating curve are known. If the y-axis is chosen as the rotation axis, we obtain $\bar{x}$, while of the x-axis is taken as the rotation axis, we get $\bar{y}$.

EXAMPLE. Find the center of gravity of one quadrant of a circle (Fig. 2-40):

$$L = \pi r/2, \quad A = 2\pi r^2,$$

$$2\pi\bar{x}(\pi r/2) = 2\pi r^2, \quad \bar{x} = 2r/\pi.$$

Since the bisector of the angle in the first quadrant is an axis of symmetry, it follows that $\bar{y} = \bar{x}$.

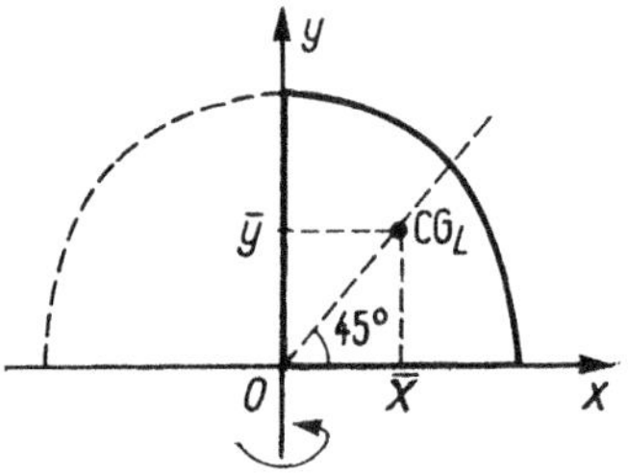

FIG. 2-40

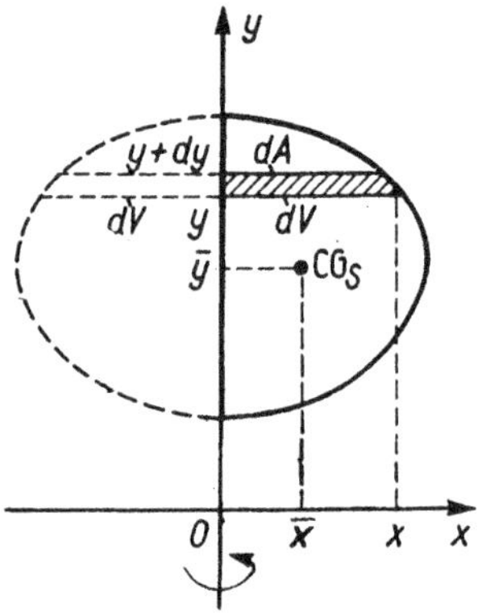

FIG. 2-41

FIG. 2-40. Center of gravity of a quarter circle.

FIG. 2-41. The second theorem of Pappus.

Second theorem of Pappus. Again, let $y = f(x)$ be the equation of the generating curve of a figure of rotation (Fig. 2-41). If the generating curve rotates about the y-axis, it generates a figure of revolution. Consider an element of volume dV cut from the figure by two horizontal planes at elevations y and $y + dy$. Then

$$dV = \pi x^2\, dy, \quad V = \pi \int x^2\, dy,$$

where x is the abscissa of the generating curve at $y = f(x)$.

From Eqn. (16-2), we have:

$$\bar{x} = \frac{1}{A}\int x\, dA. \tag{16-3}$$

We can choose as an element of area dA, the rectangle of length x and width dy. The moment of this area with respect to the origin of coordinates is obtained by multiplying its surface weight by the moment arm of its center of gravity, $x/2$. Then,

$$\int x\, dA = \int \frac{x}{2}(x\, dy) = \int \frac{x^2}{2}\, dy.$$

Combining this equation with Eqn. (16-3), we get

$$\bar{x} = \frac{1}{2A}\int x^2\,dy.$$

Comparing this with the formula for the volume V, we see that

$$V = 2\pi\bar{x}A,$$

which is the second theorem of Pappus. The product $2\pi\bar{x}$ is the length of path traveled by the center of gravity CG_S of the area of the generating surface in the rotation about the y-axis.

By analogy, we get for rotation about the x-axis, $V' = 2\pi\bar{y}A$.

Second Theorem of Pappus. The volume of a figure of rotation equals the product of the area of the generating surface and the length of path traveled by the center of gravity of this area in one revolution about the rotation axis.

From this law we can calculate the volume of a figure of rotation when the area of the generating surface and the position of its center of gravity are known. Conversely, from the known volume of a figure of revolution of known generating surface area, we can compute the position of the center of gravity of the generating surface.

Example. Consider one quadrant of an ellipse (Fig. 2-42). The volume of the half-ellipsoid obtained by its revolution about the x-axis is $(2\pi/3)\,ab^2$.

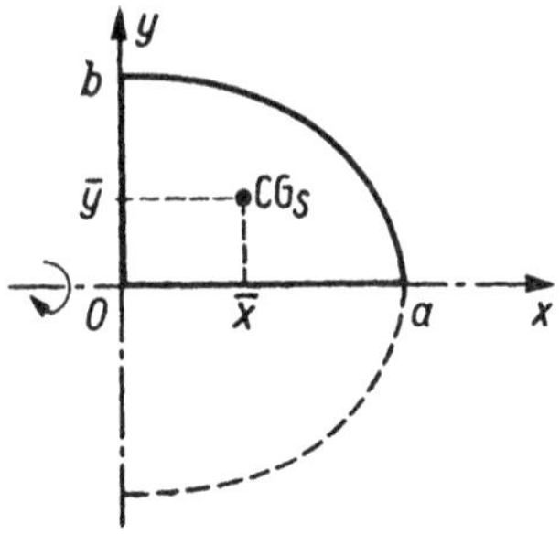

Fig. 2-42. Center of gravity of one quadrant of an ellipse.

The surface of the quarter-ellipse is $(\pi/4)\,ab$. From the second theorem of Pappus,

$$\frac{2\pi}{3}ab^2 = 2\pi\bar{y}\frac{\pi}{4}ab,$$

and hence $\bar{y} = 4b/3\pi$. Similarly,

$$\bar{x} = \frac{4a}{3\pi}.$$

17. The Three Types of Equilibrium

We can distinguish between three different types of equilibrium, called *stable*, *unstable*, and *neutral*. All three satisfy the conditions

$$\sum \boldsymbol{F}_i = 0, \quad \text{and} \quad \sum \boldsymbol{M}_i = 0.$$

However, only one of these types of equilibrium exhibits the property we intuitively associate with the concept of equilibrium, that of returning to the equilibrium state after being slightly displaced from it. This is called stable equilibrium.

An equilibrium state is called *stable* if small displacements give rise to forces or moments which bring the system back to its original equilibrium position. This type of equilibrium always exists when the center of gravity (CG) of a rigid body is vertically below the point N about which it is suspended. Any rotation about N (see Fig. 2-43) then gives rise to a restoring moment.

A system is in *unstable* equilibrium if a small displacement gives rise to forces or moments which increase the displacement still further. This type of equilibrium exists when the center of gravity of a rigid body is vertically above the point about which it is pivoted (Fig. 2-44). The smallest rotation about the pivot N creates a moment which increases the initial rotation.

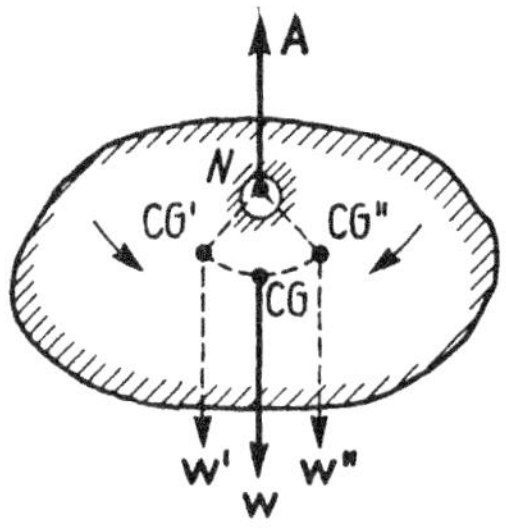

Fig. 2-43

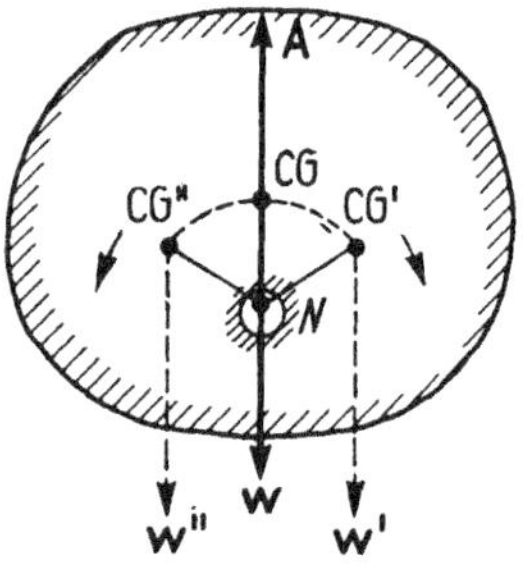

Fig. 2-44

Fig. 2-43. Stable equilibrium.

Fig. 2-44. Unstable equilibrium.

A system is in *neutral* equilibrium if a small displacement gives rise to no new forces or moments. This type of equilibrium exists when the center of gravity of a rigid body always remains vertically below, above, or at the point of support.

For example, a sphere on a horizontal plane and a railroad car on horizontal rails are in neutral equilibrium.

It is important and interesting to consider whether an equlilibrium state is stable or unstable when an object is supported on legs and the center of gravity lies above the supporting surface. We define the *tipping lines* of the body as the external tangents to the points of contact between the body and the supporting surface. If the body is tipped by a small angle about a possible line of tipping, then, in general, a restoring moment is set up. However, if as a result of the tipping, the center of gravity reaches a point vertically above the tipping line, the equilibrium becomes unstable. Depending on the nature of some accidental disturbance, the object may then return to its original equilibrium state or move to a new state of stable equilibrium. For example, a three-legged stool has three tipping lines (Fig. 2-45); a table has four tipping lines (Fig. 2-46); and a pentagonal prism has five tipping lines (Fig. 2-47). The law of moments shows that a rigid body is more stable,

FIG. 2-45

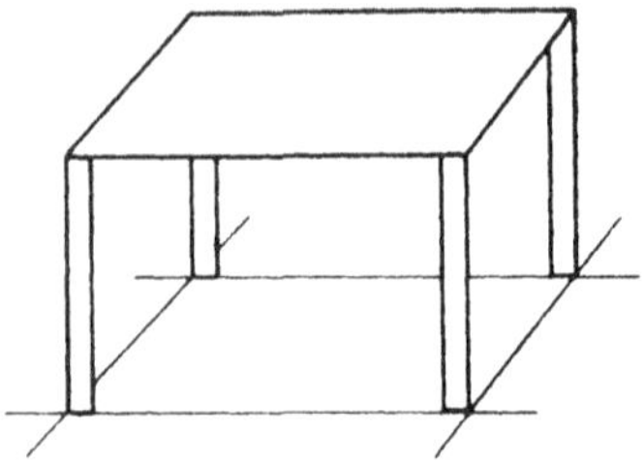

FIG. 2-46

FIG. 2-45. Tipping lines of a three-legged stool.

FIG. 2-46. Tipping lines of a table.

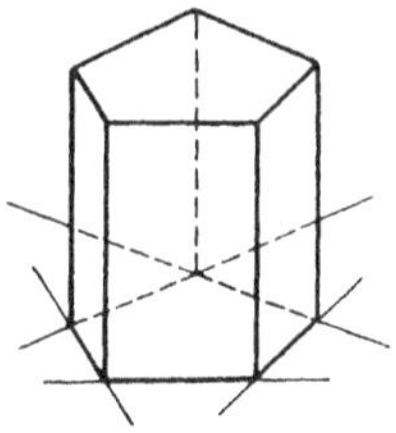

FIG. 2-47. Tipping lines of a pentagonal prism.

(a) the greater the horizontal distance from the center of gravity to the nearest (most dangerous) tipping line, (b) the lower the center of gravity, and (c) the larger the weight. These facts are expressed by the concept of *stability*. The stability of a body with respect to a specified tipping line is measured by the horizontal force, through the center of gravity, which is necessary to balance the moment of the body's weight.

EXAMPLE. Consider the stability of a rectangular pyramid (Fig. 2-48). From the top and side views of Fig. 2-48, we see that

$$S_b \frac{h}{4} = w \frac{a}{2}.$$

Hence, the stability with respect to tipping line b is

$$S_b = \frac{2wa}{h}.$$

Relative to the tipping line a,

$$S_a \frac{h}{4} = w \frac{b}{2}, \quad S_a = \frac{2wb}{h}.$$

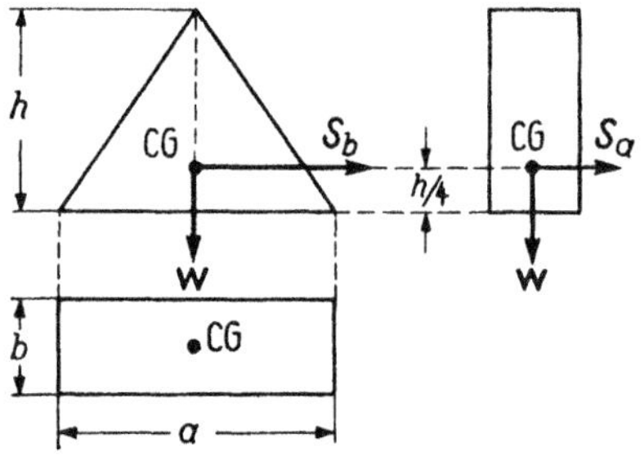

FIG. 2-48. Stability of a rectangular pyramid.

CHAPTER 3

DYNAMICS OF A PARTICLE

18. Coordinate Systems and Fundamental Concepts of Dynamics

The subject of dynamics deals with the motion of bodies, and with the causes of their motion. The concepts of *rest* and *motion* have a physical significance only with respect to a specified observer or reference system. Ordinarily, motion is referred to an observer at rest relative to the surface of the earth. The earth's surface then becomes the reference system. In the following discussion, we shall for the most part make use of a coordinate system at rest with respect to the earth.

Having agreed on the choice of a reference system, we can say that a particle is at rest if its coordinates x, y, and z relative to this reference system do not change with time. It is in motion relative to the coordinate system if one or more of its coordinates change with time.

Together with the concepts of rest and motion in dynamics, the concept of a *particle*, or a *point mass*, must also be introduced. This physical concept, which can only be approximately realized in practice, makes it possible for us to postpone the consideration of the rotational motion of an extended body. The dynamics of systems of particles will be applied later to the motions of extended bodies, and this limitation will be removed. It is much more convenient, however, to begin the development of the concepts of dynamics by considering only a particle.

Definition. A *particle*, or *point mass*, is defined as a body whose dimensions are vanishingly small, but whose mass is finite.

Later in this chapter, when we speak of spheres, carts, locomotives, and so on, we shall think of the entire mass of these systems as concentrated at a point at the center of gravity. Instead of the center of gravity, we prefer to speak of the *center of mass* of the system. The concept of center of mass will be defined when we take up the mechanics of systems of particles in Chapter 4, at which time the reason for preferring it to the center of gravity will be made clear.

19. Kinematics

Kinematics is the study of the geometrical properties of motion. It does not deal with the forces which bring about the motion, so that the concept of mass has no role in kinematics.

19-1 The Path of a Particle

When a particle moves with respect to a coordinate system, it passes through a continuous series of points of space with the passage of time.

DEFINITION. The *path* of a particle moving with respect to a coordinate system is defined as the totality of all points through which the particle passes in the course of time.

The motion of a particle is known if its coordinates are known for any arbitrary value of the time t. This will be the case if the coordinates x, y, and z of the particle are given as functions of the time:

$$x = f_1(t), \quad y = f_2(t), \quad z = f_3(t). \tag{19-1}$$

These three functions specify the *trajectory* of the particle in parametric form. The three equations may be combined into a single vector equation by considering the coordinates x, y, and z as the components of a vector $\boldsymbol{r}$, extending from the origin of the coordinate system to the position of the particle (Fig. 3-1). Equations (19-1) can then be written in the form

$$\boldsymbol{r} = \boldsymbol{r}(t).$$

This concise notation contains exactly the same information as the three component equations (19-1). The magnitude of the vector $\boldsymbol{r}$ is given by

$$r = \sqrt{(x^2 + y^2 + z^2)},$$

and it equals the distance from the origin to the particle.

The three direction cosines of the vector are

$$\cos\alpha = x/r, \quad \cos\beta = y/r, \quad \cos\gamma = z/r.$$

They satisfy the relation $\cos^2\alpha + \cos^2\beta + \cos^2\gamma = 1$.

In a short time interval Δt, the coordinates of the particle change by Δx, Δy, and Δz, and the particle moves a long a small element Δs of its path (Fig. 3-2). The vector $\Delta \boldsymbol{s}$, tangent to the path, is called the *displacement* or

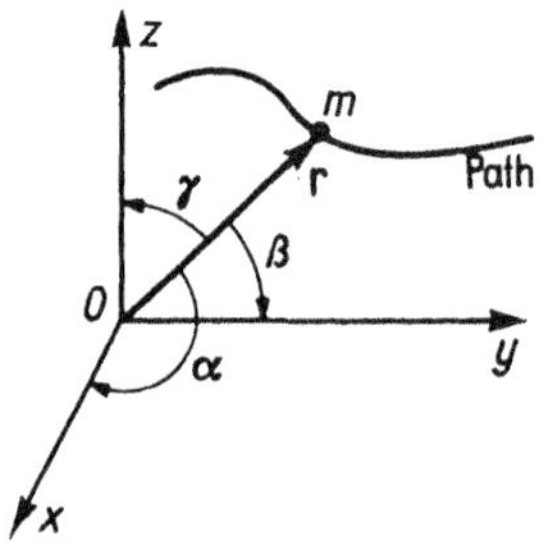

FIG. 3-1

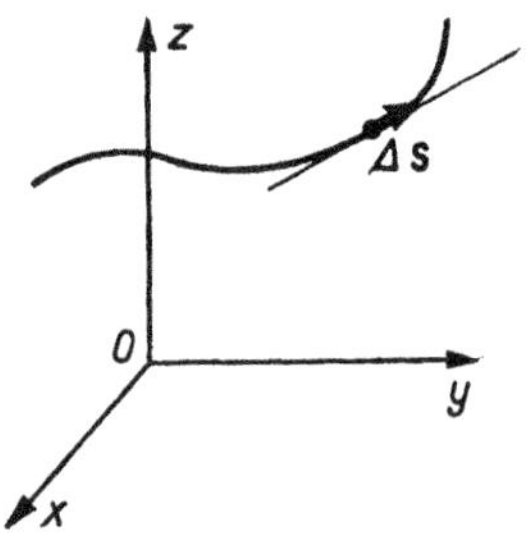

FIG. 3-2

FIG. 3-1. Path of a point mass (or particle).

FIG. 3-2. Displacement (element of path length) Δs.

element of path length. It has a unique sense, determined by the direction in which the particle moves as time elapses. The algebraic sum of the magnitudes of the displacement is the total *path length s.*

19-2 Velocity

The motion of a particle along a given path can take place in many different ways. Additional concepts are therefore necessary for a complete description of the motion.

DEFINITION. The *average velocity* $\mathbf{v}$ of a particle over the time interval Δt is defined as the quotient of the displacement $\Delta\mathbf{s}$ and the time Δt in which the particle travels this distance:

$$\mathbf{v} = \Delta\mathbf{s}/\Delta t.$$

The *instantaneous velocity* $\mathbf{v}(t)$ of a particle is defined as the limit, as Δt goes to zero, of the quotient of the path element $\Delta\mathbf{s}$ and the time Δt in which the particle travels this distance. This limiting quotient is called a derivative in differential calculus:

$$\mathbf{v}(t) = \lim_{\Delta t \to 0} \Delta\mathbf{s}/\Delta t = d\mathbf{s}/dt.$$

This vector equation comprises the following pieces of information:

(1) The *magnitude* of $\mathbf{v}$, often called the *speed*, equals the quotient of the magnitude of $\Delta\mathbf{s}(d\mathbf{s})$ and the magnitude of $\Delta t(dt)$:

$$v = \Delta s/\Delta t \quad \text{(magnitude equation)}.$$

(2) The *direction* of the vector $\mathbf{v}$ is the same as the direction of $\Delta\mathbf{s}(d\mathbf{s})$. The time interval $\Delta t(dt)$ is a scalar and does not affect the direction. The velocity vector is always tangent to the path. Velocity $\mathbf{v}$ is a *derived* quantity, and includes both the concepts of distance and of time. Its dimensions are $[v] = [LT^{-1}]$, and its units are: cgs system, 1 cm/sec; mks system, 1 m/sec; fps system, 1 ft/sec.

A graph of the magnitude of the velocity $\mathbf{v}$ of a particle as a function of time t is called the *velocity–time* (abbreviated *v-t*) *diagram* of the motion of the particle (Fig. 3-3). It follows from the definition $v = \Delta s/\Delta t$ that the increase of the path length, Δs, in a short time interval Δt, is $\Delta s = v\Delta t$.

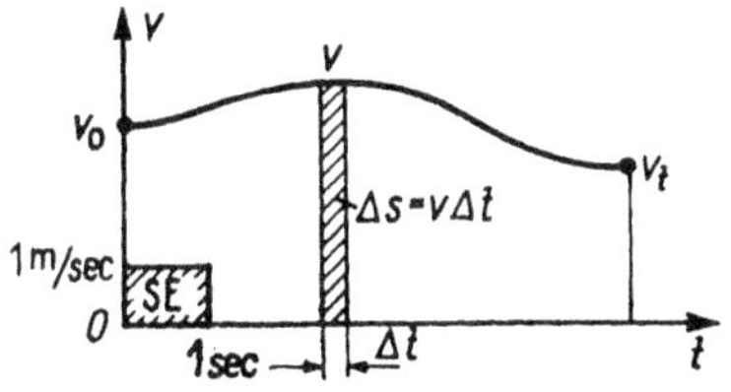

FIG. 3-3. Velocity–time diagram. The unit of surface area SE corresponds to one meter of path length.

The shaded element of area $v\Delta t$ in Fig. 3-3 represents the increase in path length Δs. The total length of path s is obtained by scalar addition of all path elements, that is,

$$s = \sum_i \Delta s_i = \sum_i v_i \Delta t_i .$$

Using the methods of calculus for the above, we would say

$$v = ds/dt, \quad ds = v\,dt,$$

$$s = \int ds = \int v(t)\,dt.$$

The path length is therefore represented geometrically by the area under the curve on the velocity–time graph. Thus can be measured by graphical or analytical integration if we are given $v(t)$. The measured area under the curve, say in cm^2, is divided by the area in cm^2 of the surface element (SE) and multiplied by the unit of path length which this element represents. The result is independent of the choice of units in which the area is measured.

19-3 Acceleration

The velocity of a moving particle tells us the rate at which its position is changing. If the velocity is known at all instants, the distance of the particle from a known starting point can be found. However, the velocity $\boldsymbol{v}$ is, in general, not constant in either magnitude or direction. This fact is described in terms of an additional quantity called the *acceleration.*

DEFINITION. The *average acceleration* $\boldsymbol{a}$ of a particle over the time interval Δt is defined as the quotient of the vector change in velocity $\Delta\boldsymbol{v}$ and the time interval in which this change has taken place:

$$\boldsymbol{a} = \Delta\boldsymbol{v}/\Delta t.$$

The *instantaneous acceleration* $\boldsymbol{a}(t)$ of a particle with respect to a fixed coordinate system is defined as the limit as Δt approaches zero of the quotient of the vector change in velocity, $\Delta\boldsymbol{v}$, and the time interval Δt in which this change takes place:

$$\boldsymbol{a}(t) = \lim_{\Delta t \to 0} \Delta\boldsymbol{v}/\Delta t = d\boldsymbol{v}/dt.$$

Since velocity is a vector quantity, changes in velocity include changes in direction as well as in magnitude. The acceleration has the dimensions $[a] = [LT^{-2}]$, and the units: cgs system, 1 cm/sec^2; mks system, 1 m/sec^2; fps system, 1 ft/sec^2.

Properties of acceleration. Acceleration is obtained by taking the difference between two velocity vectors and therefore is itself a vector. Dividing the vector $\Delta\boldsymbol{v}(d\boldsymbol{v})$ by the scalar $\Delta t(dt)$ changes only the magnitude. To determine the acceleration of a moving particle, it is necessary to find the change in velocity in a time interval Δt. This can be done graphically from a diagram of the path by constructing the velocity vectors at two neighboring points

P_1 and P_2. If the two velocity vectors are drawn with a common origin, we obtain a polar diagram [Fig. 3-4 (b)] from which $\Delta \boldsymbol{v}$ can be easily determined.

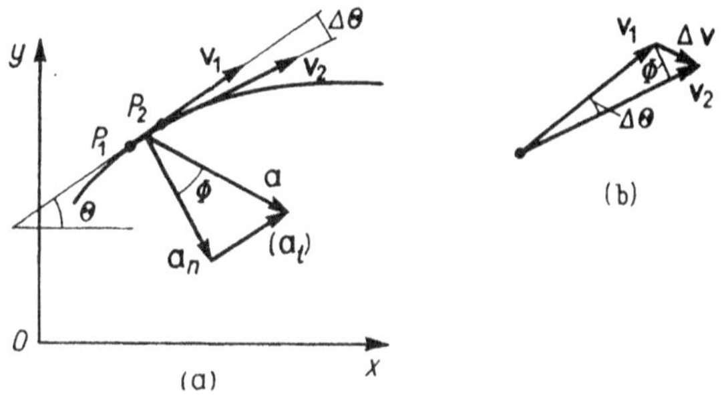

FIG. 3-4 (a). Path of a point mass in a plane, showing its velocity and acceleration. (b) Polar diagram of velocity vectors for determining $\Delta \boldsymbol{v}$.

From this construction, we obtain the vector $\Delta \boldsymbol{v} = \boldsymbol{a}\Delta t$, which determines the direction of the acceleration vector $\boldsymbol{a}$. The line of action of $\boldsymbol{a}$ passes through the point of intersection of the tangents to the path at P_1 and P_2. The magnitude of $\boldsymbol{a}$ is $a = \Delta v/\Delta t$.

It is often helpful to resolve the acceleration into two rectangular components: (a) the tangential component $\boldsymbol{a}_t$ in the direction of the tangent to the path; (b) the normal component $\boldsymbol{a}_n$ in the direction at right angles to the path.

The motion of a particle is completely determined if we know its acceleration, together with the initial position and initial velocity. From these three quantities, the position and velocity at any time can be found, as we shall show in the following sections.

20. The Most Important Cases in Kinematics

20-1 Uniform Rectilinear Motion

In *uniform rectilinear motion,* the velocity vector remains constant in magnitude and direction, and the average and instantaneous velocities are

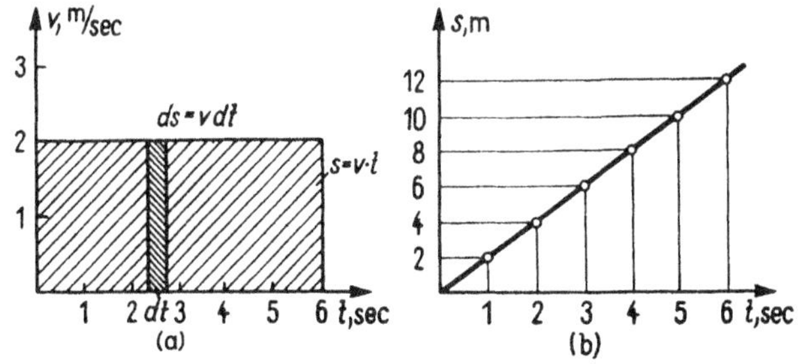

FIG. 3-5. Uniform rectilinear motion. (a) Velocity–time diagram. (b) Path length–time diagram.

equal. The path is a straight line in the direction of the vector $\boldsymbol{v}$. The v-t diagram is given in Fig. 3-5(a). Motion of this sort is illustrated, for example, by a sphere falling in a viscous liquid such as glycerin. In time dt, the particle covers a distance $ds = vdt$ along its path, shown in Fig. 3-5(a) as the small shaded rectangle. The total length of path covered in time t is the sum of all such areas and increases linearly with time [Fig. 3-5(b)] according to the relation $s = vt$.

20-2 Rectilinear Motion with Constant Vector Acceleration (pure tangential acceleration)

This case occurs frequently in practice. Since the magnitude and direction of $\boldsymbol{a}$ are constant, motion can take place along a straight line *only* if the acceleration vector $\boldsymbol{a}$ is parallel to the velocity vector, that is, if the acceleration lies along the direction of the path (Fig. 3-6). We then speak of

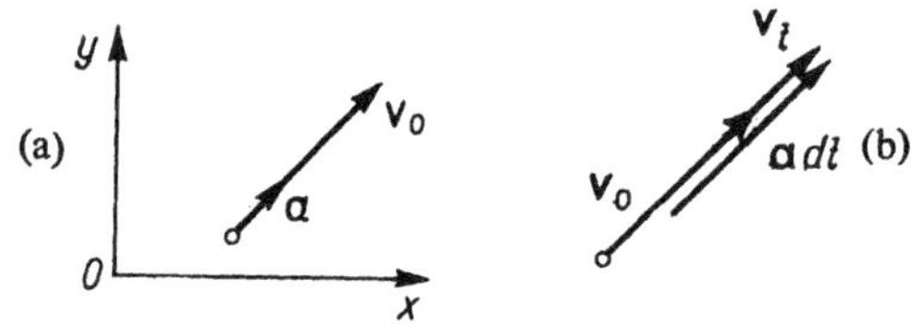

FIG. 3-6. Accelerated rectilinear motion. (a) Velocity and acceleration at $t = 0$. (b) Polar diagram of velocity and acceleration.

uniformly accelerated motion. The velocity changes in magnitude only. From the definition of acceleration as $\boldsymbol{a} = d\boldsymbol{v}/dt$, it follows that

$$d\boldsymbol{v} = \boldsymbol{a}\,dt.$$

In equal time intervals dt, the change in the magnitude of the velocity is the same and is given by adt. The total change of velocity in time t is therefore at; that is, the velocity increases uniformly with time. In an acceleration-time (a-t) diagram, the change in velocity is represented by a small rectangle (Fig. 3-7). If we represent the initial velocity by v_0, the velocity v_t at any time t is

$$v_t = v_0 + at. \tag{20-1}$$

To calculate the distance travelled, we return to the definition of velocity,

$$v = ds/dt.$$

The displacement $ds = vdt$ is represented by the shaded strip in the v-t diagram [Fig. 3-8(a)]. The distance s_t covered in time t is the sum of all elements ds from time 0 to time t. This sum is represented in the diagram by the area of a trapezoid, and the distance moved is therefore

$$s_t = \frac{v_0 + v_t}{2}t = v_0 t + \frac{1}{2}at^2. \tag{20-2}$$

FIG. 3-7

FIG. 3-8

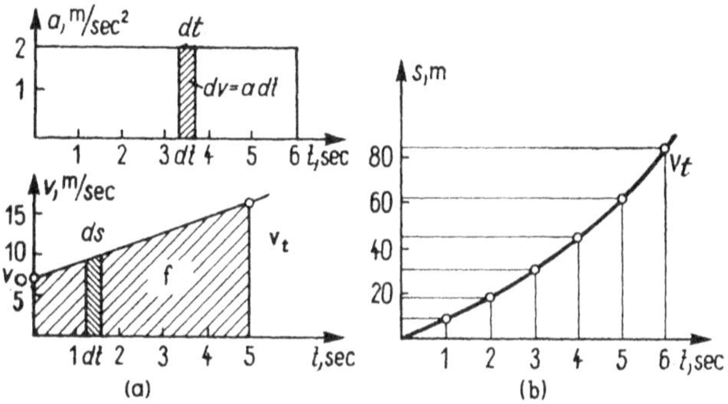

FIG. 3-7. Acceleration–time diagram for uniform accelerated rectilinear motion. Δv is the velocity increment in time Δt.

FIG. 3-8. Uniformly accelerated rectilinear motion. (a) Velocity–time diagram. (b) Path length–time diagram.

These two equations suffice to answer all kinematical questions regarding the motion. Another equation can be obtained by eliminating t between them:

$$v_t^2 = v_0^2 + 2as_t. \tag{20-3}$$

This particular case of constant acceleration is important because it describes the motion of a free particle moving in a vertical direction near the surface of the earth. Experimental study of such motion shows that the acceleration due to the earth's gravitational field is constant and directed vertically downward. At latitudes in the neighborhood of 45°, the magnitude of the acceleration is about 981 cm/sec², 9.81 m/sec², or 32.2 ft/sec². This particular acceleration is called the *acceleration due to gravity* and is represented by g. Also, to a first approximation, the starting and stopping of most vehicles can be considered as motion with constant acceleration. Here, however, the acceleration is not known and must be computed from given data.

EXAMPLE. A railway train has an initial velocity v_0 of 80 mi/hr. The brakes are applied for 20 sec, producing a constant negative acceleration. At the end of this time the velocity is 50 mi/hr. Compute the acceleration and the distance covered:

$$v_0 = 80 \text{ mi/hr} = 117.3 \text{ ft/sec},$$

$$v_t = 50 \text{ mi/hr} = 73.3 \text{ ft/sec},$$

$$t = 20 \text{ sec}.$$

From Eqn. (20-1),

$$a = \frac{44.0 \text{ ft/sec}}{20 \text{ sec}} = 2.20 \frac{\text{ft}}{\text{sec}^2}.$$

From Eqn. (20-2),

$$s_t = \frac{(117.3 + 73.3) \text{ ft/sec}}{2} \times 20 \text{ sec} = 1906 \text{ ft}.$$

20-3 Uniform Circular Motion
(pure normal acceleration)

Figure 3-9(a) represents a particle moving on the circumference of a circle of radius r with a velocity $\boldsymbol{v}$ of constant magnitude. This is an accelerated motion, since the direction of the velocity is continually changing.

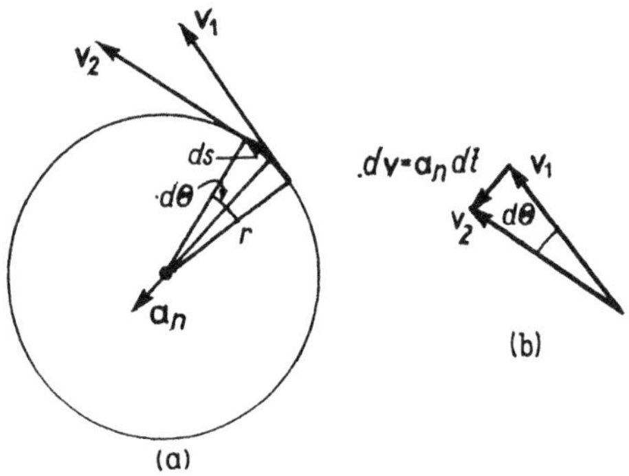

FIG. 3-9. Uniform circular motion.

In equal time intervals dt, the direction of the velocity vector $\boldsymbol{v}$ turns through equal angles $d\theta$ (expressed in radians). The radius vector of the moving point, in the same time intervals, rotates through the same angles $d\theta$. It follows that the magnitude v of the velocity can be written as

$$v = ds/dt = r\,d\theta/dt = r\omega. \tag{20-4}$$

The quantity $\omega = d\theta/dt$ is called the *angular velocity* of the particle. The speed v equals the product of the radius and the angular velocity.

DEFINITION. The *angular velocity* ω of a particle moving uniformly in a circle is defined as the quotient of the angle $d\theta$ through which the radius vector of the particle turns divided by the time dt that elapses during the motion:

$$\omega = d\theta/dt.$$

It has the dimensions $[\omega] = [T^{-1}]$, and the units 1 radian sec^{-1} in all systems.

To calculate the acceleration, we note that since the magnitude of the velocity remains constant, the change in velocity is always perpendicular to the velocity itself. The acceleration is therefore directed along the radius toward the center of the circle. The motion is one of pure *normal* acceleration and is also called a *radial* acceleration. Its magnitude is easily derived with the help of the polar diagram, Fig. 3-9(b), from which the relation

$$dv = v\,d\theta = a_n\,dt$$

is obtained. Then using Eq. (20-4) it follows that

$$a_n = v\,d\theta/dt = v\omega = r\omega^2 = v^2/r. \tag{20-5}$$

The radial acceleration is therefore proportional to ω^2 and also to v^2.

Motion in a circle has a property we have not previously encountered, namely, that of being *periodic* in time. The motion repeats itself in all kinematic aspects after the lapse of a certain time T. This time T is called the *period.* It corresponds to the duration of one complete revolution, and is a scalar quantity. Another concept, called the *frequency*, is related to the period.

DEFINITION. The *frequency* f of a uniform circular motion is defined as the number of complete revolutions per unit time. Frequency is also a scalar quantity. It has the dimensions $[f] = [T^{-1}]$, and the units (in all systems) 1 revolution (or cycle) $\sec^{-1}$.

The relation between f and T follows directly from their definitions. If a particle makes f revolutions per second, the period or elapsed time of one revolution is

$$T = 1/f.$$

Also, the relation between the angular velocity ω and the frequency f follows at once from the definition of the latter. If the frequency is f, the radius vector turns through an angle $2\pi f$ in unit time, and hence

$$\omega = 2\pi f.$$

Uniform circular motion plays an important role in the rotation of all rigid bodies about fixed axes. It is applied to a number of practical problems such as transmissions, gear systems, and rolling wheels. The concept of "rolling without slipping" is of importance in many problems of this sort. A wheel is said to roll without slipping if the point of contact of the wheel is instantaneously at rest, relative to the surface on which it rolls. In this case, the distance moved by a point on the circumference of the wheel with respect to the center of the wheel equals the distance which the point of contact moves along the surface. When a wheel rolls on a plane surface without slipping, the peripheral velocity relative to the center is equal in magnitude to the velocity of the center of the wheel relative to the surface.

EXAMPLE. 1. The center of gravity of an automobile moves in a circle of radius $R = 150$ ft with a velocity $v = 30$ ft/sec (Fig. 3-10). The wheel separation, s, is 4.0 ft. (a) What is the angular velocity Ω of the automobile, as seen from the center of curvature C? Since $v = R\Omega$

$$\Omega = \frac{v}{R} = \frac{30\text{ ft/sec}}{150\text{ ft}} = 0.2\,\frac{\text{rad}}{\text{sec}}\;(\simeq 11.5^\circ\ \text{sec}^{-1}).$$

(b) The radius r of each wheel is 1 ft. Find the peripheral velocities and the angular velocities of the inner and outer wheels.

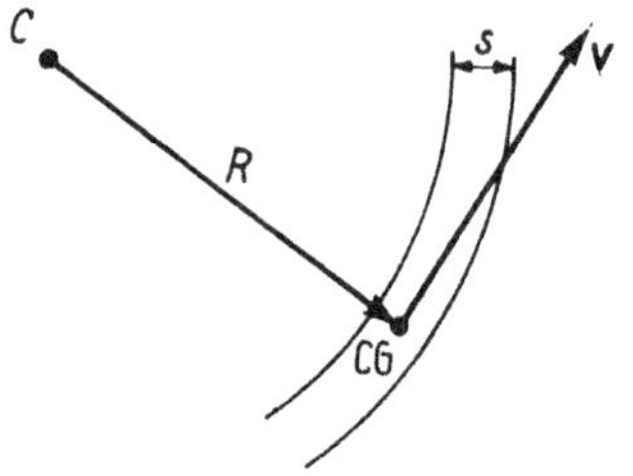

FIG. 3-10. Auto travelling on a curve.

Inner wheels:

$$v_i = \left(R - \frac{s}{2}\right)\Omega = \frac{(2R - s)\,v}{2R}$$

$$= \frac{(300 - 4)\,30}{300} = 29.6\,\frac{\text{ft}}{\text{sec}}.$$

$$\omega_i = \frac{v_i}{r} = \frac{29.6\ \text{ft/sec}}{1\ \text{ft}} = 29.6\,\frac{\text{rad}}{\text{sec}},$$

$$f_i = \frac{\omega_i}{2\pi} = 4.71\,\frac{\text{rev}}{\text{sec}}.$$

Outer wheels:

$$v_0 = \frac{(2R + s)\,v}{2R} = 30.4\,\frac{\text{ft}}{\text{sec}}$$

$$\omega_0 = \frac{v_0}{r} = 30.4\,\frac{\text{rad}}{\text{sec}},$$

$$f_0 = \frac{\omega_0}{2\pi} = 4.84\,\frac{\text{rev}}{\text{sec}}.$$

(c) Find the radial acceleration of the automobile as a whole, and the radial acceleration of points on the peripheries of the inner and outer wheels:

$$a_{\text{auto}} = \frac{v^2}{R} = \frac{(30\ \text{ft/sec})^2}{150\ \text{ft}} = 6\,\frac{\text{ft}}{\text{sec}^2},$$

$$a_i = \frac{v_i^2}{r} = 876\,\frac{\text{ft}}{\text{sec}^2},$$

$$a_0 = \frac{v_0^2}{r} = 924\,\frac{\text{ft}}{\text{sec}^2}.$$

EXAMPLE. 2. *Kinematics of the moon's orbit.* The velocity of the moon in its orbit, and its radial acceleration, can be computed from the moon's period ($T = 27.3$ day), and its distance from the earth ($r_{av} = 384{,}400$ km),

$$v = r\omega = \frac{2\pi r}{T} = \frac{2\pi \times 384{,}400\ \text{km}}{27.3 \times 24 \times 3600\ \text{sec}} = 0.93\,\frac{\text{km}}{\text{sec}},$$

$$a_r = \frac{v^2}{r} = \frac{(0.93 \times 10^5\ \text{cm/sec})^2}{3.84 \times 10^{10}\ \text{cm}} = 0.23\,\frac{\text{cm}}{\text{sec}^2}.$$

The radial acceleration of 0.23 cm/sec^2 means that the velocity changes in each second by 0.23 cm/sec, in a radial direction. Thus the velocity remains constant in magnitude but is deviated in direction in such a way that the moon remains in a circular orbit.

20-4 Representation of Angular Velocity ω as a Vector

In every case of rotation about an axis, the axis defines a direction in space. This suggests the possibility of representing an angular velocity ω by a vector along the axis of rotation, whose length corresponds to the magnitude of $\boldsymbol{\omega}$. The right-hand rule (Section 9-2) can be used to define the sense of the vector, since by this rule the sense of the vector is uniquely related to the direction of rotation. With the help of the vector $\boldsymbol{\omega}$, the velocity of any point in a rotating rigid body can be represented by a vector product.

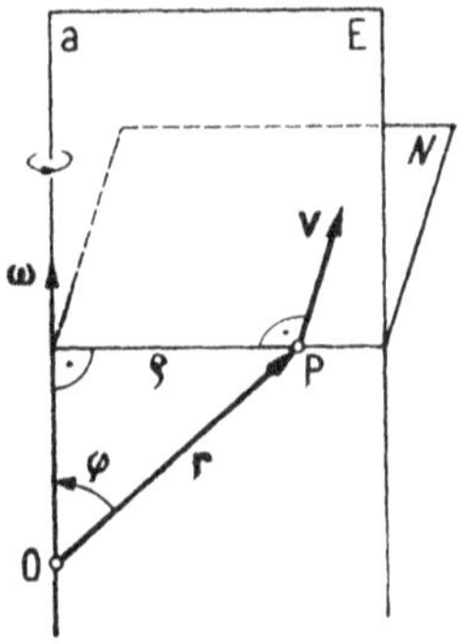

FIG. 3-11. Diagram showing how the velocity $\boldsymbol{v}$ is the vector product of $\boldsymbol{r}$ and $\boldsymbol{\omega}$.

In Fig. 3-11, *Oa* is the axis of rotation of a rigid body and *P* is an arbitrary point in the body whose velocity we wish to find. As the body rotates, *P* moves in a circle of radius ϱ. Let $\boldsymbol{r}$ be the radius vector drawn to *P* from an arbitrary origin *O* on the rotation axis. Then

$$\varrho = r \sin \varphi,$$

and the magnitude of the velocity of point *P* is

$$v = \omega\varrho = \omega r \sin \varphi .$$

The direction of $\boldsymbol{v}$ is perpendicular to the plane *E* which contains the vectors $\boldsymbol{\omega}$ and $\boldsymbol{r}$. It is therefore possible to represent $\boldsymbol{v}$ by the vector product $\boldsymbol{v} = \boldsymbol{\omega} \times \boldsymbol{r}$.

20-5 Uniformly Accelerated Circular Motion

Motion in a circle with constant angular acceleration is called uniformly accelerated circular motion.

DEFINITION. *Angular acceleration* α is defined as the limiting quotient of the change in angular velocity $d\omega$ divided by the time interval dt in which the change takes place:

$$\alpha = d\omega/dt.$$

It has the dimensions $[\alpha] = [T^{-2}]$, and the units (all three systems) 1 radian $\sec^{-2}$.

From this definition of α, it follows that

$$d\omega = \alpha\, dt,$$

and if α is constant, we have

$$\omega_t = \omega_0 + \alpha t,$$

where ω_0 is the angular velocity at time $t = 0$. Using the relation $\omega_t = d\theta/dt$, we find that the angle θ turned through in any time t is (as in Section 20-2)

$$\theta_t = \omega_0 t + \tfrac{1}{2}\alpha t^2.$$

The relations between α, ω, and θ correspond exactly to those between a, v, and s given in Eqns. (20-1), (20-2), and (20-3). Equation (20-3) becomes

$$\omega_t^2 = \omega_0^2 + 2\alpha\theta_t.$$

The magnitude of the tangential acceleration at the perimeter of the circle is

$$a_{tg} = r\alpha.$$

To a first approximation, the starting and stopping of wheeled vehicles, motors, generators, etc., are motions that can be considered to take place with constant angular acceleration.

20-6 Trajectory of a Projectile in Vacuum

Using the methods of vector algebra, we can resolve any arbitrary motion into its components parallel to the coordinate axes. Superposition of these components then gives the original motion again.

Motion of a projectile over a horizontal surface. A particle moving freely in a vacuum is acted on by only one force, its weight. This is of constant magnitude and is directed vertically downwards. From Newton's second law, the acceleration is then also constant and vertically downward.

Let us construct a coordinate system with the x-axis horizontal and the y-axis vertical (Fig. 3-12). The particle is projected from the origin at an angle of elevation α relative to the x-axis and with an initial velocity $\boldsymbol{v}_0$. The motion is to be resolved into x- and y-components.

In the x-direction,

$$a_x = dv_x/dt = 0.$$

But if $dv_x = 0$, then

$$v_x = \text{const} = v_0 \cos\alpha,$$

and

$$x = v_0 t \cdot \cos\alpha. \tag{20-6}$$

The horizontal component of velocity therefore remains constant, and the particle moves in the x-direction with uniform rectilinear motion.

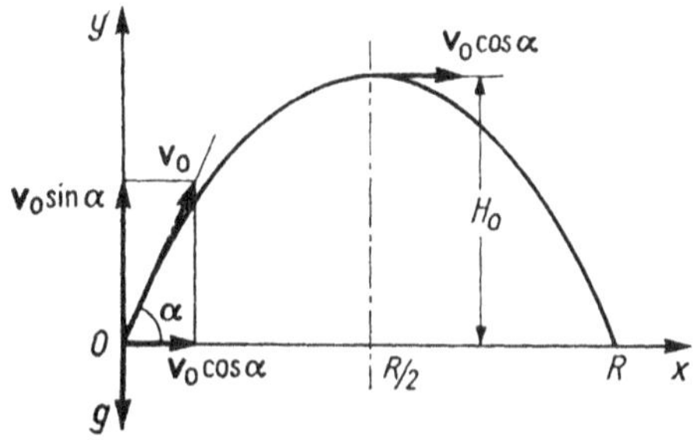

FIG. 3-12. Trajectory over a horizontal surface.

In the y-direction,

$$a_y = -g.$$

The motion in the y-direction is uniformly accelerated, with a constant acceleration $-g$ and an initial velocity $v_0 \sin\alpha$. Therefore

$$v_y = v_0 \sin\alpha - gt,$$

$$y = v_0 t \sin\alpha - \tfrac{1}{2}gt^2. \tag{20-7}$$

Equations (20-6) and (20-7) describe completely all kinematical aspects of the motion. Any details of the motion can be obtained from them, as shown below.

The parametric equations of the trajectory are

$$x = v_0 t \cdot \cos\alpha, \quad y = v_0 t \cdot \sin\alpha - \tfrac{1}{2}gt^2.$$

The explicit equation of the trajectory is obtained by eliminating t:

$$y = \frac{x \sin\alpha}{\cos\alpha} - \frac{gx^2}{2v_0^2 \cos^2\alpha} = x \tan\alpha - \frac{gx^2}{2v_0^2 \cos^2\alpha}. \tag{20-8}$$

The trajectory is therefore a second-degree parabola with its axis vertical and it opens downward. It intersects the x-axis at the points $x = 0$ (the starting point), and, from the equation $\tan\alpha - gx/2v_0^2 \cos^2\alpha = 0$, at

$$x = \frac{v_0^2 \sin 2\alpha}{g}.$$

From these values of x, the horizontal range R is

$$R = \frac{v_0^2 \sin 2\alpha}{g},$$

which is a maximum when $\alpha = 45°$.

The time required to reach the highest point is called the ascent time, T_a. The time between the two crossings of the x-axis is called the flight time, T_f.

The ascent time can be found from the fact that at the highest point $v_y = 0$:

$$v_y = v_0 \sin\alpha - gT_a = 0, \quad T_a = \frac{v_0 \sin\alpha}{g}.$$

The flight time can be found from the constant x-velocity and the horizontal range:

$$R = \frac{2v_0^2 \sin\alpha \cos\alpha}{g} = v_0 T_f \cos\alpha, \quad T_f = \frac{2v_0 \sin\alpha}{g} = 2T_a.$$

To calculate H_0, the maximum height, we insert the time of ascent into Eqn. (20-7):

$$H_0 = v_0 \sin\alpha \frac{v_0 \sin\alpha}{g} - \frac{gv_0^2 \sin^2\alpha}{2g^2} = \frac{v_0^2 \sin^2\alpha}{2g}.$$

The x-coordinate of the highest point follows from the time of ascent and Eqn. (20-6):

$$x_{H_0} = v_0 \cos\alpha \frac{v_0 \sin\alpha}{g} = \frac{1}{2}\frac{v_0^2 \sin 2\alpha}{g} = \frac{R}{2}.$$

The apex of the trajectory therefore lies at the midpoint of the horizontal range, which is obvious from the symmetry of the parabolic path.

Motion of a projectile over an inclined plane. Let us construct a coordinate system, as in Fig. 3-13, with the x-axis horizontal, the y-axis vertical, and

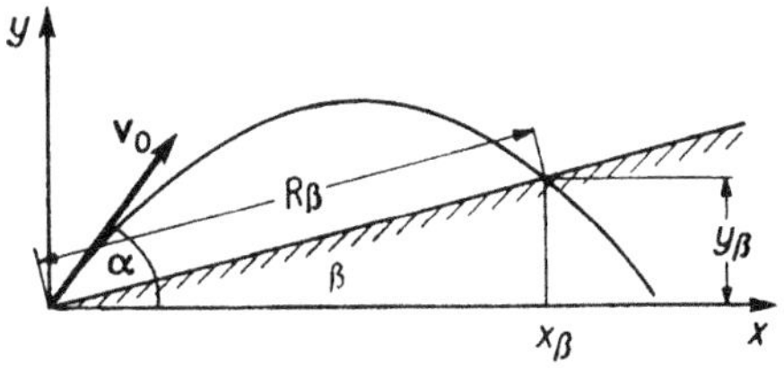

FIG. 3-13. Trajectory over an inclined plane.

the projection point at the origin. The angle of elevation is α and the angle of inclination of the plane is β. To compute the range R_β, along the plane, we note that

$$x_\beta = v_0 t \cos\alpha.$$

From Eqn. (20-8) and Fig. 3-13, it follows that

$$y_\beta = x_\beta\left(\tan\alpha - \frac{gx_\beta}{2v_0^2 \cos^2\alpha}\right) = x_\beta \tan\beta.$$

Hence

$$\tan\beta = \tan\alpha - \frac{gx_\beta}{2v_0^2 \cos^2\alpha},$$

or

$$x_\beta = \frac{2v_0^2 \cos^2 \alpha}{g}(\tan \alpha - \tan \beta).$$

The range on the inclined plane is therefore

$$R_\beta = \frac{x_\beta}{\cos \beta} = \frac{2v_0^2 \cos^2 \alpha(\tan \alpha - \tan \beta)}{g \cos \beta}$$

or

$$R_\beta = \frac{2v_0^2 \cos \alpha \sin(\alpha - \beta)}{g \cos^2 \beta}.$$

21. Newton's Laws

Newton's laws are the foundation of the science of mechanics, and hence are the backbone of all physics. A significant portion of physical science consists of the logical consequences of these laws. They were first published in 1687 by Sir Isaac Newton (1642–1727) in his work *Philosophiae Naturalis Principia Mathematica.*

The concepts of Newton were so inspired and his laws were so fruitful that not until 1905 did certain difficulties in connection with the electrodynamics of moving media require their reformulation. This was done by Albert Einstein, in an equally gifted way, using his theory of relativity. The work of Einstein contains the laws of Newton as a special case.

Every study of dynamics therefore begins today with the discussion of Newton's laws. In classical physics, these are considered as postulates which are valid without exception, and which have been confirmed by hundreds of years of experimental proof.

Newton's first law, sometimes known as the *law of inertia,* was anticipated by the work of Galileo (1638), but was set by Newton within a wider framework. In the first law, Newton establishes the concept of dynamically equivalent coordinate systems, and recognizes the concept of acceleration as the distinguishing characteristic of every change of motion.

The *second law* defines the concept of mass and relates it quantitatively to the concepts of acceleration and force.

The *third law* makes possible the transition from a single particle to a system of many particles.

These three fundamental laws will now be considered in detail.

21-1 Newton's First Law

Every body continues in a state of rest or of uniform rectilinear motion (inertial motion) unless it is acted upon by some external force.

Changes of motion consist exclusively of changes in the magnitude or direction of the velocity vector. Both of these correspond to an acceleration.

The characteristic quantities of a motion (path, position, velocity, and acceleration) can be specified only with respect to some coordinate system. It is evident, however, that three of these, the path and the position and velocity vectors, cannot be inherent properties of the motion, since they depend on the choice of coordinate system.

EXAMPLE. A railway car moves with constant velocity v_B relative to the earth. An observer A sits in the car and an observer B stands beside the track. How does each describe the motion of a particle which moves vertically upward inside the car with a constant velocity v_A relative to the car?

From the standpoint of observer A, the path of the particle is a vertical straight line; the displacement ds is directed vertically upward and its magnitude is $v_A\,dt$; the velocity vector v_A points vertically upward.

For observer B, the path is a straight line but is inclined at an angle α with the horizontal (Fig. 3-14), where $\tan \alpha = v_A/v_B$; the displacement $ds = \sqrt{[(v_A)^2 + (v_B)^2]}\,dt$. The velocity vector is in the same direction as the path, and its magnitude is $v = \sqrt{[(v_A)^2 + (v_B)^2]}$.

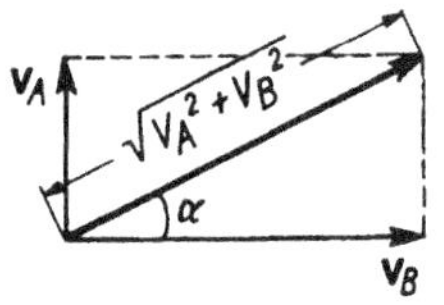

FIG. 3-14. Velocity of a moving point inside a moving vehicle, as seen by an outside observer at rest.

Neither observer tells us anything intrinsic about the motion when he describes it in this way. These are statements which depend only on the standpoint of the observer. A third description of the motion, fully equivalent to these two, would be given by a third observer moving relative to the others with uniform rectilinear motion. The descriptions of the same motion, with respect to the path, the displacement vector, and the velocity vector, are different for different observers, and their statements depend on the relative state of motion of the observer and the particle.

The situation is quite different for the acceleration. If a particle falls freely in the railway car, under the influence of its weight, its acceleration is that of free fall, and the two observers describe it as follows.

Observer A: The path is a straight line directed vertically downward; the acceleration is vertically downward and its magnitude is g.

Observer B: The path is a parabola having an initial horizontal tangent and a vertical axis; the acceleration is vertically downward and its magnitude is g.

This means, therefore, that there are quantities in the world which are the same in two coordinate systems which are moving relative to each other with constant velocity.

Coordinate systems in which force-free bodies experience no acceleration are called *inertial systems*. Furthermore, all coordinate systems moving with uniform rectilinear motion relative to an inertial system are also inertial systems.

All observers who carry out kinematic investigations with respect to an inertial system will observe the same acceleration for a given motion. Hence, when we speak with Newton of a "change of motion", and wish to determine it objectively, this acceleration must be measured only by observers at rest with respect to some inertial system. Only through the concept of inertial systems does the first law aquire an objective meaning. *All inertial systems are dynamically equivalent.*

21-2 Newton's Second Law

A force is recognizable by the fact that it can produce a deformation of a solid body, such as a coil spring. Newton's second law makes an additional statement regarding the effect of a force. This law states: *Every force acting on a body produces an acceleration. The acceleration is proportional to the force and has the same direction as the force*:

$$\boldsymbol{F} = m\boldsymbol{a}.$$

Thus, according to Newton's second law, force has a second property by which it can be recognized, namely, as the cause of acceleration. The second law consists of more than just the defining equation of force; it contains a physical statement about force. The following remarks may be made regarding the second law:

(1) The equation relating force and acceleration is a vector equation and thus is a statement of the three component equations

$$F_x = ma_x, \quad F_y = ma_y, \quad F_z = ma_z.$$

(2) The equation $\boldsymbol{F} = m\boldsymbol{a}$ implies two things: (a) The direction of $\boldsymbol{F}$ is the same as the direction of $\boldsymbol{a}$, and (b) the magnitude of $\boldsymbol{F}$ is $F = ma$.

The proportionality factor between a force and the acceleration produced by it is called the *mass* of the body. It is a scalar and can be considered as the resistance offered to acceleration.

In addition to the more obvious properties of constancy of mass and impenetrability, experiments have shown that matter also resists acceleration and attracts other matter. Despite this complex character, Newton recognized that there dwelt in every body a "quantity" which was independent both of the specific material of which it was composed and of its location in space and time. This "quantity" is of the same nature in metallic copper, in crystalline diamond, and in liquid water. It can be determined by measuring the acceleration $\boldsymbol{a}$ produced by a force $\boldsymbol{F}$ (as observed in an inertial coordinate system). We designate this "quantity" as the *mass* of the body.

Newton himself distinguished four essential properties of matter.† (a) Matter is *impenetrable*; that is, a point in space cannot be occupied simultaneously by two bodies. Any attempt to place a body at a point already occupied by another body leads to the production of forces (Newton's

† An element that has more than one property is never suitable as a fundamental element for axiomatic systems. Matter is a complex concept.

third law). (b) Every body resists acceleration in proportion to its mass; mass has *inertia.* (c) The mass of a quantity of matter is an invariant, that is, it is independent of position and of state of motion (law of conservation of mass). This statement is valid only for velocities that are small compared with the velocity of light c ($c = 3 \times 10^{10}$ cm/sec). For a velocity v of the same order of magnitude as c, the mass depends on v (see Section 24). (d) Every body exerts a force of attraction on every other body in proportion to the product of their masses (gravitation).†

Because of its invariance with position, time, and state of motion, mass is selected as a fundamental unit of the system of measurement. This has resulted in the adoption of a standard of mass. Mass has the dimension $[M]$, and the unit, in the cgs system, 1 gm (gram-mass). This equals one one-thousandth part of the standard kilogram-mass located in Paris. In the mks system, its unit is 1 kgm (kilogram-mass). This is the mass of the "kilogram des archives". In the fps system, its unit is 1 lbm (pound-mass). The United States standard of mass is the avoirdupois pound, 1 lbm = 453.59 gm.

Having defined the unit mass, we find that the unit of force in any system of measurement is uniquely determined by Newton's second law, $\boldsymbol{F} = m\boldsymbol{a}$.

DEFINITION. The unit force is that force which gives unit acceleration to unit mass. The dimensions of force are $[MLT^{-2}]$, and the units are: cgs system, 1 gm · cm · sec^{-2} (called 1 *dyne*); mks system, 1 kgm · m · sec^{-2} (called 1 *newton*); fps system, 1 lbm · ft · sec^{-2} (called 1 *poundal*).
Note: 1 newton = 1 kgm · m · sec^{-2} = 10^3gm · 10^2 cm · sec^{-2} = 10^5 dynes.

From the experimental fact that in vacua at a given place all bodies fall with the same constant acceleration, we can conclude with the help of the second law that a constant force acts on a freely falling body in a vacuum. This force is called the *weight* $\boldsymbol{w}$ of the body, that is, the force with which the mass of the rotating earth attracts the mass of the body. The second law then takes the form

$$\boldsymbol{w} = m\boldsymbol{g},$$

where $\boldsymbol{g}$ is the local acceleration due to gravity.

Precise measurements show that the magnitude of the acceleration due to gravity depends on position and varies with geographic latitude and with elevation above sea level. There are also small local variations in the magnitude. The acceleration due to gravity is approximately constant along a circle of latitude. Average values at sea level are:

At the equator:	$g = 978.04$ cm/sec^2 = 32.088 ft/sec^2.
At 45° latitude:	$g = 980.62$ cm/sec^2 = 32.173 ft/sec^2.
At the poles:	$g = 983.22$ cm/sec^2 = 32.257 ft/sec^2.

† Translators' Note: See Note 6, Appendix, of Cajori's revised translation of Newton's *Principia*, Univ. Calif. Press, 1946.

It follows, therefore, that the weight of one and the same mass varies with its location. For example, 1 gm weighs 978 dynes at the equator and 983 dynes at the poles.

There is another system of units, often used in engineering, called the *engineering* system, or the *gravitational* system of units, in which the weight of a standard mass is used for the definition of the unit of force.

DEFINITION. In a *gravitational* system of units, the unit of force is that force with which the earth attracts a standard unit of mass at a point where the acceleration due to gravity has the standard value.

In countries where the metric system is commonly used, the gravitational unit of force is the force acting on a standard kilogram mass at a place where the acceleration due to gravity is 9.80665 m/sec^2. This force is called 1 kilogram-force (1 kgf). Similarly the force on a standard gram mass is 1 gram-force (1 gf). In the United States and England, the gravitational unit of force is the force acting on a standard pound mass at a place where g has the value 32.176 ft/sec^2. This force is called 1 pound-force (1 lbf). (These force units are also called 1 pound-weight, 1 kilogram-weight, and 1 gram-weight, which are abbreviated 1 lbwt, 1 kgwt, and 1 gwt.)

It is easily seen that 1 lbf = 32.176 poundals, 1 kgf = 9.80665 newtons = 980,665 dynes, and 1 gf = 980.665 dynes.

The fact that in the gravitational systems a unit of force is chosen as a fundamental unit, instead of a unit of mass, leads to the consequence that the units of mass in these systems have unfamiliar values. The mass units are uniquely defined by the equation $\boldsymbol{F} = m\boldsymbol{a}$. If the pound-force is taken as the force unit, the corresponding mass unit is defined as follows:

DEFINITION. The unit of mass in the *gravitational* system is that mass which is given unit acceleration by the application of unit force.

Thus, in the United States and England, the gravitational unit of mass is that mass which is accelerated 1 ft/sec^2 by 1 pound-force. This unit is called a *slug*.

Let us calculate the mass of the standard pound body, expressed in slugs. When such a body falls freely at a point where g has its standard value, the force acting on it is 1 lbf, by definition, and its acceleration is 32.176 ft/sec^2. Hence using $F = ma$, we have

$$m = \frac{F}{a} = \frac{1 \text{ lbf}}{32.176 \text{ ft/sec}^2} = \frac{1}{32.176} \text{ slug}.$$

Since the mass of the body is also 1 pound-mass, we have the following equations:

$$1 \text{ lbm} = 1/32.176 \text{ slug} = 0.03108 \text{ slug},$$

$$1 \text{ slug} = 32.176 \text{ lbm}.$$

That is, a mass of 1 slug is 32.176 times as great as the mass of the standard pound body.

If the kilogram force is taken as the force unit, the corresponding unit of mass in the metric gravitational system is that mass which is accelerated 1 m/sec^2 when acted on by a force of 1 kilogram-force. This unit is called 1 *technical mass unit*, abbreviated TMU (1 TMU = 9.81 kgm).

We now consider two important concepts which are related to mass and weight.

DEFINITION. The *density* ϱ of a homogeneous body is defined as the quotient of its mass and its volume,

$$\varrho = m/V.$$

Its dimensions are $[ML^{-3}]$, and its units are: cgs system, 1 gm/cm^3; mks system, 1 kgm/m^3; fps system, 1 lbm/ft^3.

EXAMPLE. Water at 4°C:

$$\varrho = 1 \text{ gm/cm}^3 = 1000 \text{ kgm/m}^3 = 62.4 \text{ lbm/ft}^3.$$

Since density is the quotient of two scalars, it is also a scalar and is independent of position. (See Section 66-3, for a discussion of the variation of density with temperature.)

From the definition of specific weight (Section 16-2) as the ratio of the weight of a body to its volume, we see that the specific weight, ϱ, depends on position, since the weight depends on position. Specific weight has the dimensions $[MLT^{-2}]/L^3 = [ML^{-2}T^{-2}]$, and the units: cgs system, 1 dyne/cm^3; mks system, 1 newton/m^3; fps system, 1 poundal/ft^3.

EXAMPLE. Water at 4°C and 45° latitude:

$$\varrho = 980.62 \text{ dynes/cm}^3 = 9806.2 \text{ newton/m}^3 = 2007.4 \text{ poundal/ft}^3.$$

21-3 Newton's Third Law

The first and second laws of Newton are extended and completed by *Newton's third law*. Its significance has already been demonstrated in the statics of a system of particles, where it plays an important role in relating the forces acting on different parts of the system. Newton's third law is: *Whenever one body exerts on a second a force* $\mathbf{F}_1$ *(action), the second body exerts on the first a force* $\mathbf{F}_2$ *(reaction). The action and reaction have the same magnitude, are oppositely directed, and have the same line of action.*

EXAMPLE. A body of mass m is attracted by the earth with a force $w = mg$ (its weight). This force is exerted by the mass of the earth on the mass of the body. The third law states that a force of the same magnitude but oppositely directed is exerted by the body on the earth. The weight acts on the body and imparts to it the aceleration due to gravity, g. The reaction acts on the earth and accelerates it toward the body. Because of the large mass of the earth, the acceleration imparted to it is very small.

Detailed discussions of the significance of Newton's third law in problems in statics were given in Section 9-5, and it is unnecessary to repeat them here. Derivation of the center of mass law, in connection with the mechanics of a system of particles, will provide an opportunity to further demonstrate the utility of Newton's third law (Section 43-1).

22. Application of Newton's Laws

22-1 Equations of Motion

Before we can derive the equations of motion of a body acted on by given forces, it is necessary to choose a coordinate system. This is usually taken as being at rest relative to the earth's surface, with axes oriented so that they are parallel to as many of the forces as possible. If F_{xi}, F_{yi}, and F_{zi} are the components of the ith force, and x, y, and z the coordinates of a particle, then Newton's second law gives, for the three coordinate directions, the equations

$$\sum_i F_{xi} = ma_x, \quad \sum_i F_{yi} = ma_y, \quad \sum_i F_{zi} = ma_z.$$

These three simultaneous equations always have a unique solution if the initial coordinates x_0, y_0, and z_0, and the initial velocity $\mathbf{v}_0$, are given. Solutions can be obtained by elementary methods, however, only in those cases in which the forces are constant.

Example. 1. An elevator of mass $m = 1800$ lbm travels upward. It starts from rest ($v_0 = 0$), moves with constant acceleration, and reaches its final velocity $v_t = 10$ ft/sec after traveling a distance $s_1 = 6.0$ ft. The local acceleration due to gravity is $g = 32.2$ ft/sec². Compute the tension T_1 in the elevator cable.

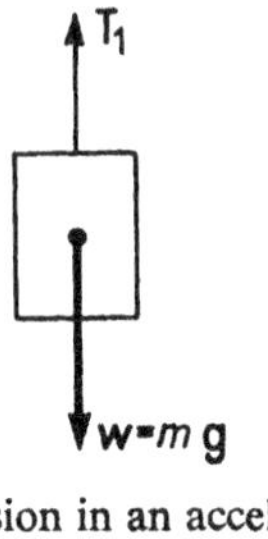

Fig. 3-15. Cable tension in an accelerating elevator.

The coordinate system (Fig. 3-15) is chosen so that the forces are in the y-direction. Newton's second law then reduces to the equation

$$T_1 - mg = ma_y.$$

For the given data, we obtain for the acceleration, using Eqn. (20-3),

$$a_y = \frac{v_t^2 - v_0^2}{2s_1} = 8.33 \text{ ft/sec}^2.$$

For the tension, we obtain

$$T_1 = m(g + a_y) = 72{,}900 \text{ poundals}.$$

After the initial period of acceleration, $a_y = 0$ and the tension is

$$T_2 = mg = 57{,}900 \text{ poundals}.$$

In engineering units

$$m = 55.94 \text{ slugs}, \quad T_2 = mg = w = 1800 \text{ lbf}.$$

EXAMPLE. 2. A sled of mass $m = 180$ kgm, on a level frictionless road, is acted on for 6 sec by a horizontal force P of 25 kgf = 245.2 newtons (Fig. 3-16). The initial velocity $v_0 = 0$. At the end of this time, a, v, and s have the following values (mks system):

$$P = ma_x, \quad a_x = \frac{P}{m} = 1.36 \text{ m/sec}^2,$$

$$v_t = a_x t = 8.17 \text{ m/sec}, \quad s_t = \tfrac{1}{2}a_x t^2 = 24.5 \text{ m}.$$

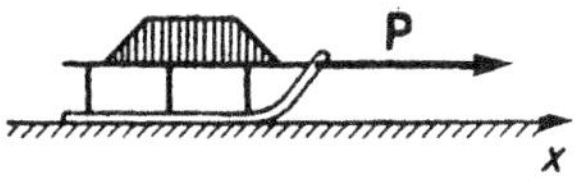

FIG. 3-16. Sliding without friction.

EXAMPLE. 3. A sphere of mass $m = 24$ gm and radius $r = 1$ cm moves with constant tangential velocity $v = 1.2$ m/sec around the inside surface of a frictionless cylindrical bowl of radius $r_1 = 81$ cm (Fig. 3-17). What forces are exerted by the bowl on the sphere?

In the radial direction, we have

$$\sum F_{r_i} = -F_1 = m\frac{v^2}{r_1 - r},$$

$$F_1 = -4.32 \times 10^3 \text{ dynes}.$$

In the y-direction, $a_y = 0$, and hence

$$\sum F_{y_i} = 0 = -mg + F_2, \quad F_2 = mg.$$

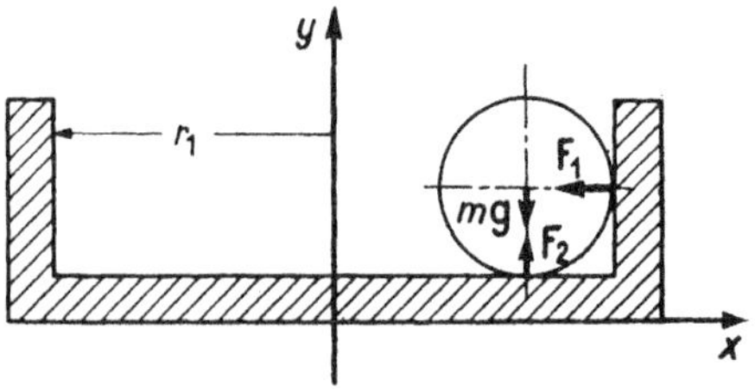

FIG. 3-17. Sphere revolving in a bowl.

22-2 D'Alembert's Principle. Inertial Forces

Newton's second law can always be written in the form

$$\boldsymbol{F} - m\boldsymbol{a} = 0.$$

This equation, according to D'Alembert, can be given the following interpretation. The force $\boldsymbol{F}$, acting on the mass m, produces an acceleration $\boldsymbol{a}$ such that the vector $m\boldsymbol{a}$ has the same magnitude and direction as the vector $\boldsymbol{F}$. The vector $-m\boldsymbol{a}$ is then in the opposite direction to $\boldsymbol{F}$. It is called the *inertial force* $\boldsymbol{I}$ and represents the inertial resistance of the mass to acceleration:

$$\boldsymbol{I} = -m\boldsymbol{a}.$$

The inertial force is not a true force as we have defined this term, for it cannot produce an acceleration; it is only brought into existence by a true force. It has, however, the same units as a true force (dyne, newton, poundal). Introducing this newly defined quantity in Newton's second law, we have

$$\boldsymbol{F} + \boldsymbol{I} = 0.$$

This equation expresses the so-called *d'Alembert condition of equilibrium* or *d'Alembert's principle*.

The equation asserts that if we extend the system of external forces acting on a body (having a resultant $\boldsymbol{F}$) to include the inertial force $\boldsymbol{I}$, then the true forces and inertial forces together form a system in *apparent equilibrium.* The equilibrium is only apparent, because here one cannot speak of an equilibrium state in the sense in which equilibrium is used in statics. Such an apparent equilibrium is called *d'Alembert equilibrium*, and it can be used advantageously in the solution of many problems in dynamics, which can then be treated by the graphical methods that we have developed for handling problems in statics. It is customary to represent the inertial force $\boldsymbol{I}$ by a *dotted* vector.

The subject of *analytical mechanics* makes use of Newton's second law chiefly in the form of equations of motion such as those developed in Section 22-1. In *elementary* mechanics, however, the d'Alembert form is easier and more straightforward for problems involving the forces acting on a body, since one can then use the simpler condition of equilibrium in graphical or analytical form.

It is necessary to emphasize that the inertial forces in a given coordinate system cannot be considered as ordinary forces. If they were ordinary forces, the sum of all forces would be zero and no acceleration would result. Furthermore, in every coordinate system moving relative to the given system with *constant* velocity, the inertial forces $\boldsymbol{I}$ must also be considered only as apparent forces. On the other hand, there are coordinate systems in which these inertial forces must be considered as real forces, capable of producing acceleration. Such systems, however, are *accelerated* relative to the original system. It is inadvisable to make use of such coordinate systems, since an observer in such a system is subject to effects which distort his observations. Newton himself defined the concept of force with reference only to an inertial system, and in no case will it be necessary for us to make use of an accelerated coordinate system.

Applications of d'Alembert's principle are manifold. Examples include circular motion, the problem of reaction forces in constrained motion, and the determination of the banking angle of a road or railway track.

22-3 Inertial Forces in Circular Motion

Suppose that a particle of mass m moves on a level frictionless surface with constant velocity v and describes in this plane a circle of radius r (Fig. 3-18). By means of a cord, the particle is kept at a constant distance r from a pivot point.

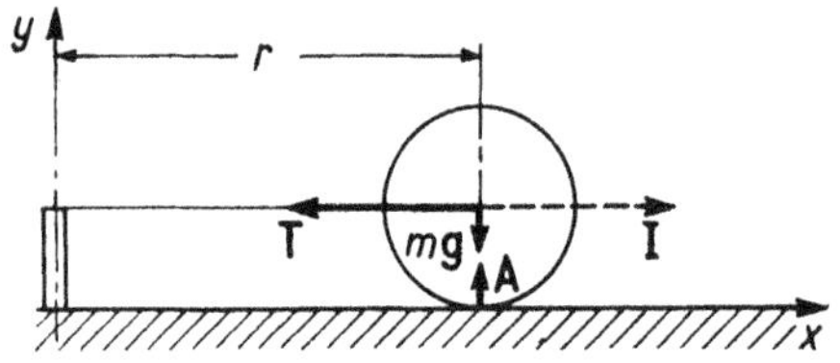

FIG. 3-18. Inertial force for circular motion.

The particle is acted on by the forces mg, T, and A, and by an inertial force I, which replaces $-m\boldsymbol{a}_{\mathrm{rad}}$:

$$\sum F_y = 0 = A - mg, \quad A = mg,$$
$$\sum F_r = 0 = I - T, \quad T = I = m(v^2/r).$$

The *inertial force* $\boldsymbol{I}$, which results from the circular motion, is of magnitude mv^2/r and is directed radially outward.

The real force $\boldsymbol{T}$ exerted by the cord on the particle is directed inward, from the particle toward the center of the circle, and is called a *centripetal* force. It has the same magnitude as $\boldsymbol{I}$:

$$T = m(v^2/r) = mr\omega^2.$$

We now discuss the equilibrium of the cord (Fig. 3-19). Two forces $\boldsymbol{T}'$ and $\boldsymbol{P}$, act on the cord, which is assumed to be weightless; $\boldsymbol{T}'$ is the force

FIG. 3-19. The cord in equilibrium.

exerted on the cord by the particle. The force $\boldsymbol{T}'$ pulls outward on the cord, and is the reaction to the force $\boldsymbol{T}$. Its magnitude is therefore

$$T' = m(v^2/r).$$

The force $\boldsymbol{P}$ is the force exerted by the pivot on the cord. Since the cord is in equilibrium, the magnitude of this force is also

$$P = m(v^2/r).$$

In this discussion, the inertial force $\boldsymbol{I}$ acting at the center of mass of the sphere has not been considered as a real force. Although this force does not exist from the standpoint of an observer at rest, it is nevertheless often referred to as a real force, and is called a *centrifugal force*. To speak of it in this way is incorrect. If we wish to consider the centrifugal force as a real force, this can be done correctly only if we rotate with the moving particle (see Section 27).

The controversy regarding the real existence of centrifugal forces is only a question of the standpoint of the observer, and becomes pointless as soon as we state clearly the coordinate system to which the motion is referred. Much more important is the recognition that a reconciliation of the two points of view is possible only through an extension of the concept of force. In the mechanics of Newton, forces are defined only for inertial coordinate systems. But in rotational motion, and from the viewpoint of an observer who partakes of this motion, certain radially directed accelerations occur for which no cause can be found by an inertial observer, who does not share the accelerated motion. The concept of *inertial forces*, introduced by d'Alembert, first showed how to make the transition from inertial systems to accelerated systems, and made it possible to define the concept of force in an accelerated coordinate system. We shall return to this question in Section 27.

22-4 Influence of the Earth's Rotation on Weight

The weight of a body is defined as the force with which the body is attracted by the earth. All bodies on the earth partake of the earth's rotation and as a consequence are accelerated toward the earth's axis. In considering the d'Alembert equilibrium of a body subject to this acceleration, we must realize that a body at rest relative to the surface of the earth is acted on by an inertial force whose magnitude is negligible for most engineering purposes, but which must be taken into account in work of high precision.

Consider a body of mass m at a point where the latitude is φ and the acceleration due to gravity is $\boldsymbol{g}$. If the earth were not rotating, the weight of the body would be $\boldsymbol{w}_0 = m\boldsymbol{g}_0$, where $\boldsymbol{g}_0$ is the acceleration due to gravity on a nonrotating earth. The line of action of $\boldsymbol{w}_0$ lies along the radius of the earth, assuming it to be spherical. The body rotates with the earth, however, and in the course of each day moves along the circumference of a circle of radius ϱ (Fig. 3-20), where $\varrho = R \cos \varphi$. The body is therefore acted on by an inertial force directed radially outward from the center of rotation D. The magnitude I of the inertial force is

$$I = m\varrho\omega^2 = mR\omega^2 \cos \varphi .$$

The vector $\boldsymbol{I}$ has a component

$$I_1 = mR\omega^2 \cos^2 \varphi$$

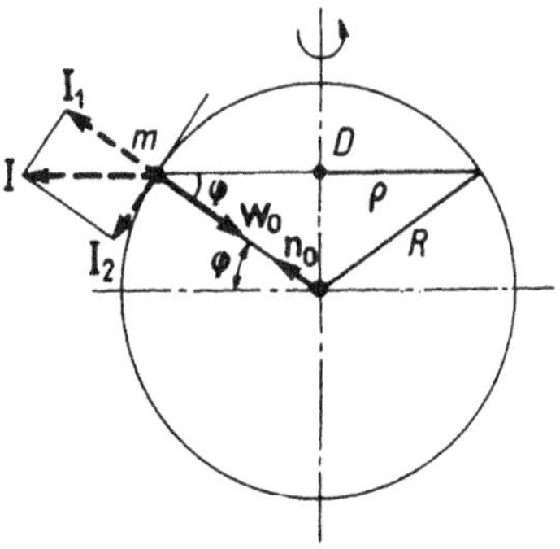

FIG. 3-20. Effect of the earth's rotation on weight. $\boldsymbol{I}$ is drawn much too large relative to $\boldsymbol{w}_0$. At latitude, 45°, $\boldsymbol{w}_0$ is about 600 times larger than $\boldsymbol{I}$.

in the direction of the radius, and a component toward the south of

$$I_2 = mR\omega^2 \cos\varphi \sin\varphi = \tfrac{1}{2}mR\omega^2 \sin 2\varphi .$$

Since the component $\boldsymbol{I}_2$ is normal to $\boldsymbol{w}_0$ and is small compared with it, it has little effect on the magnitude of $\boldsymbol{w}$, and the resultant force $\boldsymbol{P}$ on the rotating body is, to a first approximation,

$$\boldsymbol{P} = \boldsymbol{w}_0 - \boldsymbol{I}_1 ,$$

and the weight $\boldsymbol{w}$ is

$$\boldsymbol{P} = \boldsymbol{w} = m(g_0 - R\omega^2 \cos^2\varphi)\,\boldsymbol{n}_0 = m\boldsymbol{g}_{\mathrm{exp}} .$$

The experimentally determined acceleration $\boldsymbol{g}_{\mathrm{exp}}$ has the magnitude

$$g_{\mathrm{exp}} = g_0 - R\omega^2 \cos^2\varphi .$$

The value of $\boldsymbol{g}_{\mathrm{exp}}$ varies with latitude, both as a result of the flattening of the earth and of the earth's rotation.

22-5 Reaction of a Roadbed

When a vehicle such as a railway car or an automobile is forced to move in a curved path with a given velocity, forces must be exerted on the vehicle by the rails or the surface of the road. Only those cases in which the path is circular can be treated by elementary methods.

EXAMPLE. A cart of mass m moves with velocity v in a circle of radius r on a level surface. Let s represent the spacing between the wheels and let h be the vertical height of the center of gravity (Fig. 3-21). To calculate the force which must be exerted by friction at the points of contact between wheels and road, and to determine the limiting conditions for motion without slipping, it is convenient to use d'Alembert's principle.

Reference to Fig. 3-21 shows that the cart is acted on by the five real forces $\boldsymbol{w}$, $\boldsymbol{A}_1$, $\boldsymbol{A}_2$, $\boldsymbol{R}_1$, and $\boldsymbol{R}_2$, and by the inertial force $\boldsymbol{I}$. The forces $\boldsymbol{w}$ and $\boldsymbol{I}$ are known in magnitude and direction.

In the case of motion without slipping $\boldsymbol{A}_1$ and $\boldsymbol{A}_2$ are directed vertically upward, and their magnitudes are unknown; $\boldsymbol{R}_0 = \boldsymbol{R}_1 + \boldsymbol{R}_2$ is directed horizontally, and its

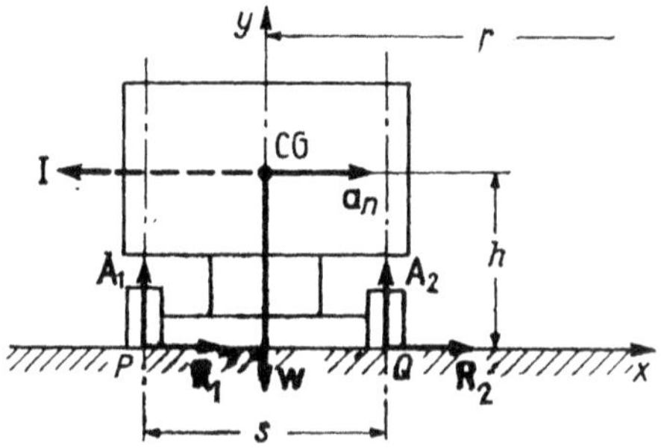

FIG. 3-21. Moving vehicle on a curve.

magnitude is unknown. Since the friction forces are not fully developed, the forces $\boldsymbol{R}_1$ and $\boldsymbol{R}_2$ cannot be found separately, although the force $\boldsymbol{R}_0$ can be determined. The conditions for translational equilibrium give

$$\sum F_{x_t} = 0 = R_0 - I, \quad R_0 = I = m\frac{v^2}{r},$$

$$\sum F_{y_t} = 0 = -w + A_1 + A_2, \quad w = A_1 + A_2.$$

The condition for rotational equilibrium gives, with respect to point P,

$$\sum M_t = 0 = -w\frac{s}{2} + A_2 s + Ih, \quad A_2 = \frac{ws - 2Ih}{2s},$$

with respect to point Q,

$$\sum M_t = 0 = w\frac{s}{2} - A_1 s + Ih, \qquad A_1 = \frac{ws + 2Ih}{2s}.$$

As a numerical example, let $v = 25$ ft/sec, $m = 2400$ lbm $\cong 75$ slugs, $r = 150$ ft, $s = 5$ ft, $h = 2$ ft. Then

$$I = R_0 = 75 \times \frac{625}{150} = 312.5 \text{ lbf},$$

$$A_1 = 1325 \text{ lbf}, \quad A_2 = 1075 \text{ lbf}.$$

In the limiting case of no slipping, the friction forces can be computed separately:

$$R_1 = \mu_0 A_1, \quad R_2 = \mu_0 A_2.$$

The conditions of equilibrium give

$$\sum F_{x_t} = 0 = -I + R_1 + R_2, \quad I = R_1 + R_2 = \mu_0(A_1 + A_2),$$

$$\sum F_{y_t} = 0 = -w + A_1 + A_2, \quad w = A_1 + A_2.$$

With respect to the center of gravity CG,

$$\sum M_t = 0 = -A_1\frac{s}{2} + A_2\frac{s}{2} + (R_1 + R_2)\,h.$$

From these equations, the unknown forces A_1, A_2, R_1, and R_2 are calculated to be

$$A_1 = w\left(\frac{1}{2} + \frac{\mu_0 h}{s}\right), \quad A_2 = w\left(\frac{1}{2} - \frac{\mu_0 h}{s}\right),$$

$$R_1 = \mu_0 w\left(\frac{1}{2} + \frac{\mu_0 h}{s}\right), \quad R_2 = \mu_0 w\left(\frac{1}{2} - \frac{\mu_0 h}{s}\right).$$

The limiting velocity for which the vehicle will not skid is found from the equation

$$I = m(v^2/r) = R_1 + R_2 = \mu_0 w = \mu_0 mg,$$

from which

$$v_{max} = \sqrt{(\mu_0 gr)}.$$

22-6 Banked Curves

When a vehicle travels in a curved path on a plane surface, the vehicle's inertia makes it tend to go off on a tangent to the curve, rather than follow the prescribed path. The prescribed motion is maintained by forces exerted at the points of contact of wheels and track, either by friction, as for an automobile or bicycle or, in the case of a railway car, by forces exerted against the flanges of the wheels by the rails. As a consequence, large forces of sliding friction can be developed between the roadbed or the rails and the wheels. By inclining the roadbed at an angle α to the horizontal, the required centripetal force can be obtained as a component of the normal reaction force exerted by the rails. The angle of inclination depends on the velocity of the vehicle and on the radius of curvature of the path. When the angle has its correct value, the vehicle can round the curve with neither the inner nor the outer flange in contact with the rails.

At the correct angle of banking, and neglecting rolling friction, three real forces and one inertial force act on the vehicle (Fig. 3-22). These forces are

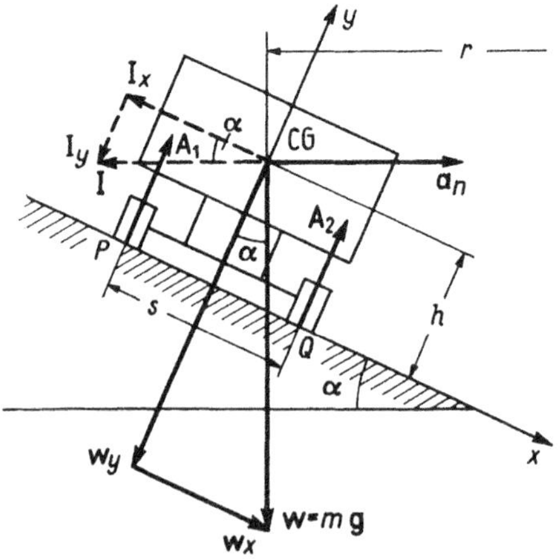

FIG. 3-22. Moving vehicle on a banked curve.

$w = mg$, which is known; A_1 and A_2, of unknown magnitude, both perpendicular to the roadbed; I (magnitude mv^2/r), acting at the center of gravity of the vehicle and directed horizontally outward.

Using the coordinate system indicated in the diagram, we have

$$\sum F_{x_i} = 0 = mg \sin \alpha - m(v^2/r) \cos \alpha,$$

and hence

$$\tan \alpha = v^2/rg.$$

The banking angle α is independent of m, as would be expected, since both the weight w and the inertial force I are directly proportional to m.

22-7 Motion in a Viscous Gas or Liquid

When a body moves in a gas or a liquid, resisting forces act on it which result in part from *friction*, and in part from the formation of *vortices* or *eddies* behind the body. These two types of resistance depend in quite different ways on the velocity of the moving body and on its shape. They will be discussed more fully in the section on the hydrodynamics of viscous fluids. It will be sufficient to state here that they increase with increasing velocity, and that both are zero when $v = 0$. These properties enable us to draw a number of conclusions regarding simple types of motion. We shall assume only that a body is acted on by a constant driving force. This will be the case, for example, for a projectile or a falling body in air, and for heavy particles sinking in a liquid, as in the process of sedimentation.

If an object is released from rest in a liquid or gas, no initial resisting force acts on it. The only accelerating force is its weight w, corrected for the buoyancy of the fluid. If buoyancy is neglected, the initial motion is vertically downward, with an acceleration g. As a result of the increasing velocity, a resisting force F develops, and we know that this increases at first slowly, and then more and more rapidly as v increases. If we plot both forces in an F-v diagram, as in Fig. 3-23, it is easy to see that regardless of the exact form of the law of increase of F with velocity (type I, II, or III), a velocity will always eventually be reached at which the resisting force is equal to the driving force. Since these forces are oppositely directed, their resultant from

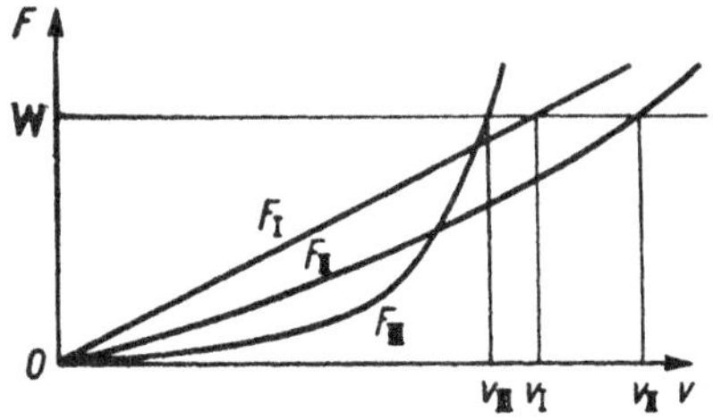

FIG. 3-23. Terminal velocity for the of case a constant driving force W and for several kinds of resistance force F.

this time on is zero, and the motion continues with constant velocity. In the initial phase of the motion, the acceleration decreases from $\boldsymbol{g}$ to zero. Only if the dependence of $\boldsymbol{F}$ on $\boldsymbol{v}$ is known, however, can we compute the distance moved in a given time and the magnitude of the final constant velocity, called the *terminal velocity*. The following are the chief cases of interest.

Vertical free fall in air. For bodies of sufficiently large specific weight, the buoyancy of the air can be neglected and the initial acceleration is $\boldsymbol{g}$. If the height of fall is sufficiently great, the terminal velocity is reached when the resisting force equals the weight $\boldsymbol{w}$. For a given density of air, and a given shape and size of the falling body, the terminal velocity will be greater the larger the weight $\boldsymbol{w}$. For particles of low specific weight, the buoyancy of the air must be taken into account.

This type of motion is of importance in Millikan's method of measuring the charge of an electron (see Vol. II).

Motion of a projectile in air. During the time the projectile is rising, the resisting force exerts on it a downward component F_1. The downward acceleration is therefore greater than g, and it decreases gradually until the highest point is reached. At this point, $v_y = 0$ and the acceleration equals g. The maximum height reached is therefore less than it would be in vacuum (Fig. 3-24).

The horizontal motion is likewise retarded, since the resisting force has a horizontal component F_2. The horizontal component of velocity decreases asymptotically from the initial value $v_0 \cos \alpha$ to zero. The trajectory has a vertical asymptote.

Sedimentation in liquids. When a particle falls in a stationary undisturbed liquid whose density is less than that of the particle, the resultant of the weight of the particle and of the buoyant force is vertically downward. The particle starts to fall, and as a result, a resisting force develops which increases with increasing v. After a time, the velocity of fall becomes constant. The terminal velocity is greater the greater the difference of density between

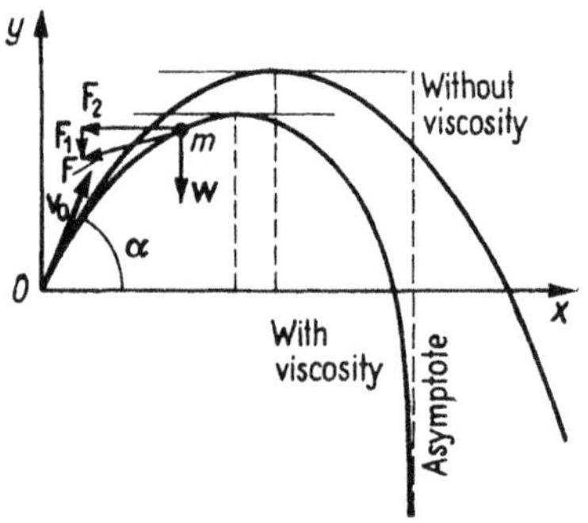

FIG. 3-24. Projectile in air.

the particle and the liquid. It depends also on the shape of the particle and increases with increasing ratio of volume to surface area. This ratio is significant because the driving force is proportional to the volume, while the resisting force is proportional to the surface area. In a suspension of particles of the same form, the larger particles will fall more rapidly than the smaller, since the ratio increases with particle size. This effect is called *sedimentation*. Bubbles of gas rising in a liquid show an analogous behavior. In a rotating system, an outward sedimentation of the heavier particles occurs. By rotating the system is a centrifuge, the rate of sedimentation can be enormously increased over that in the earth's gravitational field.

23. Sliding and Rolling Friction

23-1 Sliding Friction

In Section 14 we considered the behavior of friction forces up to the limiting case of sliding. Figure 3-25 shows a body on a plane surface, acted

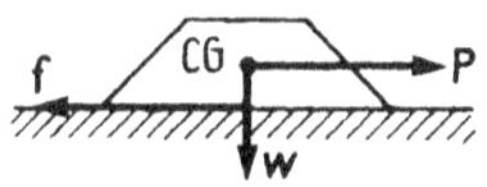

FIG. 3-25. Static friction.

on by a horizontal force P. As the force P is gradually increased from zero, the body at first remains at rest, since the force of static friction, f, increases in the same way as the force P until it reaches its maximum value. It was found by Coulomb that this maximum value is proportional to the normal force N, that is,

$$f_{max} = \mu_0 N.$$

If the force P is increased still further, we find by experiment that the body starts to accelerate, and that the coefficient of friction decreases from μ_0 (the coefficient of *static* friction) to a somewhat smaller value μ, called the *coefficient of sliding, or kinetic friction.* This coefficient remains nearly constant up to velocities of the order of 30 ft/sec.

We can apply Newton's second law to this situation. The force of sliding friction becomes one of the set of forces acting on the body. Its direction at every instant is opposite to that of the velocity.

EXAMPLE. A sled slides down an inclined plane with an angle of inclination α under the influence of its weight w and the force of sliding friction. The coefficient of sliding friction is μ. For what slope angle α_1 will the motion be one of constant velocity (Fig. 3-26)?

Using d'Alembert's principle, we have

$$\sum F_{x_i} = -w \sin \alpha + \mu N + ma_x = 0,$$

$$ma_x = w \sin \alpha - \mu N;$$

$$\sum F_{y_i} = -w \cos \alpha + N = 0,$$

$$N = w \cos \alpha = w_n.$$

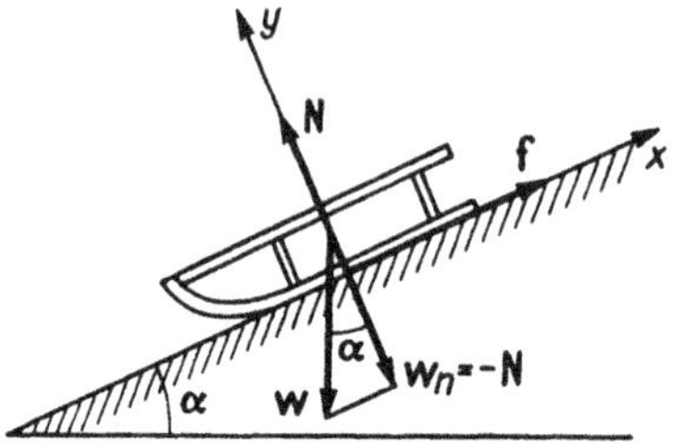

FIG. 3-26. Sliding with friction on an inclined plane.

The acceleration for any arbitrary value of α is therefore

$$a_x = g(\sin\alpha - \mu\cos\alpha).$$

The motion will be uniform when $a_x = 0$, that is, when

$$\sin\alpha_1 = \mu\cos\alpha_1 \quad \text{or} \quad \tan\alpha_1 = \mu.$$

If for example $\mu = 0.06$, then $\alpha_1 = 3.5°$.

23-2 Rolling Friction

Friction enters into the mechanics of rolling in two separate ways. First, there must be a force of static friction in order that any rolling at all may take place. This requirement is independent of whether the wheel or the surface is completely rigid or elastic (Section 50), and it involves no dissipation of energy. Secondly, if either the wheel or the surface or both are elastic, there will be a dissipative torque called the *rolling friction moment.*

A wheel can roll on a surface only if a force f of static friction acts on the wheel at the point of contact between the wheel and the surface [Fig. 3-27(a)]. If this force did not act, the wheel would only slide over the surface. On the other hand, when the maximum force of static friction is exceeded, the wheel can slide and roll simultaneously, as is often observed when an automobile accelerates or decelerates rapidly on a slippery road.

When a wheel rolls on a surface, both the wheel and the surface are elastically deformed as a result of the weight and its reaction. As a consequence, when the wheel is in motion, the line of action of the reaction force

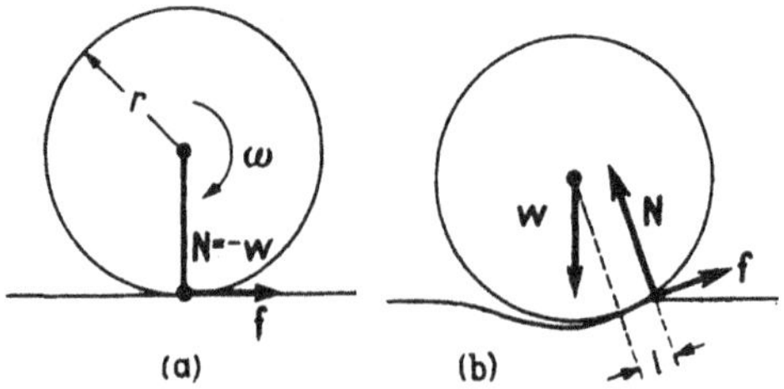

FIG. 3-27. Rolling friction.

exerted on the wheel by the surface no longer passes through the center of the wheel [Fig. 3-27 (b)]. Thus the reaction force N has a lever arm l about the center of the wheel and exerts a torque about the wheel axis. This torque is called the *rolling friction moment* M_r, and it opposes the motion of the wheel.

It is found experimentally that to a first approximation the rolling friction moment is independent of the radius of the wheel and of its velocity. We have set above that

$$M_r = lN.$$

However, it is more usual to replace l by μ_r, the *coefficient of rolling friction*, giving the equation

$$M_r = \mu_r N.$$

The coefficient μ_r depends on the materials of the wheel and the surface, and has the dimensions of length. Its magnitude, for steel rolling on steel, is of the order of 0.05 cm.

24. Momentum

Newton's second law, $\boldsymbol{F} = m\boldsymbol{a}$, has been formulated thus far in terms of the concept of acceleration. This statement, however, is more restricted than the law as originally given by Newton. In the form above, it is applicable only to the motion of a body of constant mass. The original formulation of Newton, however, is valid for a body of variable mass. Only in such form is the law compatible with the more recent developments in mechanics using the theory of relativity.

Newton prefaced the statement of his laws by two definitions. In the first definition, Newton makes it clear that he considered "quantity of matter" to mean the same thing as "mass." The second definition states: "The quantity of motion is the measure of the same, arising from the velocity and quantity of matter conjointly." The concept which he referred to as "quantity of motion" and defined as the product of the mass and the velocity, we now call the *momentum*, $\boldsymbol{p}$. Momentum is a vector, in the same direction as the velocity:

$$\boldsymbol{p} = m\boldsymbol{v}.$$

The dimensions of momentum are $[p] = [MLT^{-1}]$, and its units are: cgs system, 1 gm · cm/sec; mks system, 1 kgm · m/sec; fps system, 1 lbm · ft./sec.

Newton formulated his second law as follows:

"*The time rate of change of momentum is proportional to the external force and takes place in the direction of this force.*"

Or expressed mathematically,

$$d\boldsymbol{p}/dt = \boldsymbol{F}.$$

When stated in this way, Newton's second law can be called the law of change of momentum or, more briefly, the *momentum law*. The time rate of change of momentum can be expanded in the form,

$$dp/dt = m\,(dv/dt) + v(dm/dt).$$

For a body of constant mass, this form reduces to the second law as previously stated, since if m is constant,

$$dm/dt = 0 \quad \text{and} \quad dp/dt = m(dv/dt) = ma.$$

Newton's second law, in the form of the momentum law, remains valid for the motion of a body of *variable* mass, a problem which is of the greatest importance in the field of modern physics. The special theory of relativity, which is verified by many experimental measurements, predicts that mass depends on velocity. If m_0 represents the mass of a body at rest and c the velocity of light in vacuum, the mass m_v of the body when its velocity is v is given by

$$m_v = \frac{m_0}{\sqrt{[1 - (v/c)^2]}}.$$

The mass increases with increasing velocity, as shown in Fig. 3-28. This variability of mass becomes significant only for velocities which are as yet

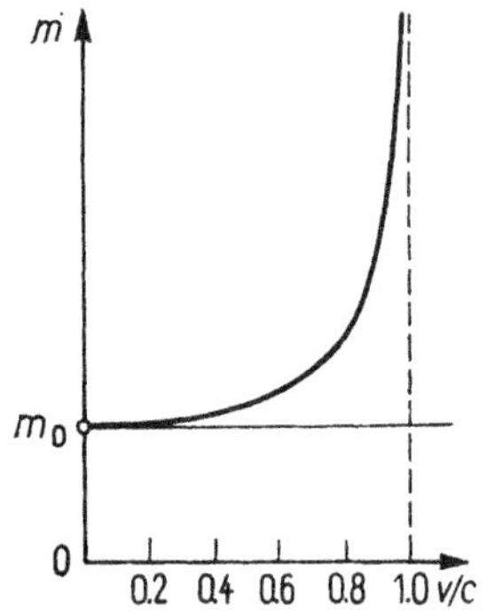

FIG. 3-28. Increase in mass as a function of v/c.
c = velocity of light = 3×10^{10} cm/sec.

unattainable by bodies of macroscopic size, and hence we can neglect it in our considerations. However, the variability has been observed for particles of subatomic size. The concept of mass thus loses its absolute significance, since it need not have the same value for two observers moving uniformly relative to each other.

Newton's first law, the law of inertia, is a special case of the momentum law. If no external forces act on a body, there is no change in its momentum, and the momentum remains constant in magnitude and direction. That is, if $\boldsymbol{F} = 0$, then $d\boldsymbol{p}/dt = 0$, and

$$\boldsymbol{p} = \text{const}.$$

This formulation of the inertia law is known as the *law of conservation of momentum*.

Newton's second law is one of the fundamental principles of physics. Its significance goes beyond the mere fact that it serves as the defining equation for force or mass. Its physical content lies in the assertion that *force* determines the *first derivative of momentum* with respect to time. In making this statement, Newton gave to his contemporaries and to the world a concept of the greatest significance. The widespread implications of this law will be seen many times as we carry out our exposition of physics.

25. The Law of Areas for Central Forces

Suppose that a particle moving with an arbitrary initial velocity $\boldsymbol{v}_0$ is acted on by a force $\boldsymbol{F}(r)$ which is always directed toward a fixed point. Forces of this sort are called *central* forces. Let us select the point toward which the force vector is directed as the origin of a coordinate system (Fig. 3-29). Newton's second law states that

$$\boldsymbol{F} = m(d\boldsymbol{v}/dt) = m\boldsymbol{a}.$$

When both sides of this equation are multiplied vectorially by $\boldsymbol{r}$, we get

$$\boldsymbol{r} \times \boldsymbol{F} = m\left(\boldsymbol{r} \times \frac{d\boldsymbol{v}}{dt}\right) = 0,$$

since $\boldsymbol{r}$ and $\boldsymbol{F}$ are anti-parallel.

Now

$$\frac{d}{dt}(\boldsymbol{r} \times \boldsymbol{v}) = \frac{d\boldsymbol{r}}{dt} \times \boldsymbol{v} + \boldsymbol{r} \times \frac{d\boldsymbol{v}}{dt}.$$

But

$$\frac{d\boldsymbol{r}}{dt} \times \boldsymbol{v} = 0,$$

since $d\boldsymbol{r}/dt = \boldsymbol{v}$. We therefore have the relation

$$m\left(\boldsymbol{r} \times \frac{d\boldsymbol{v}}{dt}\right) = m\frac{d}{dt}(\boldsymbol{r} \times \boldsymbol{v}) = 0,$$

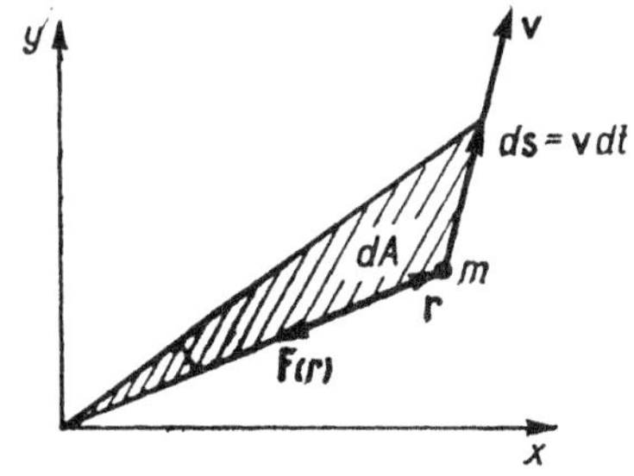

FIG. 3-29. The law of areas.

and hence

$$\boldsymbol{r} \times \boldsymbol{v} = \text{const.} \tag{25-1}$$

This equation has a simple interpretation in terms of the area swept over by the radius vector $\boldsymbol{r}$ in the time dt (Fig. 3-29). If dA is the shaded element of area, and using $d\boldsymbol{s} = \boldsymbol{v}\,dt$, we have

$$2\,dA = r\,ds \sin(\boldsymbol{r}, d\boldsymbol{s}) = rv\,dt \sin(\boldsymbol{r}, \boldsymbol{v}),$$

and hence

$$2\frac{dA}{dt} = rv \sin(\boldsymbol{r}, \boldsymbol{v}) = |\boldsymbol{r} \times \boldsymbol{v}|.$$

It follows that *for motion under the action of central forces only, the rate at which area is swept over by the radius vector is constant.* This is the *law of areas.* It plays an extensive role in the theory of the orbits of planets and comets, and in the scattering of electrically charged particles by radially symmetric electric fields, as, for example, the scattering of alpha particles by atomic nuclei. The law was drived independently by D. Bernoulli, in 1745, and by L. Euler in 1746.

It also follows from Eqn. (25-1), since the vector product is a constant for the motion, that the orbit of the particle lies in a plane. At perihelion (closest point of approach to the sun or other force center) the magnitude $|\boldsymbol{r}|$ of the radius vector is *smaller* than at aphelion (farthest point of the orbit from the force center). Therefore the velocity at perihelion is *greater* than at aphelion.

26. Newton's Law of Gravitation

Newton discovered that the motions of the planets and their satellites could be calculated from the second law, provided it was assumed that forces of attraction existed between them which were proportional to the product of their masses and inversely proportional to the square of the distance between their centers of gravity. This force of attraction is called a *gravitational* force. According to Newton's third law, the mutual forces between any two bodies are equal in magnitude and opposite in direction, and Newton postulated that their line of action lies along the line joining the centers of gravity. Experimental measurements first performed by Cavendish have verified this law in all of its particulars. The magnitude of the force is given by the equation

$$F_{1,2} = G\frac{m_1 m_2}{r^2}.$$

If we take the standpoint that the three fundamental units of mass, length and time suffice for the description of all mechanical processes, we are not free to assign an arbitrary value to G, and its magnitude must be determined by experiment. The factor G is called the *gravitational constant.* If

$m_1 = m_2 = 1$ gm, and $r = 1$ cm, then

$$|F_{1,2}| = |F_{2,1}| = G.$$

That is, the magnitude of G equals the force (in dynes) with which a mass of 1 gm attracts another mass of 1 gm when their centers of gravity are 1 cm apart. Experimental measurements of G give

$$G = 6.67 \times 10^{-8} \text{ gm}^{-1} \text{ cm}^3 \text{ sec}^{-2}$$
$$= 6.67 \times 10^{-11} \text{ kgm}^{-1} \text{ m}^3 \text{ sec}^{-2}.$$

The dimensions of G are $[G] = [M^{-1}L^3T^{-2}]$, and it units are: cgs system, 1 $\text{cm}^3/\text{gm} \cdot \text{sec}^2$; mks system, 1 $\text{m}^3/\text{kgm} \cdot \text{sec}^2$; fps system, 1 $\text{ft}^3/\text{lbm} \cdot \text{sec}^2$.

It may be remarked that the value of G can be obtained only from laboratory measurements of the force between known masses at a known separation. The force is measured with the help of a sensitive torsion balance, since its magnitude, for any two masses that can be used, is extremely small.

EXAMPLE.

$$m_1 = 10^6 \text{ gm}, \quad m_2 = 10^3 \text{ gm},$$
$$r = 1 \text{ m}, \quad F = 6.68 \times 10^{-3} \text{ dyne}.$$

Gravitational forces are significant only for masses of cosmic magnitude. They have no observable influence on the motions of atoms and molecules. The following section illustrates how the masses of cosmic bodies can be determined.

26-1 The Gravitational Attraction of the Earth

A body of mass m_B at the surface of the earth is attracted by the earth with a force given by

$$F_{E,B} = G\frac{m_B m_E}{R_E^2},$$

where m_E is the mass of the earth and R_E is its radius. From the definition of the term *weight*, the force $F_{E,B}$ is identical with the weight of the body, except for the small correction resulting from the earth's rotation.

If the only force on the body is its weight (that is, if it is falling freely in vacuum), then according to Newton's second law the body will be accelerated with an acceleration g_0, the acceleration due to gravity on a nonrotating earth (cf. Section 22-4). The maximum correction to the observed acceleration g because of the earth's rotation is about 0.3 per cent, and at a latitude of 45°, $g_0 = 982.36$ cm/sec².

Combining the law of gravitation and the second law, we obtain

$$G\frac{m_B m_E}{R_E^2} = m_B g_0,$$

and hence

$$g_0 = G\frac{m_E}{R_E^2}. \tag{26-1}$$

The acceleration g_0 is independent of the mass of the body. All bodies fall with the same acceleration at the same point on the earth's surface.

The calculation of g_0 is significant in a number of respects: The value of g_0 depends on the distance of a point on the earth's surface from the center of the earth. To a first approximation, the earth is an oblate ellipsoid of revolution. The poles are therefore nearer the center than is the equator, and g_0 at the poles should be greater than at the equator. This is confirmed by the following experimental values:

$$g_{eq} = 978.03\ \text{cm/sec}^2, \quad g_{0,\,eq} = 981.42\ \text{cm/sec}^2,$$

$$g_{pole} = 983.09\ \text{cm/sec}^2, \quad g_{0,\,pole} = 983.09\ \text{cm/sec}^2.$$

The value of g_0 decreases with increasing distance above the earth's surface. Its value at an elevation h is

$$g_h = \frac{Gm_E}{(R_E + h)^2} \approx \frac{Gm_E}{R_E^2}\left(1 - \frac{2h}{R_E}\right) = g_0\left(1 - \frac{2h}{R_E}\right),$$

if $h \ll R_E$.

An example taken from measurements at the Physikalischtechnischen Reichsanstalt in Berlin† will illustrate this variation of g with h. Two bodies each of mass 1 kgm, were suspended at different levels from the pans of an equal-arm balance. The measured value of the difference between their weights, for a difference in level of 1 m, was $\Delta w = 0.295$ mgf. The difference as computed from the law of gravitation, with $g_{Berlin} = 981.3\ \text{cm/sec}^2$ and $R_E = 6370$ km, is $\Delta w = 0.314$ mgf.

The mass of the earth can be computed if g_0, R_E, and G are known. We have from Eqn. (26-1),

$$m_E = g_0 R_E^2/G.$$

For $g_0 = 982.3\ \text{cm/sec}^2$ and $R_E = 6370$ km, we obtain

$$m_E = 5.98 \times 10^{27}\ \text{gm}.$$

The mean density of the earth is therefore

$$\bar{\varrho} = m_E/V_E = 5.52\ \text{gm/cm}^3.$$

This is considerably larger than the density of the materials composing the earth's crust, so the interior of the earth must be of much higher density.

It must be pointed out that the cancellation of m_B from the left and right sides of Eqn. (26-1) rests on the assumption that this quantity is the same in both the gravitational attraction of a body and the opposition offered by it to an acceleration (equality of gravitational and inertial mass). This

† Karl Scheel, *Grundlagen der praktischen Metronomie*, Braunschweig, 1911, p. 111.

postulate has been verified by experiments performed by von Eötvös, using a very sensitive rotation balance.†

26-2 The Earth–Moon System

The moon revolves around the earth in an orbit which is very nearly circular. The average radius of the orbit is $R_M = 384{,}000$ km. The mutual attraction of the two bodies can be computed from the law of gravitation (Fig. 3-30):

$$F_{E,M} = G\frac{m_E m_M}{R_M^2} = m_M a_M = m_M R_M \omega^2,$$

where ω is the angular velocity of the moon in its orbit. Since the mass of the moon cancels, the angular velocity and hence the period T can be computed if m_E and R_M are known. We have

$$\omega = \sqrt{(Gm_E/R_M^3)},$$

$$T = 2\pi/\omega = 2\pi\sqrt{(R_M^3/Gm_E)} = 27.4 \text{ days}.$$

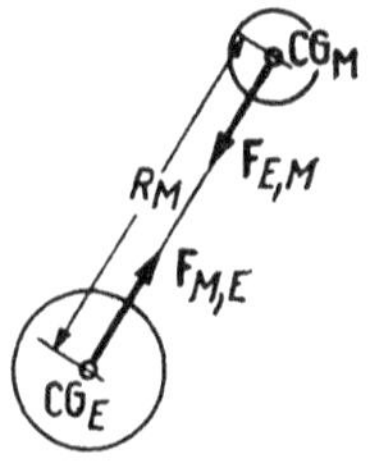

FIG. 3-30. The earth–moon system.

26-3 The Tides

Newton's third law requires that the moon attract the earth with a force of the same magnitude as that which the earth exerts on the moon. Since the mass of the earth is 81 times as great as that of the moon, its acceleration is only 1/81 as great. The motion of the earth resulting from this acceleration will be discussed in the chapter on the dynamics of rigid bodies. For the present, wc will only mention that this force draws the movable water masses of the oceans toward the moon. A mountain of water builds up on the side of the earth facing the moon. In each rotation of the earth, this mountain of water crosses every meridian and produces the so-called *flood tides*. The valleys of water, which are displaced 90° in longitude (~6 hr) from the peaks, are called the *ebb tides*. It will be explained in Section 42-3 why two flood tides occur daily at each meridian.

The sun, as well as the moon, exerts a force of attraction on the earth. This force is, in general, in a different direction from that of the moon and

† von Eötvös, *Die Naturwissenschaften*, **6**, 445 (1918); **7**, 149, 327 (1919); and more recently, R. H. Dicke, *Scientific American*, **205**, 84 (1961).

combines with it into a resultant which is a maximum at the times of the new moon and the full moon, and is a minimum when the moon is in its first and third quarters. The influence of the sun on the tides is almost exactly half as great as that of the moon.

26-4 Kepler's Laws

Kepler succeeded in unifying the astronomical observations which had previously been made on the motions of the planets. It remained for Newton, through his law of gravitation, to give the physical basis underlying the laws of Kepler. Kepler's laws are:

(1) The planets move in elliptical orbits having the sun at one focus. (This law cannot be derived by elementary methods.)

(2) The radius vector from the sun to a planet sweeps over equal areas in equal times. This is the law of areas for central forces (Section 25), and it implies that the force exerted on a planet by the sun lies along the line joining them.

(3) The square of the period of a planet's rotation about the sun is proportional to the cube of the major axis of its orbit.

The third law, which was considered by Kepler as clear evidence of a harmonic universe, can easily be derived by elementary methods from the law of gravitaion if we replace the elliptical orbits by circles. Let m_p be the mass of a planet, R_p its distance from the sun, and m_s the mass of the sun. Neglecting any disturbing effects of the other planets, the gravitational force F on the planet is

$$F = G\frac{m_p m_s}{R_p^2}.$$

This force provides the centripetal force for the planetary motion:

$$F = m_p R_p \omega_p^2 = \frac{m_p R_p 4\pi^2}{T_p^2}.$$

It follows from these two equations that

$$T_p^2/R_p^3 = 4\pi^2/Gm_s = \text{const},$$

which is Kepler's third law. The mass of the sun can be computed if T_p and R_p are known for any planet.

27. Transformation to Accelerated Coordinate Systems

In order to present the meaning of D'Alembert inertial forces in a new light, let us imagine an observer A in a completely enclosed box which is very far from all gravity-exerting masses and is in force-free, uniform rectilinear motion. The observer A has in the box a number of test objects of small mass which have arbitrary velocities relative to the box and on which he proceeds to make measurements. To give our ideas a concrete basis let us think of the observer as attached to and moving with one wall of the box. He exerts no mechanical forces on the test objects. If a test object is at rest relative to the observer, it remains in this state. If it is in motion relative to the box, this motion is rectilinear and uniform until the object collides with the walls or with another test object. When this occurs, the test object is accelerated as a result of contact forces in the collision. After the collision is over, the motion is again rectilinear and uniform.

We now impart to the box (but not to the test objects) an acceleration, by applying an external mechanical force in the direction of the z-axis. We can think of this force as being exerted on the box by an external observer B. Observer B draws the following conclusions: The box, and with it observer A, experiences an acceleration in the z-direction of magnitude a_z. The test objects in the interior of the box remain in a state of uniform, rectilinear, inertial motion.

Observer A, inside the box, comes to different conclusions. It appears to him that all objects, which formerly moved about inside the box with uniform rectilinear motion, now have an acceleration $-a_z$ toward the floor of the box. (If originally at rest relative to the box, they fall vertically; if originally in motion, they move in parabolas.) After striking the floor, they bounce up and again fall, with an acceleration $-a_z$. Observer A asserts that his test objects are in a gravitational field, and therefore fall with an acceleration $-a_z$.

There is no way of finding out, from experiments performed inside the box, whether the box is in a gravitational field (that is, is at rest or in inertial motion relative to a gravitationally attracting mass), or whether the box is acted on by an accelerating force which acts on it alone, but not on the test objects, and which is opposite in direction to the presumed gravitational force.

After communicating with each other, the two observers must agree to postulate the following *principle of equivalence*: A coordinate system in *inertial* motion relative to a gravitationally attracting mass is equivalent to an *accelerated* coordinate system which is isolated from any gravitationally attracting mass, provided that the accelerating force has the same magnitude as the gravitational force and is oppositely directed. It is thus seen that gravitational forces can never be *screened* or *shielded* by a wall or a filter, which, in fact, is an essential property of forces of this kind. This fact that gravitational forces cannot be "screened out" is an important difference between gravitational forces on the one hand and electrical and magnetic forces on the other, since both of the latter can be "screened out" by electrical or magnetic shielding.

For observer A, in the box, the accelerating force which appears to act gravitationally on his test objects is a completely real force. However, according to observer B, outside the box, the test objects experience no force at all; the gravitational force is only a simulated one, since to him the objects appear to move inertially, as before. The downward acceleration seen by A is simulated by the upward acceleration of the box. The external observer B can, however, predict the motion that will be observed within the box (by means of Newton's law), if he adds to the forces on all objects within it a force of magnitude $I = -ma_z$. But this simply says that Newton's laws can also be used in an accelerated coordinate system to compute motions and solve problems involving forces, even though the concept of force is defined only for inertial systems. It is only necessary to add to the forces acting in the accelerated system an inertial force $\boldsymbol{I} = -m\boldsymbol{a}_z$, where $\boldsymbol{a}_z$ is the acceleration of the observer who is not moving with inertial motion.

Thus, the principle of equivalence can be restated in the following form: An accelerated reference frame can be treated as an inertial frame and Newton's laws can then be applied, provided that we postulate a gravitational field that acts in the accelerated frame. The gravitational acceleration and the acceleration of the frame have the same magnitude, but opposite directions.

In this sense, the concept of d'Alembert inertial forces acquires a deeper significance. d'Alembert's principle is nothing more than a *transformation* of the observer's standpoint from an *inertial* coordinate system to an *accelerated* coordinate system. This transformation can be used to study the motion of point masses and rigid bodies, but not of bodies which have internal motions. Inertial forces, in an inertial coordinate system, must be considered as "apparent" forces, since they cannot produce an acceleration. From the standpoint of an accelerated observer, however, they are to be considered as "real" forces, at least insofar as he disclaims any knowledge of the acceleration of his coordinate system. An observer moving with the system ordinarily denies that he has any acceleration, and must then postulate a gravitational field which corresponds to the inertial force $\boldsymbol{I}$ in magnitude and direction.

For the case of uniform circular motion, the situation is as follows: An external observer in inertial motion considers that a single real force acts on the revolving body (see Section 22-3). It is the force exerted by the cord on the revolving body and is directed toward the center of rotation (the centripetal force). For an observer moving with the revolving body, two real forces act on it: the centripetal force exerted by the cord and, in addition, an inertial force $\boldsymbol{I}$, of the same magnitude as the centripetal force and directed radially outward (the "centrifugal force"). The two forces produce equilibrium (an *apparent equilibrium*) for the observer moving with the body (see Section 41).

Albert Einstein, in his general theory of relativity, has pointed out the true significance of the relation between gravitation and inertia. To understand these ideas of Einstein, a completely new orientation toward mechanics is necessary. In particular, the Newtonian concepts of *absolute space* and *time* must be abandoned. Space and time are influenced by the masses which exist in them. The space–time continuum in which physical processes take place is not rigid, but contracts or expands depending on its contents.

28. Work and Energy

The introduction of the concepts of *work* and *energy* makes possible a great simplification in the formulation of physical laws and the description of physical processes. These quantities are an important part of the structure of mechanics, and, in fact, of all physical science.

28-1 The Concept of Work

It requires an effort to displace a body on which another force is acting. Bodies do not displace themselves spontaneously, as we know from common experience.

DEFINITION. The *work* W done by a constant force $\boldsymbol{F}$, when its point of application moves through a distance s, is defined as the scalar product of the force $\boldsymbol{F}$ and the displacement $\boldsymbol{s}$:

$$W = \boldsymbol{F} \cdot \boldsymbol{s} = F_s \cos \varphi,$$

where φ is the angle between $\boldsymbol{F}$ and $\boldsymbol{s}$.

Since $F \cos \varphi$ is the component of the force in the direction of the displacement, F_s, we can also define work as the force component in the direction of the displacement, multiplied by the displacement, $F_s s$ (not simply "work equals force times distance").

From its definition, work is a scalar quantity. If F_s and s are in the same direction, the work done by the force is *positive* ($\cos \varphi > 0$). We say that work is done *on* the system on which the force acts; the force or the object producing the force supplies the work.

EXAMPLE 1. When a body falls, positive work is done by the weight of the body. When we stretch a spring by muscular effort, positive work is done *by* the force $\boldsymbol{F}_{HS}$ which our hand exerts on the spring (Fig. 3-31).

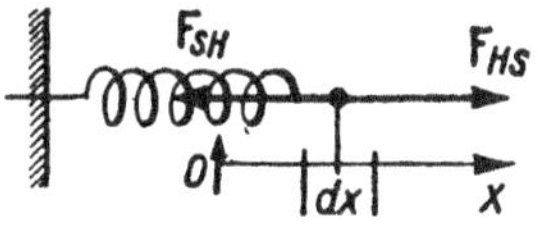

FIG. 3-31. Work in stretching a spring.

If $\boldsymbol{F}_S$ and s are directed oppositely, the work done by the force is negative ($\cos \varphi < 0$). We say in this case that the system on which the force acts is doing work.

EXAMPLE 2. When we stretch a spring, negative work is done by the force $\boldsymbol{F}_{SH}$ which the spring exerts on our hand (Fig. 3-31).

When the force component is not constant, the work must be expressed in differential form:

$$dW = F_s\,ds = \boldsymbol{F} \cdot d\boldsymbol{s}.$$

The total work done in a finite displacement $\boldsymbol{s}$ is the integral of dW:

$$W = \int dW.$$

Work has the dimensions $[W] = [ML^2T^{-2}]$, and the units: cgs system, 1 dyne-cm $=$ 1 gm $\cdot$ cm^2 sec^{-2} $=$ 1 erg; mks system, 1 newton-m $=$ 1 kgm $\cdot$ m^2 $\cdot$ sec^{-2} $=$ 1 joule; fps system, 1 ft-poundal $=$ 1 lbm $\cdot$ ft^2 $\cdot$ sec^{-2}.

In the cgs system, the unit of work is 1 dyne-cm, called, for brevity, 1 erg. An erg is the work done by a constant force of 1 dyne when its point of application is displaced 1 cm in the direction of the force. One erg is very nearly equal to the work done when a body weighing 1 mgmf is lifted 1 cm.

In the mks system, the unit 1 newton-m is called, for brevity, 1 joule. The erg and the joule are related by the equation

$$1 \text{ joule} = 1 \text{ newton-m} = 10^7 \text{ dyne-cm} = 10^7 \text{ ergs}.$$

REMARK 1. The performance of work is always associated with a *force* and a *displacement* of its point of application. In an electric motor, for example, the displacement is produced by electromagnetic forces. When a man does work, the displacement is produced by muscular forces.

REMARK 2. In all instances in which work is done, it is important to distinguish between two systems which exert forces on each other. In every such process, two quantities of work are done, having equal magnitudes and opposite signs. To illustrate this, let us consider a simple process in which work is done. Suppose that one end of a spring is attached to a wall and the spring is stretched by a muscular force exerted on its other end. Two forces act on the spring: (a) The wall exerts a force on the spring. The point of application of this force is not displaced in the stretching process and no work is done by it. (b) A muscular force is exerted on the spring. This force does positive work on the spring. The reaction to this force is exerted on the hand of the person stretching the spring. The work done by the reaction force is negative, since its direction is opposite to the displacement of its point of application (Fig. 3-31).

REMARK 3. The work done by a force is zero if the component of force in the direction of the displacement is zero. This is the case when the force itself is zero, and also when it is perpendicular to the displacement. The work done by a force is also zero if the displacement is zero.

REMARK 4. When a force $\boldsymbol{F}$ acts on a body, and work is done by the force in a displacement $d\boldsymbol{s}$, where the force $\boldsymbol{F}$ makes an angle φ with $d\boldsymbol{s}$, a force of constraint is always necessary. The constraining force has such a magnitude that its resultant with the force F is in the same direction as ds. The work done is thus

$$dW = F_s\,ds = F\cos\varphi\,ds,$$

since no work is done by the force of constraint. This justifies the original definition of work as the product of the component of a force in the direction of the displacement of its point of application and the displacement.

28-2 The Most Important Forms of Work

Work done in lifting a body. When a body of mass m is lifted slowly in vacuum, two forces act on the body (Fig. 3-32), its weight mg and the muscular force F exerted by the person lifting it. If the velocity is constant, these forces are in equilibrium; that is $|F| = |mg|$. To lift the body, work must be done by the person lifting it. If the point of application of F is displaced upward a distance h, the work done on the body by muscular effort is $W = Fh = mgh$.

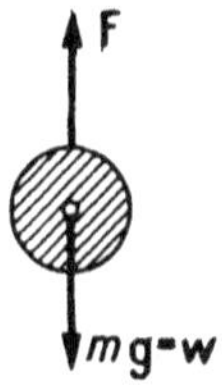

FIG. 3-32

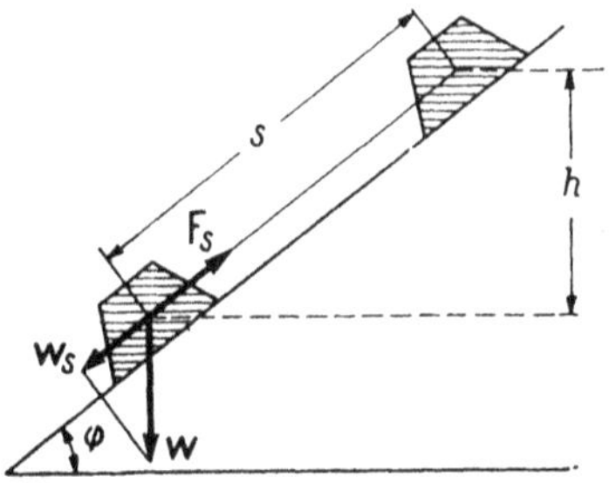

FIG. 3-33

FIG. 3-32. Forces on a rising weight.

FIG. 3-33. Work of moving a weight up an inclined plane.

When a body is moved up a frictionless inclined plane with constant velocity (Fig. 3-33), the work done is

$$W = F_s s = mgs\sin\varphi = mgh.$$

The work done in raising the body depends only on the difference in elevation between the initial and final positions. It is independent of the length or configuration of the path between these points. By means of an inclined plane, the work can be done by a smaller force than when the body is lifted vertically.

Work done in free fall. When a body falls freely, the only force acting on it is its weight. Work is done on the falling body by its weight mg, since this force is in the same direction as the displacement. The work done when the body descends through a height h is $W = mgh$.

This work done on the body by its weight is stored up in the body as kinetic energy (see Section 28-4). That is, the body can do work on some other body in the process of being brought to rest.

Work done in stretching a spring. Let a spring be slowly stretched a distance dx by muscular effort. The work done by the stretching force (Fig. 3-31) is

$$dW = F_{HS}\, dx,$$

where F_{HS} is the muscular force exerted on the spring. If the elastic limit of the spring is not exceeded, the force is given by the linear equation $F = kx$, where x is the extension of the spring beyond its unstressed length. The muscular work done in an extension dx (Fig. 3-34) is

$$W = kx \cdot dx,$$

which corresponds to the area of the narrow vertical strip in Fig. 3-34. The total work done when the elongation is increased from zero to x equals the area of the shaded triangle, or

$$W = \int W = \int_0^x kx \cdot dx = \tfrac{1}{2}kx^2.$$

The work increases with the square of the elongation.

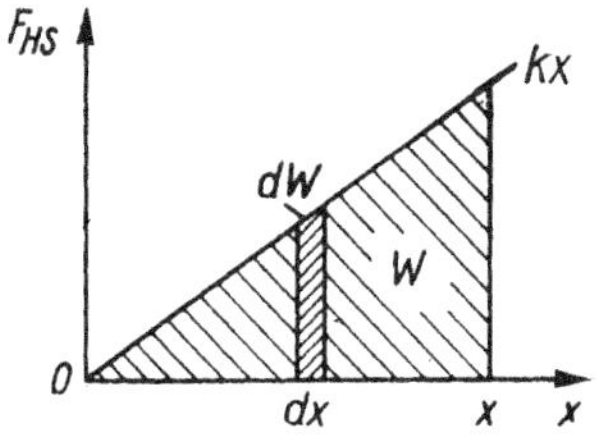

FIG. 3-34. Work of stretching a spring.

Work done in accelerating a body. Work must be done to increase the velocity of a body of mass m from zero to a value $\boldsymbol{v}$. To compute this work, we consider the special case in which the acceleration is constant. The body is then acted on by a constant force $\boldsymbol{F}$ until its velocity is $\boldsymbol{v}$, where $\boldsymbol{F} = m\boldsymbol{a}$.

From Eqn. (20-3), the corresponding displacement of the body is

$$s = \tfrac{1}{2}(v^2/a).$$

The work done by the force F is therefore

$$W = Fs = ma\,\tfrac{1}{2}(v^2/a) = \tfrac{1}{2}mv^2.$$

The work necessary to increase the velocity of a body from zero to $\boldsymbol{v}$ is proportional to the square of v. For a given body, the work depends only on the final velocity and not on the acceleration or the accelerating force. The same amount of work would be required if the force varied in any arbitrary way during the accelerating process, and the above result, although derived for a special case, is, in fact, perfectly general.

Work against sliding friction. When a body slides (with constant velocity) through a distance s along a plane surface for which the coefficient of sliding friction is μ, an external force must be exerted on it (Fig. 3-35). The frictional force f has the value $f = \mu N$, where N is the normal force exerted on the body by the surface. Let F be the external force. Then, since the body is in equilibrium, $F = f = \mu N$, and the work done by the force F is

$$W = fs = \mu N s .$$

The force F must do this amount of work on the body against the force of friction.

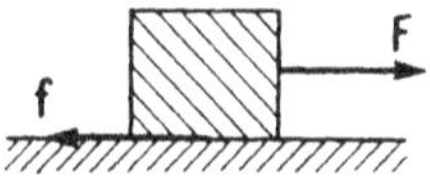

FIG. 3-35

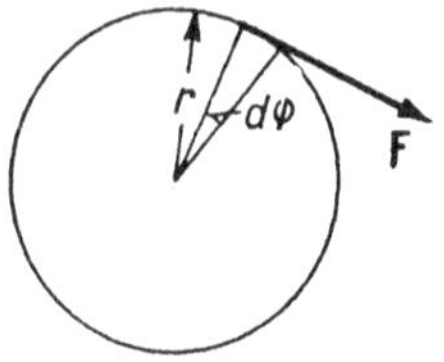

FIG. 3-36

FIG. 3-35. Work in the case of sliding friction.

FIG. 3-36. Work by a torque.

Work done by a torque. Figure 3-36 represents a wheel mounted on a fixed axis. A force F, at a distance r from the axis, is exerted on the whee perpendicular to the radius. The moment of the force about the axis (called the *torque*) is

$$M = Fr .$$

When the wheel turns through the angle $d\varphi$, the point of application of the force moves a distance $ds = r\,d\varphi$, and the work done is

$$W = F\,ds = Fr\,d\varphi = M\,d\varphi .$$

Work against rolling friction. Consider a wheel rolling on a plane surface for which the rolling friction torque is given by (cf. Section 23-2)

$$M_r = \mu_r N ,$$

where N is the normal reaction of the surface and μ_r is the coefficient of rolling friction. The work done on the wheel by this torque is therefore

$$W = M_r\,d\varphi = \mu_r N\,d\varphi ,$$

where $d\varphi$ is the angle through which the wheel turns. Since the wheel rolls without slipping $r\,d\varphi = ds$, where ds is the displacement of the center of the wheel. The work is therefore $dW = (\mu_r/r)\,N\,ds$.

28-3 Graphical Determination of Work

A graph of the force component F_s as a function of the path length s is called a *force-displacement diagram* (Fig. 3-37). When the units of force and distance along the respective axes are specified, the amount of work

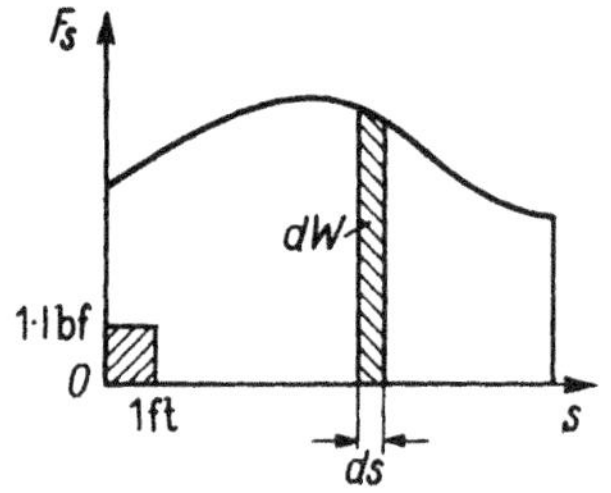

FIG. 3-37. Graphical deteımination of work.

represented by the diagram is determined. Thus the shaded area adjacent to the origin in Fig. 3-37 represents 1 ft · lbf of work. When the path length s increases by ds, the work done is $dW = F_s\,ds$. The work dW is therefore represented by the area of the narrow vertical strip, since its height is F_s and its width is ds. The area is to be considered positive if ds is in the positive direction along the s-axis and if F_s is positive. If the motion is backward, and if F_s is positive, the area is negative. The total work is represented by the algebraic sum of all elements of area under the graph of F_s.

28-4 Energy

First we define energy.

DEFINITION. The *energy* of a system is defined as its capacity to do work.

Energy is equivalent to work, and is therefore a scalar. It is expressed in the same units as work, that is, ergs, joules, or foot · poundals, but we represent it by the symbol E.

In order that a system shall be able to do work, it is necessary that work shall first be done on it. A body that has been lifted, a spring that has been stretched, and a body that has been accelerated have all acquired energy as a result of work that has been done on them. They now have the ability to do work on some other system.

We distinguish between two principal forms of energy, kinetic and potential. The energy of motion, or *kinetic energy* E_k, of a mass is the amount of work it can do as a result of the work done in accelerating it. A moving mass m therefore has a kinetic energy given by

$$E_k = \tfrac{1}{2}mv^2.$$

The energy of position, or *potential energy* E_p, is the work a system can do as a result of the work done on it in changing the position either of portions of the system or of the system as a whole. Potential energy exists in many different forms.

(1) *Gravitational energy*. When a body of mass m is lifted through a vertical height h, where h is small compared with the earth's radius, it acquires a potential energy $E_p = mgh$.

(2) *Elastic energy*. An elastic body which has been deformed can do work in returning to its original shape. For a spring which has not been strained beyond the elastic limit, the stored elastic potential energy is $E_p = \frac{1}{2}kx^2$. Other forms of elastic energy, in the Hooke's law region, are the energy of a bent beam, of a twisted rod, etc.

(3) *Chemical energy*. Substances, or mixtures of substances, are able to do work as a result of rearrangements of the atoms among the molecules of the substance. Burning is one means used to liberate this energy.

(4) *Electrical energy*. Electrical charges have the ability to do work as a result of the mutual forces between them.

(5) *Mass*. The mass of a body represents a highly concentrated source of energy. It can be utilized in part for doing work as a result of rearrangements of nucleons among the atomic nuclei by means of nuclear fission or fusion reactions.

29. The Principle of Conservation of Energy

29-1 Introduction

Mankind has attempted for hundreds of years to devise a machine which would do work, that is, deliver energy, without requiring that work be done on it. This hypothetical device has been given the inappropriate name *perpetual motion* machine. Such a name is misleading. What we desire is not a machine which will continue to *move* forever without requiring that work be done on it. We have such systems before us daily, in the form of the planets and their satellites, which have been moving in their orbits for millions of years without any observable reduction in their speeds. What we would really like is a machine which will continue to do *work* for us, without our having to supply *energy* to it.†

The term *machine* refers here to a device which periodically performs a *cyclic process* (see Section 82-1). This means that at the completion of the process the machine has returned to its original state in all respects. It can be proved by the principles of Newtonian mechanics that in *purely mechanical processes*, the total amount of work done *on* a machine in a cyclic process is exactly equal to the total work done *by* the machine. Processes which are

† Such a machine is called a *perpetual motion machine of the first kind. Perpetual motion machines of the second kind* are discussed in Section 86-1.

not purely mechanical will be excluded in the considerations of this section. Friction, for example, is of this sort, since heat is always developed when frictional forces are present. The processes of lifting a body, stretching a spring, or accelerating a body are purely mechanical, although that of stretching a spring can be included only for completely elastic deformations.

For centuries, inventors have sought without success for a machine which would continuously do work, but on which work would not have to be done. By the middle of the last century the experimental evidence had become extraordinarily abundant that energy could be neither created nor destroyed by the use of a machine. The statement of this fact is the principle of *conservation of energy*. It should be noted, that a general formulation of the energy concept must go further and include in the energy balance not only mechanical, but all other forms of energy as well.

The search for a general formulation of the law of conservation of energy has shown that this law holds without exception in all branches of natural science. The apparent exceptions, such as those which occur in mechanics when frictional forces are present, are explained when we include the heat developed as a result of friction. This conversion of mechanical energy to heat energy will be considered later in connection with the first law of thermodynamics.

As a result of a wealth of experimental evidence, the principle of conservation of energy is accepted today as a postulate that is universally valid. It ranks with Newton's laws as one of the foundations of physics, and is true even in the atomic domain, where the latter must be modified. This principle forms a basis for interpreting the results of all scientific investigations.

29-2 Formulation of the Principle of Conservation of Energy

Closed systems. A closed system is one which exchanges no work or energy with its surroundings. Such a system is completely isolated from the rest of the world; hence the principle of conservation of energy is expressed as follows.

The total energy of the system neither increases nor decreases. The energy content of the system remains constant and consists of the sum of its potential and kinetic energy:

$$\sum (E_p + E_k) = \text{const.}$$

Energy of one form can be transformed *quantitatively* into another form.

Open systems. A system is said to be *open* if it exchanges work and energy with its surroundings. In this case the law of conservation of energy takes the following form: The increase in the internal energy of the system is equal to the work done *on* it by external forces.

When a closed system performs a cyclic process, the initial and final states of the system are the same in all respects and the sum of the changes in energy of all parts of the system must be zero at all times, since the total energy is always a constant:

$$\sum dE_i = 0 \quad \text{(for all times } t\text{)}.$$

If an open system performs a cyclic process, the initial and final energies of the system are equal, since all parts of the system have returned to their initial state, but the total energy may well vary during the course of the process:

$$E_{\text{initial}} - E_{\text{final}} = 0.$$

The energy difference, ΔE, between two states of an open system, one state being reached from the other along a prescribed path s, equals the algebraic sum of the amounts of work done on or by the system along this path:

$$\Delta E = E_2 - E_1 = \sum_i W_i.$$

This is true only if we allow nothing but mechanical processes to take place. In forming this sum, work done on the system is considered positive and work done by the system is considered negative. This convention of signs is appropriate, because work done on a system increases its ability to do work, or its energy. The same convention is used in thermodynamics, where the work done by a system is considered negative and the work done on a system is positive.

The fact that the net change in energy is zero in a cyclic process for a purely mechanical, open system leads to an important law:

If a purely mechanical, open system is taken from any arbitrary initial state A to any other arbitrary end state B (Fig. 3-38), the difference between

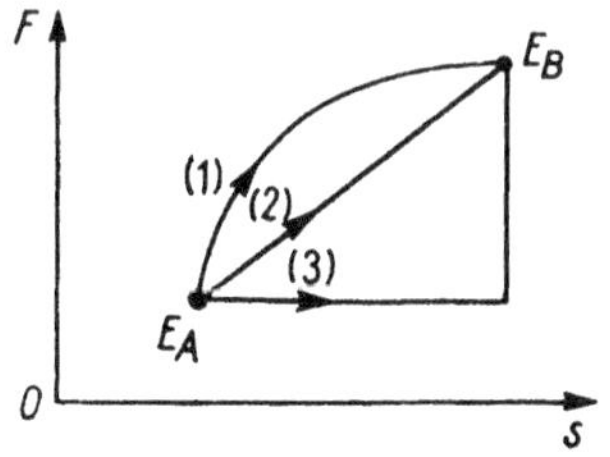

FIG. 3-38. The difference in energy between two states is independent of the path from one state to the other.

its energies in the two states is completely independent of the process by which it is taken from the first state to the second:

$$E_B - E_A = \sum_i W_i = \text{const}$$

for all paths between A and B.

To prove this law, we first make the assumption that it is false. We can then take the system along path 1, in Fig. 3-38, from A to B, and return it from B to A along another path such as 2 or 3. Then for the complete process, $\sum_{i} W_i \neq 0$, and we have performed a cyclic process in which the energy of the system has increased or decreased, which contradicts the principle of conservation of energy. Hence the assumption was false and the law as stated is correct. This law, which is obviously true for *all* closed systems, can be extended to any open system by a proper consideration of all the various forms of energy (see Chapter 10).

The *energy* of an open system is therefore a quantity whose value depends only on the state of a system, and whose change between two states is independent of the path. Quantities having this property are called *potentials*, and much use is made of them in advanced dynamics and in thermodynamics.

29-3 Application of the Law of Conservation of Energy

The total energy of a *closed* system, $\sum (E_p + E_k)$, remains constant at all times. If we determine the energy at any two points of a process, such as the initial point and the endpoint, it will have the same value. The mathematical expression of this fact is the so-called *energy equation*, which can be used to compute one unknown quantity for the process.

If the system is *open*, the amounts of work done on or by the system must be included in the energy equation. If the system does work against a frictional force, the initial energy equals the final energy plus the frictional work.

EXAMPLE 1. *The motion of a projectile.* A body of mass m is projected from the origin of a coordinate system at an angle of departure α and with an initial velocity v_0 (Fig. 3-39)

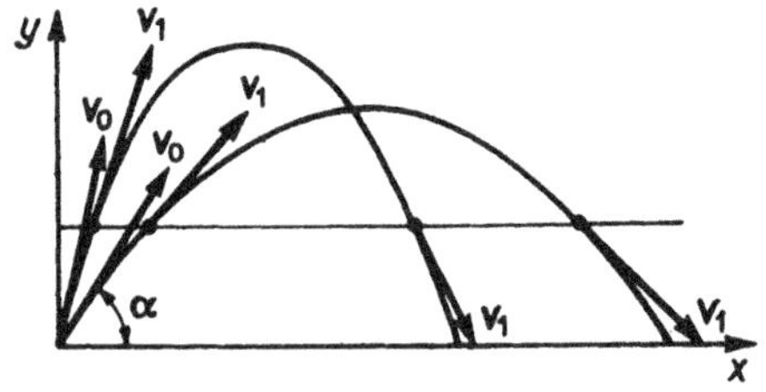

FIG. 3-39. Motion of a projectile in a vacuum.

To calculate the velocity at any height h_1, we set up the energy equation. Let us consider the potential energy to be zero at the origin. Then

$$\text{Initial position:} \quad E_k = \tfrac{1}{2}mv_0^2, \quad E_p = 0,$$

$$\text{Final position:} \quad E_k = \tfrac{1}{2}mv_1^2, \quad E_p = mgh_1.$$

From the energy law, the total energy must be the same in both positions. That is,

$$\tfrac{1}{2}mv_0^2 = \tfrac{1}{2}mv_1^2 + mgh_1.$$

The only unknown is the velocity v_1, and hence

$$v_1 = \sqrt{(v_0^2 - 2gh_1)}.$$

The velocity is independent of the mass and of the angle of departure α. Also, the velocity is the same, at the same elevation, on both the ascending and descending portions of the trajectory.

EXAMPLE 2. A particle slides without friction on the inside of a vertical circular track of radius R. To complete the entire circular motion without leaving the track, it must be given a certain minimum velocity v_0 at the lowest point. We wish to compute this velocity (Fig. 3-40).

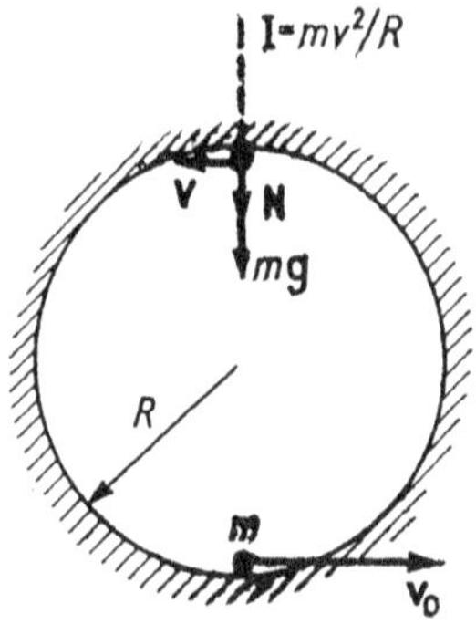

FIG. 3-40. Sliding on a vertical circular track without falling.

The critical point of the motion is the highest point of the circular path. If the particle passes this point without leaving the track, the remainder of the path will be completed safely. The real forces on the particle at the highest point are its weight mg and the reaction of the track, N. The inertial force I equals mv^2/R, where v is the velocity. Considering the state of the particle as one of d'Alembert equilibrium, we have $I = N + mg$.

The particle will not leave the track at the highest point provided that any reaction force N exists at this point. But if $N \geqq 0$, then $I \geqq mg$ or $mv^2/R \geqq mg$.

The required velocity v_0 can now be computed from the energy law and the requirements that $I \geqq mg$:

$$E_0 = \tfrac{1}{2}mv_0^2 = \text{const} = \tfrac{1}{2}mv^2 + 2mgR.$$

Since mv^2/R must be greater than mg, or $mv^2 \geqq mgR$, it follows from the energy equation that

$$\tfrac{1}{2}mv_0^2 \geqq 5mgR/2,$$

and that

$$v_0 \geqq \sqrt{(5gR)}.$$

EXAMPLE 3. A railway train travelling at 60 mi/hr passes the point Q in Fig. 3-41 and encounters a 12 $^0/_{00}$ grade (i.e., $\tan\alpha = 0.012$). The power is cut off at point Q. The kinetic energy of the train enables it to travel a certain distance up the grade. The combined effects of air resistance and sliding and rolling friction are equivalent to a coefficient of sliding friction $\mu = 0.005$.

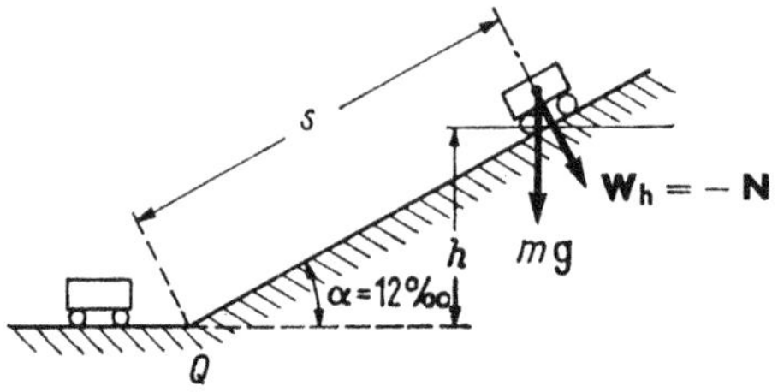

FIG. 3-41. Train on an incline.

We first find the distance s moved up the grade. The initial kinetic energy, $\frac{1}{2}mv_0^2$, is transformed into potential energy and frictional work. The energy equation is

$$\tfrac{1}{2}mv_0^2 = mgh + \mu Ns = mgs \sin \alpha + \mu mgs \cos \alpha .$$

The unknown distance s is therefore

$$s = \frac{v_0^2}{2g(\sin \alpha + \mu \cos \alpha)} = 970 \text{ ft}.$$

The distance is independent of the mass, since all quantities in the energy equation are proportional to the mass.

Next we find the velocity v' when the train rolls back to Q. It follows from the law of conservation of energy that

$$\tfrac{1}{2}mv_0^2 = \tfrac{1}{2}mv'^2 + 2\mu mgs \cos \alpha$$

from which

$$v' = \sqrt{(v_0^2 - 4\mu gs \cos \alpha)} = 84.4 \text{ ft/sec} = 57.5 \text{ mi/hr}.$$

30. Power and Efficiency

We define the quantity *power* in the following way:

DEFINITION. The *power P* developed by a system is defined as the quotient of the work done by the system and the time interval in which the work is done:

$$P = dW/dt.$$

Power has the dimensions $[P] = [ML^2T^{-3}]$, and the units: cgs system, 1 erg/sec; mks system, 1 joule/sec = 1 watt; fps system, 1 ft · lbf/sec.

It follows from the definitions of the erg and the joule that 1 watt = 10^7 ergs/sec. Units of more convenient size in engineering are 1 kilowatt (kw) = 10^3 watts, 1 megawatt (mw) = 10^6 watts, 1 horsepower (hp) = 550 ft · lbf/sec = 33,000 ft · lbf/min. It is easy to show that 1 hp = 746 watts. A unit of work that is commonly used is the *kilowatt · hour* (kwh):

$$1 \text{ kwh} = 10^3 \times 3600 \text{ watt} \cdot \text{sec} = 3.6 \times 10^6 \text{ joules}.$$

The *useful work output* of a machine is always less than the *energy input* because of work done against friction, in making noise, etc.

DEFINITION. The *efficiency* η of a machine is defined as the ratio of the useful work output to the energy input (or as the ratio of the useful power output to the power input):

$$\eta = \frac{\text{useful work output}}{\text{energy input}}.$$

To apply the energy equation to problems involving power, the change in energy per unit time and the work done per unit time (power) must be equated. The energy change between two neighboring states of the system is divided by the time interval dt required to go from one state to the other:

$$\frac{dE}{dt} = \frac{E_2 - E_1}{dt} = \frac{dW}{dt} = P.$$

If the power P is positive, work is done on the system and its energy is increased. If the power is constant, the preceding equation can be written for a finite time t as

$$\frac{E_2 - E_1}{t} = \frac{W}{t} = P.$$

EXAMPLE 1. A turbine with 70 per cent efficiency receives a flow of 12,400 litre/min of water, which descends through a vertical height of 18 m. The power available at the entrance of the turbine is

$$P_{\text{in}} = gh(\Delta m/\Delta t).$$

The ratio $\Delta m/\Delta t$ is the mass of water flowing into the turbine per unit time. The power input is therefore

$$P_{\text{in}} = 36{,}450 \text{ watts}.$$

The power output is

$$P_{\text{out}} = \eta P_{\text{in}} = 25{,}515 \text{ watts} = 25.5 \text{ kw}.$$

EXAMPLE 2. An automobile weighing 4000 lbf travels on a level road at a velocity of 60 mi/hr. The effective coefficient of friction including air resistance at this velocity is $\mu = 0.08$ and the efficiency of the motor is $\eta = 70$ per cent. The motor utilizes the chemical-potential energy E_{chem} of gasoline + oxygen. An amount of energy ηE_{chem} is delivered by the motor to the drive shaft and is completely converted to work against friction. We wish to compute the horsepower input required to drive the automobile at a constant velocity of 60 mi/hr:

$$\eta \frac{\Delta E_{\text{chem}}}{\Delta t} = \eta P_{\text{in}} = \frac{\mu mg\, \Delta s}{\Delta t} = \mu mgv,$$

$$P_{\text{in}} = \frac{\mu mgv}{\eta} = 40{,}228 \frac{\text{ft} \cdot \text{lbf}}{\text{sec}} = 73.1 \text{ hp}.$$

EXAMPLE 3. The energy output of the sun results in a decrease in its mass. According to the Einstein equation, the energy equivalent of a mass m is

$$E = mc^2,$$

where c is the velocity of light, 3×10^{10} cm/sec. What is the decrease of mass of the sun per second if every square centimeter of the earth's surface receives radiant energy from the sun at a rate $Q = 1.92$ cal/min $= 1.92 \times 4.185$ joules/min? The distance from the earth to the sun is 1.495×10^8 km:

$$\Delta mc^2 = 4\pi r^2 Q, \quad \Delta m = 4\pi r^2 Q/c^2 = 4.18 \times 10^{12} \text{ gm/sec}.$$

For comparison, the mass of the sun is 1.97×10^{33} gm.

EXAMPLE 4. Water flows into the intake of a hydroelectric plant at an elevation of 600 ft above the turbines. Each pound-mass of water does 570 ft · lbf of work on a turbine. The efficiency η_2 of the turbine is 80 per cent, and that of the generators η_3 is 95 per cent. (a) What is the mechanical efficiency η_1 of the falling water? (b) What is the overall efficiency η of the installation? (c) What is the power output at the generator terminals, if 2×10^4 lbm of water flow through the turbine per second?

$$\eta_1 = \frac{570}{600} = 0.95,$$

$$\eta = \eta_1 \eta_2 \eta_3 = 0.72,$$

$$P = mgh\eta = 2.78 \times 10^8 \text{ ft} \cdot \text{lbf/sec} = 505{,}000 \text{ hp} = 377{,}000 \text{ kw}.$$

31. HARMONIC MOTION

Periodic oscillatory motions occur in many natural processes. A study of such motion adds nothing new to our knowledge of the principles of mechanics. Nevertheless, a special treatment of this subject is worth while, because of the great importance of oscillatory motion in physics and engineering.

Any motion that is periodic in time is called *oscillatory*. At the end of a time interval called the *period T*, a quantity which characterizes the motion, the body returns to its initial position. The oscillation is described mathematically by the expression for the coordinates of the body as a function of time. We shall restrict the discussion at first to linear oscillations, which can be described by a single coordinate.

Sinusoidal or *harmonic* oscillations are of particular importance. According to a theorem developed by Fourier, every periodic oscillation can be represented as a superposition of properly chosen sinusoidal oscillations. Harmonic oscillations are therefore of fundamental significance.

31-1 Kinematics of Harmonic Oscillations

We shall make use of a graphical method to begin our study of harmonic motion. Suppose that a point p moves with constant tangential speed v in a circle of radius r (Fig. 3-42). Let the motion be resolved into two-component motions, parallel to the axes of a coordinate system whose origin is at the center of the circle. We now consider the motions of the projections of the endpoint of the radius vector to p on the x- and y-axes, represented by the points p_x and p_y in Fig. 3-42. The velocity and acceleration of these points are equal to the components of the velocity v and the acceleration a of the particle p, where $v = r\omega$, $a = r\omega^2$, and ω is the angular velocity of

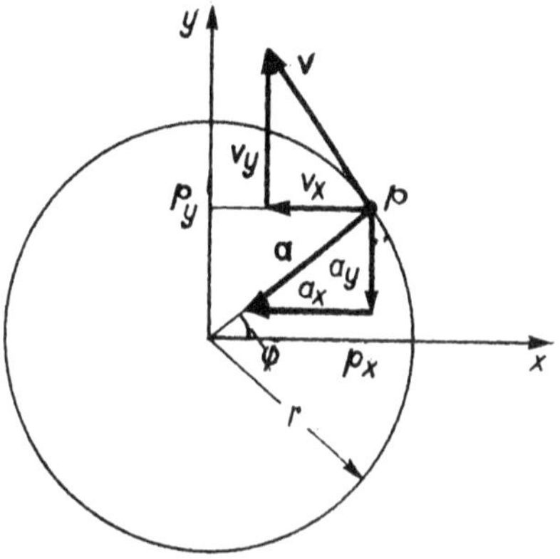

FIG. 3-42. Circular motion related to harmonic oscillation.

the circular motion. Let φ be the angle between the x-axis and the radius drawn to the particle p at any time t. Its magnitude is $\varphi = \omega t$. The motions of p_x and p_y are listed in Table 3-1.

TABLE 3-1

Motion of p_x	Motion of p_y
$x = r\cos\varphi = r\cos\omega t$	$y = r\sin\varphi = r\sin\omega t$
$v_x = -v\sin\varphi = -r\omega\sin\omega t$	$v_y = v\cos\varphi = r\omega\cos\omega t$
$a_x = -a\cos\varphi = -r\omega^2\cos\omega t$	$a_y = -a\sin\varphi = -r\omega^2\sin\omega t$

We now disregard the circular motion and consider the component x- and y-motions independently. We now call r $(=A)$ the *amplitude* of oscillation and ω $(= 2\pi f)$ the *angular frequency* of oscillation. The amplitude A is the maximum displacement from the midpoint of the motion. The frequency f is the number of complete oscillations per second. It follows that the period $T = 1/f$. Therefore

$$T = 1/f = 2\pi/\omega.$$

In terms of these quantities, the equations of motion can be written in a form which is completely independent of the circular motion considered above (Table 3-2).

TABLE 3-2

x-motion	y-motion
$x = A\cos\omega t$	$y = A\sin\omega t$
$v_x = -A\omega\sin\omega t$	$v_y = A\omega\cos\omega t$
$a_x = -A\omega^2\cos\omega t = -\omega^2 x$	$a_y = -A\omega^2\sin\omega t = -\omega^2 y$

An oscillation having the kinematic properties listed in Table 3-2 is called a *harmonic* oscillation.

REMARK 1. The total length of path covered by an oscillating body is not ordinarily of any significance. The quantity in which we are interested is the instantaneous coordinate x or y, which is a periodic function of the time having maximum magnitude A [Fig. 3-43(a)].

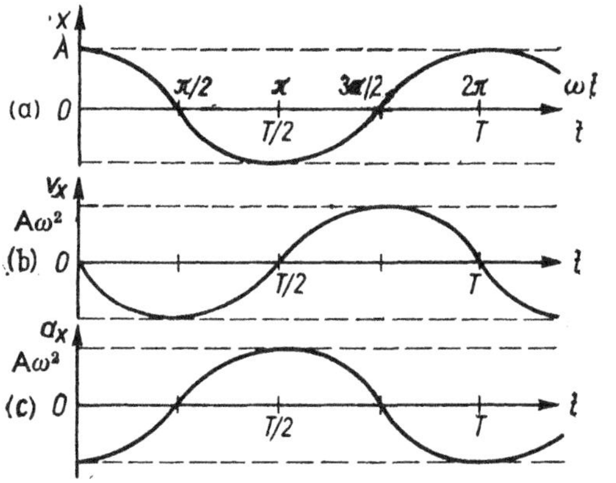

FIG. 3-43. Position, velocity, and acceleration in a harmonic oscillation.

REMARK 2. The velocity varies with time, with a period equal to that of the coordinate. At the midpoint of the oscillation, v has its maximum magnitude $A\omega$; at the endpoints of the path the velocity is zero [Fig. 3-43(b)].

REMARK 3. The acceleration is also a periodic function of the time, with the same period as the coordinate. It has its maximum value $A\omega^2$ at the endpoints of the path and is zero at the midpoint. It is important to note that at every point of the path, the *direction* of the acceleration is toward the midpoint, which is the point of equilibrium, and is proportional to the distance from this point [Fig. 3-43(c)].

31-2 Forces in Harmonic Motion

Every acceleration is caused by a force. Knowing the acceleration, we can compute the force on a body in harmonic motion from Newton's second law. The force F_x at any coordinate x is given by

$$F_x = ma_x = -m\omega^2 x,$$

where m is the mass of the body.

Harmonic motion therefore results from the action of a *restoring* force whose magnitude is directly proportional to the displacement from the point of equilibrium, which is here the origin (Fig. 3-44). Any force which is proportional to the coordinate x and is oppositely directed will produce harmonic motion along the x-axis. Forces of this type are given by the equation $F = -kx$, where k is a constant.

The angular frequency ω can be expressed in terms of the force as follows:

$$F = -kx = -m\omega^2 x, \quad \omega = \sqrt{(k/m)}.$$

Thus, the angular frequency, which is the most important characteristic of harmonic motion, is determined by k and m. These quantities, together with the amplitude A, completely determine the motion.

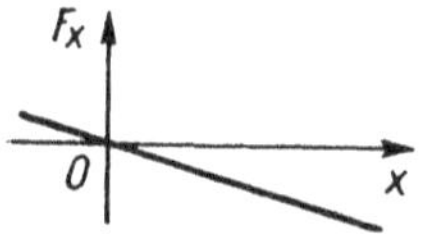

FIG. 3-44

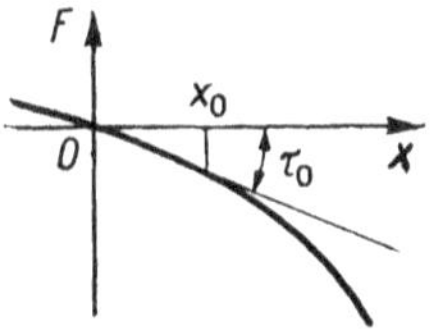

FIG. 3-45

FIG. 3-44. Linear restoring force in harmonic motion.

FIG. 3-45. Quasiharmonic force function.

Forces of the sort described above occur in many instances. For example, suppose that a particle of mass m is attracted toward a fixed point by a force F which is some arbitrary function of x, that is,

$$F = -f(x), \quad f(0) = 0.$$

In general, this force will *not* result in harmonic motion. However, *for small values of* x, the curve $F = -f(x)$ can be approximated by its tangent, as in Fig. 3-45, so that when x is small the law of force takes the form

$$F = -x \tan \tau_0 .$$

Hence, if the amplitude is small, the particle oscillates with harmonic motion. The force constant k then equals $\tan \tau_0$ and the angular frequency ω is given by

$$\omega = \sqrt{(\tan \tau_0/m)}.$$

A restoring force of this sort is called *quasi-harmonic*, in contrast to one which is strictly harmonic. The forces which bind molecules to their positions in a crystal are of this sort, and the molecular oscillations are harmonic if the amplitude is small. The same is true of the oscillations of a pendulum, and of the vibrations of strings and bars.

32. EXAMPLES OF LINEAR RESTORING FORCES

32-1 Oscillation of a Body Suspended from a Spring

At this point, we shall not consider the mechanism which leads to a linear force law when a spring is stretched (Section 57). However, it is an experimental fact that when a spring is stretched or compressed, it exerts on its surroundings a restoring force which is directly proportional to x, provided that the elastic limit of the spring is not exceeded. That is, $F = -kx$.

The proportionality constant k is called the *force constant* or the *spring constant*. It is numerically equal to the magnitude of the force required to stretch the spring by a unit distance, and can be determined by measuring the change in length of the spring when a body of known weight is suspended from it.

It is necessary to distinguish between a so-called *compression* spring and a *prestressed* spring. A compression spring is one in which the coils do not touch one another when no load is applied to the spring. Such a spring can be stretched or compressed without having its coils touch. For a compression spring, the same law of force applies to both elongation and shortening. In a prestressed spring, the coils press against one another when no stretching force acts on the spring. The spring does not start to elongate until the stretching force exceeds a certain minimum value F_0 (see Fig. 3-46) called the buffer force. For forces greater than F_0, the law of force is linear, as shown in Fig. 3-46.

If a body of mass m is hung from a compression spring, the spring is extended and the body executes harmonic oscillations about an equilibrium position. If a braking force is exerted on it, it comes to rest with the spring extended beyond its unloaded length (Fig. 3-47). In this position, two forces act on the body:

$$w = mg \quad \text{and} \quad F = -k\Delta x = -kx_0.$$

Since the body is in equilibrium, the magnitudes of these forces are equal. The equilibrium postion is therefore determined by the condition $x_0 = mg/k$.

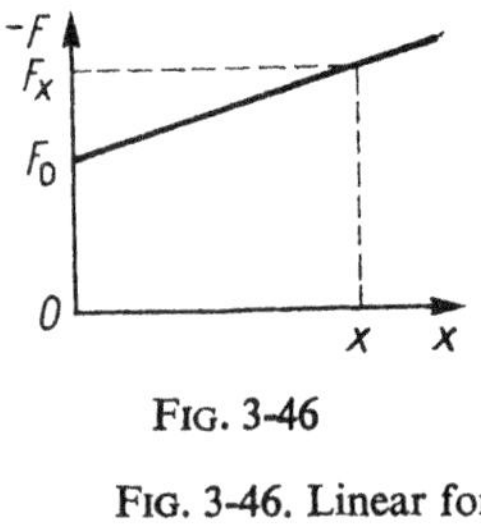

FIG. 3-46

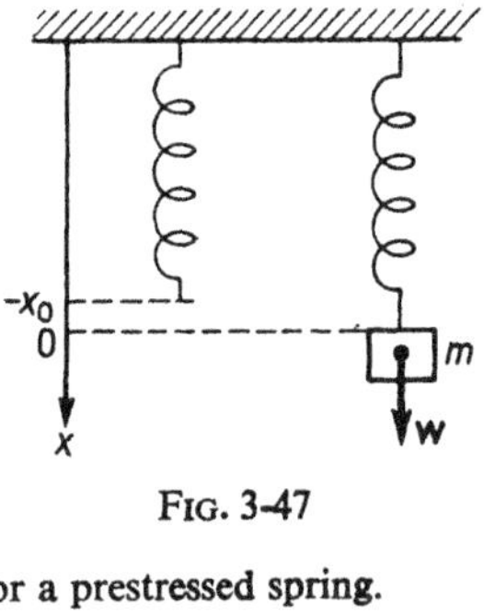

FIG. 3-47

FIG. 3-46. Linear force function for a prestressed spring.

FIG. 3-47. Equilibrium position for a loaded spring.

If the body is displaced vertically and released, it oscillates about this equilibrium position. Let us consider the equilibrium position as the origin. When the body is at a distance x below the origin, the elongation of the spring, S, is

$$S = x_0 + x,$$

and the resultant force on it is

$$F = mg - k(x_0 + x) = -kx.$$

The body is therefore acted on by a linear restoring force, independent of the load m provided, as always, that the elastic limit of the spring is not exceeded.

The angular frequency is

$$\omega = \sqrt{(k/m)},$$

which is valid only if the mass of the spring itself can be neglected, since the spring also takes part in the oscillatory motion. All springs have some mass; hence the experimentally measured value of ω is always smaller than that computed from the spring constant, as above.

EXAMPLE 1. A body of mass $m = 150$ gm is suspended from a spring of force constant $k = 42{,}000$ dynes/cm (Fig. 3-48). The elongation of the spring, when the body hangs at rest, is $x_0 = 3.54$ cm. The body is now given an initial velocity $v_0 = 40$ cm/sec. We wish to compute the amplitude, maximum acceleration, frequency, and the period of the ensuing motion. Frictional forces and the mass of the spring can be neglected. We have

$$\omega = \sqrt{(k/m)} = 16.7 \text{ sec}^{-1},$$

$$f = \omega/2\pi = 2.67 \text{ sec}^{-1},$$

$$T = 1/f = 0.374 \text{ sec}.$$

The amplitude A can be found from the velocity at the midpoint:

$$A = v_0/\omega = 2.39 \text{ cm}.$$

The maximum acceleration occurs at the ends of the path, and is

$$a_{max} = \pm A\omega^2 = \pm 668 \text{ cm/sec}^2.$$

EXAMPLE 2. A body of mass m, moving with a vertical velocity v_0, is suddenly attached to a spring of force constant k (Fig. 3-49). The body executes a harmonic oscillation whose midpoint is at a distance mg/k below the end of the unloaded spring. The lowest point reached in the oscillatory motion determines the amplitude A. Let us call the

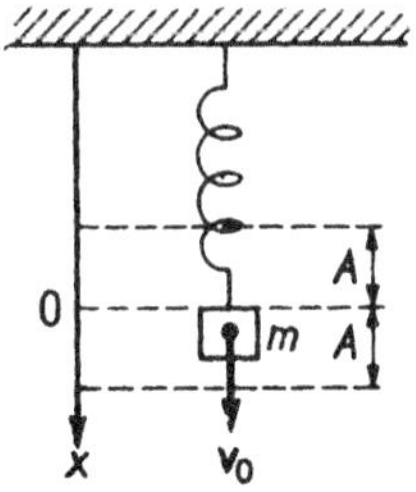

FIG. 3-48

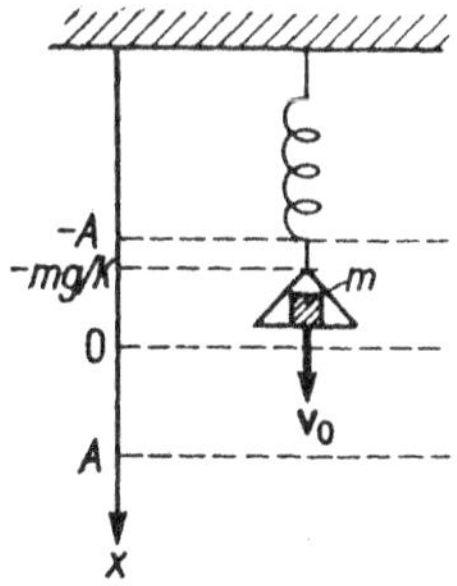

FIG. 3-49

FIG. 3-48. Vertically oscillating loaded spring.

FIG. 3-49. Conservation of energy applied to a loaded spring.

potential energy zero at the lowest point. The initial energy of the system is

$$E_0 = mg(A + mg/k) + \tfrac{1}{2}mv_0^2.$$

The energy at the lowest point (see Section 28-4) is

$$E_1 = \tfrac{1}{2}k(mg/k + A)^2.$$

From the principle of conservation of energy, the energies E_0 and E_1 are equal, so

$$mg(A + mg/k) + \tfrac{1}{2}mv_0^2 = \tfrac{1}{2}k(mg/k + A)^2.$$

For brevity, let $u = mg/k + A$. Then

$$\tfrac{1}{2}ku^2 - mgu - \tfrac{1}{2}mv_0^2 = 0,$$

and

$$u = \frac{mg \pm \sqrt{(m^2g^2 + kmv_0^2)}}{k}.$$

The amplitude A is therefore

$$A = \pm \frac{\sqrt{(m^2g^2 + kmv_0^2)}}{k}.$$

The amplitude is independent of the direction of the initial velocity v_0.

32-2 The Ideal Pendulum

An *ideal* pendulum, also called a *mathematical* pendulum, consists of a particle of mass m attached to one end of a weightless inextensible cord of length l. The other end of the cord is fixed in position. When the particle is displaced from its equilibrium position (see Fig. 3-50) it is acted on by a

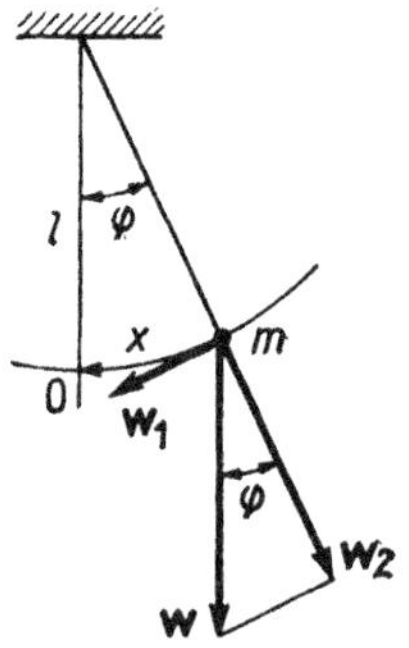

FIG. 3-50. The ideal pendulum.

restoring force given by the tangential component of its weight. Let x be the length of arc from the mass to the equilibrium point. Then

$$F = w_1 = -w \sin \varphi = -w \sin (x/l).$$

Since this force is not proportional to the distance x, the resulting motion is not harmonic. However, if the displacement x is small compared with the length of the pendulum, then we can replace $\sin(x/l)$ by x/l, obtaining as the law of force

$$F = -mgx/l.$$

We see that for small x the force law is linear. For small displacement, then, the pendulum oscillates with harmonic motion and

$$-mgx/l = -m\omega^2 x, \quad \omega^2 = g/l.$$

The frequency f and the period T are therefore

$$f = \omega/2\pi = (1/2\pi)\sqrt{(g/l)}, \quad T = 1/f = 2\pi\sqrt{(l/g)}.$$

The period is independent of the mass of the pendulum. The reason for this is that the gravitational mass of the pendulum, which gives rise to the restoring force, is equal to its inertial mass.

Since all of the forces acting on the pendulum lie in a vertical plane through the supporting cord, the pendulum continues to move in this plane. In order for this statement to be strictly correct, the motion must be referred to an inertial coordinate system. Let us therefore construct such a system whose origin is at the center of the earth, but which does not partake of the earth's rotation. (The motion of the earth around the sun can be considered as uniform and rectilinear for the length of time needed to perform the experiment.) It is clear that the gravitational field of the earth, which moves along with the earth in its motion around the sun, will not alter the plane of the pendulum motion, but that the surface of the earth will rotate with respect to this invariable plane. Hence, if a pendulum that is free from any disturbing forces is allowed to swing for a long period of time, the plane of oscillation appears to rotate slowly in a direction opposite to that of the earth's rotation. Actually, the plane of oscillation of the pendulum remains fixed and the earth's surface rotates with respect to it in the same direction as the earth's rotation. Such a pendulum is called a *Foucault pendulum*, since this experiment was first performed by Foucault to demonstrate the rotation of the earth.

The effect of the acceleration due to gravity on the period of a pendulum can be demonstrated by the so-called Mach pendulum. This consists of a rigid pendulum (Fig. 3-51) which can oscillate about the axis $a - a$. The angle α of the sloping oscillation plane E can be changed by changing the inclination of the axis $a - a$. The effective gravitational force acting on the pendulum *in its plane of oscillation E* is the component of its weight, $mg_1 = mg\cos\alpha$. The restoring force at an angular displacement φ from the line f in the plane E (Fig. 3-52), where f passes through the point of equilibrium and the axis, is

$$F = -mg\cos\alpha\sin(x/l).$$

If $x \ll l$, then

$$F = -(mgx\cos\alpha)/l = -m\omega^2 x,$$

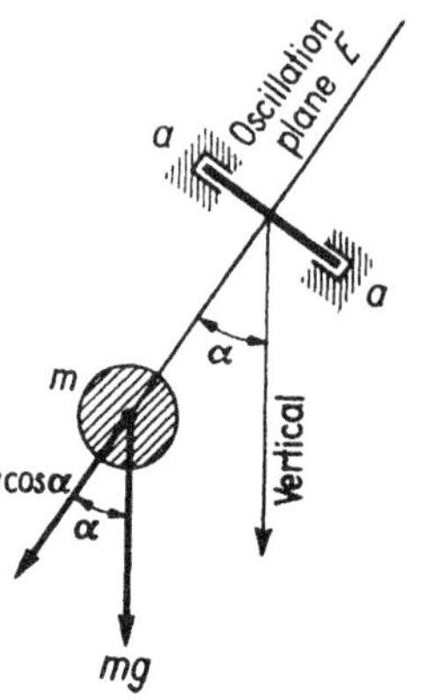

FIG. 3-51

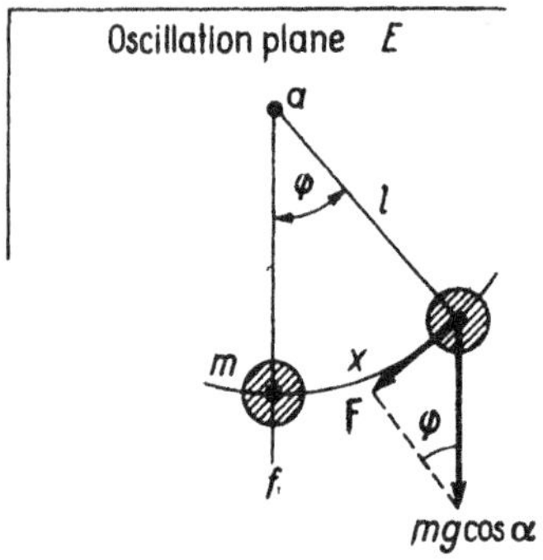

FIG. 3-52

FIG. 3-51. A Mach pendulum.

FIG. 3-52. Force on a Mach pendulum.

and the pendulum oscillates with harmonic motion of angular frequency

$$\omega = \sqrt{(g \cos \alpha / l)}.$$

The angular frequency ω depends on the component of the acceleration due to gravity in the oscillation plane.

When the displacement x of a pendulum is not small compared with l, $\sin (x/l)$ cannot be replaced by x/l. The restoring force F_1 is then smaller than the harmonic force (see Fig. 3-53), and the vibrations are no longer harmonic. For angular amplitudes up to $\varphi = 0.1$ rad, that is, $\varphi = 5.7°$, the period differs from that of a true harmonic motion by less than 1%. Hence, if differences of 1% can be tolerated, the motion can be considered harmonic. The period, within this range of amplitudes, is independent of the amplitude. With increasing amplitude, the discrepancy increases rapidly, and the periodic time varies with the amplitude. This situation is the chief obstacle to the construction of accurate pendulum clocks. However, Bessel has shown that satisfactory isochronism (i.e., the independence of period and amplitude) can be attained by attaching the upper end of the pendulum

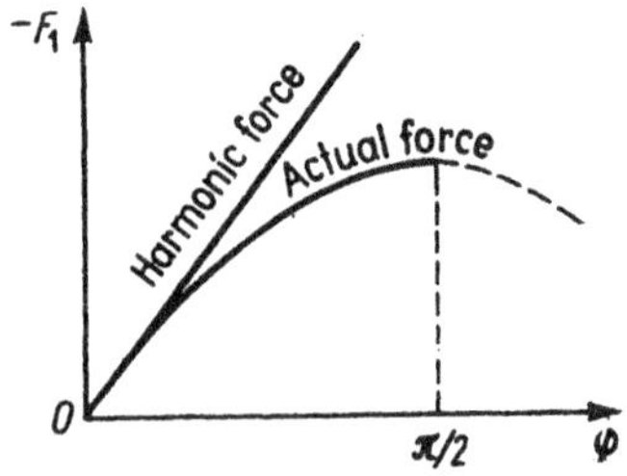

FIG. 3-53. Restoring force on a pendulum for large angle oscillations.

to a short flat spring whose length must be properly chosen. The added restoring force of the spring increases with increasing amplitude, and if properly designed, the total restoring force becomes very nearly harmonic.

33. Damped Harmonic Motion

33-1 Decrease of Amplitude

Every oscillatory motion is damped by friction forces. The amplitude A decreases from one oscillation to the next. It is often found experimentally that the ratio of the amplitudes of two consecutive oscillations is a constant. This commonly occurring case, called *exponential damping*, will now be considered in more detail. It is characterized by a *damping force* which is opposite in direction to the velocity and increases in direct proportion to the velocity.

Let A_n be the amplitude of any arbitrary oscillation and A_{n+1} that of the next oscillation. The ratio A_n/A_{n+1} is called the *damping ratio*, or the *decrement* δ:

$$\delta = A_n/(A_{n+1}).$$

This ratio, readily determined by experiment, is sufficient for describing this special type of oscillation. If A_0 is the initial amplitude and A_n that after n complete oscillations, then

$$A_0/A_n = \delta^n, \quad \text{and} \quad \delta = \sqrt[n]{(A_0/A_n)} > 1.$$

Taking logarithms of both sides of the first equation, we get

$$\ln A_n = \ln A_0 - n \ln \delta. \tag{33-1}$$

If we construct a semilogarithmic plot of A_n as a function of the number of oscillations (or of the time), we obtain a straight line, as in Fig. 3-54. In

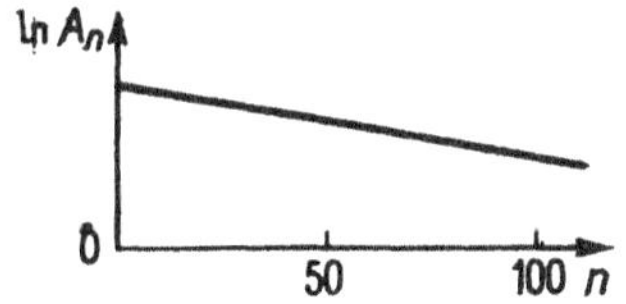

FIG. 3-54. Amplitude decay for exponential damping (semilog plot).

place of the number n, it is more convenient to make use of the continuous variable, time, given by $t = nT$, where T is the period of the oscillations. Since

$$\delta = e^{\ln \delta},$$

we can write for A_n

$$A_n = A_0\delta^{-t/T} = A_0e^{-t \ln \delta/T} \equiv A_0e^{-\varrho t}.$$

That is, A is an exponential function of t.

The equation of an exponentially damped oscillation (see Fig. 3-55) is

$$y = A_0 e^{-\varrho t} \sin \omega t.$$

Here, $e^{-\varrho t}$ is the *damping term,* $\varrho = \ln \delta / T$ is the *damping constant,* and $\ln \delta = \varrho T$ is the *logarithmic decrement* of the oscillation.

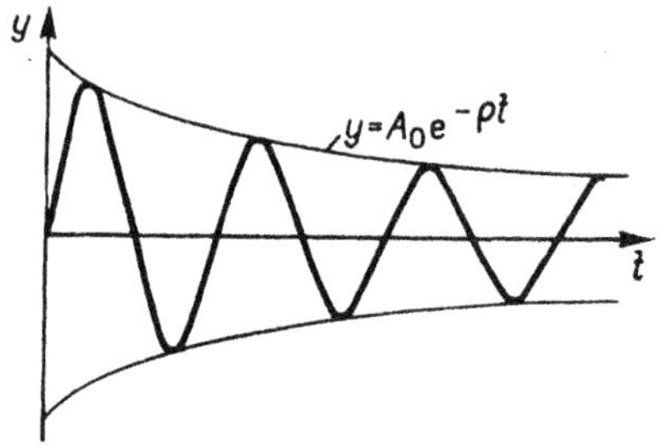

FIG. 3-55. Oscillations with exponential damping.

The *relaxation time* τ_e of the oscillation is defined as the time in which the amplitude decreases to $1/e$ of its original value:

$$A(\tau_e) = A_0/e = A_0 e^{-\varrho \tau_e}, \quad \tau_e = 1/\varrho = T/\ln \delta.$$

The *half-life* $\tau_{1/2}$ is the time in which the amplitude decreases to one-half its original value:

$$A(\tau_{1/2}) = A_0/2 = A_0 e^{-\varrho \tau_{1/2}}, \quad \tau_{1/2} = \ln 2/\varrho = 0.69/\varrho = 0.69\, \tau_e.$$

The quantities τ_e and $\tau_{1/2}$ are both readily determined by measuring the damping ratio δ and the period T.

EXAMPLE. A damped torsion pendulum has an initial amplitude $A_0 = 2.36$ rad and a period $T = 3.54$ sec. The amplitude $A = A_0 \delta^{-n}$ is shown in Fig. 3-56 as a function of the number of complete oscillations. From the measured value for 20 oscillations, $A_{20} = 1.17$, we find for δ the mean value

$$\delta = \sqrt[20]{(A_0/A_{20})} \doteq 1.0352,$$

and hence $\ln \delta = 0.03465$. The damping constant is

$$\varrho = \ln \delta / T = 9.79 \times 10^{-3}\ \text{sec}^{-1}.$$

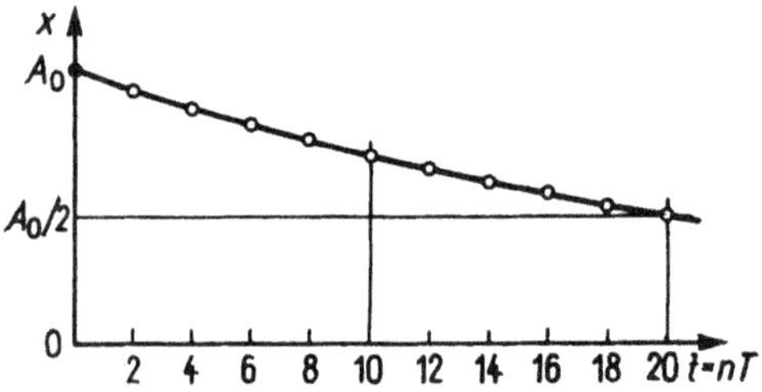

FIG. 3-56. Amplitude decay of an exponentially damped oscillation, linear plot. $T = 3.54$ sec, $\delta = 1.0352$, $A_0 = 2.36$, $A_{20} = 1.17$.

The oscillation is therefore given by the equation

$$A = 2.36\, e^{-9.79 \times 10^{-3} t} \cos\left(2\pi \frac{t}{3.54}\right).$$

For the relaxation time τ_e and the half-life $\tau_{1/2}$, we obtain

$$\tau_e = \frac{1}{\varrho} = \frac{10^3}{9.79} = 101.2 \text{ sec}, \quad \tau_{1/2} = \tau_e \ln 2 = 70.8 \text{ sec}.$$

This is in good agreement with the value $\tau_{1/2} = 70.8$ sec, as found from the graph in Fig. 3-56.

A semilogarithmic plot of the amplitude as a function of the number of oscillations is useful in the graphical treatment of the decay in amplitude (Fig. 3-57). If the damping is exponential, then A is a linear function of n:

$$\ln A = \ln A_0 - n \ln \delta.$$

The semilogarithmic plot is therefore a simple test for the existence of logarithmic damping.

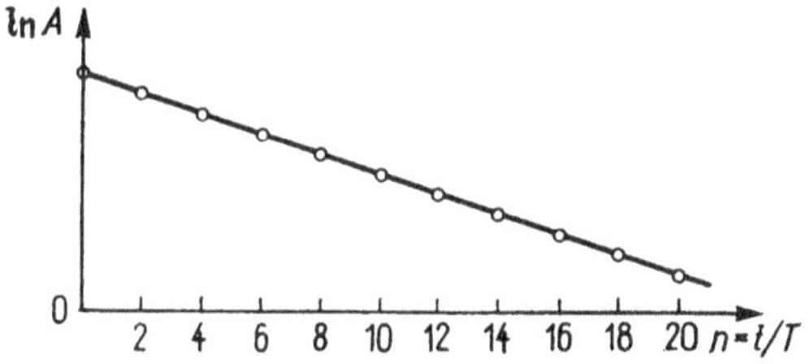

FIG. 3-57. Semilog plot of the amplitude decay in Fig. 3-56.

34. Phase Relations in Harmonic Motion

A harmonic oscillation in which the displacement is zero when $t = 0$ is represented mathematically by the equation $x = A \sin \omega t$. If the displacement is a maximum at $t = 0$, the motion is represented by $x = A \cos \omega t$. If at time $t = 0$ the displacement has some arbitrary value x_1 between 0 and A, the motion is given by

$$x = A \sin(\omega t + \varphi).$$

For a given amplitude A, φ is determined by the initial condition equation

$$x = x_1 = A \sin \varphi \quad (t = 0).$$

The most general equation of harmonic motion is therefore

$$x = A \sin(\omega t \pm \varphi).$$

The argument of the sine function, $(\omega t + \varphi)$, or $(\omega t - \varphi)$ is called the *phase* of the oscillation. Only when two harmonic oscillations are considered simultaneously does it make sense to speak of a *phase difference* Δ. The phase difference $\Delta_{1,2}$ between two oscillations given by

$$x_1 = A \sin(\omega t + \varphi_1) \quad \text{and} \quad x_2 = B \sin(\omega t + \varphi_2)$$

is defined as the difference between the phases of the two oscillations:

$$\Delta_{1,2} = \omega t + \varphi_1 - \omega t - \varphi_2 = \varphi_1 - \varphi_2.$$

The phase difference $\Delta_{1,2}$ of the first oscillation relative to the second has the opposite sign to the phase difference $\Delta_{2,1}$ of the second relative to the first.

A positive phase difference of the first relative to the second means that the first precedes the second in time; the first oscillation reaches its maxima and minima at earlier times than the second, since the argument of the first is always greater than that of the second. To determine the phase difference, both oscillations must be written in terms of the same harmonic function (sine or cosine) and must have the same sign (e.g., positive).

EXAMPLE 1. Consider $x_1 = -A \cos \omega t$, $x_2 = B \sin(\omega t + \pi/2)$. Writing both as sine functions with the same sign, we have

$$x_1 = A \sin(\omega t + 3\pi/2), \quad x_2 = B \sin(\omega t + \pi/2).$$

Then

$$\Delta_{1,2} = \omega t + \frac{3\pi}{2} - \omega t - \frac{\pi}{2} = \pi.$$

The oscillation x_1 *leads* the oscillation x_2 in phase by 180°.

EXAMPLE 2.

$$x_1 = A \cos \omega t = A \sin(\omega t + \pi/2),$$

$$x_2 = -A \sin(\omega t - \varphi) = A \sin(\omega t + \pi - \varphi),$$

$$\Delta_{1,2} = \omega t + \frac{\pi}{2} - \omega t - \pi + \varphi = \varphi - \frac{\pi}{2} = -\left(\frac{\pi}{2} - \varphi\right).$$

Oscillation x_2 *lags* oscillation x_1 by $90° - \varphi$.

35. COMPLEX NUMBERS, PHASORS, AND SUPERPOSITION OF HARMONIC OSCILLATIONS ALONG ONE AXIS

It is shown in the elementary theory of complex numbers that complex numbers can be represented by points in a plane (using a rectangular coordinate system). These numbers are, in some respects, similar to the vectors which extend from the origin to the points. Their behavior is similar to that of vectors with respect to the operations of multiplication by a real number, addition, and subtraction. The fact that the correspondence between complex numbers and vectors does not extend to the formation of products demonstrates the fundamentally different nature of the two quantities.

There are certain physical quantities which behave mathematically exactly like vectors and complex numbers with respect to the operations mentioned above. The mathematical representation of such a quantity is called a *phasor*. Phasors are uniquely determined by a magnitude, a unit, and a phase angle (but not, as for a vector, a direction in space) and they can be represented graphically in a rectangular coordinate system called a phasor diagram. A phasor is described by an equation of the form

$$y = A \sin \varphi .$$

It is evident that a harmonic oscillation can be represented by a phasor if φ is a linear function of the time.

For the graphical representation of the phasor

$$y_1 = A \sin \alpha ,$$

in a rectangular coordinate system (x, y), we lay off a line of length A from the origin in the direction making an angle α with the positive x-axis (Fig. 3-58). The projection of the phasor on the y-axis equals the instantaneous value $y_1 = A \sin \alpha$. If the phase angle α increases uniformly with

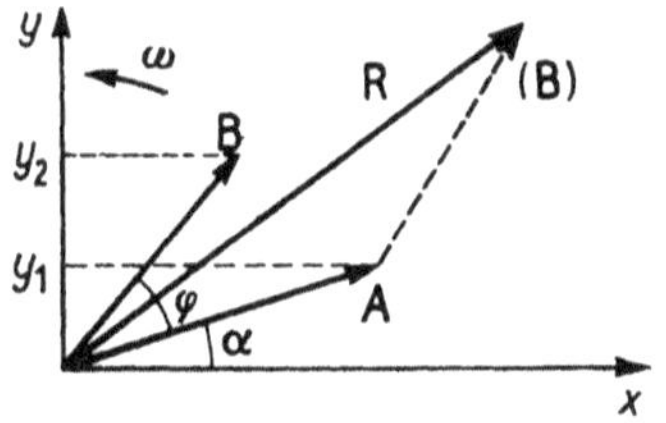

FIG. 3-58

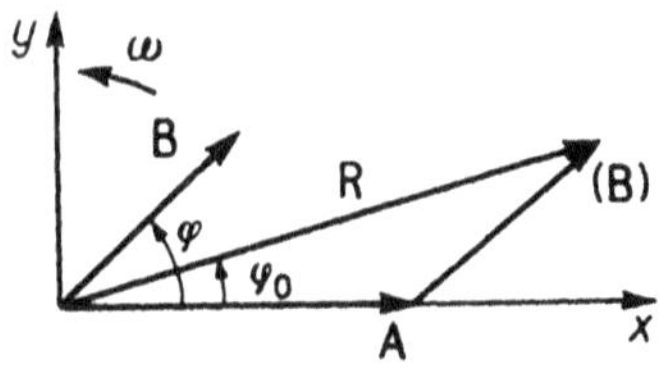

FIG. 3-59

FIG. 3-58. Diagram of two phasors ***A*** and ***B*** and their resultant ***R***.

FIG. 3-59. Addition of two phasors.

time, i.e., if $\alpha = \omega t$, the phasor rotates uniformly about the origin in the positive sense with an angular velocity ω. A second phasor in the same diagram, given by

$$y_2 = B \sin (\omega t + \varphi) ,$$

rotates with the same angular velocity, the phase difference between the phasors remaining constant:

$$\Delta_{1,2} = \omega t - (\omega t + \varphi) = -\varphi .$$

Phasors can be added according to the same rule as for ordinary vectors, and we obtain a resultant R, which rotates with the same angular velocity. The phase differences between the resultant and the components remain constant and can be determined from any instantaneous position of the phasors.

The resultant of two phasors of the same angular velocity, having unequal amplitudes and differing in phase φ, is given by (Fig. 3-59)

$$R^2 = (A + B \cos \varphi)^2 + B^2 \sin^2 \varphi .$$

Also

$$\tan \varphi_0 = \frac{B \sin \varphi}{A + B \cos \varphi}$$

gives the phase difference φ_0 between the resultant phasor and A_1. Phasors of the same angular velocity can be added and subtracted graphically, in the same way as ordinary vectors. The magnitude of the resultant and its relative phase angle can be found from the diagram. Harmonic vibrations of the same frequency can be superposed in this way.

For phasors of *unequal* angular velocity, the problem is more complicated and is treated by a different method in the next section. Phasor diagrams can be used only in the superposition of oscillating quantities of the same frequency. There, however, they find extensive application in the theories of alternating currents and three-phase currents (see Volume II).

It was shown by Fourier that it is possible to represent (almost) any periodic function $f(t)$ by a sum of the harmonic functions, sine and cosine (a Fourier series):

$$f(t) = A_0 + \sum_{n=1}^{n=\infty} A_n \sin n\omega t + \sum_{n=1}^{n=\infty} B_n \cos n\omega t.$$

Pure harmonic motion therefore occupies an extremely important position, since any periodic motion, however complicated, can be represented by a superposition of harmonic motions. We shall make use of this theorem in the study of acoustics (see Volume II).

36. Beats

The phenomenon of *beats* represents a particularly important case of the superposition of harmonic oscillations. Beats result from the superposition of two oscillations of almost equal frequency:

$$x_1 = A_0 \sin \omega_1 t = A_0 \sin \omega t,$$

$$x_2 = A_0 \sin \omega_2 t = A_0 \sin (\omega + \delta)\, t,$$

where δ is very small compared with ω.

The resultant oscillation is

$$x_0 = x_1 + x_2 = 2A_0 \sin\left(\frac{2\omega + \delta}{2}\, t\right) \cos\left(\frac{\delta t}{2}\right).$$

If $\delta \ll \omega$, then to a first approximation

$$x_0 = 2A_0 \sin \omega t \cos (\delta t/2).$$

The term $\cos (\delta t/2)$ is periodic, varying very slowly with the time t. The quantity $2A_0 \cos (\delta t/2)$ can therefore be considered as a slowly varying amplitude. As time goes on, it decreases from a magnitude $2A_0$ to zero, and increases again to $2A_0$. The time interval between these maximum values, or the *beat period* τ, is given by

$$\delta\tau/2 = \pi, \quad \text{or} \quad \tau = 2\pi/\delta.$$

This period is long compared with the fundamental period T of the oscillations, where $T = 2\pi/\omega$. The *beat frequency* f_b, given by

$$f_b = \frac{1}{\tau} = \frac{\delta}{2\pi} = \frac{\omega_2 - \omega_1}{2\pi} = f_2 - f_1,$$

determines the number of beats per unit time and equals the difference between the frequencies of the component oscillations. The resultant oscillation can be described as rapid and of very nearly the same frequency as the superposed oscillations,

$$\frac{2\omega + \delta}{2} \cong \omega \cong \omega + \delta$$

whose amplitude fluctuates slowly (Fig. 3-60).

If the amplitudes of the component oscillations are equal (as in the example above) the resultant amplitude periodically becomes zero. For components of unequal amplitude, the resultant fluctuates between the sum and the difference of the individual amplitudes.

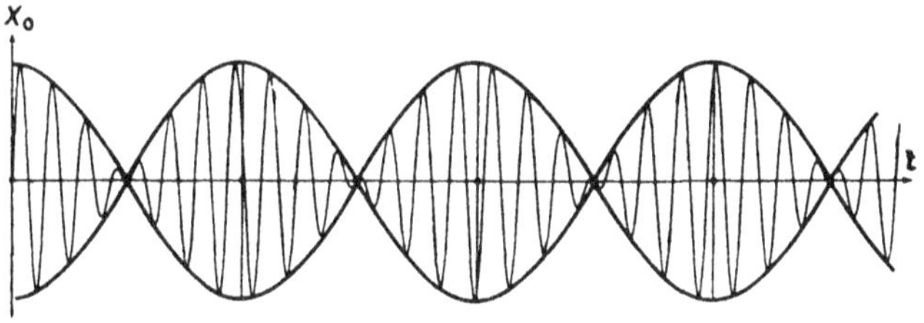

FIG. 3-60. Beats.

37. Superposition of Harmonic Oscillations along two Axes. Lissajous Figures

Suppose that a particle is acted on by two independent harmonic forces, parallel respectively to the x- and y-axes of a rectangular coordinate system. The vectors $\boldsymbol{x}$ and $\boldsymbol{y}$ of the resulting oscillation combine vectorially to form the radius vector $\boldsymbol{r}$ to the position of the particle, and since both $\boldsymbol{x}$ and $\boldsymbol{y}$ vary with time, the motion of the tip of the vector $\boldsymbol{r}$ represents the path of the particle in the x, y-plane.

In only a few cases can the resultant motion be expressed in a simple form. The simplest case is that in which the frequencies of both oscillations are the same. The path of the particle is then a straight line, a circle, or an ellipse.

The path is a straight line when the component oscillations are in phase (Fig. 3-61). Then

$$x = A \sin \omega t, \quad y = B \sin \omega t.$$

Eliminating t, we get

$$y/x = B/A = \text{const},$$

which is the equation of a straight line.

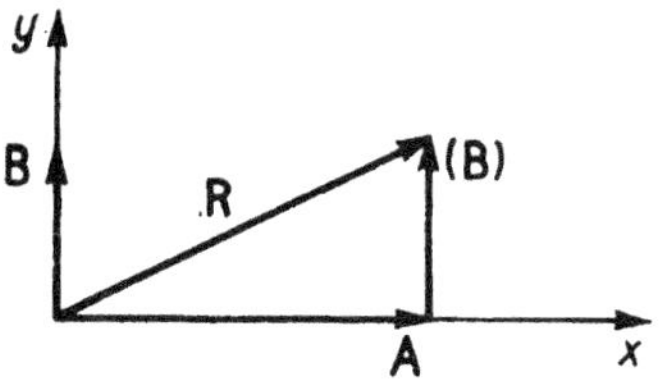

FIG. 3-61. Superposition of two oscillations having the same phase.

The path is an ellipse when there is a phase difference between the component vibrations. Then

$$x = A \sin \omega t, \quad y = B \sin (\omega t + \varphi).$$

As before, the equation of the path is found by eliminating t. To do this, we expand the term $\sin (\omega t + \varphi)$ and eliminate the terms containing t with the help of the first equation. Thus

$$y = B(\sin \omega t \cos \varphi + \cos \omega t \sin \varphi)$$

$$= B\left[\frac{x}{A} \cos \varphi + \sin \varphi \sqrt{\left(1 - \frac{x^2}{A^2}\right)}\right],$$

$$Ay - Bx \cos \varphi = B \sin \varphi \sqrt{(A^2 - x^2)}.$$

Squaring this equation gives

$$A^2y^2 + B^2x^2 \cos^2 \varphi - 2AB\, xy \cos \varphi = B^2(A^2 - x^2) \sin^2 \varphi,$$

or

$$B^2x^2 + A^2y^2 - 2ABxy \cos \varphi - A^2B^2 \sin^2 \varphi = 0.$$

This is the equation of an ellipse with its center at the origin. The major axis is inclined to the x-axis at an angle ψ, where ψ has the constant value given by

$$\tan 2\psi = \frac{2AB \cos \varphi}{B^2 - A^2}.$$

This type of motion is executed by a pendulum having an arbitrary initial velocity $\mathbf{v}$, whose line of action does not pass through the origin (the equilibrium position of the pendulum). We shall encounter an analogous situation again in optics in connection with elliptically polarized light.

FIG. 3-62. Lissajous figures for several frequency and phase ratios. The phase differences in the five columns are 0, $\pi/4$, $\pi/2$, $3\pi/4$, and π respectively.

The elliptical path becomes a circle if both amplitudes are equal and if the phase difference is 90°. Then

$$x = A\cos\omega t, \quad y = A\sin\omega t, \quad x^2 + y^2 = A^2,$$

which is the equation of a circle.

If the amplitudes are *unequal*, the path is elliptical in this case also. The major and minor axes lie along the coordinate axes and their magnitudes are $2A$ and $2B$, respectively.

The general case of the superposition of two harmonic motions in arbitrary directions x, y, of unequal amplitude, different frequency, and with an arbitrary phase difference, results in a so-called *Lissajous figure*. In parametric form, the equation of the path is

$$x = A\sin\omega_1 t, \quad y = B\sin(\omega_2 t - \varphi).$$

The equation of the path is, in general, very complicated. The following general remarks may, however, be made.

(1) The motion is periodic if the ratio of the frequencies is a rational number; that is, if it equals the ratio of two integers n_1 and n_2. Then

$$\frac{\omega_1}{\omega_2} = \frac{n_1}{n_2} = \frac{2\pi/T_1}{2\pi/T_2} = \frac{T_2}{T_1},$$

and the period of the motion, τ, is

$$\tau = n_1 T_1 = n_2 T_2 .$$

(2) If the ratio of the frequencies is irrational, the path does not close on itself and the motion is not periodic. In the course of time, if the motion is not damped, the particle covers (i.e., comes infinitely close to every point of) the entire area of a parallelogram whose sides are the doubled amplitudes of the component oscillations.

Figure 3-62 shows the form of the Lissajous figures for some simple frequency ratios and for several phase differences. The x- and y-axes are perpendicular, and the amplitudes of the motions are equal for all the figures.

The ratio of the component frequencies can easily be determined from the Lissajous figure. When the x-motion, for example, makes one complete oscillation, the Lissajous figure touches the bounding parallelogram once along a side parallel to the y-axis. The frequency ratio ω_x/ω_y equals the ratio of the number of points of contact at the y- and x-sides, respectively. If there is a point of rest at one corner of the parallelogram, this is to be counted as 1/2 for each side. The figure is then said to be *degenerate*.

38. Forced Oscillations and Resonance

A *forced oscillation* is one in which a system is acted on by a periodic external force called the *driving force*. A *free* oscillation is one in which the only forces are those of the system itself. Free oscillations of real objects are always damped.

In discussing forced oscillations it is necessary to distinguish between two frequencies: (1) the frequency of the driving force, the so-called *driving frequency* f; and (2) the *natural frequency* f_0 of the system.

We consider first a system with a single natural frequency, for example, a body suspended from a spring. The natural frequency of such a system, from Section 32-1, is

$$f_0 = (1/2\pi)\sqrt{(k/m)}.$$

Now let the body be acted on by a periodic driving force given by $F_1 = B \sin \omega_1 t$. This can be done, for instance, by suspending the spring-mass system from a cord, passing the cord over a pulley, and attaching it to a pin set in a disk at a distance r from the axis of rotation. If the angular velocity ω_1 of the motor turning the disk is constant, the cord executes periodic oscillations of amplitude r (see Fig. 3-63).

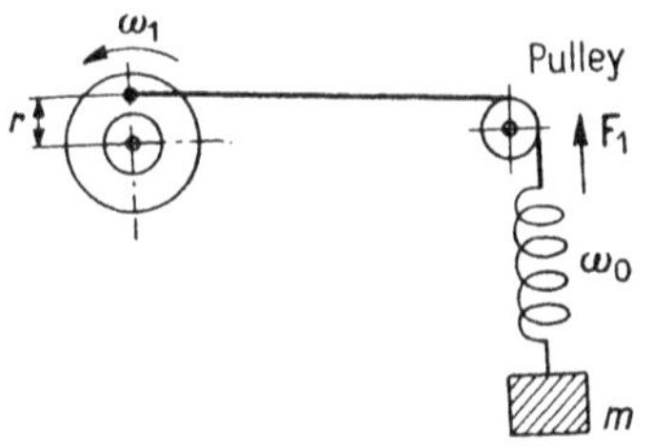

FIG. 3-63. Example of a forced oscillation.

The cases of practical importance are those in which the spring-mass system oscillates with damped motion. We shall assume that the damping force $\boldsymbol{R}$ (air resistance, for example) is proportional to the instantaneous velocity and is oppositely directed:

$$\boldsymbol{R} = -\alpha \boldsymbol{v}.$$

With sufficiently large damping, the initial free vibration, $x = A \sin \omega_0 t$, dies out in a short time, and only the forced oscillation remains.

The forced oscillation can be analyzed by the application of Newton's second law. Let the system be acted on by a sinusoidal driving force of constant amplitude and controllable frequency. Then the following forces act on the system:

(1) The driving force $F_1 = B \sin \omega_1 t$,

(2) The elastic restoring force $F_0 = -kx = -m\omega_0^2 x$, where ω_0 is the natural frequency of the undamped system, $\omega_0 = \sqrt{(k/m)}$,

(3) The damping force $R = -\alpha \boldsymbol{v}$.

It can be shown that the forced oscillation of x is harmonic and has the same frequency ω_1 as the driving force, but differs from it in phase. Thus we may write $x = A \sin(\omega_1 t - \varphi)$. Then

$$v = dx/dt = A\omega_1 \cos(\omega_1 t - \varphi), \quad a = dv/dt = -A\omega_1^2 \sin(\omega_1 t - \varphi).$$

Newton's second law, $F_0 + R + F_1 = ma$, then leads to the following equation for the forced oscillation:

$$\begin{aligned} -mA\omega_1^2 &\sin(\omega_1 t - \varphi) \\ &= -m\omega_0^2 A \sin(\omega_1 t - \varphi) - \alpha A\omega_1 \cos(\omega_1 t - \varphi) + B \sin\omega_1 t, \end{aligned}$$

which can be written in the form

$$mA(\omega_1^2 - \omega_0^2)\sin(\omega_1 t - \varphi) - \alpha A\omega_1 \cos(\omega_1 t - \varphi) + B \sin\omega_1 t = 0. \tag{38-1}$$

The amplitude A of the oscillation, and its phase angle φ, can both be determined from this equation. For brevity, let

$$p = mA(\omega_1^2 - \omega_0^2), \quad \text{and} \quad r = \alpha A\omega_1,$$

and expand the terms $\sin(\omega_1 t - \varphi)$ and $\cos(\omega_1 t - \varphi)$. Equation (38-1) then gives

$$\begin{aligned} p \sin(\omega_1 t)\cos\varphi - p\cos(\omega_1 t)\sin\varphi &- r\cos(\omega_1 t)\cos\varphi \\ &- r\sin(\omega_1 t)\sin\varphi + B\sin(\omega_1 t) = 0, \end{aligned}$$

or

$$(p\cos\varphi - r\sin\varphi + B)\sin\omega_1 t - (p\sin\varphi + r\cos\varphi)\cos\omega_1 t = 0.$$

This equation can be satisfied for all values of t only if the coefficients of the time-dependent terms $\sin\omega_1 t$ and $\cos\omega_1 t$ are both equal to zero. This leads to the two equations:

$$p\cos\varphi - r\sin\varphi = -B, \quad \text{and} \quad p\sin\varphi + r\cos\varphi = 0,$$

from which the two unknowns A and φ can be determined. By squaring and adding the equations, we get $p^2 + r^2 = B^2$, that is,

$$m^2A^2(\omega_1^2 - \omega_0^2)^2 + \alpha^2A^2\omega_1^2 = B^2,$$

and hence

$$A = \frac{B}{\sqrt{[m^2(\omega_0^2 - \omega_1^2)^2 + \alpha^2\omega_1^2]}}.$$

It follows from the second equation that

$$\tan\varphi = -\frac{r}{p} = \frac{\alpha\omega_1}{m(\omega_0^2 - \omega_1^2)}.$$

The amplitude A and the phase angle φ of the forced oscillations both depend on the frequency ω_1 of the driving force. The more closely the driving frequency approaches the natural frequency of the system, the larger the amplitude of the forced oscillations. In Fig. 3-64(a) the amplitude A is plotted as a function of the ratio ω_1/ω_0, for different values of α. The curve $A(\omega_1/\omega_0)$ is called the *resonance curve* of the system. For an undamped system ($\alpha = 0$), the amplitude becomes infinite when $\omega_1 = \omega_0$. This infinite amplitude for an undamped system is characteristic of the phenomenon of resonance between the driving force and the natural vibration. The phenomenon of resonance occurs repeatedly throughout the entire field of physics. It shows that the response of an oscillating system to a driving force is greater, the closer the driving frequency approaches the natural frequency of a system.

The shape of the resonance curve depends on the *damping constant* α [Fig. 3-64(a)]. The smaller the damping, the sharper is the resonance curve

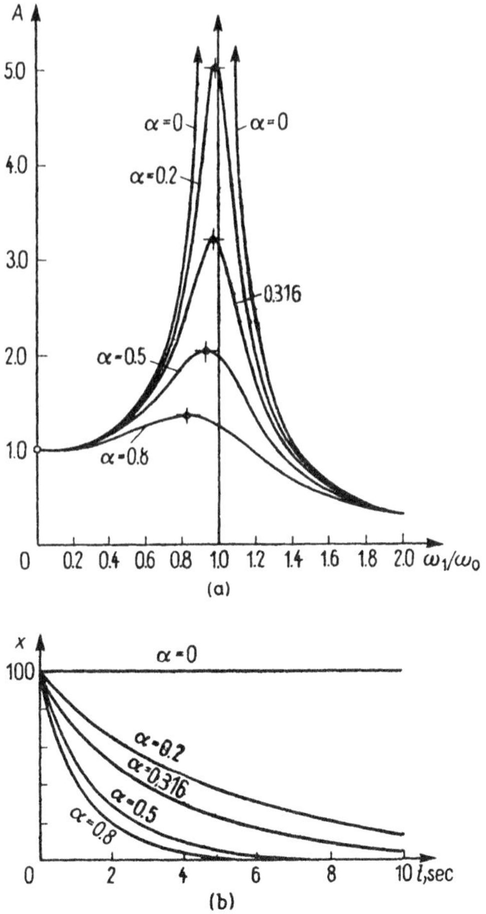

FIG. 3-64. (a) Resonance curves for various values of the damping constant α. (b) Amplitude decay of damped free oscillations.

and the more pronounced the resonance. For a given value of α, the peak of the resonance curve occurs at that frequency ω_1 for which the quantity $y = m^2(\omega_0^2 - \omega_1^2)^2 + \alpha^2\omega_1^2$ is a minimum, since $A = B/\sqrt{y}$. The condition for a minimum is

$$dy/d\omega_1 = -2m^2(\omega_0^2 - \omega_1^2)\,2\omega_1 + 2\alpha^2\omega_1 = 0,$$

and the *resonance frequency* is therefore†

$$\omega_{res} = \sqrt{(\omega_0^2 - \alpha^2/2m^2)}.$$

For a damped system, the resonance frequency is not the same as the natural frequency ω_0, but is somewhat smaller, and the amplitude at resonance remains finite.

Oscillations that may originally have been present in the system die out with time, and their duration decreases with increasing values of the damping constant α. Fig. 3-64(b) shows the decrease with time of these oscillations for different values of α. Such oscillations which die out with time are called *transients.*

The phase differency φ by which the forced oscillations lag the driving force is given by

$$\tan\varphi = \frac{\alpha\omega_1}{m(\omega_0^2 - \omega_1^2)}.$$

As ω_1 increases from zero, the phase difference φ increases from 0 to π. The value $\pi/2$ is reached somewhat beyond the maximum of the resonance curve. Figure 3-65 is a graph of φ as a function of ω_1/ω_0.

For sufficiently small damping, the amplitude of the oscillation shows the phenomenon of beats (until the transient dies out). The beat frequency equals

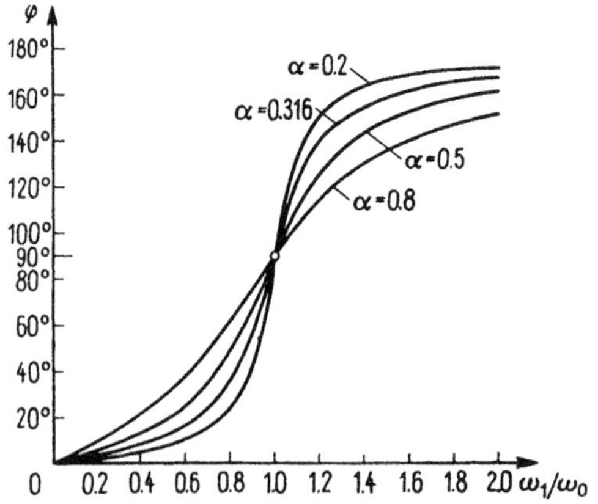

FIG. 3-65. Phase shift of a forced, damped oscillation.

† The second root, $\omega_1 = 0$, shows that the resonance curve has a horizontal tangent at $\omega_1 = 0$.

the difference between the driving frequency and the natural frequency. For ω_0 and ω_1 nearly equal, the initial amplitude of the beats can be very large. In the limit $\alpha \to 0$ and $\omega_1 \to \omega_0$, the period of the beats becomes infinite. If, at the start, the driving force F_1 and the velocity have the same direction, they remain in phase for a relatively long time. During this time, the force F_1 does work on the system; the energy of the system increases and the amplitude of the oscillation increases. When a phase difference of π between the force F_1 and the velocity has been reached, energy is withdrawn from the system and the amplitude decreases. These remarks follow from the fact that the energy of the oscillation is proportional to the square of the amplitude. Once the transient dies out, the phase difference between F_1 and the oscillation remains constant. The continued supply of energy to the system maintains the amplitude at a maximum value determined by the damping.

Certain peculiar, but extremely important, resonance processes occur when several oscillatory systems are coupled together so that they can exert forces on each other. A simple example is two pendulums of equal frequency, coupled together by a weak spring (Fig. 3-66). Assume that pendulum I is

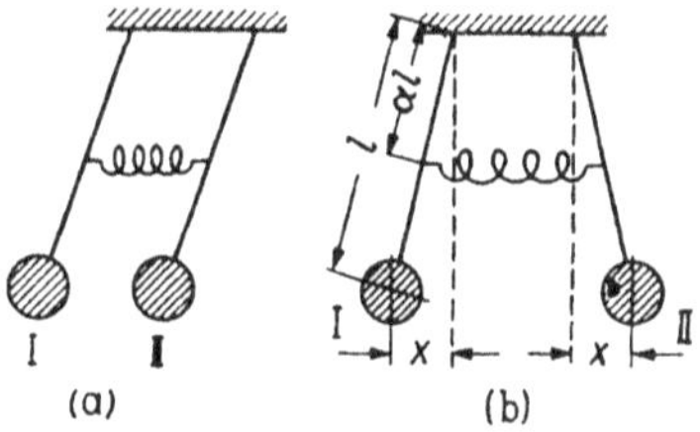

FIG. 3-66. The two fundamental modes of oscillation of two coupled pendulums.

initially at rest and pendulum II is in motion. The spring transmits energy from pendulum II to pendulum I until pendulum II is brought to rest and pendulum I has acquired all the energy of the system. This pendulum then proceeds to supply energy to the other until the latter is again vibrating with its original amplitude. The energy is transferred back and forth between the two pendulums. When one has its maximum amplitude, the other is at rest.

Such a pendulum system can also oscillate without any energy transfer, and can do so in two different ways:

(a) Both pendulums are originally given equal displacements in the same direction, so that the spring is not stretched. This so-called *first fundamental mode of oscillation* has the same frequency ω_0 as that of a single pendulum [Fig. 3-66(a)].

(b) Both pendulums are originally given equal displacements in opposite directions [Fig. 3-66(b)]. The restoring force arising from the weight of the pendulum is then augmented by the elastic force $F = -2k\alpha x$, where x is the displacement of the pendulum bob and α is a constant which depends on the position of the points at which the spring is attached. For small displacements, the total force on one pendulum is

$$F_{total} = -mg\frac{x}{l} - 2k\alpha x = -m\left(\frac{g}{l} + \frac{2k\alpha}{m}\right)x,$$

and the angular frequency is

$$\omega_1 = \sqrt{[(g/l) + 2k\alpha/m]}.$$

This is called the *second fundamental mode of oscillation*, and it has a somewhat higher frequency than the first because of the coupling of the spring.

A transfer of energy takes place if the two pendulums are "out of tune", that is, if they have different frequencies.† Processes analogous to those with coupled pendulums occur in many instances. For example, the phenomenon of homopolar binding of molecules is explained in quantum theory by a similar process of energy exchange.

39. The Laws of Collision

When two bodies of mass m_1 and m_2 collide, contact forces are developed whose variation with time cannot be determined by elementary methods. Calculation of the resulting acceleration is difficult. On the other hand, Newton's third law states that at every instant of the process the force $\boldsymbol{F}_{1,2}$ which the first body exerts on the second is equal in magnitude and opposite in direction to the force $\boldsymbol{F}_{2,1}$ which the second exerts on the first. This fact enables us to draw an important conclusion regarding collisions, with the help of Newton's second law in the form $\boldsymbol{F} = d\boldsymbol{p}/dt$.

Let $\boldsymbol{p}_1$ and $\boldsymbol{p}_2$ represent the momenta of the bodies before the collision. During the collision process, oppositely directed forces of equal magnitude act on each body, and at every instant produce equal and opposite changes in momentum. That is

$$d\boldsymbol{p}_1/dt = \boldsymbol{F}_{2,1} = -\boldsymbol{F}_{1,2} = -d\boldsymbol{p}_2/dt,$$

and hence $d\boldsymbol{p}_1/dt = -d\boldsymbol{p}_2/dt$. There is no change in the momentum of the system as a whole. The initial momentum $\boldsymbol{p}_1 + \boldsymbol{p}_2$ remains constant, or is *conserved*. Let us indicate by primes the values of quantities after the collision. Then

$$\boldsymbol{p} = \boldsymbol{p}_1' + \boldsymbol{p}_2' = \boldsymbol{p}_1 + \boldsymbol{p}_2,$$

that is, the total momentum $\boldsymbol{p}$ of the system remains constant.

This statement holds for systems of any number of bodies moving in any arbitrary way, provided that the only forces acting are internal forces

† See A. Sommerfeld, *Lectures on Theoretical Physics*, Vol. I, *Mechanics*, p. 106, Academic Press, Inc., New York, 1952.

whose reactions also act on bodies of the system. Two limiting cases of collision processes can be treated in a particularly simple way.

39-1 The Completely Inelastic Collision

A completely inelastic collision is one in which the colliding bodies stick together after the collision and hence have the same final velocity. In this case, the total momentum determines the velocity after the collision.

For an inelastic collision in which all velocities are along the same line,

$$m_1 v_1 + m_2 v_2 = (m_1 + m_2)\, v', \quad v' = \frac{m_1 v_1 + m_2 v_2}{m_1 + m_2}.$$

This case is illustrated by the ballistic pendulum.

There is a loss of kinetic energy in an inelastic collision, given by

$$\Delta E_k = \tfrac{1}{2} m_1 v_1^2 + \tfrac{1}{2} m_2 v_2^2 - \tfrac{1}{2}(m_1 + m_2)\, v'^2.$$

Inserting the expression for v' from the preceding equation, we get

$$\Delta E_k = \frac{m_1 m_2}{2(m_1 + m_2)} (v_1 - v_2)^2.$$

The "lost" kinetic energy is converted to heat.

39-2 The Elastic Collision

An elastic collision is one in which both kinetic energy and momentum are conserved. It represents a limiting case which is never completely realized in practice. Since both momentum and kinetic energy are conserved, we have (conservation of momentum)

$$m_1 \boldsymbol{v}_1 + m_2 \boldsymbol{v}_2 = m_1 \boldsymbol{v}_1' + m_2 \boldsymbol{v}_2'$$

and (conservation of kinetic energy)

$$\tfrac{1}{2} m_1 v_1^2 + \tfrac{1}{2} m_2 v_2^2 = \tfrac{1}{2} m_1 v_1'^2 + \tfrac{1}{2} m_2 v_2'^2.$$

Since elastic collisions are of great significance in the kinetic theory of gases and in atomic physics, we shall consider some common types of such collisions in more detail.

Collision of two particles of mass m_1 and m_2, whose centers move on the same straight line (central collision). From the conservation of momentum, we have

$$m_1 v_1 + m_2 v_2 = m_1 v_1' + m_2 v_2'.$$

Collecting terms which refer to the same particle, we get

$$m_1(v_1 - v_1') = m_2(v_2' - v_2). \tag{39-1}$$

From the conservation of kinetic energy, we have

$$\tfrac{1}{2} m_1 v_1^2 + \tfrac{1}{2} m_2 v_2^2 = \tfrac{1}{2} m_1 v_1'^2 + \tfrac{1}{2} m_2 v_2'^2,$$

or

$$m_1(v_1^2 - v_1'^2) = m_2(v_2'^2 - v_2^2). \tag{39-2}$$

From Eqns. (39-1) and (39-2) we can draw some conclusions regarding the *relative velocities* of the particles, $(v_2 - v_1)$ and $(v_2' - v_1')$, before and after the collision. Dividing the left and right sides of Eqn. (39-2) by Eqn. (39-1), we get

$$v_1 + v_1' = v_2' + v_2,$$

or

$$v_2 - v_1 = -(v_2' - v_1'). \tag{39-3}$$

That is, the relative velocity of the particles is reversed in direction as a result of the collision, but is unaltered in magnitude.

The velocities after the collision can now be found from Eqns. (39-1) and (39-3):

$$m_1v_1 + m_2v_2 = m_1v_1' + m_2v_2', \quad -v_1 + v_2 = v_1' - v_2'.$$

From these equations, we have

$$v_1' = \frac{m_1 - m_2}{m_1 + m_2}v_1 + \frac{2m_2}{m_1 + m_2}v_2, \tag{39-4}$$

$$v_2' = \frac{2m_1}{m_1 + m_2}v_1 + \frac{m_2 - m_1}{m_1 + m_2}v_2. \tag{39-5}$$

The expressions are symmetrical with respect to the indices 1 and 2.

Some special cases. We consider the case where the colliding particles have the same mass; that is, $m_1 = m_2$. This case occurs, for example, in the collision of two molecules of the same kind, moving on the same straight line; in the central collision of an α-particle with a helium atom; and in a central neutron-proton collision. The velocities after the collision, from Eqns. (39-4) and (39-5), are

$$v_1' = v_2, \quad v_2' = v_1.$$

Thus, the particles interchange velocities as a result of the collision. If the velocity of the second particle was zero before the collision, that of the first is zero after the collision. This fact plays an important part in the game of pool and in the slowing down of neutrons by means of collisions with hydrogen nuclei.

The exchange of velocities can readily be demonstrated by means of two pendulums of an elastic material such as ivory, having equal masses (Fig. 3-67). Let the first be displaced to the left and the second be at rest. When the first pendulum is released, it gives its entire velocity to the second during the collision, and comes to rest itself. After the next collision, the second pendulum remains at rest. Its velocity has been transferred to the first pendulum. After several collisions, the system as a whole acquires a pendulum motion, since the collisions are not completely elastic.

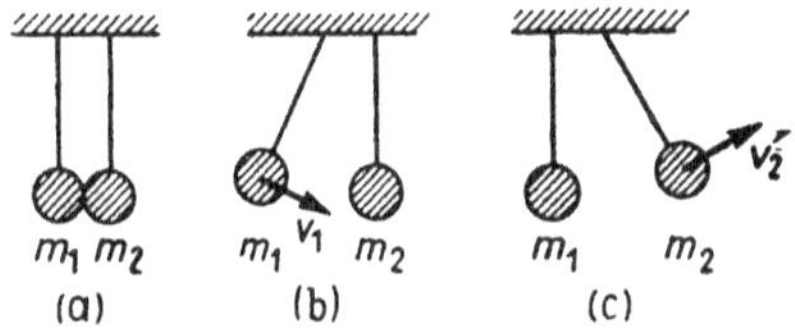

FIG. 3-67. Exchange of velocities in an elastic collision, (a) at rest, (b) before collision, (c) after collision.

We now consider central collisions between particles of unequal mass, one of which is initially at rest. Here, $m_1 \neq m_2$, $v_2 = 0$. From Eqns. (39-4) and (39-5), the velocities after the collision are

$$v_1' = \frac{m_1 - m_2}{m_1 + m_2} v_1, \quad v_2' = \frac{2m_1}{m_1 + m_2} v_1.$$

The corresponding kinetic energies are

$$E_{k1}' = \frac{m_1}{2}\left(\frac{m_1 - m_2}{m_1 + m_2}\right)^2 v_1^2 = \left(\frac{m_1 - m_2}{m_1 + m_2}\right)^2 E_{k1},$$

$$E_{k2}' = \frac{m_2}{2}\,\frac{4m_1^2}{(m_1 + m_2)^2} v_1^2 = \frac{4m_1 m_2}{(m_1 + m_2)^2} E_{k1},$$

where E_{k1} is the initial kinetic energy of the first particle and E_{k2}' is the energy given to the second particle by the first in the collision. The ratio

$$\frac{E_{k2}'}{E_{k1}} = \frac{4m_1 m_2}{(m_1 + m_2)^2}$$

is the fraction of the energy of the first particle which is transferred to the second. This ratio is plotted in Fig. 3-68 as a function of the ratio m_2/m_1. The maximum energy transfer results when $m_2 = m_1$, for which $E_{k2}' = E_{k1}$. That is, in a central collision between particles of equal mass, the energy of the first is completely transferred to the second. For any other mass ratio, the energy transfer is smaller. For very large and very small mass ratios,

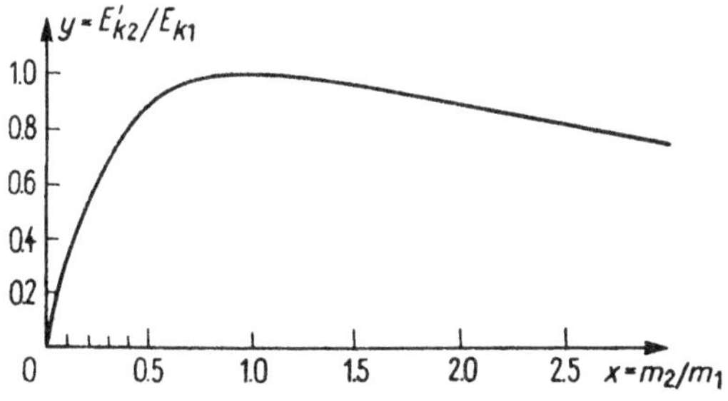

FIG. 3-68. Energy transfer in a central collision as a function of relative mass.

only a small amount of energy is transferred. These extreme mass ratios are illustrated by the collisions of molecules with electrons, on the one hand, and with rigid walls on the other.

An important problem in the kinetic theory of gases is that of the elastic collision of a particle with a stationary wall. Suppose that a molecule of mass m and velocity v collides perpendicularly with a stationary wall ($m_{\text{wall}} \to \infty$, $v_2 = 0$). The velocities after the collision, from Eqns. (39-4) and (39-5), are

$$v_1' = -v_1, \quad v_2' = 0.$$

The molecule rebounds from the wall with a velocity whose magnitude is unaltered. It loses no kinetic energy in the collision. The wall acquires no kinetic energy, since it does not move. On the other hand, the impulse of the force exerted on the wall, which equals the change in momentum of the particle, is

$$\text{impulse} = 2m_1 v_1 .$$

If the molecules strikes the wall at an angle α, as in Fig. 3-69, its velocity can be resolved into components perpendicular and parallel to the wall.

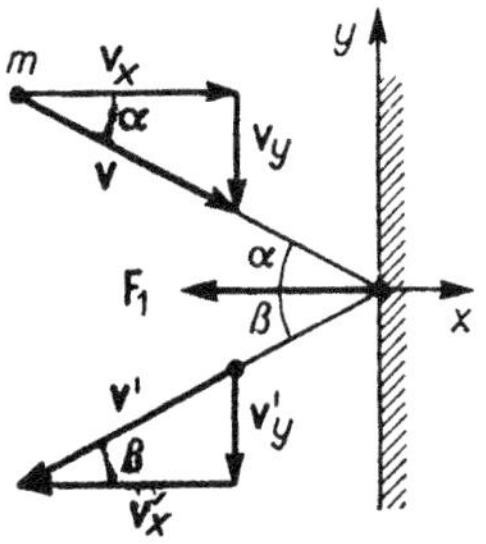

FIG. 3-69. Elastic collision with a stationary wall.

Since the wall can exert no tangential forces, the component parallel to the wall is not altered by the collision: $v_y' = v_y$. However, the component perpendicular to the wall is reversed in direction: $v_x' = -v_x$. Hence we can conclude that the *angles of incidence* and of *reflection* are equal in such an elastic collision:

$$\tan\alpha = \frac{|v_y|}{|v_x|} = \frac{|v_y'|}{|v_x'|} = \tan\beta;$$

that is, $\alpha = \beta$.

In order to understand the nature of the processes that take place in a steam engine or an internal combustion engine, where a gas expands against a piston, it is necessary to study elastic collisions with a *moving* wall. We shall postpone this question, however, until the concept of relative motion has been discussed.

Noncentral elastic collision of two spheres. We shall consider only the special case in which $m_1 \neq m_2$, $v_2 = 0$. The collision is like that of two particles, insofar as no rotational motion is set up. The line joining the midpoints of the spheres at the instant of contact is called the *collision axis* (Fig. 3-70). The *contact plane* (CP) is perpendicular to the collision axis.

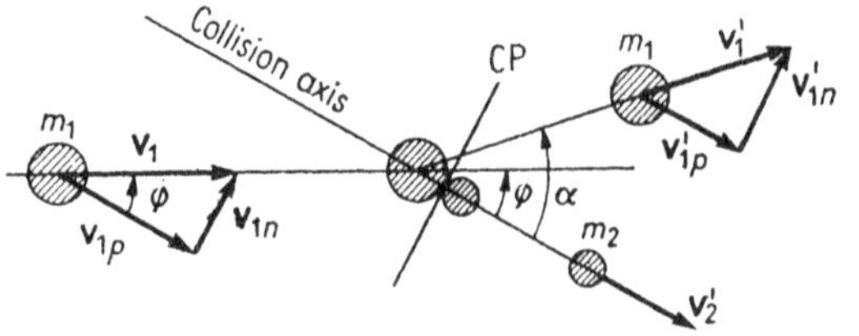

FIG. 3-70. A noncentral collision.

The *collision angle* φ is the angle between the collision axis and the incident velocity of the first sphere. Let us resolve the velocity of the incident sphere into components parallel and perpendicular to the collision axis. The component parallel to the axis will be represented by the subscript p, the perpendicular or normal component by the subscript n. The velocity normal to the collision axis is unaltered by the collision, since no forces act in this direction. Along the collision axis, on the other hand, we have a central collision. The velocity components after the collision, from Eqns. (39-4) and (39-5), are therefore

$$v'_{1p} = \frac{m_1 - m_2}{m_1 + m_2} v_{1p} = \frac{m_1 - m_2}{m_1 + m_2} v_1 \cos\varphi,$$

$$v'_{1n} = v_{1n} = v_1 \sin\varphi,$$

$$v'_{2p} = \frac{2m_1}{m_1 + m_2} v_{1p} = \frac{2m_1}{m_1 + m_2} v_1 \cos\varphi,$$

$$v'_{2n} = 0.$$

The kinetic energies after the collision are

$$E'_{k1} = \frac{m_1}{2}\left[\left(\frac{m_1 - m_2}{m_1 + m_2}\right)^2 v_1^2 \cos^2\varphi + v_1^2 \sin^2\varphi\right]$$

$$= \left[\left(1 - \frac{4m_1m_2}{(m_1 + m_2)^2}\right)\cos^2\varphi + \sin^2\varphi\right] E_{k1}$$

$$= \left[1 - \frac{4m_1m_2}{(m_1 + m_2)^2}\cos^2\varphi\right] E_{k1},$$

$$E'_{k2} = \frac{m_2}{2}\,\frac{4m_1^2}{(m_1 + m_2)^2} v_1^2 \cos^2\varphi = \frac{4m_1m_2}{(m_1 + m_2)^2} E_{k1}\cos^2\varphi.$$

Thus, $E'_{k1} + E_{k2} = E_{k1}$, which shows that the energy is conserved. The ratio of the transferred energy to the original energy of the incident particle is

$$\frac{E'_{k2}}{E_{k1}} = \frac{4m_1 m_2}{(m_1 + m_2)^2} \cos^2 \varphi .$$

If the masses are equal, the ratio reduces simply to $\cos^2 \varphi$, that is, $E'_{k2} = E_{k1} \cos^2 \varphi$.

The collision of a neutron with a proton is an example of this latter case. If neutrons of kinetic energy E_n collide with protons, the recoil energy E_p acquired by a proton when the collision angle is φ is

$$E_p = E_n \cos^2 \varphi .$$

Since all collision angles are possible, the recoil energies in a large number of collisions are distributed continuously between zero and E_n.

40. Relative Motion. Collisions with a Moving Wall

40-1 Relative Motion

In a coordinate system $O(x, y)$ the position of a particle A is determined by the vector $\boldsymbol{r}_0$ (Fig. 3-71). With respect to a second coordinate system

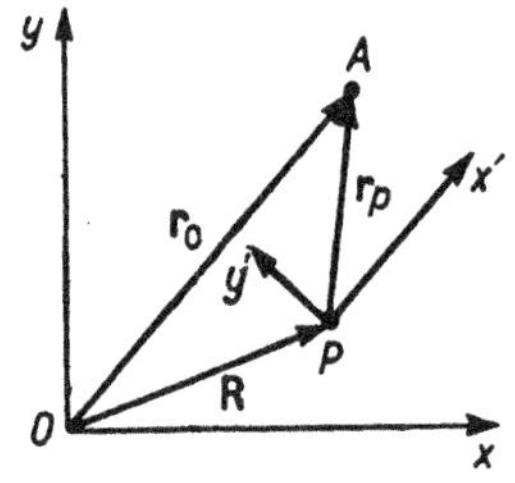

FIG. 3-71. Position of a particle relative to two-coordinate systems.

$P(x', y')$, the position of the particle is determined by the vector $\boldsymbol{r}_p$. If $\boldsymbol{R}$ is the vector from O to P, then we have the relation $\boldsymbol{r}_0 = \boldsymbol{R} + \boldsymbol{r}_p$.

Now consider the coordinate system $P(x', y')$ and the particle A to be in motion relative to the system $O(x, y)$ with a constant velocity $\boldsymbol{V}$. For an observer who is moving with the system $P(x', y')$, the position of particle A does not change. On the other hand, an observer at rest in the coordinate system $O(x, y)$ observes a change in the position of the particle. From his point of view, the particle moves with the same velocity as does the moving system, namely, $\boldsymbol{V}$ (Fig. 3-72). The same is true for all points of $P(x', y')$.

Now let particle A move with a velocity $\boldsymbol{v}_p$ *relative* to the moving coordinate system $P(x', y')$. The velocity $\boldsymbol{v}_0$ of the particle relative to the coordinate system $O(x, y)$ is then the vector sum of $\boldsymbol{V}$ and $\boldsymbol{v}_p$ (Fig. 3-73): $\boldsymbol{v}_0 = \boldsymbol{V} + \boldsymbol{v}_p$.

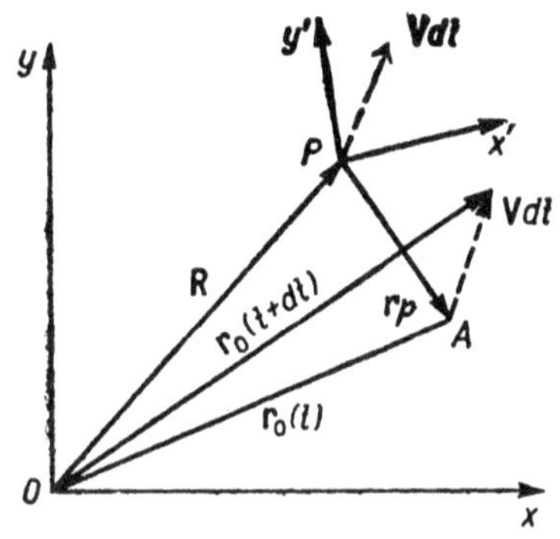

FIG. 3-72. Two-coordinate systems in relative motion.

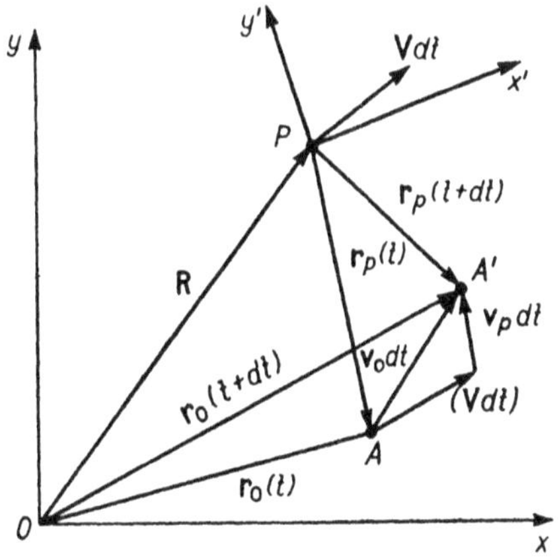

FIG. 3-73. Velocity of particle A relative to the coordinate system $O(x, y)$ which is at rest.

EXAMPLE 1. A ship moves on a lake with a velocity $V = 15$ ft/sec relative to an observer on the shore. A man walks across the deck of the ship with a velocity $v_p = 6$ ft/sec (Fig. 3-74). The velocity of the man relative to an observer on the shore is $v_0 = V + v_p = 16.2$ ft/sec.

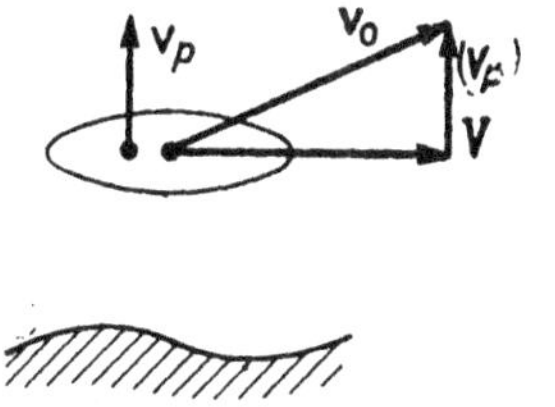

FIG. 3-74. Relative velocity of a man on a ship with respect to the shore.

EXAMPLE 2. An aircraft wishes to fly due east with a speed relative to the earth of $v_0 = 150$ mi/hr. The air is moving relative to the earth with a speed $V = 40$ mi/hr in a northeast direction (Fig. 3-75). We wish to find the velocity v_p of the aircraft relative to the air and the angle α by which the plane must head off its true course. The moving coordinate system in this case is the moving air. From the cosine law (Fig. 3-75) we get

$$v_p = \sqrt{(v_0^2 + V^2 - 2v_0 V \cos 45°)} = 125 \text{ mi/hr}.$$

The angle α can be found from the sine law:

$$\sin\alpha = \frac{v_p}{V}\sin 45° = 0.226, \quad \alpha \cong 13°.$$

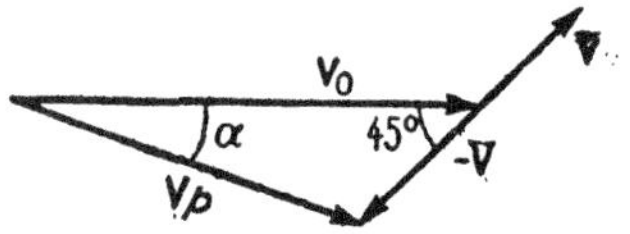

FIG. 3-75

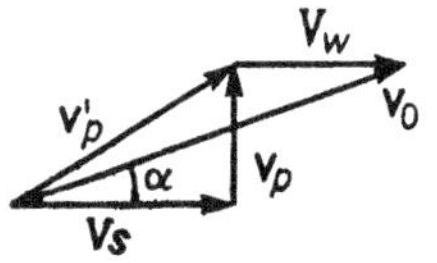

FIG. 3-76

FIG. 3-75. Relative velocity of a plane and the air.

FIG. 3-76. Passenger walking on a moving ship. $\alpha = \tan^{-1} 2.7/6.7 = 22°$, $v_p = 4.4$ m/sec, $v_0 = 7.2$ m/sec.

EXAMPLE 3. A ship travels down a river with a speed $V_s = 3.5$ m/sec relative to the flowing water. The speed of the flowing water is $V_w = 3.2$ m/sec relative to the shore. The ship and the water move parallel to each other. A passenger on the ship walks at right angles across the ship with a speed $v_p = 2.7$ m/sec relative to the ship. We wish to find the velocity v_0 of the passenger relative to the shore (Fig. 3-76).

Let us first find the velocity v_p' of the passenger relative to the water. The ship then becomes the moving coordinate system and $v_p' = V_s + v_p$. Next, we find the velocity of the passenger relative to the shore. The water now becomes the moving coordinate system and $v_0 = V_w + v_p'$.

40-2 Collisions with a Moving Wall

In Fig. 3-77, a particle of mass m, moving with a velocity v_0, collides at an angle α with a wall moving with a velocity c relative to the coordinate system $O(x, y)$. To analyze the collision process with the moving wall, it

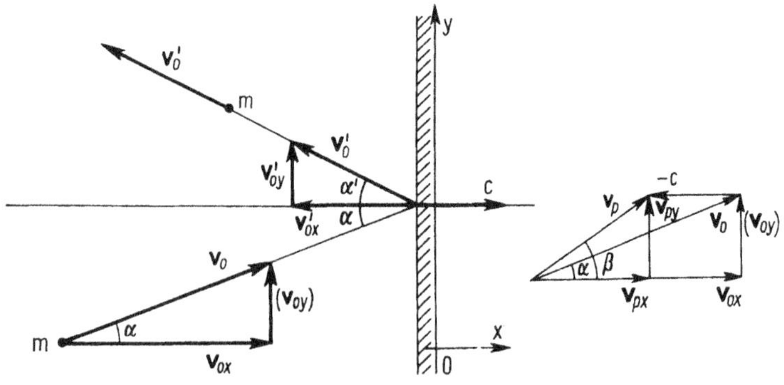

FIG. 3-77. Collision with a moving wall.

is advantageous to describe the collision from the standpoints of two different observers. The first is an external observer at rest relative to the coordinate system $O(x, y)$. The second moves with the wall and hence is at rest relative to it. From his viewpoint, the collision takes place with a stationary wall, which has already been discussed in Section 39-2.

The conclusions of the two observers are as follows:

1. Incident velocity of the particle, and its components:

$$\left.\begin{aligned} &\boldsymbol{v}_0 \\ &v_{0x} = v_0 \cos\alpha \\ &v_{0y} = +v_0 \sin\alpha \end{aligned}\right\} \text{external observer;}$$

$$\left.\begin{aligned} &\boldsymbol{v}_p = \boldsymbol{v}_0 - \boldsymbol{c} \\ &v_{px} = v_0 \cos\alpha - c \\ &v_{py} = +v_0 \sin\alpha \end{aligned}\right\} \text{observer moving with wall.}$$

The y-components appear the same to both observers, but the x-component appears to the observer moving with the wall to be decreased by an amount c.

2. Velocity of the wall: to the external observer, $\boldsymbol{c}$ is parallel to x-axis. The observer moving with the wall is not aware of its motion.

3. Angle between the normal to the wall and

$$\left.\begin{aligned} &\boldsymbol{v}_0: \\ &\tan\alpha = \left|\frac{v_{0y}}{v_{0x}}\right| \end{aligned}\right\} \text{external observer}$$

$$\left.\begin{aligned} &\boldsymbol{v}_p: \\ &\tan\beta = \left|\frac{v_{py}}{v_{px}}\right| = \left|\frac{v_{0y}}{v_{0x} - c}\right| \end{aligned}\right\} \text{observer moving with wall.}$$

4. The collision process for the external observer is to be calculated. For the observer moving with the wall, it is an elastic collision with a stationary wall. The component v_{px} is reversed in direction. The angle of incidence β is equal to angle of reflection β'.

The collision can now be readily analyzed by the external observer, making use of the conclusions of the moving observer. The velocity of the particle after the collision, relative to the moving observer, is $\boldsymbol{v}_p'$ and its components are $-(v_{0x} - c)$ and v_{0y}.

The velocity of the particle relative to the external observer after the collision is $\boldsymbol{v}_0' = \boldsymbol{c} + \boldsymbol{v}_p'$.

The components of $\boldsymbol{v}_0'$ are

$$v_{0x}' = c - (v_{0x} - c) = -(v_{0x} - 2c) = -(v_0 \cos\alpha - 2c),$$

and

$$v_{0y}' = v_{0y} = +v_0 \sin\alpha.$$

Hence for the external observer, the magnitude of the velocity $\boldsymbol{v}_0'$ after the collision is

$$v_0' = \sqrt{[(v_{0x}')^2 + (v_{0y}')^2]} = \sqrt{[(v_0 \cos\alpha - 2c)^2 + v_0^2 \sin^2\alpha]}$$
$$= v_0 \sqrt{[1 - (4c/v_0^2)(v_0 \cos\alpha - c)]}.$$

A collision with the moving wall can occur only if the speed of the wall, c, is less than $v_0 \cos\alpha$. The velocity after collision is less than that before the collision, that is, the kinetic energy of the particle has decreased. The decrease in energy is

$$\Delta E = E_k - E_k' = 2mc(v_0 \cos\alpha - c),$$

where E_k and E_k' are the kinetic energies before and after the collision, respectively. The decrease in energy results from the work done on the moving wall during the finite time of the collision. It is because of the decrease of energy that a gas can do work in an expansion process. The temperature of the gas decreases.

The angle of reflection α', for the external observer, is given by

$$\tan\alpha' = \frac{|v_{0y}'|}{|v_{0x}'|} = \frac{v_0 \sin\alpha}{v_0 \cos\alpha - 2c} = \frac{\tan\alpha}{1 - (2c/v_0 \cos\alpha)}.$$

It follows that the angle of reflection α', is larger than the angle of incidence α. For the observer on the moving wall, the reflection angle equals the observed angle of incidence and has a magnitude that is between α and α'.

If the wall moves in the negative x-direction, opposite to the motion of the particle, it is easily seen that the velocity after the collision is greater than that before the collision, and that the angle of reflection is smaller than the angle of incidence. This corresponds to the *compression* of a gas in a cylinder with a movable piston. The temperature of the gas increases, since work is done on the gas during compression.

The processes described above are of the greatest importance in connection with the problem of obtaining mechanical work from heat through the expansion of a gas against a moving piston. We shall return to them later in the study of thermodynamics (Chapter 10).

41. Relative Accelerations. Inertial Forces in Relative Motion

The relation between the velocities $\boldsymbol{v}_0$ and $\boldsymbol{v}_p$ for translational motion of a moving coordinate system has been discussed in the preceding sections. In this section, the relation will be extended to include rotational motion of the coordinate system; in addition, the relation between the accelerations $\boldsymbol{a}_0$ and $\boldsymbol{a}_p$ will be considered for both types of motion.

To this end, we must first investigate the expressions for the time derivative of a vector, relative to two different coordinate systems. We consider the following situation. The origin P of the coordinate system $P(x', y')$

remains in coincidence with the origin O of the coordinate system $O(x, y)$, and $P(x', y')$ rotates in the plane of $O(x, y)$ about O with a constant angular velocity ω. At some time t_0, the x- and x'-axes coincide, as do the y- and y'-axes [Fig. 3-78 (a)].

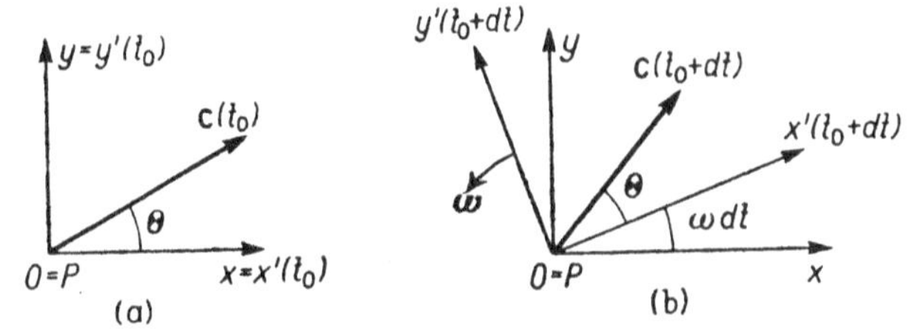

FIG. 3-78. Stationary and rotating coordinate systems with the same origin. (a) State at time t_0, (b) state at time $t_0 + dt$. The rotating system has turned through an angle $\omega\, dt$.

Let $\boldsymbol{c}$ be a vector having a constant magnitude and direction in the coordinate system $P(x', y')$. The change $(d\boldsymbol{c})_P$ of this vector in the rotating system $P(x', y')$ is zero [Fig. 3-78 (b)]. With respect to the fixed coordinate system $O(x, y)$, however, the direction of $\boldsymbol{c}$ changes, that is $(d\boldsymbol{c})_O$ is not equal to zero. The time variation of the vector $\boldsymbol{c}$ is therefore different relative to the rotating and to the fixed systems.

To obtain the general relation between the two time variations (derivatives), we consider the case in which $\boldsymbol{c}$ is variable in magnitude and direction in both coordinate systems. At the times t and $t + dt$, the situation [in the $O(x, y)$ system] is depicted in Fig. 3-79. The change in $\boldsymbol{c}$, in any coordinate system, can be measured when the positions of the tip of the vector $\boldsymbol{c}$, namely, $S(t)$ and $S(t + dt)$, are known at the times t and $t + dt$. The vector from $S(t)$ to $S(t + dt)$ is then the change in the vector $\boldsymbol{c}$.

In Fig. 3-79, the points $S(t)$ and $S(t + dt)$ in the $O(x, y)$ system are shown and their difference is the change $(d\boldsymbol{c})_O$ in the vector $\boldsymbol{c}$ in this system. If we

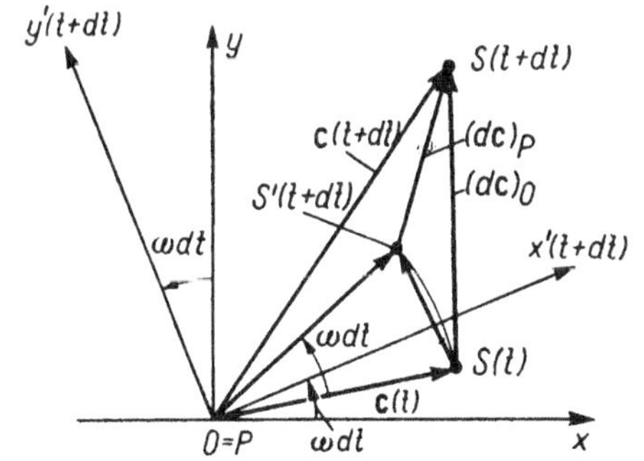

FIG. 3-79. Relating the time derivatives of a vector in a stationary and a rotating coordinate system.

wish to obtain the change $(d\mathbf{c})_P$ with respect to the rotating system $P(x', y')$, we must note that the point $S(t)$, fixed in relation to $P(x', y')$, has undergone a displacement from $S(t)$ to $S'(t + dt)$ in $O(x, y)$ as a result of the rotation of $P(x', y')$. The vector $(d\mathbf{c})_P$ from $S'(t + dt)$ to $S(t + dt)$ then represents the change in $\mathbf{c}$ in the rotating system.

If the rotation of the $P(x', y')$ system is described by a constant angular velocity vector $\boldsymbol{\omega}$,† which is perpendicular to the plane of the diagram and directed toward the reader, the displacement from $S(t)$ to $S'(t + dt)$ is given by the vector product $(\boldsymbol{\omega} \times \mathbf{c})\, dt$. We then have

$$(d\mathbf{c})_O = (d\mathbf{c})_P + (\boldsymbol{\omega} \times \mathbf{c})\, dt.$$

Division by dt gives

$$\left(\frac{d\mathbf{c}}{dt}\right)_O = \left(\frac{d\mathbf{c}}{dt}\right)_P + \boldsymbol{\omega} \times \mathbf{c}. \tag{41-1}$$

This is the desired relation between the time derivatives of the same vector $\mathbf{c}$ in the two coordinate systems, when $\mathbf{c}$ varies in both systems. It was derived here only for one special relative motion of the two systems, but it holds for any sort of relative rotational motion, provided that we understand by $\boldsymbol{\omega}$ the *instantaneous angular velocity*.

From this equation we now derive two important kinematical relations. The position vectors $\mathbf{r}_0$ and $\mathbf{r}_p$ of a particle, in the systems $O(x, y)$ and $P(x', y')$ respectively, are the same vector, provided that the origins of the two systems coincide. It follows from Eqn. (41-1) that

$$\left(\frac{d\mathbf{r}_0}{dt}\right)_O = \left(\frac{d\mathbf{r}_p}{dt}\right)_P + \boldsymbol{\omega} \times \mathbf{r}_p. \tag{41-2}$$

But

$$\left(\frac{d\mathbf{r}_0}{dt}\right)_O = \mathbf{v}_0 \quad \text{and} \quad \left(\frac{d\mathbf{r}_p}{dt}\right)_P = \mathbf{v}_p.$$

Equation (41-2) therefore gives the relation between the velocities of a particle in a *fixed* and in a *rotating* coordinate system, when the particle has a velocity $\mathbf{v}_p$ in the rotating system.

We now introduce the concept of the *instantaneous relative velocity*, $\mathbf{v}_I(t)$. This is the velocity with respect to the stationary system of a point fixed in the moving system which coincides with the particle at the time t. The vector $\mathbf{v}_I$ has already been discussed in the preceding section, where it was denoted by $\mathbf{V}$. There, however, it had the same value for all points of the moving system, whereas it now depends on the particular point under consideration in the moving system. Its value changes with time, not only because the moving system may change its angular velocity, but because the particle can move with respect to the moving system.

† In this section the angular velocity $\boldsymbol{\omega}$ will be considered a vector, as discussed in Section 20-4.

In the case where the motion of the moving system is one of rotation only, $\boldsymbol{v}_I = \boldsymbol{\omega} \times \boldsymbol{r}_p$, and we obtain from Eqn. (41-2),

$$\boldsymbol{v}_O = \boldsymbol{v}_p + \boldsymbol{v}_I. \tag{41-3}$$

Thus, we see that this equation from Section 40-1 is also valid for rotational motion. If $P(x', y')$ is moving with translational as well as rotational motion, it can be shown that the instantaneous velocity of a point in the moving system relative to the fixed system is $\boldsymbol{v}_I = \boldsymbol{V} + \boldsymbol{\omega} \times \boldsymbol{r}_p$, where $\boldsymbol{V}$ is the translational velocity of P relative to O, $\boldsymbol{\omega}$ is the instantaneous angular velocity of $P(x', y')$, and $\boldsymbol{r}_p$ is the position vector of the point in the moving system. This, the most general statement of Eqn. (41-3) is

$$\boldsymbol{v}_O = \boldsymbol{v}_p + \boldsymbol{V} + \boldsymbol{\omega} \times \boldsymbol{r}_p. \tag{41-3'}$$

EXAMPLE. A particle moves radially outward on a turntable which is rotating with constant angular velocity, as in Fig. 3-80. The radial velocity $\boldsymbol{v}_p$ and the tangential velocity $\boldsymbol{v}_I$ combine vectorially to give the resultant velocity $\boldsymbol{v}_O$.

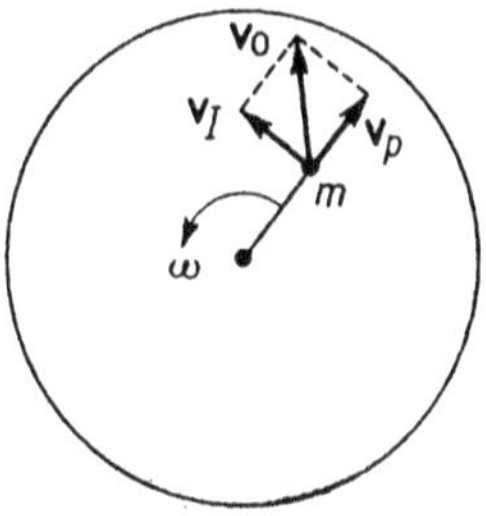

FIG. 3-80. Motion of a particle on a rotating plate.

The relation between the accelerations $\boldsymbol{a}_O$ and $\boldsymbol{a}_p$, in the two coordinate systems, can also be derived from Eqn. (41-1). These accelerations are given by

$$\boldsymbol{a}_O = \left(\frac{d\boldsymbol{v}_O}{dt}\right)_O \quad \text{and} \quad \boldsymbol{a}_p = \left(\frac{d\boldsymbol{v}_p}{dt}\right)_P.$$

The acceleration $\boldsymbol{a}_O$ can be found from Eqn. (41-3):

$$\boldsymbol{a}_O = \left(\frac{d\boldsymbol{v}_O}{dt}\right)_O = \left(\frac{d\boldsymbol{v}_p}{dt}\right)_O + \left(\frac{d\boldsymbol{v}_I}{dt}\right)_O.$$

Introducing for $\boldsymbol{c}$ in Eqn. (41-1) the relative velocity $\boldsymbol{v}_p$, we get

$$\left(\frac{d\boldsymbol{v}_p}{dt}\right)_O = \left(\frac{d\boldsymbol{v}_p}{dt}\right)_P + \boldsymbol{\omega} \times \boldsymbol{v}_p = \boldsymbol{a}_p + \boldsymbol{\omega} \times \boldsymbol{v}_p. \tag{41-4}$$

Hence the relation between the accelerations is given by

$$\boldsymbol{a}_O = \boldsymbol{a}_p + \left(\frac{d\boldsymbol{v}_I}{dt}\right)_O + \boldsymbol{\omega} \times \boldsymbol{v}_p. \tag{41-5}$$

As examples, we now consider two important special cases.

Coordinate system $P(x', y')$ moving with translational accelerated motion. In this case, $\boldsymbol{\omega} = 0$. Furthermore, all points of $P(x', y')$ have the same velocity $\boldsymbol{v}_I$ relative to the coordinate system $O(x, y)$. The acceleration of $P(x', y')$ with respect of $O(x, y)$ is $(d\boldsymbol{v}_I/dt)_O$ which we represent by $\boldsymbol{a}_I$. From Eqn. (41-5), we have

$$\boldsymbol{a}_0 = \boldsymbol{a}_p + \boldsymbol{a}_I. \qquad (41\text{-}6)$$

In the coordinate system $O(x, y)$, Newton's second law applies, that is, $\boldsymbol{F} = m\boldsymbol{a}_0$, where $\boldsymbol{F}$ is the resultant of all forces acting on a particle of mass m. Inserting the expression for $\boldsymbol{a}_0$ from Eqn. (41-6) it follows that

$$\boldsymbol{F} = m\boldsymbol{a}_p + m\boldsymbol{a}_I. \qquad (41\text{-}7)$$

An observer at rest in $P(x', y')$ who has no knowledge of the outside world can measure only the acceleration $\boldsymbol{a}_p$ of the particle relative to $P(x', y')$. On the other hand, he also finds (perhaps with a spring balance) that the resultant force on the particle is $\boldsymbol{F}$, which is not equal to $m\boldsymbol{a}_p$. (This depends on how he defines $\boldsymbol{F}$. If he defines $\boldsymbol{F}$ as equal to $m\boldsymbol{a}_p$, then the argument doesn't hold.) He concludes, therefore, that there is a force acting on the particle which he cannot detect in any other way, and which he computes from Eqn. (41-7), written in the form $m\boldsymbol{a}_p = \boldsymbol{F} - m\boldsymbol{a}_I$, to be equal to $-m\boldsymbol{a}_I$. In the system $O(x, y)$, on the other hand, an observer at rest asserts that this is only an apparent force. We shall speak of it as an additional force resulting from the relative motion, and represent it by $\boldsymbol{Z}$:

$$\boldsymbol{Z} = -m\boldsymbol{a}_I.$$

The meaning of this extra force is therefore that Newton's second law holds for the acceleration of the particle relative to $P(x', y')$ only if one adds the additional force $\boldsymbol{Z} = -m\boldsymbol{a}_I$ to the force that would be considered to act by an observer in the $O(x, y)$ system. This extra force is a generalization of the D'Alembert inertial force. For if we consider a coordinate system attached to the particle and moving with it, then

$$\boldsymbol{a}_p = 0, \quad \boldsymbol{a}_I = \boldsymbol{a}_0, \quad \text{and} \quad \boldsymbol{Z} = -m\boldsymbol{a}_0.$$

The extra force $\boldsymbol{Z}$, in this case, is therefore the same as the D'Alembert inertial force.

Rotating coordinate system. We make the same assumptions regarding the moving system $P(x', y')$ as at the beginning of the section in the derivation of Eqn. (41-1). In this case, as seen from the system $O(x, y)$, all points of $P(x', y')$ do not have the same velocity, and the term $(d\boldsymbol{v}_I/dt)_O$ in Eqn. (41-5) must be investigated further.

Figure 3-81 shows the fixed coordinate system $O(x, y)$, the moving system $P(x', y')$ at the times t and $t + dt$, and the vectors $\boldsymbol{r}_p(t)$ and $\boldsymbol{r}_p(t + dt)$ to the points $S(t)$ and $S(t + dt)$ at which the particle is located at the times t and $t + dt$. If we think of the point $S(t)$ as a fixed in $P(x', y')$, then after a time dt, we label the fixed point as $S'(t + dt)$ with position vector $\boldsymbol{r}_p'(t + dt)$.

The vector from S' to $S(t + dt)$ is $(\boldsymbol{dr}_p)_P$ (see Fig. 3-81). Thus

$$\boldsymbol{r}_p(t + dt) - \boldsymbol{r}'_p = (\boldsymbol{dr}_p)_P. \tag{41-8}$$

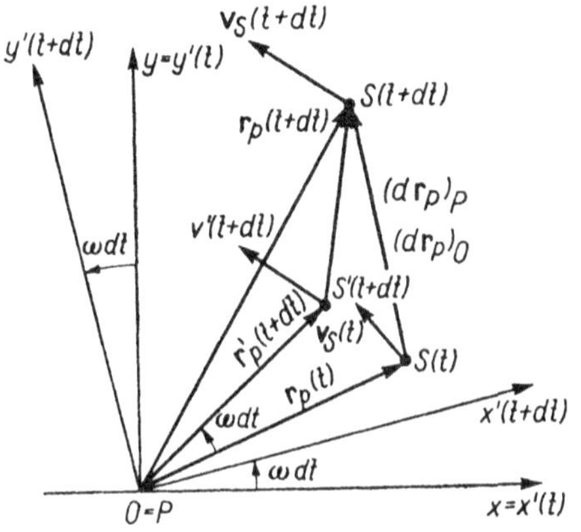

FIG. 3-81. Derivation of the Coriolis acceleration.

Figure 3-81 also shows the velocity vectors of the points $S(t)$, $S(t + dt)$, and $S'(t + dt)$. The velocity of $S'(t + dt)$ is designated by $\boldsymbol{v}'$. These velocities are

$$\boldsymbol{v}_S(t) = \boldsymbol{\omega} \times \boldsymbol{r}_p(t),$$

$$\boldsymbol{v}_S(t + dt) = \boldsymbol{\omega} \times \boldsymbol{r}_p(t + dt), \tag{41-9}$$

$$\boldsymbol{v}'(t + dt) = \boldsymbol{\omega} \times \boldsymbol{r}'_p(t + dt). \tag{41-10}$$

By definition, the velocities $\boldsymbol{v}_I(t)$ and $\boldsymbol{v}_I(t + dt)$ of points in $P(x', y')$ relative to $O(x, y)$ are

$$\boldsymbol{v}_I(t) = \boldsymbol{v}_S(t), \quad \boldsymbol{v}_I(t + dt) = \boldsymbol{v}_S(t + dt).$$

Therefore

$$(\boldsymbol{dv}_I)_O = \boldsymbol{v}_S(t + dt) - \boldsymbol{v}_S(t). \tag{41-11}$$

We now introduce the concept of the *moving system's acceleration*, $\boldsymbol{a}_I(t)$, defined as the acceleration, relative to the coordinate system $O(x, y)$, of that particular point of $P(x', y')$ at which the particle is located at time t. That is,

$$\boldsymbol{a}_I = \frac{\boldsymbol{v}'(t + dt) - \boldsymbol{v}_S(t)}{dt}.$$

If to the right-hand side of Eqn. (41-11), we now add and subtract $\boldsymbol{v}'$, we get

$$(\boldsymbol{dv}_I)_O = [\boldsymbol{v}_S(t + dt) - \boldsymbol{v}'(t + dt)] + [\boldsymbol{v}'(t + dt) - \boldsymbol{v}_S(t)].$$

Dividing through by dt, and realizing that the second term on the right is simply $\boldsymbol{a}_I$, we get

$$\left(\frac{\boldsymbol{dv}_I}{dt}\right)_O = \frac{\boldsymbol{v}_S(t + dt) - \boldsymbol{v}'(t + dt)}{dt} + \boldsymbol{a}_I.$$

We now use Eqns. (41-9) and (41-10) to substitute in the first term on the right:

$$\frac{\boldsymbol{v}_S(t+dt)-\boldsymbol{v}'(t+dt)}{dt}=\frac{\boldsymbol{\omega}\times\boldsymbol{r}_p(t+dt)-\boldsymbol{\omega}\times\boldsymbol{r}_p'(t+dt)}{dt}$$

$$=\boldsymbol{\omega}\times\frac{\boldsymbol{r}_p(t+dt)-\boldsymbol{r}_p'(t+dt)}{dt}.$$

But using Eqn. (41-8), we see that this is equal to $\boldsymbol{\omega}\times(d\boldsymbol{r}_p/dt)_P=\boldsymbol{\omega}\times\boldsymbol{v}_p$. Finally, then, we have

$$\left(\frac{d\boldsymbol{v}_I}{dt}\right)_O=\boldsymbol{a}_I+\boldsymbol{\omega}\times\boldsymbol{v}_p.$$

We see, therefore, that the time derivative in the fixed system $O(x, y)$ of the velocity of a point in the moving system is not equal to the moving system's acceleration.

Introducing the last equation into Eqn. (41-5) we have

$$\boldsymbol{a}_0=\boldsymbol{a}_p+\boldsymbol{a}_I+2\boldsymbol{\omega}\times\boldsymbol{v}_p.$$

In this case a term $2\boldsymbol{\omega}\times\boldsymbol{v}_p$ is added to the sum of the accelerations $\boldsymbol{a}_p$ and $\boldsymbol{a}_I$. It has the dimensions of an acceleration, and is called, after its discoverer, the *Coriolis* acceleration:

$$\boldsymbol{a}_c=2\boldsymbol{\omega}\times\boldsymbol{v}_p.$$

Half of the Coriolis acceleration arises from the difference between the time derivatives of $\boldsymbol{v}_p$ with respect to $O(x, y)$ and $P(x', y')$; the other half from the difference between $(d\boldsymbol{v}_I/dt)_O$ and $\boldsymbol{a}_I$. Finally, then, we have

$$\boldsymbol{a}_0=\boldsymbol{a}_p+\boldsymbol{a}_I+\boldsymbol{a}_c. \tag{41-12}$$

The acceleration relative to $O(x, y)$ is given by the sum of these terms, where $\boldsymbol{a}_p$ depends on the motion of the particle relative to the moving system $P(x', y')$, $\boldsymbol{a}_I$ depends both on the motion of $P(x', y')$ relative to $O(x, y)$ and on the position of the particle in $P(x', y')$, but not on its velocity $\boldsymbol{v}_p$, while $\boldsymbol{a}_c$ depends both on $\boldsymbol{\omega}$ and $\boldsymbol{v}_p$. It is therefore possible in many instances to discover the separate influences on the acceleration $\boldsymbol{a}_0$ of the motion of $P(x', y')$ relative to $O(x, y)$, and of the motion of the particle relative to $P(x', y')$.

Let us assume again that an observer at rest in $P(x', y')$ measures the acceleration $\boldsymbol{a}_p$ of a particle and the resultant $\boldsymbol{F}$ of all forces acting on the particle. He will again find that $\boldsymbol{F}\neq m\boldsymbol{a}_p$, since from Eqn. (41-12)

$$m\boldsymbol{a}_p=m\boldsymbol{a}_0-m(\boldsymbol{a}_I+\boldsymbol{a}_c)=\boldsymbol{F}-m(\boldsymbol{a}_I+\boldsymbol{a}_c).$$

Newton's second law therefore applies in its usual form only if the observer in the rotating system adds an extra force to the force $\boldsymbol{F}$. The extra force consists of the sum of two terms. One is the extra force already introduced and given by $\boldsymbol{Z}=-m\boldsymbol{a}_I$. In the case of uniform rotation, $\boldsymbol{a}_I$ is directed toward the center, or is centripetal, so the force Z is outward and is centrifugal. In this case, Z will be recognized as the *centrifugal inertial force*. The second part of the added force is

$$C = -ma_c = -2m\boldsymbol{\omega} \times \boldsymbol{v}_p,$$

called the *Coriolis force*. Since the vector product $\boldsymbol{\omega} \times \boldsymbol{v}_p$ is perpendicular to $\boldsymbol{v}_p$, the same is true of the Coriolis force. Its effect is to change the direction of the velocity $\boldsymbol{v}_p$ without altering its magnitude.

EXAMPLE 1. In the northern hemisphere, winds and ocean currents are deviated toward the right.

EXAMPLE 2. A locomotive of mass $m = 100$ tons travels at 47°N latitude with a velocity of 72 km/hr (a) from south to north, (b) from west to east. Calculate the Coriolis force.

Solution. (a) Fig. 3-82 shows the meridian plane, the angular velocity $\boldsymbol{\omega}$ of the earth, and the velocity $\boldsymbol{v}_p$. The vector $\boldsymbol{\omega} \times \boldsymbol{v}_p$ points toward the west, so the Coriolis force $\boldsymbol{C} = -2m\boldsymbol{\omega} \times \boldsymbol{v}_p$ is a force lying in the tangential plane and pointing toward the east. Its magnitude is

$$C = 2m\omega v_p \sin\varphi = 21.7 \text{ kgf}.$$

This force acts on the locomotive and results in an increased reaction force of the right-hand rails against the wheel flanges. Note that if this reaction force were *not* applied, the locomotive would deviate toward the right.

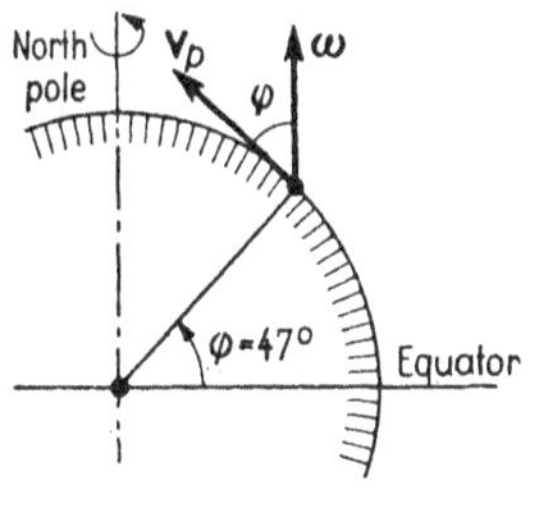

FIG. 3-82

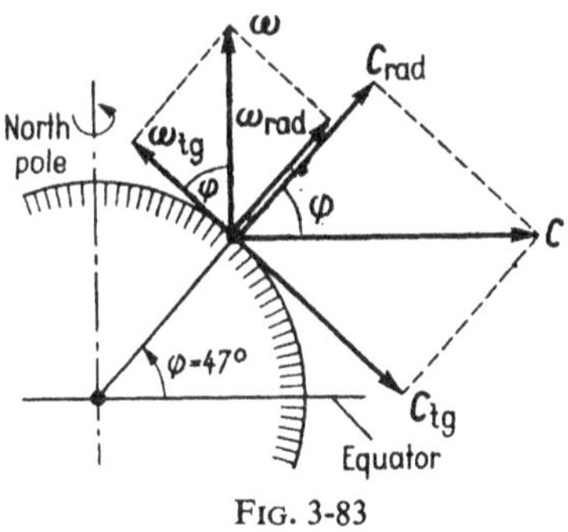

FIG. 3-83

FIG. 3-82. The Coriolis force on a locomotive traveling north.

FIG. 3-83. The Coriolis force on a locomotive traveling east.

(b) Figure 3-83 again shows the meridian plane and the angular velocity $\boldsymbol{\omega}$ of the earth, resolved this time into radial and tangential components $\boldsymbol{\omega}_{rad}$ and $\boldsymbol{\omega}_{tg}$. The velocity vector $\boldsymbol{v}_p$ points toward the east. The vector product $\boldsymbol{\omega} \times \boldsymbol{v}_p$ can be resolved into two components,

$$\boldsymbol{\omega} \times \boldsymbol{v}_p = \boldsymbol{\omega}_{rad} \times \boldsymbol{v}_p + \boldsymbol{\omega}_{tg} \times \boldsymbol{v}_p.$$

The first term gives rise to a Coriolis force $\boldsymbol{C}_{tg}$ in the tangential plane, directed toward the south and of magnitude

$$C_{tg} = 2mv_p\omega \sin\varphi = 21.7 \text{ kgf}.$$

It results in an increase in the reaction force exerted by the right-hand rails. Note that if this force were *not* applied, the locomotive would again deviate toward the right.

The second term leads to a Coriolis component C_{rad} of magnitude

$$C_{rad} = 2mv_p\omega \cos \varphi = 20.2 \text{ kgf}.$$

This component of the inertial force points vertically upward and decreases the upward reaction force exerted on the locomotive by the rails.

In both this example and the preceding one, the inertial force $\boldsymbol{Z} = -m\boldsymbol{a}_I$ acts in addition to the Coriolis force.

Equation (41-12) applies also for any arbitrary motion of the moving coordinate system $P(x', y')$. To set its interpretation in the proper light, the following remarks are in order. We have assumed throughout the preceding discussion that Newton's second law, in the form $\boldsymbol{F} = m\boldsymbol{a}_0$, holds in the coordinate system $O(x, y)$, and have then determined how an observer in the moving system must formulate this law. Let us now ask, under what circumstances will the second law in the form $\boldsymbol{F}' = m\boldsymbol{a}_p$ also apply in the moving reference system, where $\boldsymbol{F}'$ includes only those forces that would be considered to act by an observer in $O(x, y)$?

It follows from Eqn. (41-12) that this will be the case only when $P(x', y')$ moves with respect to $O(x, y)$ in uniform, rectilinear motion. [The acceleration $\boldsymbol{a}_c \equiv 0$ for any arbitrary motion relative to $P(x', y')$ if $\omega = 0$; the acceleration $\boldsymbol{a}_I \equiv 0$ for uniform rectilinear motion of $P(x', y')$.] We therefore obtain the following result:

All reference systems in which Newton's second law holds in the form $\boldsymbol{F} = m\boldsymbol{a}$ move relative to one another with inertial motion (*inertial systems*). In all systems which are accelerated relative to an inertial system, extra forces, i.e., *D'Alembert inertial forces*, must be included because of the relative motion. This can only be done, however, when the motion of the moving system is known relative to some inertial system.

CHAPTER 4

DYNAMICS OF RIGID BODIES

42. Motion of the Center of Mass; The Center of Mass Law

A *system* of particles is any collection of particles between which there are forces of any nature. If these forces are such that the positions of all the particles in the system are fixed relative to each other, the system is called a *rigid body*.

In this chapter, we shall be concerned only with those systems of particles making up a rigid body. We shall also discuss a few examples of groups of rigid bodies which themselves form nonrigid systems. In the following chapters we will consider other kinds of systems.

The first important consideration in dealing with systems of particles is the distinction between *internal* and *external* forces. *Internal* forces always occur in pairs as an *action* and its *reaction*, and are therefore oppositely directed. The resultant of the internal forces for the entire system is zero. Since an action and its reaction have the same line of action, the resultant moment of the internal forces about any reference point is also zero. The internal forces of a system can therefore be ignored, and we can fix our attention on the external forces only.

The *external* forces (that is, contact forces acting on the surface of a body, and forces which act at a distance, such as gravitation) have no reactions within the system, and can cause an acceleration of the masses making up the system. In addition, the external forces can have a resultant moment about an arbitrary reference point. The problem of the mechanics of systems can therefore be formulated through two questions. What is the acceleration produced by the external forces? What is the effect of the moment of the external forces?

The first question is answered by the *center of mass law*. The second finds its solution in the *angular momentum law*.

42-1 Formulation of the Center of Mass Law

Let $\boldsymbol{F}_i$ represent the external force acting on the ith particle of a system, and let $\boldsymbol{G}_i$ be the resultant of the internal forces exerted on this particle by all the other particles of the system. Newton's second law then leads to the following equation for the ith particle:

$$\boldsymbol{F}_i + \boldsymbol{G}_i = d\boldsymbol{p}_i/dt = \dot{\boldsymbol{p}}_i.$$

A similar equation holds for every other particle. The entire system of particles therefore obeys the equation

$$\sum_i \boldsymbol{F}_i + \sum_i \boldsymbol{G}_i = \sum_i \dot{\boldsymbol{p}}_i.$$

But $\sum_i \boldsymbol{G}_i = 0$, since the $\boldsymbol{G}_i$'s are the internal forces of the system. Therefore

$$\sum_i \boldsymbol{F}_i = \sum_i \dot{\boldsymbol{p}}_i,$$

where $\sum_i \boldsymbol{F}_i$ is the resultant $\boldsymbol{R}$ of the external forces and $\sum_i \dot{\boldsymbol{p}}_i$ is the time rate of change $\dot{\boldsymbol{p}}$ of the total momentum $\boldsymbol{p}$ of the system. Hence Newton's second law applied to a system of particles takes the form

$$\boldsymbol{R} = \dot{\boldsymbol{p}},$$

which is the *center of mass law*. That is, *the resultant of the external forces equals the time rate of change of the total momentum.*

In order to interpret this equation, we now define the *center of mass*† of the system. Let $\boldsymbol{r}_i$ be the vector from a fixed origin O to the ith particle of the system, of mass m_i. The vector $\boldsymbol{r}_O$ from the origin to the center of mass is then defined by the equation

$$m_O \boldsymbol{r}_O = \sum_i m_i \boldsymbol{r}_i, \tag{42-1}$$

where $m_O = \sum_i m_i$ is the total mass of the system. This equation is equivalent to the three component equations

$$m_O x_O = \sum_i m_i x_i, \quad m_O y_O = \sum_i m_i y_i, \quad m_O z_O = \sum_i m_i z_i.$$

The velocity $\boldsymbol{v}_O$ of the center of mass is obtained by differentiating Eqn. (42-1):

$$m_O \boldsymbol{v}_O = \sum_i m_i \boldsymbol{v}_i = \boldsymbol{p}, \tag{42-2}$$

and its acceleration $\boldsymbol{a}_O$, found by differentiating Eqn. (42-2), is

$$m_O \boldsymbol{a}_O = d\boldsymbol{p}/dt = \dot{\boldsymbol{p}}. \tag{42-3}$$

The equation $\boldsymbol{R} = \dot{\boldsymbol{p}}$ is therefore equivalent to $\boldsymbol{R} = m_O \boldsymbol{a}_O$. That is, *the center of mass* of a system of particles moves like a single particle whose mass equals the total mass of the system, and which is acted on by a single force equal to the resultant of all the external forces. This is the *law of motion of the center of mass.*

The validity of the center of mass law rests on that of Newton's third law. It does not depend on the nature of the internal forces. This law now belatedly justifies our previous treatment of the dynamics of a particle, which we applied to the motion of extended objects as if these objects were, in fact, particles. We have for a long time actually been using the concept of center of mass and the law of motion for this point.

† Daniel Bernoulli, *Traité sur le flux et le reflux de la mer*, 1740.

42-2 *Reduction to a Single Force at the Center of Mass*

In general, the line of action of the resultant external force does not pass through the center of mass. We now show how any force can be reduced to a force through the center of mass.

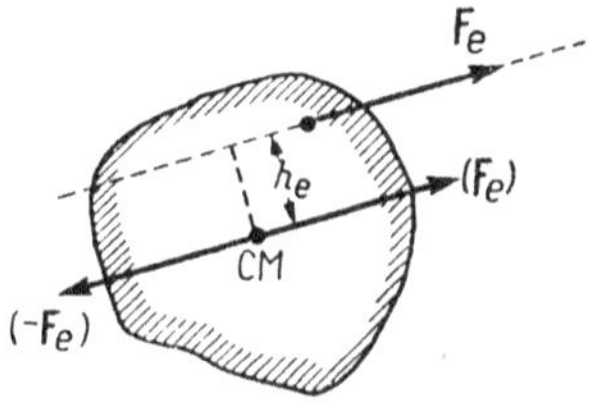

FIG. 4-1. Reduction to a force at the center of mass.

Let $\boldsymbol{F}_e$, in Fig. 4-1, represent the resultant external force acting on a system. We introduce, at the center of mass, two forces of the same magnitude as $\boldsymbol{F}_e$, one parallel and the other antiparallel to $\boldsymbol{F}_e$. Since these forces cancel one another, they have no effect on the motion of the system. The three forces can now be combined in the following way. The force $(\boldsymbol{F}_e)$ acts at the center of mass and has a magnitude F_e; the forces $\boldsymbol{F}_e$ and $(-\boldsymbol{F}_e)$ have a moment relative to the center of mass of magnitude $M = F_e h_e$, where h_e is the perpendicular distance from the line of action of $\boldsymbol{F}_e$ to the center of mass point. The effect of the external force is therefore completely equivalent to that of a force of the same magnitude acting at the center of mass plus a torque equal to that of $\boldsymbol{F}_e$ relative to the center of mass.

It again becomes evident that the mechanics of a system has to do with two motions, namely, the translation of the center of mass (center of mass law) and the rotational motion of the system (angular momentum law).

42-3 *Applications of the Center of Mass Law*

The earth–sun system. If we neglect the forces exerted on the earth by the other planets in comparison with the force exerted on it by the sun, the earth–sun system can be considered to be free from external forces. It follows that the center of mass C_0 of the system remains in a state of rest or of uniform rectilinear motion. This point can therefore be taken as the origin of an inertial reference system. By Newton's third law, the forces on the earth and sun are equal in magnitude, oppositely directed, and have the same line of action. Hence, the resultant moment about C_0 is zero, since C_0 lies on the line of action of both forces (Fig. 4-2).

If the center of mass of the earth, C_1, describes a circle (to first approximation) of radius r_1 about C_0, the center of mass of the sun, C_2, must also describe a circle of radius r_2 about C_0. Both circles are traversed in the same sense and with the same period. The radius r_2 of the path of the center of mass of the sum can be computed from the equation

$$m_{\text{earth}} r_1 \omega^2 = m_{\text{sun}} r_2 \omega^2,$$

from which

$$r_2 = r_1 \frac{m_{\text{earth}}}{m_{\text{sun}}} = 450 \text{ km}.$$

(The distance from the sun to the earth is $r_1 + r_2 = 1.5 \times 10^8$ km, and the ratio of the masses is 3.0×10^{-6}.)

The center of mass of the earth–sun system lies in the interior of the sun.

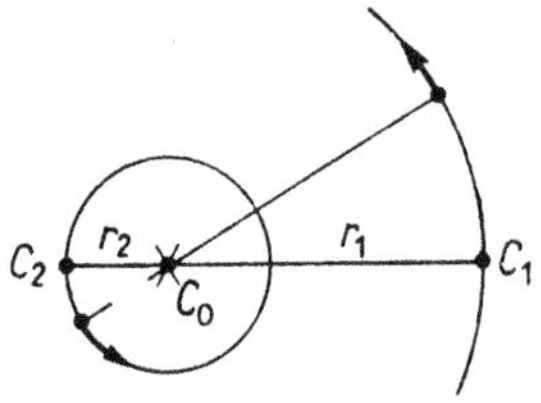

FIG. 4-2. The orbits of the earth and sun around their center of mass.

The earth-moon system. If we consider as the center of mass of the earth-moon system that point which moves in an unperturbed ellipse about the sun, and neglect the radial acceleration of the point toward the sun (about 0.4 cm/sec^2), the considerations above can be applied directly to the earth–moon system. The distance of the moon from the earth is $r_1 + r_2 = 384{,}000$ km, and the mass of the moon is 1/81 that of the earth. Thus using $r_2/r_1 = m_{\text{moon}}/m_{\text{earth}}$, we find $r_2 = 4680$ km.

The center of mass of the system, C_0, therefore lies in the interior of the earth. Hence as the earth performs its motion about the sun, points on its surface sweep out a band of width $D = 2(R_E + r_2) = 22{,}100$ km on either side of the path of the center of mass of the system (Fig. 4-3).

The motion of the earth about the center of mass of the earth–moon system is responsible for the fact that the tides have a twelve-hour period, or expressed differently, that the tide is simultaneously high at diametrically opposite points on the earth's surface. To explain this, we first neglect the rotation of the earth on its own axis, since at points of a given latitude this causes the

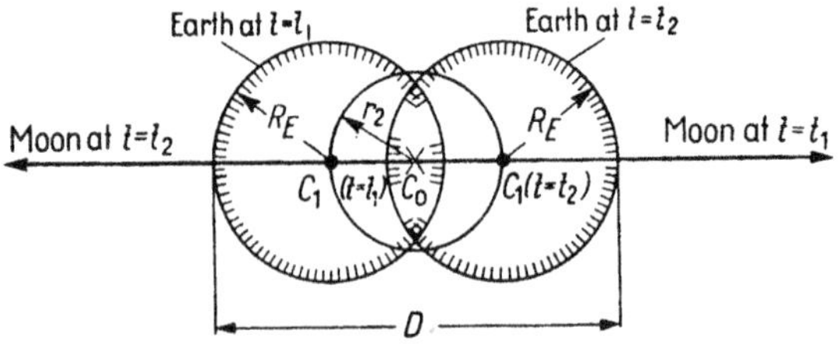

FIG. 4-3. Oscillation of the earth about the orbit of the center of mass of the earth–moon system.

same centrifugal force at all longitudes (Fig. 4-4). With this preliminary assumption, the motion of the earth is simply a rotation about the center of mass of the system C_0, with an angular velocity ω_E. The period of this motion is $T = (2\pi\omega_E)^{-1} \cong 27$ days.

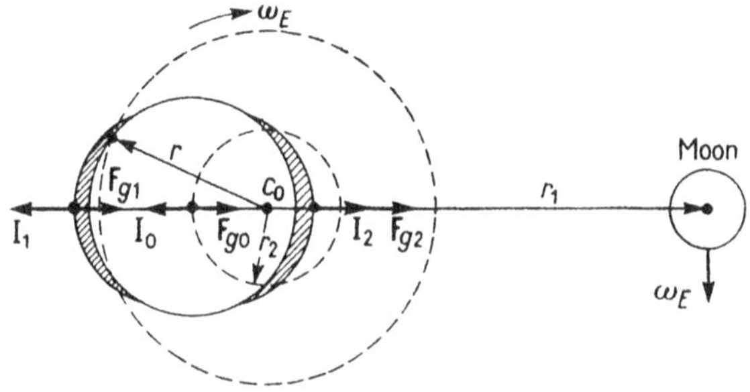

FIG. 4-4. Forces causing the tides.

Let us now choose C_0 as an origin of coordinates. This is a nearly inertial frame of reference, since we are neglecting the radial acceleration of C_0 toward the sun. For the sake of simplicity, we shall assume that the earth and moon are two dimensional, i.e., circles in their common plane of rotation about C_0. This is a good approximation, because the radii of the earth and moon are much smaller than the earth–moon separation. Then, every point of the earth will describe a circle about C_0, and all points will traverse their circles in the same time, since all move with the same angular velocity. Therefore, in addition to the gravitational force exerted by the moon, we must also consider the effect of the rotation of the earth about C_0. Then, at a point on the earth a distance r from C_0, this rotation gives rise to an acceleration of magnitude $r\omega_E^2$ directed towards C_0. The inertial force, I, on a mass m at this point is thus $mr\omega_E^2$, directed away from C_0. At the center of the earth, the inertial force and the moon's gravitational force, F_g, exactly balance (d'Alembert equilibrium). At nearly all points on the earth's surface, these two forces no longer balance. On the side of the earth closer to the moon, the gravitational and inertial forces are in the same directions, and this results in a force tending to bulge out the surface of the ocean. On the other side of the earth, the inertial and gravitational forces are in opposite directions, but since the inertial force is the greater ($mr\omega^2 > mr_2\omega^2 > mMG/R_{mM}^2$), the resultant force is directed away from the moon. Hence, on the far side of the earth, the ocean is also acted on by a force tending to bulge it outward (away from the moon). This bulging out of the ocean gives rise to the tides.

We now consider the rotation of the earth on its axis. The major effect of the motion is to carry points on the earth from the side closer to the moon to the side farthest from the moon and vice-versa. A point will be on opposite sides of a diameter of the earth once every twelve hours as the earth rotates its sides toward and away from the moon, thus giving rise to two tides per day.

Various other effects which complicate the problem of the occurrence of tides are beyond the scope of this book. These include the fact that the plane of rotation of the earth about its axis is not parallel to the plane in which the earth and moon rotate about C_0. This causes the height of the tides to vary at opposite sides of the earth. Also, the earth is not a spherical object, completely covered with water, and the frictional forces on the water due to land masses cause the tides to lag the forces producing them by several hours. Furthermore, the point C_0 is not an origin of coordinates in an inertial frame of reference, and finally, the effects of the sun's gravitational force must also be considered. This force is considerable and, among other effects, is responsible for the spring tides and neap tides that occur every month.

43. The Law of Rotational Motion for Rotation about a Fixed Axis. The Angular Momentum Law

The *translational* motion of a system of particles that form a rigid body can be treated with the methods of the mechanics of a single particle by assuming that all the mass is concentrated at the center of mass. This assumption is independent of the direction of the motion. The same simplification is not possible in the case of *rotational* motion. In order to investigate the effect of a force having a moment relative to the center of mass, we shall have to use the more sophisticated methods of the mechanics of a system of particles.

A problem of this nature that can be treated by elementary methods is the special case in which a system rotates about a *fixed axis* (q). A plane perpendicular to this axis, and intersecting it at point Q, is called a *plane of rotation*. Such a plane is shown in Fig. 4-5. All external forces are assumed parallel to this plane. Particle i, of mass m_i, is acted on by an external

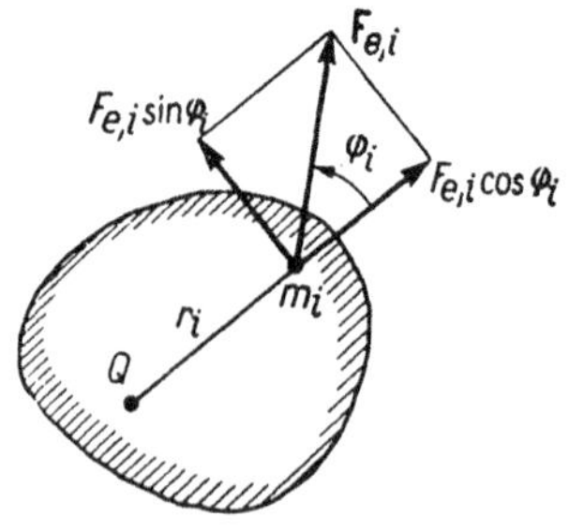

FIG. 4-5. Rotation of a rigid body about a fixed axis through the point Q. The axis of rotation is perpendicular to the plane of the figure.

force $F_{e,i}$, which we can resolve into a radial component $F_{e,i}\cos\varphi_i$ and a tangential component $F_{e,i}\sin\varphi_i$. The radial force components are balanced by reaction forces acting on the axis. The tangential components result in a torque M_q about the axis,

$$M_q = \sum r_i F_{e,i} \sin\varphi_i,$$

where r_i lies in the plane of rotation and is the moment arm of the tangential component. The force component $F_{e,i}\sin\varphi_i$ accelerates the ith particle, and hence the entire rigid body, with an angular acceleration $\dot{\omega}$. To calculate this acceleration, we apply the second law to the ith particle in the form developed in Section 42-1:

$$F_{e,i}\sin\varphi_i + G_i = m_i a_i,$$

where G_i is the tangential component of the resultant of all internal forces exerted on the ith particle by the other particles of the body. Since we are concerned with rotational motion, it is convenient to use, instead of the forces, their moments about the axis, and instead of the acceleration a_i, the angular acceleration $\dot{\omega}$. We have

$$r_i F_{e,i}\sin\varphi_i = M_{e,i}, \quad G_i r_i = M_{i,i}, \quad a_i = r_i\dot{\omega}.$$

Hence, from the second law, we have for the ith particle

$$M_{e,i} + M_{i,i} = m_i r_i^2 \dot{\omega},$$

and for the entire rigid body,

$$\sum_i M_{e,i} + \sum_i M_{i,i} = \sum_i m_i r_i^2 \dot{\omega}.$$

But $\sum_i M_{e,i}$ equals the resultant moment of the external forces about the axis, and $\sum_i M_{i,i} = 0$ since this sum represents internal forces only. Furthermore, the quantity $\dot{\omega}$ can be taken outside the summation sign, since the angular acceleration is the same for all particles of a rigid body. The remaining term, $\sum_i m_i r_i^2$, is called the *moment of inertia* of the body, I_q, with respect to the fixed axis (q). Hence the second law takes the form

$$M_q = I_q\dot{\omega}.$$

This equation is called the *law of rotational motion for a fixed axis.* It is nothing more than Newton's second law applied to rotation, and determines the rotational motion of a rigid body about a fixed axis. The law states that the resultant moment M_q of the external forces acting on the body produces an angular acceleration

$$\dot{\omega} = M_q/I_q.$$

The moment of inertia about the axis (q) is defined by the equation

$$I_q = \sum m_i r_i^2,$$

where r_i is the perpendicular distance of the ith particle from the axis. The moment of inertia about any axis is a geometrical quantity, and it can be computed from the arrangement of the particles of the body about the axis. To each of the infinite number of axes about which the body can rotate, there corresponds, in general, a different moment of inertia. The moment of inertia has the nature of a resistance to an angular acceleration, and was introduced in mechanics by Euler. It has the dimensions $[I] = [ML^2]$, and the units: cgs system, 1 gm · cm^2; mks system, 1 kgm · m^2; fps system, 1 lbm · ft^2.

The law of rotational motion can be put in a new form by introducing another quantity, called the *angular momentum* (see Fig. 4-6). This quantity

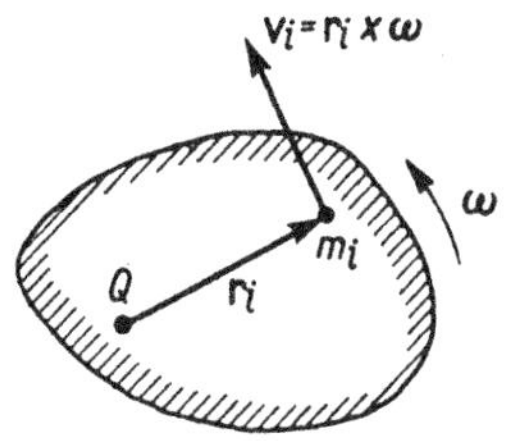

FIG. 4-6. The angular momentum about the axis (q).

is extremely useful in dealing with problems involving the rotation of a system of particles. We shall introduce the new form of the law here, although its advantages will be evident only later on.

DEFINITION. The angular momentum (also called the moment of momentum) $\boldsymbol{L}$ of a particle with respect to an axis is defined as the moment of the momentum of the particle with respect to the axis.

In the rotation of a rigid body about a fixed axis, the momentum vector $\boldsymbol{p}_i = m_i\boldsymbol{v}_i$ of a particle of mass m_i is perpendicular to the radius vector $\boldsymbol{r}_i$. The moment of this momentum is itself a vector, perpendicular to the plane of $\boldsymbol{r}_i$ and $\boldsymbol{p}_i$ (right-hand screw rule) and hence parallel to the axis. This moment is the *angular momentum* or *moment of momentum* of the particle, and its magnitude is

$$L_i = m_i r_i v_i .$$

Introducing the angular velocity ω, which is the same for all particles in a rigid body, we get $L_i = m_i r_i^2 \omega$. With the help of the vectorial representation of angular velocity (the direction of $\boldsymbol{\omega}$ is parallel to the axis or rotation), we obtain as the vector equation for the angular momentum of a particle rotating about a fixed axis,

$$\boldsymbol{L}_i = m_i r_i^2 \boldsymbol{\omega} .$$

The angular momentum $\boldsymbol{L}_q$ of the entire system is the algebraic sum of the angular momenta of the particles:

$$\boldsymbol{L}_q = \sum m_i r_i^2 \omega = \omega \sum m_i r_i^2 = I_q \omega .$$

It has the dimensions $[L] = [ML^2T^{-1}]$, and the units: cgs system, 1 gm · cm²/sec = 1 erg · sec; mks system, 1 kgm · m²/sec = 1 kp · sec; fps system, 1 lbm · ft²/sec.

For a rigid body and a fixed axis, the moment of inertia is constant. A time rate of change of angular momentum can therefore result only from an angular acceleration of the system. Hence, for this special case,

$$d\boldsymbol{L}_q/dt = I_q(d\boldsymbol{\omega}/dt) = I_q\boldsymbol{\alpha} .$$

But $I_q\boldsymbol{\alpha}$ has been shown to equal the moment $\boldsymbol{M}_q$ of the external forces, so the law of rotational motion takes the following simple form:

$$\boldsymbol{M}_q = d\boldsymbol{L}_q/dt .$$

Newton's second law for rotational motion, formulated in this way, is called the *angular momentum law*. It is the counterpart of the second law for linear motion as given in Section 24, which states: A force exerted on a particle results in a time rate of change of momentum given by $\boldsymbol{F} = d\boldsymbol{p}/dt$. The angular momentum law states: A torque exerted on a system pivoted about a fixed axis (q) results in a time rate of change of angular momentum given by $\boldsymbol{M}_q = d\boldsymbol{L}_q/dt$.

The content of the angular momentum law can be formulated in another way. If a constant torque $\boldsymbol{M}_q$ acts on a system for a time Δt, the resulting change in angular momentum is

$$\Delta\boldsymbol{L}_q = \boldsymbol{M}_q \Delta t .$$

The change in the angular momentum vector, $\Delta\boldsymbol{L}_q$ has the same direction as the torque $\boldsymbol{M}_q$.

This second form of the law of rotational motion has a wider range of validity than is apparent from the special method by which it has been derived. This will be brought out later in connection with rotational motion about axes which are not fixed. In the simple case of accelerated rotation about a fixed axis, the first form of the law is to be preferred.

43-1 Analogies between Translational Motion and Rotation about a Fixed Axis

There is a complete analogy between the translational motion of a particle and the rotation of a system of particles about a fixed axis, which is illustrated by listing the corresponding quantities and equations in Table 4-1. It is understood, of course, that $\boldsymbol{M}$ and $\boldsymbol{L}$ are referred to the same axis. In what follows, we will often drop the axis subscript when the axis is obvious.

TABLE 4-1

Translational	Rotational
Force $\boldsymbol{F}$	Torque $\boldsymbol{M}$
Acceleration $\boldsymbol{a}$	Angular acceleration $\boldsymbol{\alpha}$
Velocity $\boldsymbol{v}$	Angular velocity $\boldsymbol{\omega}$
Mass m	Moment of inertia I
Momentum $\boldsymbol{p}$	Angular momentum $\boldsymbol{L}$
Distance s	Angle φ
Kinetic energy, $\frac{1}{2}mv^2$	Kinetic energy (Section 43-2), $\frac{1}{2}I\omega^2$
Equilibrium results, if	
$\boldsymbol{F} = 0$, that is, $\boldsymbol{a} = 0$	$\boldsymbol{M} = 0$, that is, $\boldsymbol{\alpha} = 0$.
The equation of motion is	
$\boldsymbol{F} = m\boldsymbol{a}$	$\boldsymbol{M} = I\boldsymbol{\alpha}$
$\boldsymbol{F} = d\boldsymbol{p}/dt$	$\boldsymbol{M} = d\boldsymbol{L}/dt$

43-2 Kinetic Energy of a Rotating Body

The kinetic energy of a body rotating about a fixed axis is obtained by summing the energies of all particles of the body:

$$E_k = \sum_i \tfrac{1}{2}m_i v_i^2.$$

Setting $v_i = r_i\omega$, we get

$$E_k = \tfrac{1}{2}\sum_i m_i r_i^2\omega^2 = \tfrac{1}{2}I\omega^2.$$

Here also we find an exact analogy with the kinetic energy of a particle, $\frac{1}{2}mv^2$.

44. MOMENTS OF INERTIA

Every body has an infinite number of possible axes of rotation, to each of which there corresponds a moment of inertia. The moments of inertia about the axes passing through the center of mass are of particular importance, because when these are known, it is easy to calculate the moment of inertia about any parallel axis using the parallel axis theorem.

The moment of inertia about an arbitrary axis is the sum of the products of the mass of each particle by the square of the perpendicular distance of the particle from the axis,

$$I = \sum_i m_i r_i^2.$$

The *radius of gyration, k*, is the distance from the axis at which a particle of mass $m_T = \sum m_i$ should be placed in order to have the same moment of inertia as the body:

$$I = \sum m_i r_i^2 = m_T k^2.$$

The different moments of inertia of a given body are related by three general laws. These are: (1) the *parallel axis theorem*, (2) the law of the *principal moments of inertia*, and (3) *the combination law*, for combining moments of inertia.

44-1 The Parallel Axis Theorem

Let I_c be the moment of inertia of a body about an axis (c) through the center of mass (CM) (Fig. 4-7):

$$I_c = \sum_i m_i r_i^2.$$

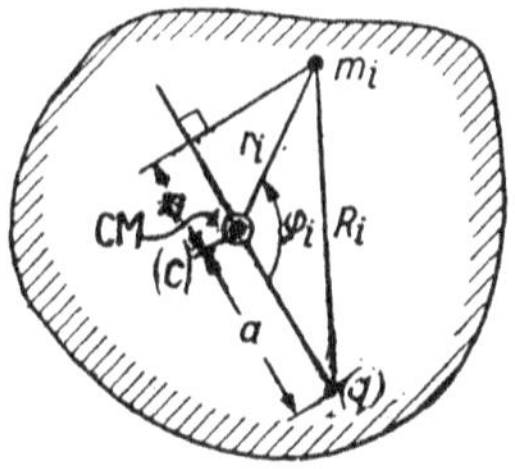

FIG. 4-7. The parallel axis theorem.

If (q) is an axis parallel to (c) and at a distance a from it, then from the cosine law,

$$R_i^2 = r_i^2 + a^2 - 2ar_i \cos \varphi_i,$$

and the moment of inertia about axis (q) is

$$I_q = \sum m_i R_i^2 = \sum m_i r_i^2 + a^2 \sum m_i - 2a \sum m_i r_i \cos \varphi_i.$$

But $r_i \cos \varphi_i = x_i$, so $\sum m_i r_i \cos \varphi_i = \sum m_i x_i = 0$, since the center of mass lies on the axis (c). We therefore obtain the *parallel axis theorem*:

$$I_q = I_c + m_O a^2.$$

The moment of inertia about an axis at a distance a from the center of mass equals the moment of inertia about a parallel axis through the center of mass, plus the total mass of the body multiplied by a^2.

It follows that the moment of inertia about an axis through the center of mass is smaller than that about any parallel axis.

44-2 Principal Moments of Inertia

Figure 4-8 represents a body in the form of a flat plate. Let us construct two perpendicular axes, x and y, through the center of mass of the body. The moment of inertia I_{xx} about the x-axis is $I_{xx} = \sum m_i y_i^2$, where y_i is the ordinate of the ith particle. The moment of inertia I_{yy} about the y-axis is $I_{yy} = \sum m_i x_i^2$, where x_i is the abscissa of the ith particle.

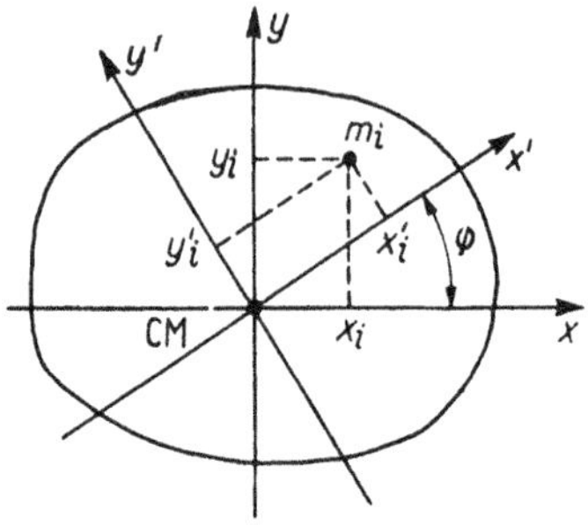

FIG. 4-8. Rotation of axes to find the principal moments of inertia.

We can now consider the quantities I_{xx} and I_{yy} as known, and ask for the moments of inertia about two new perpendicular axes x' and y', which are obtained from the old axes by a rotation through an angle φ. From the transformation equations of analytic geometry, the coordinates of a particle in the two systems are related by the equations

$$x'_i = x_i \cos\varphi + y_i \sin\varphi, \quad y'_i = y_i \cos\varphi - x_i \sin\varphi.$$

It follows that

$$x'_i y'_i = x_i y_i \cos 2\varphi + \frac{y_i^2 - x_i^2}{2} \sin 2\varphi,$$

$$x_i'^2 = x_i^2 \cos^2\varphi + y_i^2 \sin^2\varphi + 2x_i y_i \sin\varphi \cos\varphi,$$

$$y_i'^2 = y_i^2 \cos^2\varphi + x_i^2 \sin^2\varphi - 2x_i y_i \sin\varphi \cos\varphi.$$

The moments of inertia about the axes x' and y' are therefore

$$I'_{xx} = \sum m_i y_i'^2 = I_{xx} \cos^2\varphi + I_{yy} \sin^2\varphi + I_{xy} \sin 2\varphi,$$

$$I'_{yy} = \sum m_i x_i'^2 = I_{xx} \sin^2\varphi + I_{yy} \cos^2\varphi - I_{xy} \sin 2\varphi,$$

where for brevity we let $I_{xy} = \sum m_i x_i y_i = I_{yx}$. I_{xy} appears in the expressions for the moment of inertia about the new axes, and is called the *product of inertia* with respect to the axes x and y. The product of inertia with respect to the new axes is

$$I'_{xy} = \sum m_i x'_i y'_i = I_{xy} \cos 2\varphi + \frac{I_{yy} - I_{xx}}{2} \sin 2\varphi.$$

Although moments of inertia are formed from the sum of positive quantities only, products of inertia contain both positive and negative terms. For symmetrical bodies, it is possible for the product of inertia to be zero. This will be the case when the body has axial symmetry with respect to the axes of rotation. Then to every particle with coordinate (x_i, y_i) there corresponds one with coordinate $(x_i, -y_i)$, and to every particle with coordinates $(-x_i, -y_i)$ there corresponds one with coordinates $(-x_i, y_i)$. The product of inertia I_{xy} is therefore zero for an axially symmetric mass distribution.

Of special interest is the fact that even for a completely unsymmetrical distribution of particles there always exists a pair of mutually perpendicular axes for which the product of inertia is zero. This is the *law of principal moments of inertia.* To find these particular axes, it is only necessary to choose the angle φ in the expression for I'_{xy} in such a way that I'_{xy} vanishes, which will be the case if

$$\tan 2\varphi = \frac{2I_{xy}}{I_{yy} - I_{xx}}.$$

All the rotational inertia properties for rotation about the new axes x' and y' are given then by the two quantities I'_{xx} and I'_{yy}. These quantities are called the *principal moments of inertia* of the body.

44-3 The Law of Combination for Moments of Inertia

Figure 4-9 represents a body K, with center of mass CM, from which a portion K', with center of mass CM′ can be removed. Select any arbitrary

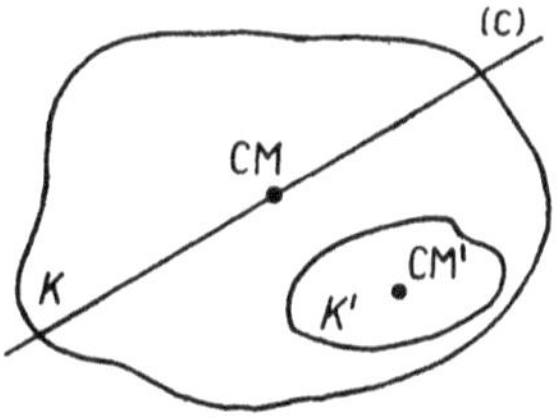

FIG. 4-9. Combination law for moments of inertia.

axis of rotation (c) through the center of mass CM. Then the moment of inertia of the entire body K, about (c), is

$$I_c = \sum_K m_i r_i^2,$$

and that of the portion K', about (c), is

$$I'_c = \sum_{K'} m_i r_i^2.$$

Since I_c includes the sum $\sum_{K'} m_i r_i^2$, it follows at once that the moment of inertia I_R of that portion of the body remaining after the portion K' has been removed is given by

$$I_R = I_c - I'_c,$$

which is the combination law.

The moment of inertia of homogeneous bodies of frequently encountered geometrical shapes, and for the most important axes through their centers of mass, can usually be computed by the methods of integral calculus. Some examples of the methods of setting up and evaluating these integrals are given in Appendix II. The moment of inertia of a body which is a combination of several simple shapes can be calculated using these integrals and the combination law. In Section 45-2, we shall describe an experimental method for obtaining moments of inertia, which is useful when the geometrical shape of a body is very complicated or when the distribution of mass is unknown.

45. Applications of the Angular Momentum Law to Rotation about a Fixed Axis

45-1 The Torsion Pendulum

Consider a body pivoted about an axis (s) and held in an equilibrium position by a torsion bar [Fig. 4-10 (a)]. Let I_s be the moment of inertia

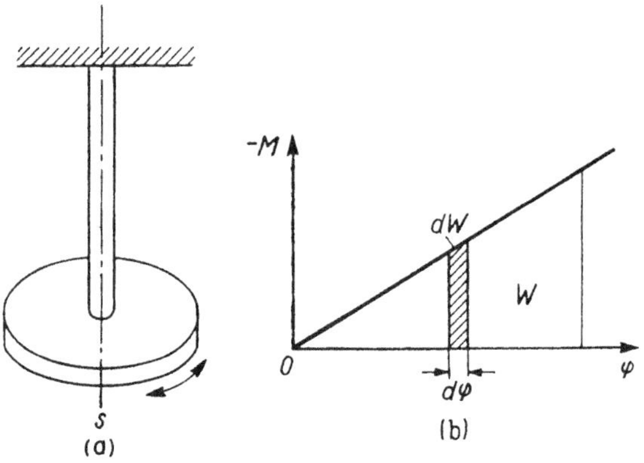

FIG. 4-10. (a) A torsion pendulum. (b) M-φ diagram for a torsion bar showing the linear relationship of the torque M and angular displacement φ.

of the body about the axis (s). Torsion bars have the property, within the proportional limit, of exerting a torque which is proportional to the angular displacement and is oppositely directed (Section 56),

$$M = -K\varphi.$$

The constant K is called the *torsion constant* and corresponds to the force constant k of a tensile spring. It has the dimensions $[K] = [ML^2T^{-2}]$, and the units: cgs system, 1 dyne · cm/rad; mks system, 1 newton · meter/rad; fps system, 1 lbm · ft/rad.

When a torsion bar is turned through an angle $d\varphi$, the work done is $dW = Md\varphi = -K\varphi\, d\varphi$. The total work done in an angular displacement from O to φ is represented by an area in an M-φ diagram [Fig. 4-10 (b)]:

$$W = -\tfrac{1}{2}K\varphi^2.$$

This work is stored as elastic potential energy of the bar:

$$E_p = +\tfrac{1}{2}K\varphi^2.$$

The angular acceleration α of the torsion pendulum produced by the bar is given by

$$-K\varphi = I_s\alpha,$$

$$\ddot{\varphi} = \alpha = -K\varphi/I_s, \qquad (45\text{-}1)$$

where I_s is the moment of inertia of the pendulum. This equation is exactly analogous to that for the acceleration of a body in harmonic motion (Section 31-2). The angular acceleration α corresponds to the linear acceleration a, and φ corresponds to x, K to k, and I_s to m. Just as in the case of a body suspended from a spring, the ensuing motion is harmonic:

$$\varphi = \varphi_0 \sin \Omega t.$$

From the analogies between the two problems, the harmonic frequency Ω is

$$\Omega = \sqrt{(K/I_s)},$$

and the period T is

$$T = 2\pi \sqrt{(I_s/K)}.$$

The motion provides a simple experimental method of determining moments of inertia. If the torsion constant K and the period T are measured, the moment of inertia is given by

$$I_s = KT^2/4\pi^2.$$

45-2 The Physical Pendulum

A *physical pendulum* is a rigid body of any arbitrary shape, pivoted about a fixed horizontal axis (a), as in Fig. 4-11. Let s be the distance from (a) to

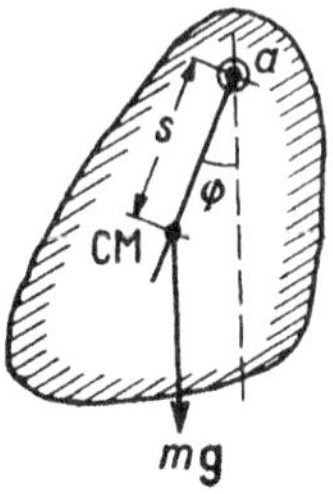

FIG. 4-11. A physical pendulum.

the center of mass CM. If the pendulum is displaced by an angle φ from its equilibrium position, it is acted on by a restoring torque $\boldsymbol{M}$ of magnitude $M = -mgs \sin \varphi$. The moment of inertia about the axis of rotation is

$$I_a = I_c + ms^2 = m(k^2 + s^2),$$

where k is the radius of gyration about the center of mass. The angular acceleration is therefore given by

$$-mgs \sin \varphi = I_a\alpha = m(k^2 + s^2)\alpha, \quad \ddot{\varphi} = \alpha = -\frac{gs \sin \varphi}{k^2 + s^2}.$$

The solution for small φ is $\varphi = \varphi_0 \sin \Omega t$, where $\Omega = \sqrt{(mgs/I_a)}$.

A *mathematical* or *ideal pendulum* consists of a point mass swinging on the end of weightless rod or cord. Let us compute the length l_0 of a mathematical pendulum having the same mass and angular acceleration as the

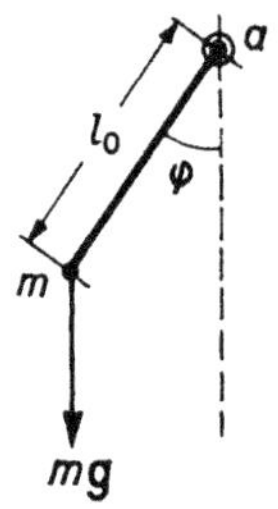

FIG. 4-12. A mathematical pendulum.

above physical pendulum. For the mathematical pendulum (Fig. 4-12), the restoring torque is

$$M' = -mgl_0 \sin \varphi,$$

the moment of inertia is

$$I'_a = ml_0^2,$$

and the angular acceleration is

$$\alpha = M'/I'_a = -g \sin \varphi / l_0.$$

Equating the angular accelerations, we get

$$l_0 = \frac{k^2 + s^2}{s} = \frac{I_a}{ms}. \tag{45-2}$$

The length l_0 is called the ***equivalent length*** of the physical pendulum. It is the length of a mathematical pendulum having the same period as the physical pendulum. If its equivalent length l_0 is known, the period of a physical pendulum for oscillations of small amplitude can be computed. We have

$$T_{\text{phys pend}} = 2\pi \sqrt{(l_0/g)} = 2\pi \sqrt{[(k^2 + s^2)/gs]} = 2\pi \sqrt{(I_a/mgs)},$$

where mg is the weight of the physical pendulum. This relation can also be used to measure experimentally the moment of inertia of a body of irregular shape.

We see from the preceding equation that the period T, for a given pendulum, depends only on the distance s from the axis of rotation to a parallel axis through the center of mass:

$$T = 2\pi \sqrt{[(k^2 + s^2)/gs]}. \tag{45-3}$$

Since this equation is quadratic in s, it follows that there are two values of s which will result in the same period. That is, for every axis (a) at a distance s_a from the center of mass, there corresponds a second axis (b) at a distance s_b for which the period is the same. It is easy to see from Eqn. (45-3) that if s_a is one solution, then $s_b = k^2/s_a$ is also a solution. From Eqn. (45-2), the sum of the two distances is

$$s_a + s_b = s_a + \frac{k^2}{s_a} = \frac{s_a^2 + k^2}{s_a} = l_0,$$

and hence equals the equivalent length of the pendulum.

A pendulum can be used to measure the acceleration due to gravity, $\boldsymbol{g}$. This must necessarily be a physical pendulum, since the mathematical pendulum is only an idealization. To make a determination of g, the equivalent length of the pendulum must be known. *Calculation* of the length is impractical, partly because of the complexity of the problem, but, more important, because of uncertainty with regard to the homogeneity of the pivoted body. It is therefore necessary to measure l_0 experimentally. This is done by using a so-called *reversible* pendulum. This is a body to which are attached two movable knife edges. The pendulum is pivoted first about one knife edge and then about the other, and the distance between them is adjusted until the period about each is the same. If the knife edges both lie on a straight line through the center of gravity, the distance between them equals the equivalent length l_0. It is therefore possible to measure l_0 with great accuracy, since the equality or inequality of the two periods can be determined with great precision. The reversible pendulum therefore makes possible very accurate measurements of $\boldsymbol{g}$. The classical measurements of $\boldsymbol{g}$ (1811–1831) by J. H. F. Bohnenberger, H. Kater, and F. W. Bessel are landmarks in the domain of high-precision measurements.

45-3 Rotation about a Fixed Axis with a Constant Frictional Torque

Suppose that a rotationally symmetrical body is given an initial angular velocity ω_0 about a fixed axis through its center of gravity, and is acted on thereafter by a constant frictional torque $\boldsymbol{M}_f$. The angular acceleration of the body is

$$\alpha = -M_f/I_s = \text{const}.$$

If t is the time required for the body to come to rest ($\omega_t = 0$), then

$$\omega_t = 0 = \omega_0 + \alpha t, \quad \text{and} \quad t = -\omega_0/\alpha = \omega_0 I_s/M_f.$$

This relation can be used for an experimental measurement of frictional torque.

The discussion in this section has been restricted to rotation about a fixed axis. In Section 46 we shall consider the application of the equation $M = I\alpha$ to rotation about axes that are not fixed. We shall show that this equation can also be applied to motion about an axis moving with translatory motion, provided only that the distance of the moving axis from the center of gravity remains constant. Before going through this derivation, we first give examples of some applications to motion in which this condition is fulfilled.

45-4 Rolling on an Inclined Plane

A cylinder of mass m and radius r rolls with a coefficient of static friction μ_0 on a plane making an angle θ with the horizontal (Fig. 4-13).

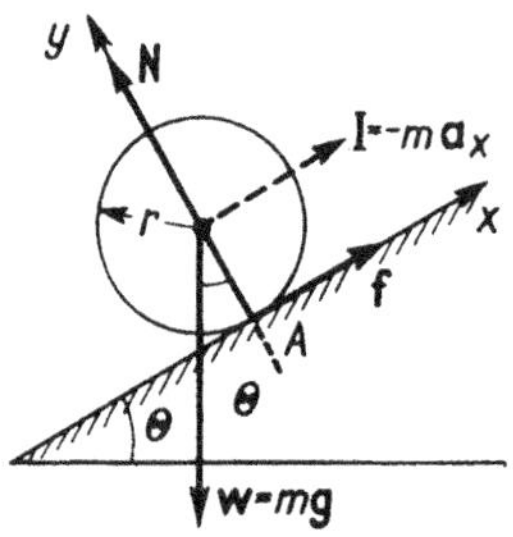

FIG. 4-13. Rolling cylinder on an inclined plane.

The motion of the center of mass in the x-direction is given by the second law in d'Alembert form (using the inertial force, $I = -ma_x$),

$$\sum F_{x_i} = 0 = -mg \sin\theta + f + ma_x, \quad f = mg \sin\theta - ma_x,$$

where f is the frictional force. In the y-direction, we have

$$\sum F_{y_i} = 0 = N - mg \cos\theta, \quad N = mg \cos\theta.$$

The law of rotational motion with respect to an axis through the center of mass gives us α_c, and hence $a_x(a_x = r\alpha_c)$, from $M_c = I_c\alpha_c$, where $M_c = fr$. The friction force f acting at the line of contact A is the cause of the angular acceleration. If there were no friction, there would be no torque, and the cylinder would slide without rolling.

The moment of inertia I_c of a cylinder is

$$I_c = \tfrac{1}{2}mr^2.$$

It follows that

$$\alpha_c = 2fr/mr^2 = 2f/mr,$$

and hence $a_x = 2f/m$. Inserting this expression for a_x in the equilibrium equation for the x-direction, we obtain $f = mg \sin\theta - 2f$, that is,

$$f = \tfrac{1}{3}mg \sin\theta.$$

The acceleration a_x is therefore

$$a_x = 2f/m = \tfrac{2}{3}g \sin\theta .$$

The factor $\frac{2}{3}$ shows that the translational motion down the inclined plane is reduced by the simultaneous rolling of the cylinder.

Now we will obtain this same result from consideration of the motion about on axis that is itself in translatory motion. This axis is the line of contact, A, between the cylinder and the plane, and is called the *instantaneous axis of rotation*. The moment about this axis is $M_A = mgr \sin\theta$. By the parallel axis theorem, the moment of inertia about A is

$$I_A = \tfrac{1}{2}mr^2 + mr^2 = \tfrac{3}{2}mr^2 .$$

Then the angular acceleration is

$$\alpha_A = M_A/I_A = (2g/3r) \sin\theta ,$$

and the translational acceleration of the instantaneous axis of rotation and hence also of the center of mass is

$$a_x = \alpha_A r = \tfrac{2}{3}g \sin\theta ,$$

as found above.

In the limiting case of rolling without slipping, the frictional force is fully developed. Hence for rolling without slipping,

$$mg \sin\theta/3 \leqq \mu_0 mg \cos\theta , \quad \text{or} \quad \tan\theta \leqq 3\mu_0 .$$

Thus while a flat object does not slide down an inclined plane until the slope angle is increased to a value given by $\tan\theta = \mu_0$, a rolling cylinder will not start to slip until the tangent of the angle of inclination is three times as great. For a sphere ($I_c = \frac{2}{5} mr^2$) the limiting angle is given by $\tan\theta = \frac{7}{2}\mu_0$.

45-5 Rolling of a Sphere on a Cylinder

In Fig. 4-14, a sphere is released from rest at the highest point of a cylindrical surface. It rolls down the surface until a certain point is reached at which it leaves the surface. The position of this point can be found from the condition that when it is reached, the normal force N exerted on the sphere by the surface becomes zero.

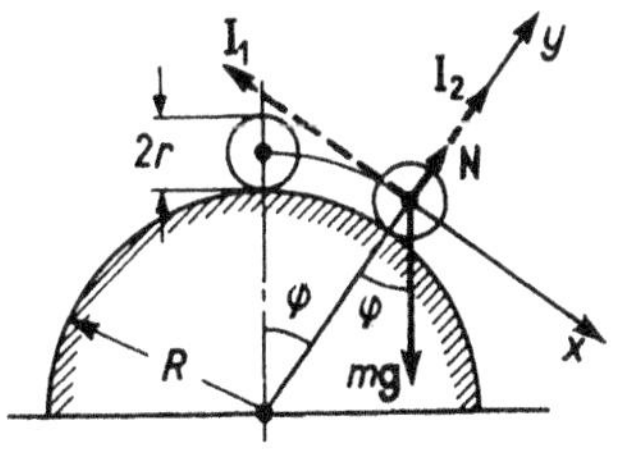

FIG. 4-14. Sphere rolling on a cylinder.

Let m be the mass of the rolling body, r its radius, and R the radius fo the cylindrical surface. Consider the motion as one of d'Alembert equilibrium. The actual forces on the sphere are its weight w and the force N exerted on it by the surface. The inertial force can be resolved into a component $I_1 = ma_t$, tangent to the path, and a component $I_2 = mv^2/(R + r)$, normal to the path. Choosing x- and y-axes as shown in Fig. 4-14, we obtain the following equation for the y-direction:

$$\sum F_y = 0 = -mg\cos\varphi + \frac{mv^2}{R + r} + N,$$

$$N = mg\cos\varphi - \frac{mv^2}{R + r}. \qquad (45\text{-}4)$$

The point at which the sphere leaves the surface cannot be found from this equation alone, since both φ and v are unknown. However, the energy equation can be used to obtain the velocity v.

Case A. Consider first a particle of radius $r = 0$, which slides on the surface without friction. From the principle of conservation of energy, we have

$$mgR(1 - \cos\varphi) = \tfrac{1}{2}mv^2, \quad v = \sqrt{[2gR(1 - \cos\varphi)]}.$$

Inserting the expression for v in Eqn. (45-4), we get (since $r = 0$)

$$N = mg[\cos\varphi - 2(1 - \cos\varphi)] = mg(3\cos\varphi - 2).$$

Hence when $N = 0$,

$$\cos\varphi = \tfrac{2}{3}, \quad \varphi = 48°11'.$$

The particle therefore leaves the surface when $\varphi = 48°11'$.

Case B. Now consider a sphere which rolls down the surface without slipping. In applying the principle of conservation of energy, the rotational energy of the sphere must also be included. If I_c is the moment of inertia of the sphere about an axis through its center, then conservation of energy tells us that

$$mg(R + r)(1 - \cos\varphi) = \tfrac{1}{2}mv^2 + \tfrac{1}{2}I_c\omega^2.$$

Since $v = r\omega$ and $I_c = \frac{2}{5}mr^2$, this equation becomes

$$mg(R + r)(1 - \cos\varphi) = \tfrac{1}{2}mv^2 + \tfrac{1}{2}(\tfrac{2}{5}mr^2)(v^2/r^2) = \tfrac{7}{10}mv^2,$$

$$v^2 = \tfrac{10}{7}(R + r)g(1 - \cos\varphi).$$

Inserting this expression for v^2 in Eqn. 45-4, we get

$$N = mg\cos\varphi - \tfrac{10}{7}mg(1 - \cos\varphi) = (mg/7)(17\cos\varphi - 10),$$

and when $N = 0$,

$$\cos\varphi = \tfrac{10}{17}, \quad \varphi = 53°\,58'.$$

We see that the angle φ is independent of the radius of the sphere. No matter how small its radius, any sphere will leave the cylinder at an angle of about 54°. Because of the rotational energy of the sphere, the problem cannot be solved by the methods of particle mechanics.

45-6 The Maxwell Wheel (or Yo-Yo)

An axially symmetrical body of mass $m = 100$ gm consists of two disks connected by an axle of radius $r = 0.5$ cm, about which is wound a thin, flexible cord (Fig. 4-15). If the upper end of the cord is kept fixed, the body

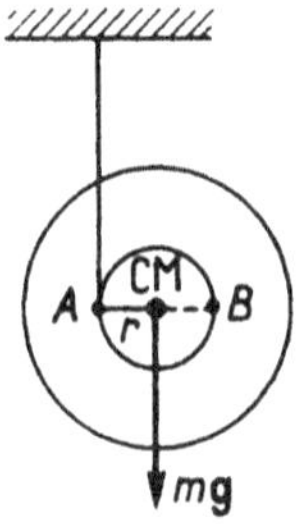

FIG. 4-15. The Maxwell wheel, or yo-yo.

rolls down the cord, reverses its motion, and then rolls up the cord. Frictional forces will be neglected. We wish to consider the kinematic and dynamic aspects of the motion, given that the radius of gyration, k, of the body is 1.5 cm and that the length of the cord is 1 m.

With respect to the instantaneous axis of rotation A, the moment of inertia and the torque are, respectively,

$$I_A = m(k^2 + r^2), \quad M_A = mgr.$$

The angular acceleration α about A, and the linear acceleration a_c of the center of mass are, respectively,

$$\alpha = \frac{M_A}{I_A} = \frac{gr}{k^2 + r^2}, \quad a_c = r\alpha = \frac{gr^2}{k^2 + r^2}.$$

The velocity of the center of mass can be expressed in terms of the distance of descent l and the acceleration a_c [see Eq. (20-2)]:

$$v_t^2 = 2a_c l, \quad v_t = r\sqrt{[2gl/(k^2 + r^2)]}.$$

The angular velocity is therefore

$$\omega_t = v_t/r = \sqrt{[2gl/(k^2 + r^2)]}.$$

When the lowest point of the path is reached, the velocity reverses from v_t to $-v_t$. Figure 4-16 illustrates a number of stages of this process. The impulse, p, of the force exerted on the body, during the reversal of its motion, is $p = 2mv_t = F\Delta t$. To a first approximation, the time interval Δt can be considered as the time required for the body to rotate through an angle π

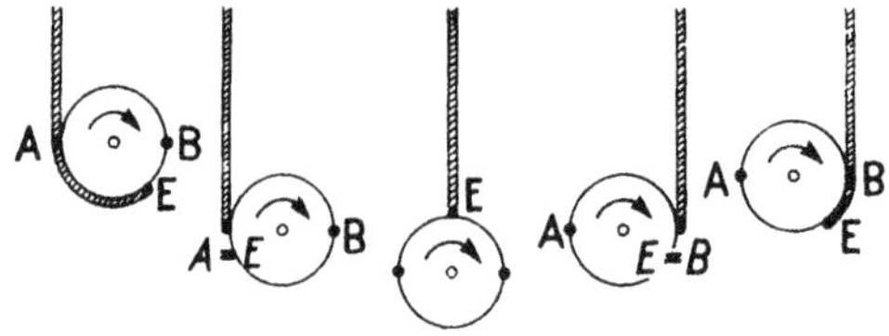

FIG. 4-16. Several steps in the velocity reversal of a Maxwell wheel.

with an angular velocity ω_t:

$$\Delta t = \pi/\omega_t .$$

The impulsive force F which the cord must exert is therefore

$$F = \frac{2mv_t\,\omega_t}{\pi} = \frac{4mgrl}{\pi(k^2 + r^2)} .$$

Introducing the numerical values given above, we find

$$\omega_t = 280 \text{ rad/sec},$$
$$\alpha = 196.2 \text{ rad/sec}^2,$$
$$a_c = r\alpha = 98.1 \text{ cm/sec}^2,$$
$$\Delta t = \pi/\omega_t = 1.12 \times 10^{-2} \text{ sec},$$
$$F = 2.50 \times 10^6 \text{ dynes} = 25 \text{ n}.$$

The velocities v_t and ω_t could have been found more simply from the law of conservation of energy. With respect to the axis A, we have at the lowest point in the motion,

$$mgl = \tfrac{1}{2}I_A\omega_A^2 = \frac{m(k^2 + r^2)}{2}\frac{v_t^2}{r^2}, \quad v_t = r\sqrt{[2gl/(k^2 + r^2)]}.$$

The law of rotational motion, in the form in which we have used it, is not applicable to all cases in which a body rotates about a moving axis. To determine the conditions under which this equation can be used, it is necessary to derive the general law of rotational motion. We shall then see under what conditions the special form can be applied and when it is necessary to use the more general relation. The general law will be needed for a study of the motion of a top and of a gyroscope.

46. The General Law of Rotational Motion

In the preceding sections, the angular momentum of a rigid body was defined with respect to an axis of rotation fixed in space, and a relation was obtained between the angular momentum and the moments of the external forces acting on the body. In this section, we shall first extend this relation to the general motion of a system of particles and then consider in more detail the method of applying the result to the plane motion of a rigid body.

46-1 *Angular Momentum with Respect to a Point*

Thus far, we have discussed only the angular momentum of a system with respect to a fixed axis of rotation. It is now necessary to define angular momentum with respect to a point. This is readily done for a single particle of a system, of mass m_k. Its position in space is defined by the position vector $\boldsymbol{r}_k$ from a fixed point O (Fig. 4-17). Its velocity is then $\boldsymbol{v}_k = d\boldsymbol{r}_k/dt$, and its momentum is $\boldsymbol{p}_k = m_k\boldsymbol{v}_k$.

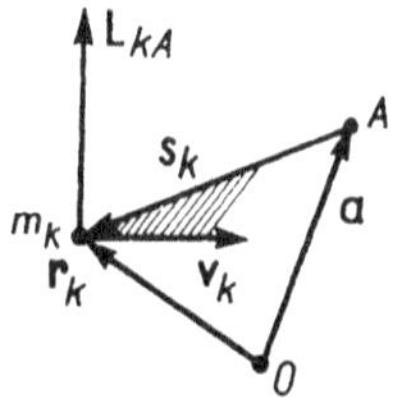

FIG. 4-17. Angular momentum of m_k relative to a point A.

Let A be any arbitrary point of space, at rest or in motion, whose position vector relative to O is $\boldsymbol{a}$. The vector from A to m_k will be designated by $\boldsymbol{s}_k$. Then

$$\boldsymbol{r}_k = \boldsymbol{a} + \boldsymbol{s}_k. \tag{46-1}$$

The *angular momentum* or *moment of momentum* of the particle with respect to A is now defined as the vector

$$\boldsymbol{L}_{kA} = \boldsymbol{s}_k \times \boldsymbol{p}_k = \boldsymbol{s}_k \times m_k\boldsymbol{v}_k = \boldsymbol{s}_k \times m_k(d\boldsymbol{r}_k/dt). \tag{46-2}$$

The angular momentum vector $\boldsymbol{L}_{kA}$ is perpendicular to the plane of $\boldsymbol{s}_k$ and $\boldsymbol{v}_k$.

The angular momentum $\boldsymbol{L}_A$ of a system of particles, relative to the point A, is defined as the vector sum of the angular momenta of the particles making up the system, relative to point A:

$$\boldsymbol{L}_A = \sum_k \boldsymbol{L}_{kA}.$$

This is a sum of vectors, all of which are, in general, different in magnitude and direction. The magnitude and direction of $\boldsymbol{L}_A$ can be determined only when the positions and velocities of all particles making up the system are known. Like the moment of a set of forces, the vector $\boldsymbol{L}_A$ (the moment of the set of momenta) depends on the choice of the reference point A.

46-2 *The Angular Momentum Law with Respect to a Moving Reference Point*

To obtain the relation between the angular momentum and the moments of the external forces, we differentiate the defining equation, Eqn. (46-2), with respect to the time. We obtain

$$\frac{d\boldsymbol{L}_{kA}}{dt} = \frac{d\boldsymbol{s}_k}{dt} \times \boldsymbol{p}_k + \boldsymbol{s}_k \times \frac{d\boldsymbol{p}_k}{dt}.$$

From Eqn. (46-1), $d\boldsymbol{s}_k/dt = d\boldsymbol{r}_k/dt - d\boldsymbol{a}/dt$, so

$$\frac{d\boldsymbol{L}_{kA}}{dt} = \frac{d\boldsymbol{r}_k}{dt} \times \boldsymbol{p}_k - \frac{d\boldsymbol{a}}{dt} \times \boldsymbol{p}_k + \boldsymbol{s}_k \times \frac{d\boldsymbol{p}_k}{dt}. \tag{46-3}$$

The first term on the right vanishes, since the vector $\boldsymbol{p}_k$, equal to $m_k\boldsymbol{v}_k$, is parallel to the vector $d\boldsymbol{r}_k/dt$ which equal $\boldsymbol{v}_k$. Furthermore, the vector $d\boldsymbol{p}_k/dt$, in the last term, equals the resultant force $\boldsymbol{R}_k$ acting on the particle ($\boldsymbol{R}_k$ is the resultant of all *external and internal* forces acting on m_k). The vector product $\boldsymbol{s}_k \times \boldsymbol{R}_k$ is the moment of $\boldsymbol{R}_k$ with respect to A, which we denote by $\boldsymbol{M}_{kA}$. Equation (46-3) then becomes

$$\frac{d\boldsymbol{L}_{kA}}{dt} = \boldsymbol{M}_{kA} - \frac{d\boldsymbol{a}}{dt} \times \boldsymbol{p}_k. \tag{46-4}$$

Writing the corresponding equations for all particles of the system and summing over all particles, we get

$$\sum_k \frac{d\boldsymbol{L}_{kA}}{dt} = \sum_k \boldsymbol{M}_{kA} - \sum_k \frac{d\boldsymbol{a}}{dt} \times \boldsymbol{p}_k.$$

Interchanging the order of summation and differentiation in the term on the left, we see that this term is simply the time rate of change of the total angular momentum of the system. Since the internal forces cancel in pairs (by Newton's third law), the first term on the right reduces to the moment of the external forces with respect to A, which we shall designate as $\boldsymbol{M}_A$. In the last term, the derivative $d\boldsymbol{a}/dt$ is a common factor in each term of the sum, so this term can be written as

$$\frac{d\boldsymbol{a}}{dt} \times \sum \boldsymbol{p}_k = \frac{d\boldsymbol{a}}{dt} \times \boldsymbol{p},$$

where $\boldsymbol{p}$ is the total momentum of the system, defined as the vector sum of the momenta of all of its particles. We therefore have

$$\frac{d\boldsymbol{L}_A}{dt} = \boldsymbol{M}_A - \frac{d\boldsymbol{a}}{dt} \times \boldsymbol{p}. \tag{46-5}$$

This relation is called the angular momentum law for the system with respect to any point A, fixed or moving. It expresses a relation between the time rate of change of angular momentum with respect to A, the moment of all external forces with respect to A, the total momentum $\boldsymbol{p}$, and the velocity $d\boldsymbol{a}/dt$ of the reference point A.

It should be noted that in this section the coordinate system of the observer is always that of the fixed point O. By the "reference point" we shall mean any point that may be selected for convenient calculation of the moments of the forces and the angular momentum. This avoids introducing the complication of a moving coordinate system.

It may be inquired whether there is any utility in introducing a moving reference point. We have done so because there are two cases in which this greatly simplifies the calculation of angular momentum and of torque:

(a) The *center of mass* of the system is selected as the reference point. We shall consider this case shortly.

(b) The introduction of the so-called *instantaneous axis* is helpful in formulating the angular momentum law for certain special cases of plane motion of a rigid body. We shall return to this question in Section 46-6.

46-3 The Angular Momentum Law with Respect to the Center of Mass

Let us select the center of mass CM of a system as the reference point A. The total momentum of the system is

$$\boldsymbol{p} = \sum_k m_k \boldsymbol{v}_k = \sum_k m_k \frac{d\boldsymbol{r}_k}{dt} = \frac{d}{dt}\sum_k m_k \boldsymbol{r}_k .$$

From the definition of the center of mass, Eqn. (42-1), the sum $\sum m_k \boldsymbol{r}_k$ is equal to $m\boldsymbol{r}_c$, where m is the total mass of the system and $\boldsymbol{r}_c$ is the position vector of the center of mass with respect to the point O. Therefore

$$\boldsymbol{p} = \frac{d}{dt}(m\boldsymbol{r}_c) = m\frac{d\boldsymbol{r}_c}{dt},$$

or, designating the velocity of CM by $\boldsymbol{v}_c$, $\boldsymbol{p} = m\boldsymbol{v}_c$. But, on the other hand, $d\boldsymbol{a}/dt$ is also equal to $\boldsymbol{v}_c$, since A was chosen to coincide with CM. Therefore in Eqn. (46-5), $d\boldsymbol{a}/dt \times \boldsymbol{p} = \boldsymbol{v}_c \times m\boldsymbol{v}_c = 0$, and this equation simplifies to

$$d\boldsymbol{L}_c/dt = \boldsymbol{M}_c \equiv \boldsymbol{M}_A . \tag{46-6}$$

This is called the angular momentum law for a system, with respect to the center of mass of the system. Writing Eqn. (46-6) in the form $d\boldsymbol{L}_c = \boldsymbol{M}_c dt$, we see that the change in angular momentum with respect to CM in the time dt is in the same direction as $\boldsymbol{M}_c$ and is of magnitude $M_c dt$. That is, the moment of the external forces attempts to set the angular momentum vector parallel to the torque vector (Fig. 4-18).

The angular momentum law with respect to the center of mass is most useful when a rigid body moves in such a way that none of its points remain at rest. With the help of the center of mass law, we first compute the motion of the center of mass. As the second step, using the angular momentum law with respect to the center of mass, we then determine the motion of the body about its center of mass.

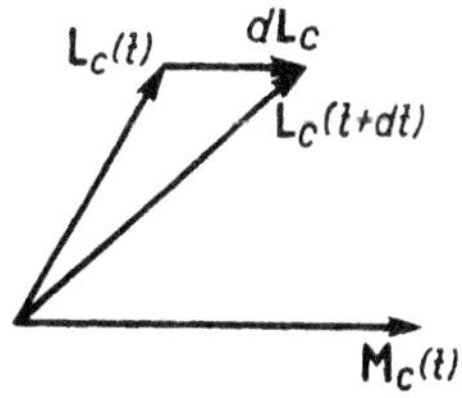

FIG. 4-18. The torque and the change in angular momentum are parallel vectors.

EXAMPLE. Let a rigid body be thrown in the air with a whirling motion. The path of the center of mass in vacuum is a parabola. Since no external torques act about the center of mass, the initial angular momentum vector remains constant in magnitude and direction.

46-4 Angular Momentum with Respect to a Fixed Point

In this case, the term $d\boldsymbol{a}/dt$ in Eqn. (46-5) is zero. Then with respect to a fixed point A,

$$d\boldsymbol{L}_A/dt = \boldsymbol{M}_A. \tag{46-7}$$

This equation is called the angular momentum law with respect to a fixed point. It has the same external form as the angular momentum law with respect to the center of mass, but of course the interpretation is different. The law can be applied in this form to any motion in which one point of a system remains at rest. This point is then selected as the reference point.

When the motion of a system can be computed from the center of mass law and the angular momentum law with respect to its center of mass, which is always possible for a rigid body, the formal agreement of Eqns. (46-6) and (46-7) shows that the motion *about* the center of mass is the same as if this point were at rest.

There are, of course, other points in addition to the center of mass, with respect to which Eqn. (46-5) simplifies to the form of Eqn. 46-6). For example, suppose the center of mass CM moves in a circle about the fixed point O, (Fig. 4-19). If we select as the reference point A any point on the position vector of CM, then $d\boldsymbol{a}/dt$ is always parallel to $\boldsymbol{v}_c$, that is, parallel to $\boldsymbol{p}$ and hence $d\boldsymbol{a}/dt \times \boldsymbol{p} = 0$. However, with the exception of the center of mass, and the instantaneous axis to be discussed in Section 46-6, such moving reference points are rarely of help in the solution of practical problems.

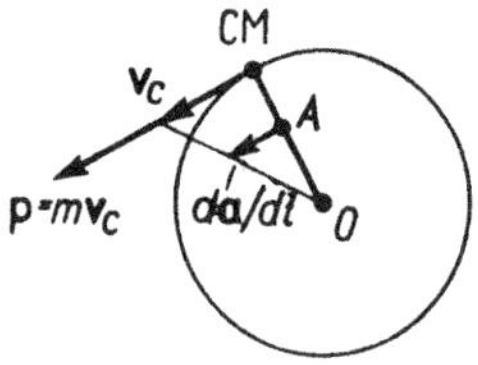

FIG. 4-19. A special reference point for which Eqn. (46-7) applies.

46-5 Angular Momentum with Respect to an Axis (*axial angular momentum*)

In formulating the general angular momentum law, the angular momentum was defined with respect to a point. In certain cases it is simpler to define the angular momentum with respect to a line, as we have already done for the special case of a rigid body rotating about a fixed axis. Between the two definitions there exists a purely geometric relation, which we shall now formulate. The result also applies to any arbitrary vector moment such as the moment of a force.

From point A of a straight line g (Fig. 4-20) construct the vector $\boldsymbol{r}$ to point B, and at B construct another vector $\boldsymbol{c}$. Construct through B the plane E_g containing the line g, and the plane E_n, normal to g. Now set up a coordinate system x, y, z, whose x-axis is the line of intersection of E_g and E_n,

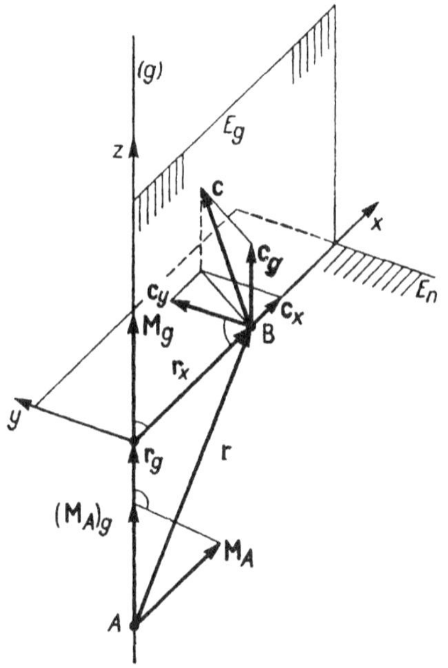

FIG. 4-20. Moment of a vector $\boldsymbol{c}$ relative to, (1) a point A, and (2) a line g.

whose y-axis is perpendicular to this line and lies in E_n, and whose z-axis coincides with the line g. The vectors $\boldsymbol{r}$ and $\boldsymbol{c}$ are now resolved into components parallel to these axes. Components of $\boldsymbol{r}$ are r_x, $r_y = 0$, and $r_z = r_g$. Components of $\boldsymbol{c}$ are c_x, c_y, and $c_z = c_g$.

DEFINITION. The *vectorial moment* $\boldsymbol{M}_A$ of the vector $\boldsymbol{c}$ with respect to the *point* A is defined as the vector product $\boldsymbol{M}_A = \boldsymbol{r} \times \boldsymbol{c}$.

The *axial moment* $\boldsymbol{M}_g$ of the vector $\boldsymbol{c}$ with respect to the *line* g is defined as the vector product $\boldsymbol{M}_g = \boldsymbol{r}_x \times \boldsymbol{c}_y$.

From Section 6-8, $\boldsymbol{M}_A$ is given by

$$\boldsymbol{M}_A = \begin{vmatrix} \boldsymbol{e}_x & \boldsymbol{e}_y & \boldsymbol{e}_z \\ r_x & 0 & r_g \\ c_x & c_y & c_g \end{vmatrix}$$

The z-component, $\boldsymbol{e}_z r_x c_y$, is therefore equal to $\boldsymbol{M}_g$, regardless of the location of A along g. If any number of vectors $\boldsymbol{c}_i$ are considered, the moment law (Chapter 2) applies.

We therefore have for the moment of any number of forces that the axial moment $\boldsymbol{M}_g$ of a set of forces, with respect to a line g through a point A, is equal to the component along g of the vectorial moment about A, $\boldsymbol{M}_A$:

$$M_g = (\boldsymbol{M}_A)_g.$$

For the angular momentum: the axial moment L_g of the momenta of all of the particles of a system, with respect to a line g through a point A, is equal to the component along g of the vectorial moment of momentum about A, $\boldsymbol{L}_A$:

$$L_g = (\boldsymbol{L}_A)_g.$$

In Section 41 it was shown that the component along any line of the time rate of change of a vector is not necessarily equal to the time rate of change of the component of the vector along that line. In the special case where the line moves with a motion of pure translation, however, the two are equal [see Eqn. (41-1)], since then $\boldsymbol{\omega} = 0$. Thus, if $\boldsymbol{\omega} = 0$, then $dL_g/dt = (d\mathrm{L}_A/dt)_g$.

Designating the g-component of the vector product in Eqn. (46-5) by $[(d\boldsymbol{a}/dt) \times \boldsymbol{p}]_g$, we can write Eqn. (46-5) as

$$\frac{dL_g}{dt} = M_g - \left(\frac{d\boldsymbol{a}}{dt} \times \boldsymbol{p}\right)_g. \tag{46-8}$$

This relation does not depend on the particular point on line g which is selected as the reference point A. It can be called the angular momentum law with respect to a line moving in pure translation. The vector $d\boldsymbol{a}/dt$ is then the velocity of this line.

If g passes through the center of mass CM, then, as explained earlier, the last term drops out and we have

$$dL_g/dt = M_g. \tag{46-9}$$

Thus, the same relation which we had in Eqns. (46-6) and (46-7) also applies to a component of the angular momentum along g, where g is a line through CM moving with pure translation. The last term in Eqn. (46-8) also vanishes if g is fixed in space. If g is *moving with respect to the body*, it is not advisable to use this relation, since the axial angular momentum and the axial moment are then not readily evaluated. However, if g is fixed in the body, that is, if the body rotates about g, we obtain the equation already derived (see Section 43), $I_g\alpha_g = M_g$, since in this case (Fig. 4-21)

$$L_g = \sum_k r_k m_k v_k = \omega_g \sum_k m_k r_k^2 = \omega_g I_g,$$

and

$$\frac{dL_g}{dt} = \frac{d}{dt}(I_g\omega_g) = I_g\frac{d\omega_g}{dt} = I_g\alpha_g.$$

It may be remarked that when a rigid body rotates about an axis fixed in space, the *vectorial* angular momentum with respect to a point on the axis is parallel to the axis only when the distribution of mass about the axis has the necessary degree of symmetry. In general, it does not lie along the axis, and rotates with the body about the axis. (One might think of a triangular plate rotating about one of its edges, and consider the angular momentum relative to one of the corners lying on the axis.) The time variation of the angular momentum vector in this case is not parallel to the axis of rotation. Then, using Eqn. (46-7), we see that the resultant moment (relative to the same corner) of all the forces acting on the body has a component perpendicular to the axis of rotation. This component must be provided by the reaction forces of the bearings on the body.

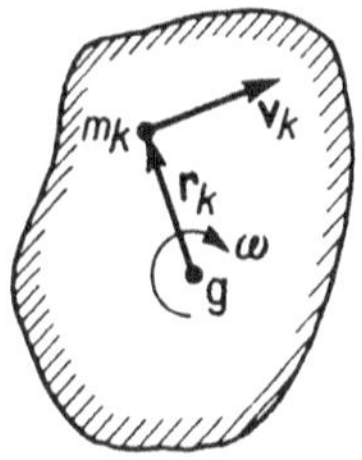

FIG. 4-21. Angular momentum for rotation about a body-fixed axis, *g*.

46-6 Applications of the Angular Momentum Law to the Plane Motion of a Rigid Body

The motion of a rigid body is said to be *planar* if all particles in the body move parallel to a fixed plane (the plane of motion). Examples of plane motion are rotation about a fixed axis and the rolling of a cylinder.

For a rotation about a line c perpendicular to the plane of motion and passing through the center of mass CM, the angular momentum law takes the form of Eqn. (46-9). In plane motion the axial angular momentum with respect to c is readily computed. We consider first the special case of pure translation. The axial angular momentum with respect to c is then (Fig. 4-22)

$$L_c = \sum r_k m_k v_k .$$

Since all points have the same velocity $\boldsymbol{v} = \boldsymbol{v}_c$,

$$L_c = v_c \sum m_k r_k .$$

But from the definition of CM, $\sum m_k r_k = 0$ and hence $L_c = 0$.

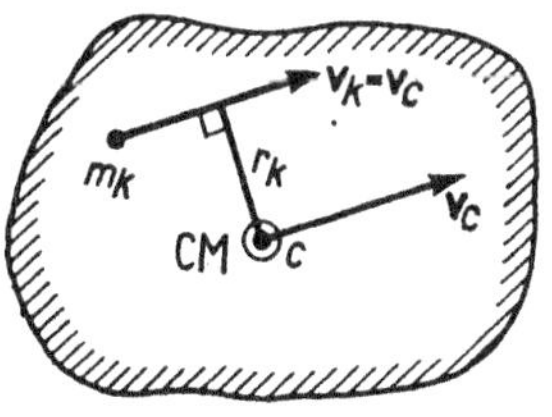

FIG. 4-22. Angular momentum in the case of pure translation.

The most general plane motion can be considered as a superposition of a translation, and a rotation about CM. The angular momentum $\boldsymbol{L}_c$ can be separated into two parts, of which the first, arising from the translation, is zero as shown above, while the second has the value $I_c \boldsymbol{\omega}_c$ (compare rotation about a fixed axis). We therefore have

$$\frac{d(I_c \boldsymbol{\omega}_c)}{dt} = I_c \frac{d\boldsymbol{\omega}_c}{dt} = I_c \boldsymbol{\alpha}_c = \boldsymbol{M}_c.$$

For any plane motion of a rigid body there exists a so-called *instantaneous axis of rotation*; that is, a line in space about which the body is instantaneously rotating. Subject to certain additional assumptions about the motion, which will be given later, the angular momentum law can be applied to this instantaneous axis of rotation in the same form as is used for a space-fixed axis. But first we will show that such an instantaneous axis actually exists.

We consider first a simple kinematical example. A hoop rolls in its own plane, without slipping, along a straight line a (Fig. 4-23). A point P on its

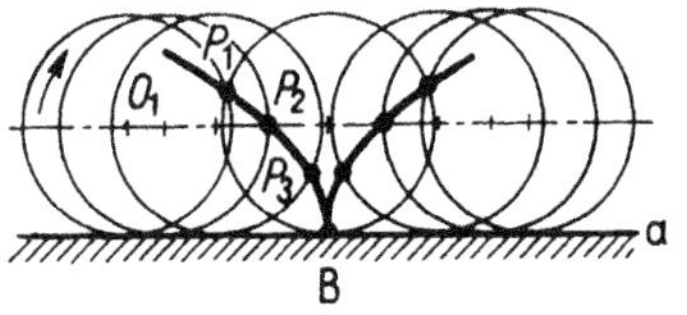

FIG. 4-23. A hoop rolling on a straight line.

periphery then moves along a cycloid. When P is at a cusp B of the cycloid, it is momentarily at rest relative to a. At this instant, therefore, the motion of the hoop is a rotation about the point of contact. The point of contact is therefore the *instantaneous center of rotation* for the motion. As the motion proceeds, this point moves along the line a. At every instant, a different point of the hoop makes contact with a and is at rest relative to a.

If a rigid wheel or cylinder rolls on a plane without slipping, it makes contact with the plane along a line in its surface. This line then becomes the *instantaneous axis of rotation*, and as the motion proceeds it is displaced along the surface with a motion of pure translation.

In the general case of plane motion, there also exists at every moment an instantaneous axis. This can be shown as follows. Consider any two points A and B (Fig. 4-24) in a plane parallel to the plane of motion, and construct at each point a line perpendicular to the velocity of the point. What is the velocity $\boldsymbol{v}_D$ of the point D at which these lines intersect? Since AD is perpendicular to $\boldsymbol{v}_A$, then in a time interval Δt the particle moves to a position A', where $A'D = AD$ to first order in Δt. In other words, the instantaneous motion of A can be considered to be a circular motion with D as the center of the circle. Because the length of the segment AD is not changing, the components of $\boldsymbol{v}_A$ and $\boldsymbol{v}_D$ in the direction of AD must be equal, and must equal zero, since $\boldsymbol{v}_A$ is perpendicular to AD. Similarly, the length of BD does not change, and the component of $\boldsymbol{v}_D$ in the direction BD must equal zero. But since points A and B can always be chosen in such a way that A, B, and D do not lie on the same straight line, it follows that the components of $\boldsymbol{v}_D$ in two different directions are both zero, and hence that $\boldsymbol{v}_D$ itself is zero. All points on a line d through D, perpendicular to the plane of motion, are also at rest, and this line is therefore the instantaneous axis.

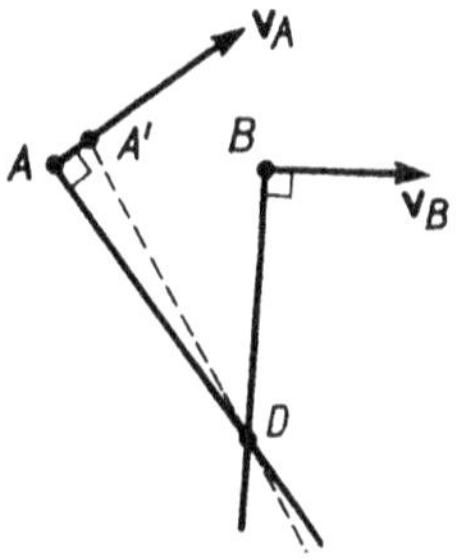

FIG. 4-24

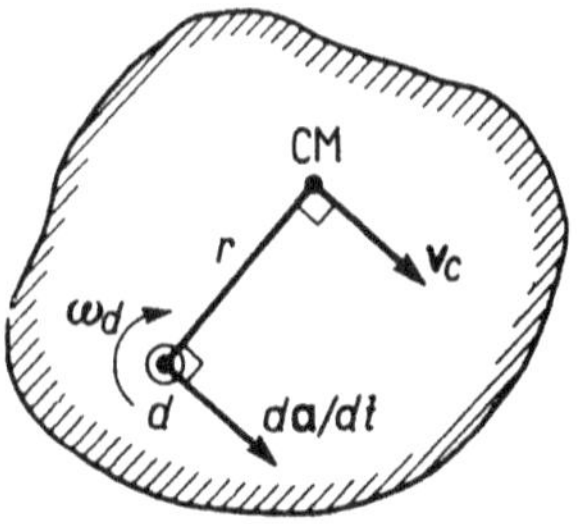

FIG .4-25

FIG. 4-24. Demonstration of the existence of instantaneous axis of rotation.

FIG. 4-25. Angular momentum in the case of plane motion.

The angular momentum law can now be expressed in the form of Eqn. (46-8), relative to the line d. Note that even though D and d are instantaneously at rest, the position of D and d is moving through the body in general, and hence the factor $d\boldsymbol{a}/dt$ in Eqn. (46-8) is not zero. However in certain special cases of plane motion, the last term in this equation vanishes (Fig. 4-25). For example, if the distance $\boldsymbol{r}$ of the center of mass from the line d is constant in magnitude (as in the case of a rolling wheel), then the velocity of the center of mass, $\boldsymbol{v}_c$, and the velocity, $d\boldsymbol{a}/dt$, of the instantaneous axis must have equal components in the direction of $\boldsymbol{r}$. Since $\boldsymbol{v}_c$ is perpendicular to $\boldsymbol{r}$, $d\boldsymbol{a}/dt$ must be perpendicular to $\boldsymbol{r}$ also. Hence $d\boldsymbol{a}/dt$ is parallel to $\boldsymbol{v}_c$ and to $\boldsymbol{p}$, and the vector product $(d\boldsymbol{a}/dt) \times \boldsymbol{p}$ is zero.

The axial angular momentum of the body with respect to d is $L_d = I_d\omega_d$. From the parallel axis theorem, $I_d = I_c + mr^2$, and for the motion described above (r = const) I_d is constant in time. It follows that

$$\frac{dL_d}{dt} = \frac{d}{dt}(I_d\omega_d) = I_d\frac{d\omega_d}{dt} = I_d\alpha_d,$$

and therefore from Eqn. (46-8), $I_d\alpha_d = M_d$.

To summarize, we can say: When a rigid body moves with plane motion, the relation $I_c\alpha_c = M_c$ holds for an axis c through its center of mass and perpendicular to the plane of motion. If the motion is such that the distance from the center of mass to the instantaneous axis d remains constant, we also have

$$I_d \alpha_d = M_d. \tag{46-10}$$

This justifies the statement made previously in Section 45-3.

EXAMPLE. A homogeneous ladder of length $2a$ leans against a vertical wall, as in Fig. 4-26. Frictional forces are negligible. The ladder starts to slide, and as long as its

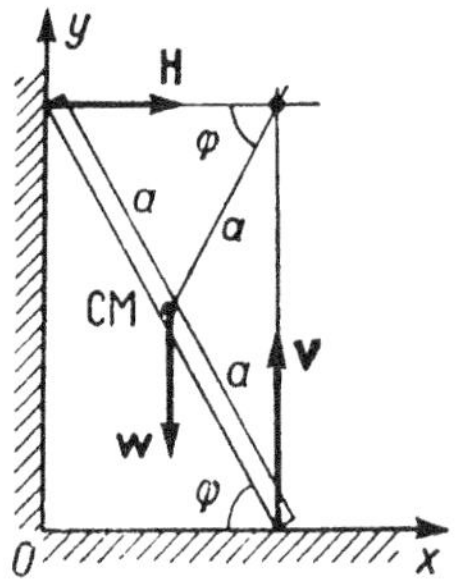

FIG. 4-26. Sliding ladder.

upper end remains in contact with the wall, the instantaneous axis of rotation passes through the point of intersection of the lines of action of the reaction force H and V. From the diagram, it is evident that the distance of the axis from the center of mass is constant. The motion can then be completely described by an angle of inclination, φ, and from Eqn. (46-10),

$$(I_c + ma^2)(d^2\varphi/dt^2) = -mga \cos \varphi.$$

The function $\varphi(t)$ can be determined from this differential equation.

The physics of the problem is more evident, however, if we do not make use of the concept of an instantaneous axis. The following equations are necessary. Angular momentum law for the center of mass:

$$I_c \frac{d^2\varphi}{dt^2} = Ha \sin \varphi - Va \cos \varphi. \tag{46-11}$$

Center of mass law:

$$m \frac{d^2 x_c}{dt^2} = H,$$

$$m \frac{d^2 y_c}{dt^2} = V - w. \tag{46-12}$$

Geometric conditions:

$$x_c = a \cos \varphi, \quad y_c = a \sin \varphi. \tag{46-13}$$

To obtain the equation for $\varphi(t)$, we compute H and V as functions of φ from Eqns. (46-12) and (45-13), and insert these expressions in Eqn. (46-11).

47. ELEMENTARY THEORY OF THE TOP

A body that moves in such a way that one of its points remains fixed in space is called a top. According to the laws of center of mass and of angular momentum, any arbitrary motion of a rigid body can be considered as a superposition of a motion *of* its center of mass and a *rotation about* its center of mass. The latter motion is the same *as if* the center of mass remained at rest and can be investigated independently of the motion of the center of mass. The molecules of a gas can be thought of as tops in this sense. In general, however, the fixed point of a top is not its center of mass. We shall therefore study the rotational motion of a top with an arbitrary fixed point, since this includes the case of a fixed center of mass as a special case.

Frequently, one point of a top makes contact with a supporting surface. This point ordinarily does not remain fixed, as for example in the case of a toy top set spinning on a smooth surface. In spite of this, it is legitimate to speak of the body as a top. To do this, we must consider separately the motion of the center of mass and the rotation about the center of mass. The rotational motion is determined by the angular momentum law. If a varying torque is exerted on the body by external forces (such as the force exerted by the supporting surface, by friction, and by the weight of the body), the angular momentum of the body changes accordingly. Its rate of change is given by Eqn. (46-5):

$$\frac{d\boldsymbol{L}_A}{dt} = \boldsymbol{M}_A - \frac{d\boldsymbol{a}}{dt} \times \boldsymbol{p}.$$

Because the vector product does not, in general, vanish, only a relatively small number of problems involving the motion of a top can be treated quantitatively using the methods of elementary mathematics. Even with the help of advanced mathematics, however, only approximate solutions can be obtained in the most general cases. But the properties of tops are so unusual that it is well worth the trouble that it takes to study them. It turns out be possible to demonstrate all of the typical properties of tops by considering only the simplest types. We shall confine ourselves to cases in which the vector product $(d\boldsymbol{a}/dt) \times \boldsymbol{p}$ is zero. However, as the following discussion will show, we must specialize even further than this if our elementary methods are to be sufficient.

To analyze the motion of a top, it is necessary to introduce three axes.

The figure axis. This term is applied to a top whose ellipsoid of inertia (see Section 47-1) with respect to a fixed point of space is an ellipsoid of rotation. The axis of rotation of the ellipsoid of inertia is then called the *figure axis* of the top. This case occurs when the distribution of mass of the top has rotational symmetry. The axis of symmetry then becomes the figure axis.

Instantaneous axis of rotation. If one point of the top is fixed in space, then at every instant there is an axis, passing through this point, about which the top is rotating. Its direction is that of the instantaneous angular velocity $\boldsymbol{\omega}$. This instantaneous axis of rotation is, in general, not fixed with respect to the top, and as result the motion of the top is a complicated one.

The angular momentum axis. The angular momentum axis is that of the angular momentum vector $\boldsymbol{L}$. From the angular momentum law for a fixed point ($d\boldsymbol{L}_A/dt = \boldsymbol{M}_A$), the time rate of change of the angular momentum vector equals the torque vector. The effect of an external torque is, in general, to change both the magnitude and direction of the angular momentum vector corresponding to the equation $d\boldsymbol{L} = \boldsymbol{M}dt$. The angular momentum is therefore the most important dynamic property of a top. The difficulty which we have in understanding the behavior of tops rests largely on the fact that angular momentum is not a quantity that we ordinarily conceive of intuitively.

Two preliminary remarks are necessary before we begin a discussion of some simple examples of the motions of tops.

47-1 Kinetic Energy of a Top. The Ellipsoid of Inertia

Let A be a point on the instantaneous axis of rotation. The velocity of the kth particle of the top is

$$\boldsymbol{v}_k = \boldsymbol{\omega} \times \boldsymbol{r}_k,$$

where $\boldsymbol{\omega}$ is the angular velocity about the instantaneous axis of rotation, and $\boldsymbol{r}_k$ is the radius vector from A to the particle (Fig. 4-27). The kinetic energy of the top is given by the sum

$$E_{\text{kin}} = \tfrac{1}{2}\sum m_k v_k^2.$$

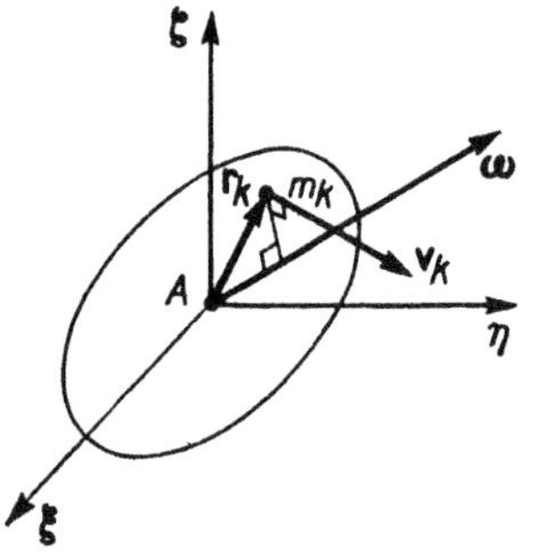

FIG. 4-27. The kinetic properties of a top.

Let the velocity $\boldsymbol{v}_k$ be resolved into components relative to a coordinate system fixed with respect to the top and having its origin at A. Let the components of $\boldsymbol{\omega}$ and $\boldsymbol{r}_k$ be represented respectively by ω_ξ, ω_η, ω_ζ and ξ_k, η_k, ζ_k. Then, as shown in Section 34, the components of $\boldsymbol{v}_k$ are

$$v_{k\xi} = \omega_\eta \zeta_k - \omega_\zeta \eta_k,$$

$$v_{k\eta} = \omega_\zeta \xi_k - \omega_\xi \zeta_k,$$

$$v_{k\zeta} = \omega_\xi \eta_k - \omega_\eta \xi_k,$$

and the kinetic energy of the top is given by

$$\begin{aligned}2E_{\text{kin}} &= \sum m_k(v_{k\xi}^2 + v_{k\eta}^2 + v_{k\zeta}^2) \\ &= \omega_\xi^2 \sum_k m_k(\eta_k^2 + \zeta_k^2) + \omega_\eta^2 \sum_k m_k(\zeta_k^2 + \xi_k^2) \\ &\quad + \omega_\zeta^2 \sum_k m_k(\xi_k^2 + \eta_k^2) - 2\omega_\xi\omega_\eta \sum_k m_k\xi_k\eta_k \\ &\quad - 2\omega_\xi\omega_\zeta \sum_k m_k\xi_k\zeta_k - 2\omega_\eta\omega_\zeta \sum_k m_k\eta_k\zeta_k.\end{aligned}$$

The sums of the right side of this equation have already been encountered in Section 44, where they were designated as the *moments of inertia* and the *products of inertia*. In terms of the symbols introduced in Section 44,

$$\begin{aligned}I_{\xi\xi} &= \sum m_k(\eta_k^2 + \zeta_k^2), \quad & I_{\eta\eta} &= \sum m_k(\zeta_k^2 + \xi_k^2), \\ I_{\zeta\zeta} &= \sum m_k(\xi_k^2 + \eta_k^2), \quad & I_{\xi\eta} &= \sum m_k\xi_k\eta_k, \\ I_{\eta\zeta} &= \sum m_k\eta_k\zeta_k, \quad & I_{\zeta\xi} &= \sum m_k\zeta_k\xi_k,\end{aligned}$$

we have

$$2E_{\text{kin}} = I_{\xi\xi}\omega_\xi^2 + I_{\eta\eta}\omega_\eta^2 + I_{\zeta\zeta}\omega_\zeta^2 - 2I_{\xi\eta}\omega_\xi\omega_\eta - 2I_{\eta\zeta}\omega_\eta\omega_\zeta - 2I_{\zeta\xi}\omega_\zeta\omega_\xi. \qquad (47\text{-}1)$$

On the other hand, the kinetic energy of the top can also be expressed in terms of its moment of inertia I with respect to the instantaneous axis, and the instantaneous velocity $\boldsymbol{\omega}$: $2E_{\text{kin}} = I\omega^2$.

Equating the right sides of the preceding equations, we get

$$I\omega^2 = I_{\xi\xi}\omega_\xi^2 + I_{\eta\eta}\omega_\eta^2 + I_{\zeta\zeta}\omega_\zeta^2 - 2I_{\xi\eta}\omega_\xi\omega_\eta - 2I_{\eta\zeta}\omega_\eta\omega_\zeta - 2I_{\zeta\xi}\omega_\zeta\omega_\xi. \qquad (47\text{-}2)$$

This equation gives the moment of inertia I of the top, for any arbitrary axis of rotation through the point A, as will now be shown. The moments of inertia for different axes of rotation can be represented geometrically by an ellipsoid. This so-called *ellipsoid of inertia* is obtained as follows.

Consider any arbitrary axis of rotation making angles α, β, and γ with the axes ξ, η, and ζ, and lay off on this axis a distance from the point A proportional to $1/\sqrt{I}$ (Fig. 4-28). This determines a point E on the rotation

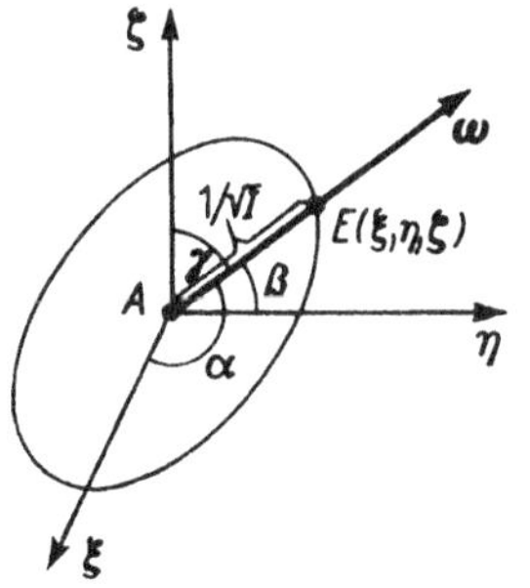

FIG. 4-28. Construction of the ellipsoid of inertia.

axis, having coordinates ξ, η, ζ, and whose distance from A is a measure of the moment of inertia about this axis. We now show that the locus of all such points is an ellipsoid. From the procedure by which E was located, it follows that

$$\cos\alpha = \xi\sqrt{I}, \quad \cos\beta = \eta\sqrt{I}, \quad \cos\gamma = \zeta\sqrt{I}. \tag{47-3}$$

When Eqn. (47-2) is divided through by ω^2, the quotients ω_ξ/ω, ω_η/ω, and ω_ζ/ω are also equal respectively to the cosines of the angles α, β, and γ. Equation (47-2) therefore takes the form

$$I = I_{\xi\xi}\cos^2\alpha + I_{\eta\eta}\cos^2\beta + I_{\zeta\zeta}\cos^2\gamma$$
$$-2I_{\xi\eta}\cos\alpha\cos\beta - 2I_{\eta\zeta}\cos\beta\cos\gamma - 2I_{\zeta\xi}\cos\gamma\cos\alpha.$$

Introducing from Eqn. (47-3) the expressions for the cosines of the angles, and dividing both sides by I, we get

$$1 = I_{\xi\xi}\xi^2 + I_{\eta\eta}\eta^2 + I_{\zeta\zeta}\zeta^2 - 2I_{\xi\eta}\xi\eta - 2I_{\eta\zeta}\eta\zeta - 2I_{\zeta\xi}\zeta\xi. \tag{47-4}$$

This is the equation of an ellipsoid and represents the previously mentioned *ellipsoid of inertia* of the top for the point A. From this ellipsoid, the moment of inertia of the top can be determined for an axis through the point A in any arbitrary direction. The distance from the midpoint A of the ellipsoid to the point of intersection of any axis with its surface is equal to $1/\sqrt{I}$, where I is the moment of inertia about the chosen axis. It is evident that the moment of inertia I is completely determined by the six quantities $I_{\xi\xi}$, $I_{\eta\eta}$, $I_{\xi\eta}$, etc.

From analytic geometry, we know that it is possible to rotate the coordinate system so that the equation of the ellipsoid assumes the normal form, in which the coefficients of the cross forms in ξ, η, and ζ are zero. (See Section 44-2 for a discussion of the two-dimensional case.) We shall define the *principal axes* of the body to be the three mutually perpendicular axes for which the ellipsoid of inertia is in normal form. By choosing the body fixed axes to coincide with the principal axes, we find that the products of inertia vanish and the equation of the ellipsoid becomes

$$I_1\xi^2 + I_2\eta^2 + I_3\zeta^2 = 1.$$

The ellipsoid of inertia is uniquely determined by the *principal moments of inertia*, I_1, I_2, and I_3. The largest and the smallest values of the moments of inertia are given by the principal moments of inertia. The largest moment of inertia corresponds to the shortest axis of the inertia ellipsoid, and the smallest moment of inertia corresponds to the longest axis.

It is possible for the ellipsoid of inertia to be an ellipsoid of revolution, or even a sphere. In the first case one principal axis is uniquely determined. All axes perpendicular to it are also principal axes. The moment of inertia about the latter axes are all equal, while for the axis of revolution of the ellipsoid, the moment of inertia is either a minimum (prolate ellipsoid) or a maximum (oblate ellipsoid). If the ellipsoid of inertia is a sphere, all axes through A are principal axes and the moment of inertia of the body is the same about all of these axes. These special cases result for bodies in which the distribution of mass has rotational or central symmetry. However, these conditions are not necessary for the degeneration of the inertia ellipsoid into a more symmetric figure.

Once the ellipsoid of inertia is known for the center of mass, all other moments of inertia can be calculated using the parallel axis theorem. The ellipsoid for the CM is called the *central ellipsoid.*

47-2 Relation between Angular Momentum and Angular Velocity

The angular momentum of a particle m_k of a top, with respect to a fixed point A is (Fig. 4-27)

$$\boldsymbol{L}_{A,k} = \boldsymbol{r}_k \times m_k \boldsymbol{v}_k = m_k(\boldsymbol{r}_k \times \boldsymbol{v}_k).$$

If $\boldsymbol{\omega}$ is the angular velocity about the instantaneous axis, the velocity $\boldsymbol{v}_k$ is $\boldsymbol{v}_k = \boldsymbol{\omega} \times \boldsymbol{r}_k$, and hence

$$\boldsymbol{L}_{A,k} = m_k[\boldsymbol{r}_k \times (\boldsymbol{\omega} \times \boldsymbol{r}_k)].$$

But from the properties of vector products,

$$\boldsymbol{r}_k \times (\boldsymbol{\omega} \times \boldsymbol{r}_k) = r_k^2 \boldsymbol{\omega} - (\boldsymbol{r}_k \cdot \boldsymbol{\omega})\, \boldsymbol{r}_k.$$

Therefore

$$\boldsymbol{L}_{A,k} = m_k r_k^2 \boldsymbol{\omega} - m_k(\boldsymbol{r}_k \cdot \boldsymbol{\omega})\, \boldsymbol{r}_k.$$

The total resultant angular momentum is obtained by summing over all particles:

$$\boldsymbol{L}_A = \boldsymbol{\omega} \sum m_k r_k^2 - \sum m_k(\boldsymbol{r}_k \cdot \boldsymbol{\omega})\, \boldsymbol{r}_k.$$

Let us again resolve $\boldsymbol{r}_k$ and $\boldsymbol{\omega}$ into components relative to axes ξ, η, ζ fixed with respect to the top. We then obtain for the ξ-component of $\boldsymbol{L}_A$

$$\begin{aligned} L_\xi &= \omega_\xi \sum m_k(\xi_k^2 + \eta_k^2 + \zeta_k^2) - \sum m_k(\xi_k\omega_\xi + \eta_k\omega_\eta + \zeta_k\omega_\zeta) \\ &= \omega_\xi \sum m_k(\eta_k^2 + \zeta_k^2) - \omega_\eta \sum m_k\xi_k\eta_k - \omega_\zeta \sum m_k\xi_k\zeta_k, \end{aligned}$$

and introducing the moments and products of inertia, we have

$$L_\xi = I_{\xi\xi}\omega_\xi - I_{\xi\eta}\,\omega_\eta - I_{\xi\zeta}\omega_\zeta. \tag{47-5}$$

Similarly,

$$L_\eta = -I_{\eta\xi}\omega_\xi + I_{\eta\eta}\omega_\eta - I_{\eta\zeta}\omega_\zeta, \tag{47-6}$$

$$L_\zeta = -I_{\zeta\xi}\omega_\xi - I_{\zeta\eta}\omega_\eta + I_{\zeta\zeta}\,\omega_\zeta. \tag{47-7}$$

These equations give the relation between the angular momentum vector $\boldsymbol{L}$ and the angular velocity vector $\boldsymbol{\omega}$. The equations are linear, but are not simple. The reason that no simple relations exist lies in the nature of the concept of moment of inertia.

For the special case in which the ξ-axis always coincides with the axis of $\boldsymbol{\omega}$, that is, if

$$\omega_\xi = \omega, \quad \omega_\eta = \omega_\zeta = 0,$$

we obtain the familiar expression for rotation about a fixed axis, $L_\xi = I_{\xi\xi}\omega$.

47-3 The Symmetrical Top. Euler's Equations

Only in a few special cases can the motion of a top about its point of contact with a supporting surface be computed by elementary methods. One such case occurs, for example, when the inertia ellipsoid of the top, with respect to the point of contact A, is an ellipsoid of rotation. The top is then said to be *symmetrical.* In the practical applications of tops, as in gyroscopes for example, use is made almost exclusively of mass distributions which are axially symmetric and for which the axis of rotation is that of the largest moment of inertia. In discussing the symmetrical top, it is convenient to introduce a rectangular coordinate system ξ, η, ζ, fixed with respect to the top, and whose axes coincide at point A with the principal axes of inertia of the top (Fig. 4-29). Let the figure axis f coincide with the ζ-axis. Then $I_{\xi\xi} = I_{\eta\eta}$, and, for brevity, let

$$I' = I_{\xi\xi} = I_{\eta\eta}, \quad I_f = I_{\zeta\zeta}.$$

The vector $\boldsymbol{\omega}$ has components, in this system, of ω_1, ω_2, and $\omega_3 = \omega_f$. The instantaneous state of motion of the top is a superposition of three rotations about the principal axes. To each of these partial rotations there corresponds a partial angular momentum, the vector sum of which equals the resultant angular momentum $\boldsymbol{L}_A$ of the top. The partial angular momentum which corresponds to ω_3 is obtained by setting $\omega_1 = \omega_2 = 0$ in Eqns. (47-5), (47-6), and (47-7). Also, since our coordinate axes are principal

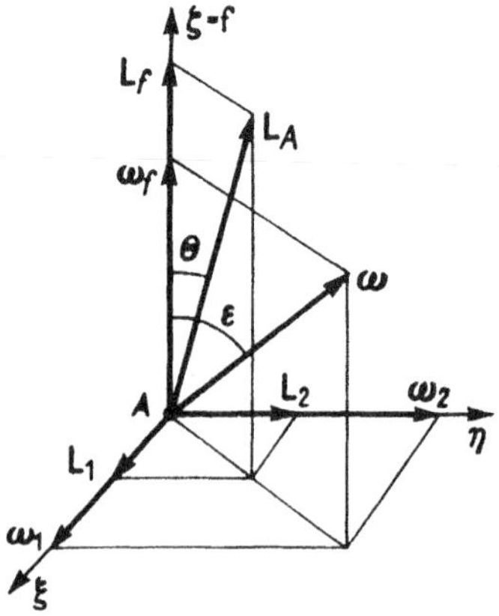

FIG. 4-29. The relative positions of the figure axis, angular momentum vector, and instantaneous axis of rotation.

axes, the products of inertia vanish and the partial angular momentum corresponding to ω_f lies along the f-axis. Both of the other partial angular momenta also lie along the corresponding principal axes. This justifies the statement made in Section 46-5 that only when a body is dynamically balanced will the angular momentum with respect to a point on the fixed axis of rotation lie along this axis, i.e., only when the axis of rotation is a principal axis of inertia.

Since the partial angular momenta lie along the coordinate axes, they give directly the rectangular components of the total angular-momentum. From Eqns. (47-5), (47-6) and (47-7), their respective magnitudes are

$$L_1 = \omega_1 I', \quad L_2 = \omega_2 I', \quad L_3 = L_f = \omega_f I_f. \tag{47-8}$$

The ξ- and η-components of $\boldsymbol{L}_A$ are proportional to the corresponding components of $\boldsymbol{\omega}$, and therefore f, $\boldsymbol{\omega}$, and $\boldsymbol{L}_A$ lie in the same plane (see Fig. 4-29). This fact will be very useful to us in describing the motion of the top. However, since $I_f \neq I'$, the direction of $\boldsymbol{L}_A$ does not coincide with that of $\boldsymbol{\omega}$.

To determine the motion of the top, it suffices, as explained in Section 46, to set up the law of angular momentum with respect to the point A. We shall do this in component form, with respect to the axes ξ, η, and ζ, which are fixed with respect to the top and which, of course, rotate with it. We consider this as our moving system P (in the sense of Section 40), moving with respect to a coordinate system O fixed in space. From Eqn. (41-2) of Section 41, we have for any arbitrary vector $\boldsymbol{r}$,

$$\left(\frac{d\boldsymbol{r}}{dt}\right)_O = \left(\frac{d\boldsymbol{r}}{dt}\right)_P + \boldsymbol{\omega} \times \boldsymbol{r}.$$

Hence, for the angular momentum vector $\boldsymbol{L}_A$,

$$\left(\frac{d\boldsymbol{L}_A}{dt}\right)_O = \left(\frac{d\boldsymbol{L}_A}{dt}\right)_P + \boldsymbol{\omega} \times \boldsymbol{L}_A.$$

From the angular momentum law,

$$\boldsymbol{M}_A = \left(\frac{d\boldsymbol{L}_A}{dt}\right)_O = \left(\frac{d\boldsymbol{L}_A}{dt}\right)_P + \boldsymbol{\omega} \times \boldsymbol{L}_A.$$

The components of $\boldsymbol{L}_A$ with respect to the moving system P are known from Eqn. (47-8), and their time derivatives are

$$\frac{dL_1}{dt} = I'\frac{d\omega_1}{dt}, \quad \frac{dL_2}{dt} = I'\frac{d\omega_2}{dt}, \quad \frac{dL_f}{dt} = I_f\frac{d\omega_f}{dt}.$$

The components of the vector product $\boldsymbol{\omega} \times \boldsymbol{L}_A$ are

$$\begin{aligned}
\omega_2 L_f - \omega_f L_2 &= \omega_2\omega_f I_f - \omega_f\omega_2 I' = (I_f - I')\,\omega_2\omega_f,\\
\omega_f L_1 - \omega_1 L_f &= \omega_f\omega_1 I' - \omega_1\omega_f I_f = -(I_f - I')\,\omega_1\omega_f,\\
\omega_1 L_2 - \omega_2 L_1 &= \omega_1\omega_2 I' - \omega_1\omega_2 I' = 0.
\end{aligned}$$

Then if M_1, M_2, and M_f are the components of $\boldsymbol{M}_A$ relative to the ξ, η, ζ system, we have the three equations

$$M_1 = I'\frac{d\omega_1}{dt} + (I_f - I')\,\omega_2\omega_f,$$

$$M_2 = I'\frac{d\omega_2}{dt} - (I_f - I')\,\omega_1\omega_f,$$

$$M_f = I_f\frac{d\omega_f}{dt}.$$

These are the so-called *Euler equations* for the motion of a symmetrical top. They represent, in fact, nothing more than Newton's second law, as formulated for the motion of the top. That the equations are more complex than those for a single particle results from the complicated character of the moment of inertia. Only a few special cases are of practical importance.

The force-free symmetrical top. This is a symmetrical top for which the resultant moment of all external forces about the center of mass is zero. Since a rigid body rotates about its center of mass as if that point were at rest, we shall consider this to be the case.

We shall assume in this special case that the top is supported in a frictionless manner so that its center of mass remains at rest. Thus in Fig. 4-29, the fixed point A is now the center of mass and the angular momentum $\boldsymbol{L}_c$ is represented by $\boldsymbol{L}_A$ in the figure. Since $\boldsymbol{M}_c = 0$, it follows that the angular momentum $\boldsymbol{L}_c$ remains constant in magnitude and direction. Since $\boldsymbol{M}_f$ also is zero, the angular velocity ω_f and the component of angular momentum $L_f = \omega_f I_f$ are also constant. Hence the angle θ in Fig. 4-29 between the vectors $\boldsymbol{L}_c(=\boldsymbol{L}_A)$ and $\boldsymbol{L}_f$ is constant also. On the other hand, we have shown in the preceding section that f, $\boldsymbol{\omega}$, and $\boldsymbol{L}_A$ always lie in the same plane for a symmetric top. It follows that

$$\boldsymbol{L}_c \cdot \boldsymbol{\omega} = L_c\omega \cos(\varepsilon - \theta).$$

Since no forces act on the body, the kinetic energy is constant. But, in general,

$$\boldsymbol{L} \cdot \boldsymbol{\omega} = 2E_{\text{kin}} = \text{const},$$

as can be seen by substituting Eqns. (47-5), (47-6), and (47-7) in $\boldsymbol{L} \cdot \boldsymbol{\omega}$ and then comparing with Eqn. (47-1). Then by expansion, we have,

$$\boldsymbol{L}_c \cdot \boldsymbol{\omega} = L_c\omega\,(\cos\varepsilon\cos\theta + \sin\varepsilon\sin\theta)$$

$$= L_f(\omega\cos\varepsilon) + L_c\sin\theta(\omega\sin\varepsilon) = \text{const}.$$

But we have just shown above that L_f, L_c, and θ are all constant. Hence ω and ε must also be constant to satisfy this equation. For an observer fixed in space, we can therefore make the following statements. The total angular momentum vector of the top $\boldsymbol{L}_c$ remains fixed in space. The angular velocity ω can be resolved into two constant components in the direction of the angular momentum and of the figure axis. The motion of the top consists of a superposition of a constant rotation about the figure axis and a uniform rotation of the figure axis about the fixed direction of the angular momentum. This motion of the figure axis is called *precession* (in this case, *force-free precession*) if $\boldsymbol{L}_A$ points along the z-axis of a space-fixed coordinate system.

The motion of the top can also be described in terms of the rolling of one cone on another. Its motion at any instant consists of a rotation about an instantaneous axis in the direction of the vector $\boldsymbol{\omega}$. As a result of this motion, the axis f is displaced perpendicular to the plane of f, $\boldsymbol{\omega}$, and $\boldsymbol{L}_c$. Since f, $\boldsymbol{\omega}$, and $\boldsymbol{L}_c$ always lie in the same plane (see Fig. 4-29) and $\boldsymbol{L}_c$ is fixed in space, the instantaneous axis must follow this motion. Hence, f and $\boldsymbol{\omega}$ both rotate about the vector $\boldsymbol{L}_c$. The vector $\boldsymbol{\omega}$ sweeps over the surface of a circular cone which is fixed in space, and is called the *polhode cone*. The motion of the top is equivalent to the rolling, on the polhode cone, of a second cone fixed with respect to the top and called the *herpolhode cone* (Fig. 4-30). The figure axis of the top lies along the axis of the herpolhode cone.

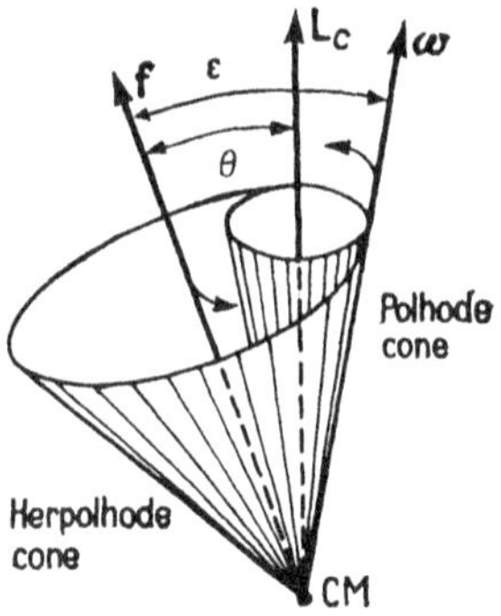

FIG. 4-30. Motion of the figure axis and the instantaneous axis of rotation of a force-free symmetrical top around the fixed total angular momentum axis. The herpolhode cone rolls on the space fixed polhode cone without sliding.

If the top is originally rotating about its figure axis, then at this instant both the angular velocity vector $\boldsymbol{\omega}$ and, from Eqn. (47-8), the angular momentum vector $\boldsymbol{L}_c$, coincide with the direction of the line f. The directions of f, $\boldsymbol{\omega}$, and $\boldsymbol{L}_c$ therefore continue to coincide, and the top rotates with a uniform angular velocity $\omega_f = \omega$ about its figure axis, which remains fixed in space. This property of a top is utilized in the *directional gyro*. It can be demonstrated by throwing a rapidly rotating toy top in the air. In the absence of air resistance, the center of mass of the top moves in a parabola. No torque acts about the center of mass of the top. At the instant at which it is thrown, the top is, to a first approximation, rotating about its figure axis, and this axis retains its direction throughout the entire motion.

The heavy, symmetrical, rapidly rotating (fast) top. When a top is supported at a point A which does not coincide with its center of gravity, the weight of the top has a moment with respect to A. If no forces act on the top other than its weight and the reaction of the supporting surface, the top is referred to as a *heavy* top. Let a symmetrical heavy top be set in rapid rotation about its figure axis, initially held at rest. If the top is now carefully released, the instantaneous axis of rotation coincides very closely with the figure axis.

For this idealized situation, the following relations apply. The figure axis, being also a principal axis of inertia, coincides with the angular momentum vector $\boldsymbol{L}_A$ and moves with $\boldsymbol{L}_A$. The change in $\boldsymbol{L}_A$, in time dt, is given by the equation $d\boldsymbol{L}_A = \boldsymbol{M}_A dt$. That is, the change in $\boldsymbol{L}_A$ is in the direction of $\boldsymbol{M}_A$. The torque $\boldsymbol{M}_A$, produced by the weight $\boldsymbol{w}$, is horizontal and perpendicular to the plane of f and $\boldsymbol{L}_A$. The change $d\boldsymbol{L}_A$ is therefore perpendicular to $\boldsymbol{L}_A$ and is horizontal (Fig. 4-31). As the vector $\boldsymbol{L}_A$ is displaced, the torque vector

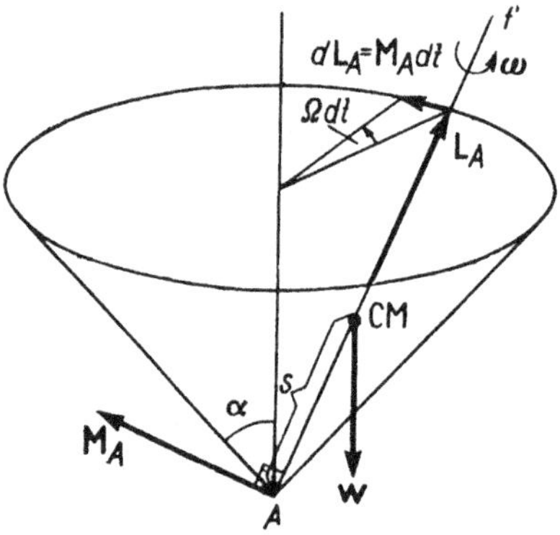

FIG. 4-31. Precession of a heavy top.

$\boldsymbol{M}_A$ rotates also, so that the torque is always perpendicular to $\boldsymbol{L}_A$. The relative positions of $\boldsymbol{M}_A, f$, and $\boldsymbol{L}_A$ are therefore the same as at the start and the original statements can be repeated. The figure axis describes a cone about a vertical line through A with an *angular velocity* Ω, and we call this motion *precession*. That is, the *heavy top* is not toppled over by its weight, but is deviated sideways. This behavior, which contradicts our intuitive feeling, makes the top a fascinating device. It is always impressive to hold a rapidly spinning top in one's hand and feel the torque developed when the axis is rotated.

The precessional velocity Ω can be found from Fig. 4-31 by computing the angular velocity Ω with which the tip of the vector $\boldsymbol{L}_A$ moves in its horizontal circular path. From the relation

$$dL_A = L_A \Omega \sin\alpha \, dt = M_A \, dt,$$

it follows that $\Omega = M_A/L_A \sin\alpha$.

If s is the distance from the point of support A to the center of gravity, then $M_A = ws \sin\alpha$, and hence

$$\Omega = ws/L_A = ws/I_f \omega_f.$$

Every top that spins on a supporting surface is acted on by frictional forces. If the top is heavy, symmetrical, and is rotating rapidly with its figure axis vertical and its angular momentum vector directed upward, then the frictional torque vector $\boldsymbol{M}_f$ is directed downward and the magnitude of the angular momentum vector decreases while its direction remains constant (Fig. 4-32):

$$dL_A = I_f d\omega = M_f dt.$$

If a top spins in a socket with its figure axis inclined from the vertical, the point of support lies at one side of both the figure axis and the angular momentum axis. The point of contact does not remain at rest and the motion of the top can be analyzed only by the use of Eqn. (46-5). However, if the socket is small, then to calculate the effect of the frictional torque on the direction of the angular momentum vector, we may, to a first approximation, apply the angular momentum law for a fixed point of contact A. The frictional torque M_f then results (Fig. 4-33) in a decrease in the magnitude of the angular velocity and a rotation of the figure axis toward the vertical.

In Fig. 4-33, point A is the point of contact, slightly displaced from the figure axis of the top, and the angular momentum vector points upward. The frictional torque vector is then perpendicular to the line ACM and can be resolved into a component M_{f_1} parallel to the figure axis, which

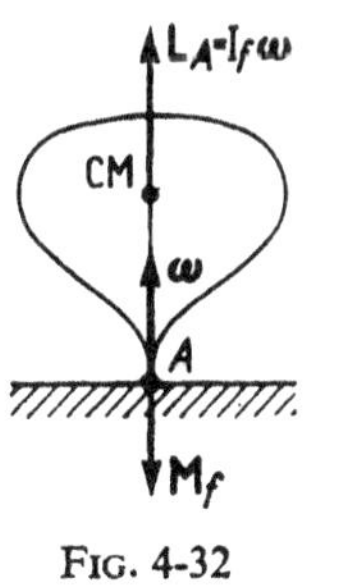

FIG. 4-32

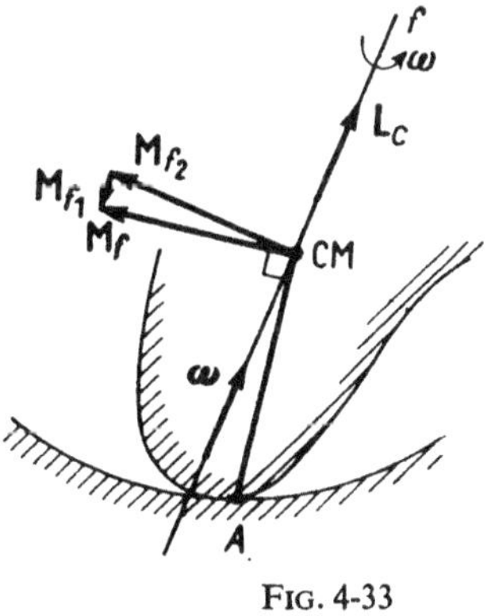

FIG. 4-33

FIG. 4-32. Symmetric top with a vertical figure axis and a frictional torque.

FIG. 4-33. Righting of a top by the frictional torque.

decreases the magnitude of the angular momentum according to the equation $dL_1 = M_{f_1}\, dt$, and a component M_{f_2} perpendicular to the figure axis, whose effect is to rotate the angular momentum vector toward the vertical according to the equation $dL_2 = M_{f_2}\, dt$. Thus we see that a fast top, whose axis is not vertical, tends to right itself.

The motion of the axis toward the vertical ceases as soon as the figure axis of the top has become vertical. When this position has been reached, the weight of the top exerts no moment about the point of support, the precessional motion ceases, and the only effect of the frictional torque is to decrease the magnitude of the rotational velocity.

If, as a first approximation, we disregarded the acceleration of the center of mass of the earth in its orbit around the sun, and also about the center of mass of the earth–moon system, the earth can be considered as a symmetrical top having a gigantic angular momentum vector $\boldsymbol{L}$ with respect to its center. The earth is acted on by relatively small external torques resulting from the gravitational forces exerted by the sun and the moon on the ellipsoidal bulges of the earth. These forces give rise to a moment because the axis of the earth is inclined to the plane of its orbit (the ecliptic) and to the plane of the orbit of revolution of the moon (Fig. 4-34). The result is a precession of the earth's axis with a period of 26,000 years, and is called the *precession of the equinoxes.*

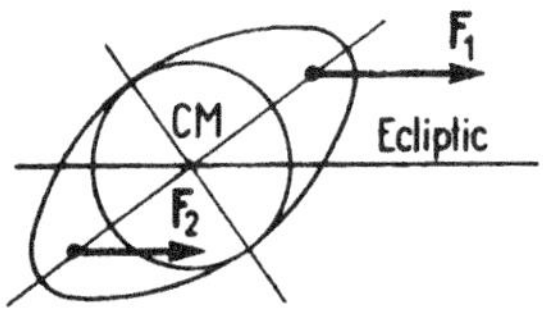

FIG. 4-34. The moment which causes the precession of the earth's axis of rotation.

48. Calculation of the Ellipsoid of Inertia of Symmetrical Bodies

It is convenient in calculating the ellipsoid of inertia to use as a reference system the coordinate system consisting of the principal axes of inertia. For symmetrical bodies, the following statements can be made regarding these axes.

(1) If a body has a plane of symmetry, this is also a symmetry plane of the inertia ellipsoid and one of the principal axes is normal to this plane. This can easily be proved as follows:

A principal axis is characterized by the vanishing of two of the products of inertia. Consider the normal to the plane of symmetry as the z-axis, and the symmetry plane as the xy-plane. Then for every particle m_k with coordinates (x, y, z), there is a symmetrical particle $m_k' = m_k$ with coordinates $(x, y, -z)$. But the product of inertia vanishes for every such pair of particles, since

$$I_{xz} = m_k xz - m'_k xz = 0, \quad I_{yz} = m_k yz - m'_k yz = 0.$$

The z-axis is therefore a principal axis.

(2) If a body has two planes of symmetry, their line of intersection is a principal axis.

This follows from the preceding discussion, according to which the normals to both of the planes of symmetry are principal axes. The third principal axis is perpendicular to the other two, and hence lies along the line of intersection of the planes of symmetry.

(3) If more than two planes of symmetry pass through a single axis, the ellipsoid of inertia has rotational symmetry.

To determine the ellipsoid of inertia with respect to a given set of principal axes (x, y, z), it is necessary to know the principal moments of inertia. Their values are given by the defining equations

$$I_1 = I_{xx} = \sum m_k(y_k^2 + z_k^2) = \sum m_k y_k^2 + \sum m_k z_k^2,$$
$$I_2 = I_{yy} = \sum m_k(x_k^2 + z_k^2) = \sum m_k x_k^2 + \sum m_k z_k^2,$$
$$I_3 = I_{zz} = \sum m_k(x_k^2 + y_k^2) = \sum m_k x_k^2 + \sum m_k y_k^2.$$

The quantities $\sum m_k x_k^2$, $\sum m_k y_k^2$, and $\sum m_k z_k^2$ are called the *moments of inertia with respect to a plane* (the Binet moments of inertia). These Binet moments of inertia serve as the starting points in setting up the integrals to evaluate the axial moments of inertia.

48-1 Some Examples

1. *The homogeneous rectangular parallelepiped* (Fig. 4-35). A homogeneous rectangular parallelepiped of density ϱ has three mutually perpendicular planes of symmetry, which intersect at its center. The principal axes are therefore determined. The Binet moments of inertia are

$$\sum m_k x_k^2 = \int_{-a/2}^{a/2} bc\varrho x^2 dx = bc\varrho \left.\frac{x^3}{3}\right]_{-a/2}^{a/2} = \tfrac{1}{12} ma^2,$$

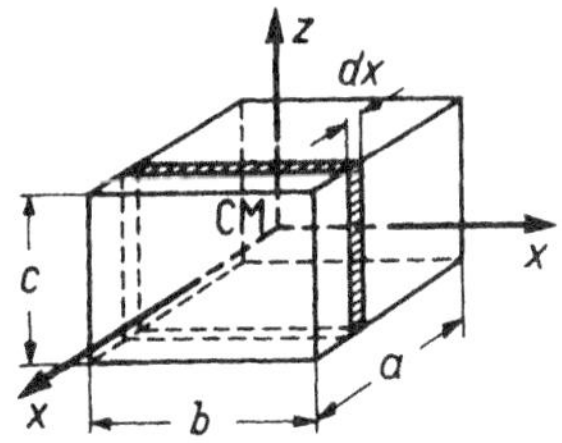

FIG. 4-35. Calculation of the inertia ellipsoid of a rectangular parallelepiped.

and, similarly,

$$\sum m_k y_k^2 = \tfrac{1}{12} mb^2, \quad \sum m_k z_k^2 = \tfrac{1}{12} mc^2.$$

The principal moments of inertia are therefore

$$I_{xx} = \frac{m}{12}(b^2 + c^2), \quad I_{yy} = \frac{m}{12}(c^2 + a^2), \quad I_{zz} = \frac{m}{12}(a^2 + b^2).$$

If $c > b > a$, then $I_{xx} > I_{yy} > I_{zz}$. The inertia ellipsoid then has its longest axis in the z-direction and its shortest axis in the x-direction.

For the special case of a cube, $a = b = c$, $I_{xx} = I_{yy} = I_{zz}$ and the inertia ellipsoid degenerates to a sphere. That is, the moment of inertia of a cube is the same about *all* axes through its center.

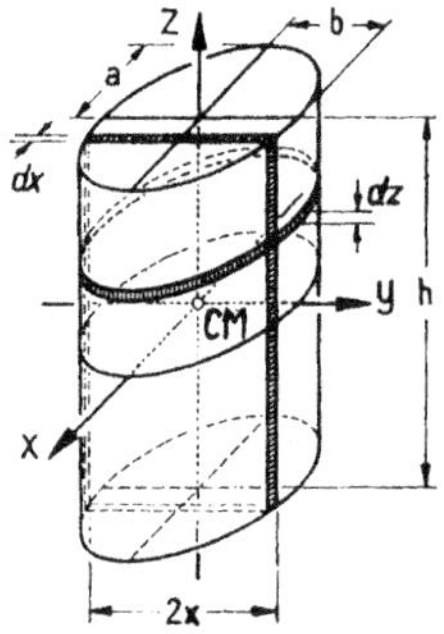

FIG. 4-36. Calculation of the inertia ellipsoid of an ellipsoidal cylinder.

The homogeneous elliptic cylinder (Fig. 4-36). The cylinder also has three mutually perpendicular planes of symmetry. The Binet moments of inertia are

$$\sum m_k x_k^2 = 2\int_0^a 2\varrho h y x^2\, dx$$

$$= 2\int_0^a \frac{2h\varrho b}{a} \sqrt{(a^2 - x^2)}\, x^2\, dx$$

$$= \frac{4h\varrho b}{a} a^4 \int_0^{\pi/2} \cos^2\varphi \sin^2\varphi\, d\varphi$$

$$= \varrho h b a^3 \int_0^{\pi/2} \sin^2(2\varphi)\, d\varphi = m\frac{a^2}{4},$$

$$\sum m_k y_k^2 = m\frac{b^2}{4},$$

$$\sum m_k z_k^2 = 2\pi ab\varrho \int_0^{h/2} z^2\, dz = m\frac{h^2}{12}.$$

The substitution $x/a = \sin\varphi$ was used in the above. The principal moments of inertia are therefore

$$I_{xx} = m\frac{b^2}{4} + \frac{mh^2}{12} = \frac{m}{12}(3b^2 + h^2),$$

$$I_{yy} = \frac{m}{12}(3a^2 + h^2), \quad I_{zz} = \frac{1}{4}m(a^2 + b^2).$$

If $h > a > b$, then I_{yy} is the largest and I_{zz} the smallest principal moment of inertia. If the cross section is a circle of radius r, then $I_{zz} = 1/2\, mr^2$.

HCl molecule (*Fig. 4-37*). The principal moments of inertia play an important role in the theory of molecular structure. The magnitudes of the principal moments of inertia can be determined from an analysis of the band spectra of the molecules of a gas and, knowing these, the geometry

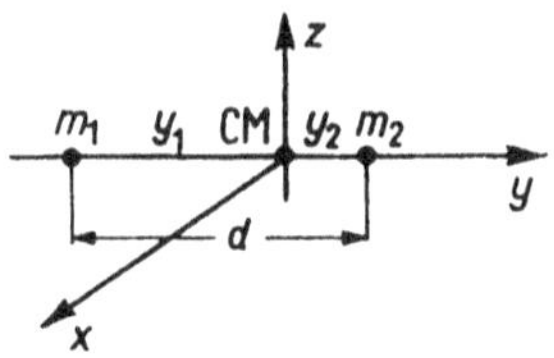

FIG. 4-37. A dumbbell molecule, such as HCl.

of the molecule can frequently be deduced. For example, the internuclear distance between the atoms forming a dumbbell-type molecule can be computed from its moment of inertia and the equation for its center of mass.

The moments of inertia of the molecule with respect to its center of mass are

$$I_{xx} = I_{zz} = m_1 y_1^2 + m_2 y_2^2; \quad I_{yy} = 0.$$

From the definition of center of mass, $m_1 y_1 = m_2 y_2$. It follows that

$$m_1 y_1^2 + \frac{m_1^2}{m_2} y_1^2 = I_{xx}, \quad y_1^2 = \frac{m_2 I_{xx}}{m_1(m_1 + m_2)}.$$

The atomic masses m_1 and m_2 can be computed from the gram-atomic weight M and Avogadro's number $N_0 = 6.025 \times 10^{23}$ molecules/mole, through the relation $m = M/N_0$. The distances y_1 and y_2 are then given by

$$y_1^2 = \frac{M_2 I_{xx} N_0}{M_1(M_1 + M_2)}, \quad y_2^2 = \frac{M_1 I_{xx} N_0}{M_2(M_1 + M_2)}.$$

For HCl, we have

$$I_{xx} = I_{zz} = 2.61 \times 10^{-40}\ \text{gm} \cdot \text{cm}^2,$$

$$M_1 = 1.01\ \text{gm}, \quad M_2 = 35.46\ \text{gm}.$$

The computed distances are

$$y_1 = 1.23 \times 10^{-8}\ \text{cm} = 1.23\ \text{Å}\ (1\ \text{Å} = 10^{-8}\ \text{cm}),$$

$$y_2 = 3.48 \times 10^{-10}\ \text{cm} = 0.035\ \text{Å}.$$

The internuclear separation of the HCl molecule is therefore

$$y_1 + y_2 = d = 1.27\ \text{Å}.$$

Note, however, that although this calculation gives the internuclear distance, it provides no information regarding the molecular "diameter", by which we mean the diameter of a rigid sphere which would behave in the same way as the molecule in intermolecular collisions. This "diameter" is mainly a function of the size of the cloud of electrons surrounding the two nuclei.

Structure of the water molecule. Optical measurements of the band spectrum of water vapor lead to the following values of the three principal moments of inertia:

$$I_1 = 1.009 \times 10^{-40}\ \text{gm} \cdot \text{cm}^2,$$

$$I_2 = 1.901 \times 10^{-40}\ \text{gm} \cdot \text{cm}^2,$$

$$I_3 = 2.908 \times 10^{-40}\ \text{gm} \cdot \text{cm}^2.$$

The nuclei in a water molecule do not lie in the same straight line (Fig. 4-38). Let us construct a coordinate system x, y, z, with origin at the center of gravity CM, having its x-axis perpendicular to the plane of the diagram, its y-axis in the plane of the diagram pointing to the right, and its z-axis pointing upward. The molecule is then symmetrical with respect to the x, z- and y, z-planes. The coordinate axes are therefore principal axes also, and the products of inertia vanish.

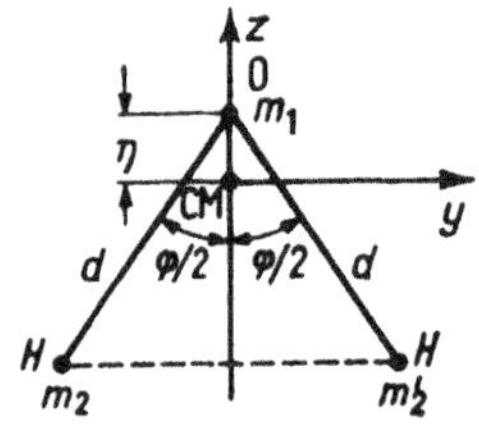

FIG. 4-38. Structure of the H_2O molecule.

In what follows, it will be convenient to represent the coordinates of all particles in a systematic way, as in Table 4-2. From the table we have the following equations: for the center of mass,

$$m_1\eta = 2m_2\left(d\cos\frac{\varphi}{2} - \eta\right);$$

TABLE 4-2. THE H_2O MOLECULE

	x	y	z
m_1 [0]	0	0	η
m_2 [H]	0	$-d \sin \frac{\varphi}{2}$	$-\left(d \cos \frac{\varphi}{2} - \eta\right)$
m_2' [H]	0	$d \sin \frac{\varphi}{2}$	$-\left(d \cos \frac{\varphi}{2} - \eta\right)$

for the moments of inertia,

$$I_{xx} = m_1\eta^2 + 2m_2 \left[d^2 \sin^2 \frac{\varphi}{2} + \left(d \cos \frac{\varphi}{2} - \eta\right)^2\right]$$

$$= m_1\eta^2 + 2m_2 (d^2 + \eta^2) - 4m_2\, d\eta \cos \frac{\varphi}{2},$$

$$I_{yy} = m_1\eta^2 + 2m_2 \left(d \cos \frac{\varphi}{2} - \eta\right)^2$$

$$= m_1\eta^2 + 2m_2 \left(d^2 \cos^2 \frac{\varphi}{2} + \eta^2\right) - 4m_2\, d\eta \cos \frac{\varphi}{2},$$

$$I_{zz} = 2m_2\, d^2 \sin^2 \frac{\varphi}{2}.$$

It will be seen from these equations that the moments of inertia are not independent. The three are related through the equation $I_{xx} = I_{yy} + I_{zz}$, an equation which, it will be seen, is verified by the experimental measurements. There are still three independent equations available, so that we may determine the parameters η, d, and $\varphi/2$. Solution of the equations above for η, d, and $\tan (\varphi/2)$ leads to

$$\eta = \sqrt{\left[\frac{2m_2(I_{xx} - I_{zz})}{m_1(m_1 + 2m_2)}\right]} = \sqrt{\left[\frac{2m_2 I_{yy}}{m_1(m_1 + 2m_2)}\right]},$$

$$d = \sqrt{\left[\frac{I_{yy}(m_1 + 2m_2) + I_{zz}m_1}{2m_1m_2}\right]},$$

$$\tan \frac{\varphi}{2} = \sqrt{\left[\frac{I_{zz}m_1}{I_{yy}(m_1 + 2m_2)}\right]}.$$

The problem has a unique solution only if we know which moment of inertia is associated with which principal axis. Since this is not the case, we must consider the two possible molecular models:

Case 1:

$$I_{xx} = I_{33}, \quad I_{yy} = I_{22}, \quad I_{zz} = I_{11}.$$

Case 2:

$$I_{xx} = I_{33}, \quad I_{yy} = I_{11}, \quad I_{zz} = I_{22}.$$

The atomic masses m_1 and m_2, from the gram-molecular weights and Avogadro's number, are

$$m_2 = m_2' = \frac{1.008}{6.02 \times 10^{23}} \text{ gm}, \quad m_1 = \frac{15.995}{6.02 \times 10^{23}} \text{ gm}.$$

The two possible models then lead to the values

Case 1:

$$\eta = 8.95 \times 10^{-10} \text{ cm}, \quad d = 9.70 \times 10^{-9} \text{ cm}, \quad \varphi = 68° \, 58'.$$

Case 2:

$$\eta = 6.52 \times 10^{-10} \text{ cm}, \quad d = 9.52 \times 10^{-9} \text{ cm}, \quad \varphi = 104° \, 34'.$$

With the help of measurements of the electrical dipole moment of the H_2O molecule, we come to the conclusion that *Case 2* is the correct model for the structure of H_2O.

CHAPTER 5

ELASTICITY

49. The General Concept of Stress

The subject of statics, as presented in Chapter 2, was based on the concept of a rigid body. This idealization of actual bodies is necessary in order to describe the mutual interactions of two bodies by means of the concept of force.

In fact, however, no truly rigid bodies exist. Every real object changes its shape when acted on by a force. A body is said to be *completely elastic* if it returns to its original shape when the forces are removed, and *completely inelastic* (or completely *plastic*) if the entire change of shape remains when the forces are removed. All actual bodies lie somewhere between these two extremes.

We shall restrict the discussion in this chapter to *isotropic* bodies. A body is said to be isotropic if its properties are independent of direction when it is in a force-free state. Although most solid bodies (the metals, for example) are microcrystalline, volume elements which are large relative to the microstructure can be considered isotropic. On the other hand, single crystals (such as crystals of rock salt or single crystals of metal) are anisotropic.

The subject of elasticity deals with the relation between the *forces* on a quasi-elastic body and the *deformations* resulting from these forces. It also determines the limits of validity of the laws which will be developed.

In studying the laws of elastic substances, it becomes evident that the actions of two systems on each other can no longer be described in terms of the elementary concept of force. In place of force, we introduce the concept of *stress*, a quantity which plays a major role in the mechanics of deformable bodies. This is done because the actions of the bodies on each other occur over extended surfaces. In contrast to rigid bodies, which make contact only at single points, elastic bodies always make contact over surfaces of finite extent.

Definition. The *stress* Σ at an element of area is defined as the quotient of the force ΔF which is exerted on the element and the area ΔA of the element:

$$\Sigma = \frac{\Delta F}{\Delta A}.$$

Its dimensions are $[\Sigma] = [ML^{-1}T^{-2}]$; and its units are: cgs system, 1 dyne/cm²; mks system, 1 newton/m²; fps system, 1 lbf/ft².

The following units are also in common use:

$$1 \text{ psi} = 1 \text{ lbf/in}^2 = 144 \text{ lbf/ft}^2,$$

$$1 \text{ atmosphere} = 14.7 \text{ lbf/in}^2 = 1.013 \times 10^6 \text{ dynes/cm}^2,$$

$$1 \text{ bar} = 10^6 \text{ dynes/cm}^2.$$

In general, the force ΔF is not perpendicular to the element of area. It can be resolved into two components, one of which is perpendicular to the element and is called the *normal* component ΔF_n, while the other lies in the plane of the element and is called the *tangential* or *shearing* component ΔF_t. Thus we have the two components of stress (Fig. 5-1):

$$\sigma = \frac{\Delta F_n}{\Delta A} = \text{normal stress}, \quad \tau = \frac{\Delta F_t}{\Delta A} = \text{shearing stress}.$$

The normal stress σ and the shearing stress τ are the components of the total stress Σ at the element of area. Since these components are associated with an element of area, they differ in a fundamental way from the components of a vector. Therefore, stress components cannot be handled in the same way as the components of a vector. Only in the case in which all stress components are referred to the same element of area is a vectorial addition of stresses possible.

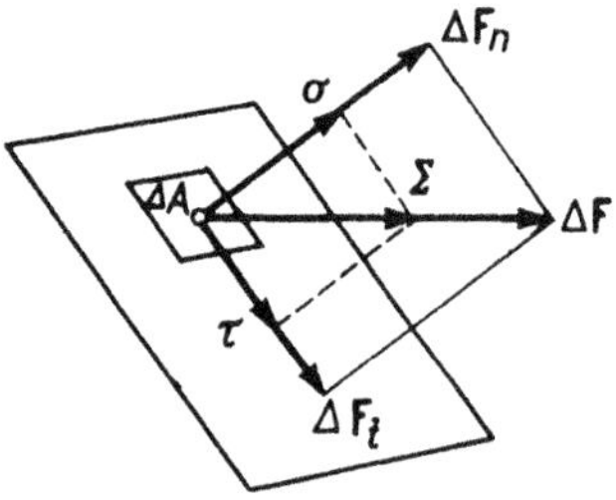

FIG. 5-1. Decomposition of a stress into its normal and shearing components.

The stresses σ and τ at a given point within an elastic body depend on the orientation of the element of area. Our first problem is to find the relation between the stresses σ and τ for different orientations of an element of area at a point within a body which is acted on by given forces. We shall show that it is possible to compute the stresses for an element in any arbitrary direction from a few known stresses acting on mutually perpendicular elements. Once the stress relations are known at every point of an elastic body, the deformation of the body can be determined from empirical laws. But first, a knowledge of the *state of stress* is necessary. We therefore begin with a study of the simplest state, that of *plane stress.*

50. The State of Plane Stress

In the elementary theory of elasticity we are often concerned with a state of stress in which all elements of area parallel to a certain plane are stress free. Such a state is known as one of *plane stress*, and it occurs only when all of the external forces acting on a body are parallel to the given plane and uniformly distributed in the direction perpendicular to the plane. That is, we are effectively considering the two-dimensional case.

To investigate the equilibrium of such a body, we construct at any point within it an infinitesimal right-angled triangular prism, as in Fig. 5-2, perpendicular to the given plane (which is taken as the plane of the diagram)

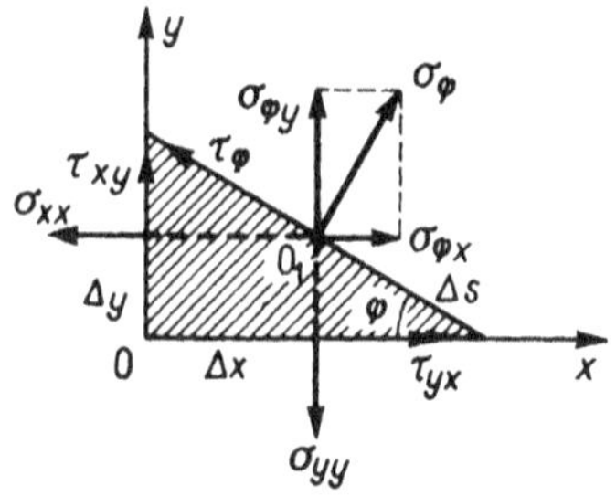

FIG. 5-2. Stress on the surfaces of a prism for a state of plane stress.

and of height Δl. The x- and y-axes of the coordinate system lie along the mutually perpendicular sides of the triangular base of the prism. The length of the hypotenuse of the triangle is Δs, and the lengths of the other sides are Δx and Δy, where

$$\Delta x = \Delta s \cos \varphi, \quad \Delta y = \Delta s \sin \varphi.$$

50-1 Designation of Stresses

In order that the symbol for a stress component shall designate the orientation of the element of area to which it refers, as well as the direction of the component itself, the symbol for a stress is given a double subscript. The first letter in the subscript indicates the direction of the normal to the element of area; the second, the direction of the stress component itself (Fig. 5-3).

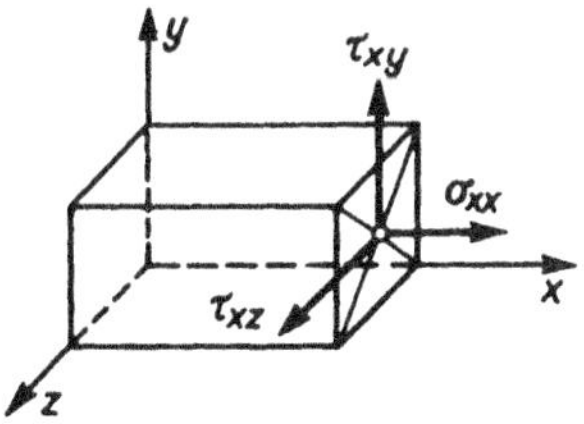

FIG. 5-3. Indices of the various stress components.

EXAMPLE. The symbol σ_{xx} represents the normal stress on an element of area which is perpendicular to the x-axis, and which is acted on by a force parallel to the x-axis. The symbol τ_{xy} is the shearing stress at an element of area which is perpendicular to the x-axis, and which results from a force in the y-direction.

50-2 Equilibrium of the Prism

For infinitesimally small volume elements, the so-called volume force (the weight of the element) can be neglected in comparison with the surface forces, because the volume forces vary with the cube of the linear dimensions, while the surface forces vary with their square. In setting up the expression for the equilibrium of the prism in Fig. 5-2 it is necessary, therefore, to consider only the surface forces. Since we are considering a state of plane stress, the triangular faces of the prism are stress free. We arbitrarily choose the directions shown for the various stresses and therefore have the following equilibrium conditions:

$$\sum F_{xi} = 0$$
$$= -\sigma_{xx}\Delta y \Delta l + \tau_{yx}\Delta x \Delta l + \sigma_\varphi \sin\varphi\, \Delta s \Delta l - \tau_\varphi \cos\varphi\, \Delta s \Delta l. \qquad (50\text{-}1)$$

Since

$$\Delta x = \Delta s \cos\varphi, \quad \Delta y = \Delta s \sin\varphi$$

it follows that

$$\sigma_\varphi \sin\varphi = \sigma_{xx} \sin\varphi - \tau_{yx} \cos\varphi + \tau_\varphi \cos\varphi. \qquad (50\text{-}2)$$

Also

$$\sum_i F_{yi} = 0 = -\sigma_{yy}\Delta x \Delta l + \tau_{xy}\Delta y \Delta l + \sigma_\varphi \cos\varphi\, \Delta s \Delta l + \tau_\varphi \sin\varphi\, \Delta s \Delta l,$$

from which

$$\sigma_\varphi \cos\varphi = \sigma_{yy} \cos\varphi - \tau_{xy} \sin\varphi - \tau_\varphi \sin\varphi. \qquad (50\text{-}3)$$

The moment equation with respect to the midpoint of the hypotenuse, O_1, gives,

$$\sum M_{O_1} = 0 = -\tau_{xy}\Delta y \Delta l \frac{\Delta x}{2} + \tau_{yx}\Delta x \Delta l \frac{\Delta y}{2},$$

or,

$$\tau_{yx} = \tau_{xy}. \qquad (50\text{-}4)$$

It follows from the last equation that *the shearing stresses on two mutually perpendicular elements of area of a body in equilibrium are equal in magnitude and that both are directed toward or both are directed away from the line of intersection of the elements.* This property of the state of stress is a consequence of the law of moments.

The stresses σ_φ and τ_φ can be computed from Eqns. (50-2) and (50-3). We find

$$\sigma_\varphi = \frac{\sigma_{yy} + \sigma_{xx}}{2} + \frac{\sigma_{yy} - \sigma_{xx}}{2} \cos 2\varphi - \tau_{xy} \sin 2\varphi \qquad (50\text{-}5)$$

and

$$\tau_\varphi = \frac{\sigma_{yy} - \sigma_{xx}}{2} \sin 2\varphi + \tau_{xy} \cos 2\varphi. \tag{50-6}$$

Thus for a state of plane stress, if the normal and shearing stresses are known for two mutually perpendicular directions at a point P in the body, then from Eqns. (50-5) and (50-6) the normal and shearing stresses σ_φ and τ_φ can be computed for any arbitrary element of area at P, in a direction given by the angle φ. A state of plane stress is completely described by giving the three stresses σ_{xx}, σ_{yy}, and $\tau_{xy} = \tau_{yx}$ at every point.

Of special significance is the angle φ_0 for which the shearing stress $\tau_\varphi = 0$. From Eqn. (50-6), when $\tau_\varphi = 0$,

$$\tan 2\varphi_0 = \frac{2\tau_{xy}}{\sigma_{xx} - \sigma_{yy}}. \tag{50-7}$$

Since this equation has two solutions in the interval $[0, 2\pi]$, namely $2\varphi_0$ and $2\varphi_0 + \pi$, it follows that at every point in the body there are two mutually perpendicular elements of area for which the shearing stress is zero. The planes in which these elements lie are called the *planes of principal stress*, and the normal stress components at these planes are called the *principal stresses* σ_1 and σ_2. They can be computed by setting $\tau_\varphi = 0$ in Eqn. (50-6) and combining this with Eqn. (50-5). The result is

$$\sigma_{1,2} = \frac{\sigma_{yy} + \sigma_{xx} \pm \sqrt{[(\sigma_{yy} - \sigma_{xx})^2 + 4\tau_{xy}^2]}}{2}. \tag{50-8}$$

Note that the principal stresses are the *maximum* and *minimum* normal stresses to be found in the material.

If the planes of principal stress are selected as those of the mutually perpendicular sides of the prism in Fig. 5-2, we then have, instead of Eqns. (50-5) and (50-6), the simpler relations,

$$\sigma_\varphi = \frac{\sigma_2 + \sigma_1}{2} + \frac{\sigma_2 - \sigma_1}{2} \cos 2\varphi,$$

$$\tau_\varphi = \frac{\sigma_2 - \sigma_1}{2} \sin 2\varphi. \tag{50-9}$$

In most practical problems, the principal stresses can be determined directly, and it is unnecessary to make use of Eqn. (50-8).

51. The Mohr Circle for the State of Plane Stress

Using Eqns. (50-9), the stresses at a point P on an element of area in any arbitrary orientation can be computed from a knowledge of the principal stresses at the point. The calculations can be represented graphically in a simple way which serves to bring out clearly the relations between the stresses. Equations (50-9) can be considered as the parametric equations of a circle of radius $r = (\sigma_2 - \sigma_1)/2$ in a rectangular coordinate system whose axes are σ_φ and τ_φ. The center of the circle lies on the σ_φ axis at an abscissa $x_M = (\sigma_2 + \sigma_1)/2$. The circle is called the *Mohr circle* (Fig. 5-4) and it

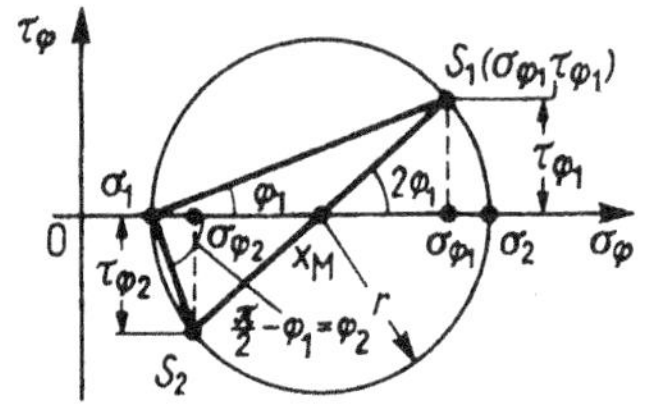

FIG. 5-4. The Mohr circle diagram for determining stresses.

is very useful in the analysis of stress distributions. The abscissa and ordinate of each point $S(\sigma_\varphi, \tau_\varphi)$ on the circle represent the normal and shearing stress σ_φ and τ_φ, respectively, at a surface element making an angle φ with the second or number 2 principal stress plane. The angle 2φ is the angle between the σ_φ axis and the radius line to the representative point on the circle. The angle between the σ_φ axis and the chord from σ_1 to the point S is φ.

It will be seen from the Mohr circle that every state of stress $S_1(\sigma_{\varphi_1}, \tau_{\varphi_1})$ can be associated with a second state $S_2(\sigma_{\varphi_2}, \tau_{\varphi_2})$ in which the shearing stress has the same magnitude but the opposite sign. This state of stress occurs on an element of area which is normal to that at which the state of stress is S_1. The normal stresses at the two elements are, in general, not equal and, as the construction shows, are located symmetrically with respect to the center of the circle.

For two elements of area inclined at angles of 45° to the principal axes of stress, the state of stress is obtained by marking off an angle of 45° at point σ_1, as in Fig. 5-5. This line intersects the circle at point S_1. The stress on

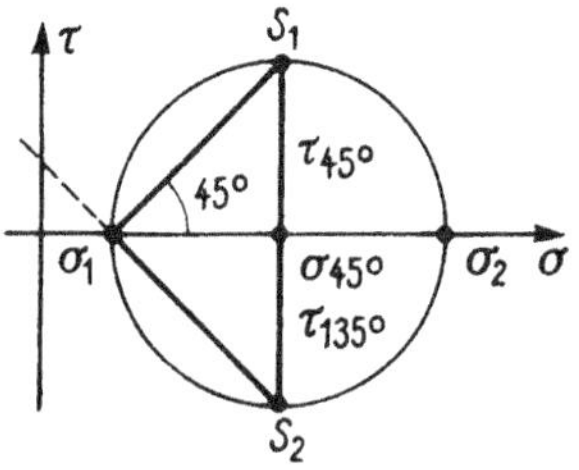

FIG. 5-5. State of stress for surface elements inclined at 45° to the principal stress planes.

the element at an angle of 135° is then represented by point S_2. The normal stresses at these elements are equal and are given by

$$\sigma_{45°} = \frac{\sigma_1 + \sigma_2}{2}.$$

The shearing stress is a maximum at these elements and the magnitude is

$$\tau_{45^\circ} = \frac{\sigma_2 - \sigma_1}{2}.$$

The following are the most important special cases of a plane stress distribution.

51-1 Hydrostatic Stress

The state of stress known as a *hydrostatic stress* results when the shearing stress, at every element of area, is zero. From Eqn. (9), this is possible only if $\sigma_1 = \sigma_2$ for all values of the angle φ. The Mohr circle then shrinks to a point. This state of stress occurs in fluids, where it is usually a hydrostatic *pressure*, so that the point lies on the negative axis of σ_φ. All elements of area experience the same normal stress, $-\sigma$.

51-2 The State of Axial Stress

A state of *axial stress* is said to exist in a body if the stress at all elements of area perpendicular to some axis is purely a normal stress, while at the same time the normal stress is zero at all elements parallel to the axis. The planes of principal stress are then determined, as are the principal stresses themselves. If the z-axis is the one referred to above, then we have

$$\sigma_2 = \sigma_z, \quad \sigma_1 = 0.$$

A beam subjected to pure tension or compression is in a state of axial stress, and the corresponding Mohr circle is shown in Fig. 5-6. For pure

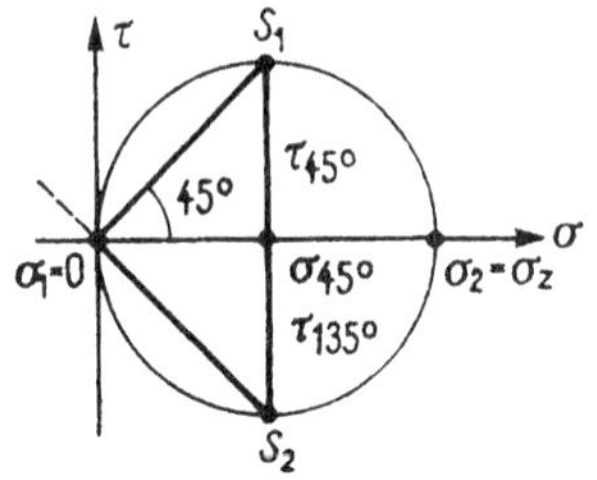

FIG. 5-6. The Mohr circle for a state of axial stress.

tension, the circle lies at the right of the origin and passes through the origin.

For $\varphi = 0$,	we have	$\sigma = \sigma_z, \tau = 0;$
$\varphi = 45^\circ$		$\sigma = \sigma_z/2, \tau = \sigma_z/2 = \tau_{max};$
$\varphi = 90^\circ$		$\sigma = 0, \tau = 0;$
$\varphi = 135^\circ$		$\sigma = \sigma_z/2, \tau = -\sigma_z/2.$

Exactly the same relations apply to a beam in pure compression along its axis, except that the Mohr circle lies at the left of the origin.

51-3 The State of Torsional Stress, or of Pure Shear

If the center of the Mohr circle lies at the origin, as in Fig. 5-7, the stress is purely normal for $\varphi = 0°$ and $\varphi = 90°$. The first is positive (tension); the second is negative (compression). The shearing stress is zero at both of these elements. At elements for which $\varphi = 45°$ and 135°, the stress is a pure shear and its magnitude is equal to the principal normal stress.

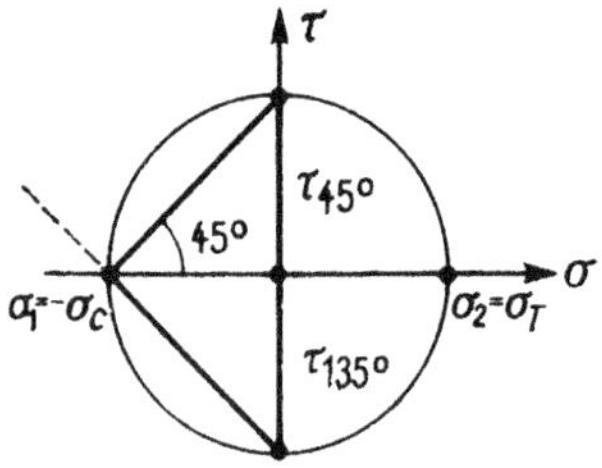

FIG. 5-7

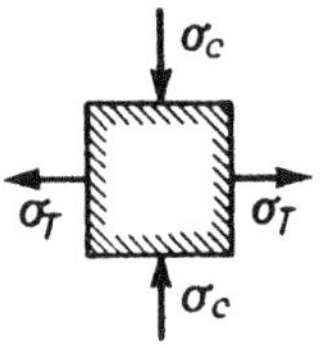

FIG. 5-8

FIG. 5-7. The Mohr circle for the torsional state of stress.

FIG. 5-8. Production of maximum shearing stress by the simultaneous application of compressive and tensile stresses. (See Fig. 5-11.)

This state of stress is made use of in certain technical processes of metal working, where the desired maximum value of shearing stress is produced as shown in Fig. 5-8 by the simultaneous applications of a compressive stress in one direction and an equal tensile stress in a perpendicular direction. If the material has a square cross-section, there is a pure shear stress along the diagonal planes.

The same state of stress also exists when a body is given a purely torsional deformation. When a rod of circular cross-section is twisted by a torque whose vector lies along the axis of the rod, all elements of area in a plane perpendicular to the axis are subjected to a pure shearing stress, since there are no external forces parallel to the axis of the rod. From the considerations given above, the stress at elements of area perpendicular to these is also a pure shear, and both stresses must point either toward or away from the line of intersection of the elements of area, as shown in Fig. 5-9. The diagonal planes of the enclosed (cross-hatched) volume element (see Fig. 5-11) are in a state of pure tension or pure compression.

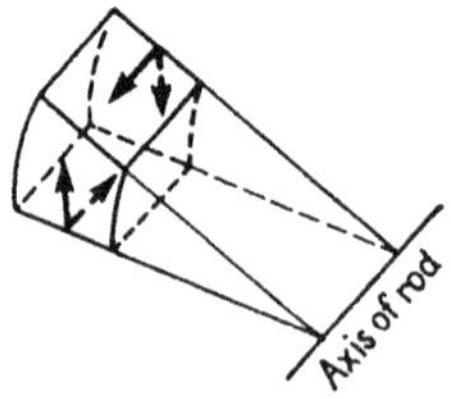

FIG. 5-9. State of stress for a plane torsional deformation.

52. Deformation and Strain

When a state of stress exists in a body, a *deformation* of the body results. The relation between the stress and the deformation can be found only by experiment and is given by a simple expression only when the stresses are less than a certain limiting value. When this is the case, the deformation process is reversible; that is, the body returns to its original form when the stresses are removed.

52-1 Hooke's Law

Hooke's law is the basis of the calculation of every deformation, as long as we remain in the region in which the deformations are reversible. If we distort an initially unstressed, isotropic material (such as a steel rod) by a tension or a compression, the principal axes of deformation coincide with the principal axes of stress. For stresses which are not too great, the deformation is directly proportional to the stress. This is a general statement of *Hooke's law*. For its precise formulation, we introduce the concept of *strain*.

The *longitudinal strain* ϵ of a rod is defined as the ratio of its deformation or change of length Δl to its original length l:

$$\varepsilon = \frac{\Delta l}{l}.$$

It follows from this definition that ε is a pure number.

Hooke's law now states that *the strain is directly proportional to the stress producing it* (see Fig. 5-10). If the normal stress in the rod is σ_{xx}, then

$$\varepsilon = \frac{\Delta l}{l}$$

$$= \frac{\sigma_{xx}}{Y}, \tag{52-1}$$

where Y is a constant, characteristic of a material, and is known as *Young's modulus*. (In some books, Young's modulus is denoted by the symbol E.) If $l = 1$ unit of length and $\sigma_{xx} = 1$ unit of stress, then

$$\Delta l = \frac{1}{Y}.$$

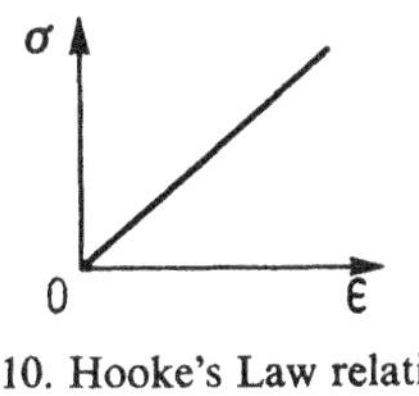

Fig. 5-10. Hooke's Law relationship.

That is, the reciprocal of Y is equal numerically to the increase in length of a rod of unit length, acted on by unit stress.

Dimensions: $[Y] = [ML^{-1}T^{-2}]$

Units: cgs system: 1 dyne/cm^2,
mks system: 1 newton/m^2,
fps system: 1 lbf/ft^2.

In engineering practice, Y is often expressed in lbf/in^2.

The significance of Young's modulus Y is not self-evident. For example, Young's modulus for steel is about 3×10^7 lbf/in^2. This means that a rod 1 ft long, subjected to a tensile stress of 1 lbf/in^2, increases in length by $1/3 \times 10^{-7}$ ft.

When a rod is stretched by a tensile stress, the transverse dimensions decrease and hence its cross-sectional area decreases. The *transverse strain* δ is defined as the ratio of the deformation Δb in a transverse dimension to the original value b of this dimension:

$$\delta = \frac{\Delta b}{b}.$$

Like ε, δ is a pure number.

According to Hooke's law, the transverse strain is also proportional to the stress σ_{xx}, although the constant of proportionality is not equal to Young's modulus. Let us write

$$\delta = \frac{\Delta b}{b} = \mu \frac{\sigma_{xx}}{Y}, \qquad (52\text{-}2)$$

where μ is a constant, characteristic of a given material. Then, since $\varepsilon = \sigma_{xx}/Y$, we have

$$\delta = \mu\varepsilon$$

or

$$\mu = \frac{\delta}{\varepsilon}$$

and the constant μ represents the *ratio* of the transverse strain to the longitudinal strain. It is called *Poisson's ratio*, and is also a pure number. The value of μ is usually between 0.30 and 0.40.

The elastic properties of an isotropic material, within the region of reversibility, are completely determined by the values of Y and μ. If the rod is compressed longitudinally, its length decreases and its transverse dimensions increase. The quantities Y and μ have the same magnitudes for compression as for tension, provided the deformation lies within the elastic limit.

In addition to longitudinal and transverse strain, we consider one more special case, that of a pure *shearing strain*. To produce a shearing strain, we apply pure shearing stresses τ_{xz} and τ_{zx} to the faces of a cube perpendicular to the x- and z-axes, while the faces perpendicular to the y-axis are stress-free. The same state of stress results in a cube contained within a larger cube, as in Fig. 5-11, if the larger cube is subjected to a tensile stress

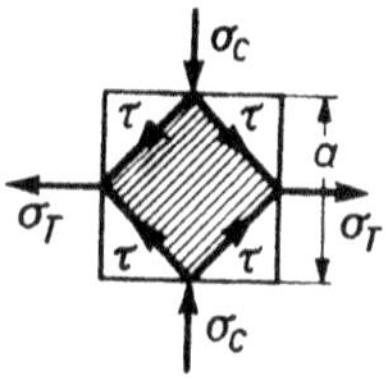

FIG. 5-11

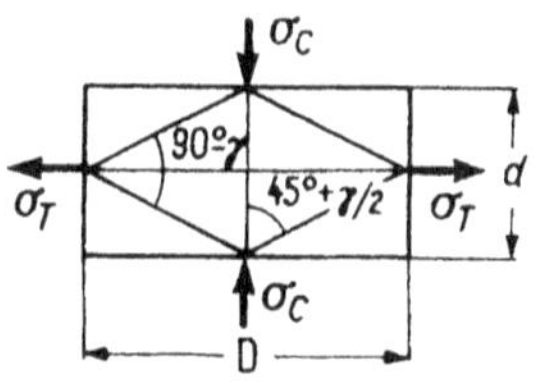

FIG. 5-12

FIG. 5-11. Shear stress.

FIG. 5-12. Change in the angles of a cube under shear stress.

along one axis and a compressive stress along a perpendicular axis. These stresses are assumed equal in magnitude:

$$\sigma_c = \sigma_T = \sigma.$$

Then, as we will show,

$$\tau_{zx} = \tau_{xz} = \tau = \sigma.$$

The enclosed cube is distorted by the shearing stresses into a rhombic prism, in which its right angles either increase or decrease by a small angle γ (Fig. 5-12). For a small stress σ, we have, from Fig. 5-12,

$$\tan\left(45^\circ + \frac{\gamma}{2}\right) = \frac{D}{d} = \frac{a + \dfrac{\sigma a}{Y} + \mu\dfrac{\sigma a}{Y}}{a - \dfrac{\sigma a}{Y} - \mu\dfrac{\sigma a}{Y}}, \tag{52-3}$$

where a is the original length of one edge of the outer cube. The quantity

$$\frac{\sigma a}{Y}(1 + \mu)$$

equals the change in length of the horizontal and vertical diagonals of the inner cube, produced by the tensile and compressive stresses on the outer cube, or by the corresponding shear stresses on the inner cube.

From the relation

$$\tan(\alpha + \beta) = \frac{\tan\alpha + \tan\beta}{1 - \tan\alpha\tan\beta},$$

we can write

$$\tan\left(45^\circ + \frac{\gamma}{2}\right) = \frac{\tan 45^\circ + \tan\dfrac{\gamma}{2}}{1 - \tan 45^\circ \tan\dfrac{\gamma}{2}}.$$

Hence from Eqn. (3), for small values of γ,

$$\frac{1+\frac{\gamma}{2}}{1-\frac{\gamma}{2}} = \frac{1+(1+\mu)\frac{\sigma}{Y}}{1-(1+\mu)\frac{\sigma}{Y}}.$$

The angle γ is therefore given by

$$\gamma = \frac{2\sigma}{Y}(1+\mu). \tag{52-4}$$

We now show that the magnitude of σ acting on the large cube is equal to that of the shearing stress τ on the small cube, provided the angle γ is small. A total shearing *force* (see Fig. 5-13) given by

$$F_{sh} = \sigma A$$

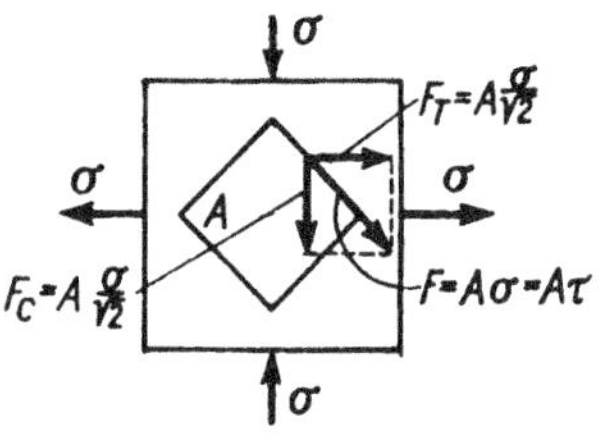

FIG. 5-13. Calculation of the magnitude of the shear stress produced by the normal stress.

is produced on the faces of the inner cube perpendicular to the plane of the figure (of area A) by the tensile and compressive stresses. The shearing stress τ is therefore

$$\tau = \frac{F_{sh}}{A} = \sigma.$$

We can therefore replace σ by τ in Eqn. (52-4), since the smaller cube is acted on only by shearing stresses. The change in angle of the cube, produced by a shearing stress τ, is therefore

$$\gamma = \frac{2(1+\mu)}{Y}\tau = \frac{\tau}{G}, \tag{52-5}$$

where

$$G = \frac{Y}{2(1+\mu)}.$$

The quantity G is called the *shear modulus* or the *torsion modulus* of the material. It has the same dimensions as Young's modulus and is expressed in the same units.

The three elastic constants Y, μ, and G satisfy the equations

$$G = \frac{Y}{2(1+\mu)} \quad \text{or} \quad \mu = \frac{Y}{2G} - 1. \tag{52-6}$$

A few general remarks may be made concerning the magnitude of μ and hence of the ratio Y/G. Let us consider a cube which is acted on by a compressive stress σ at all of its faces (Fig. 5-14). The original length a of each side of

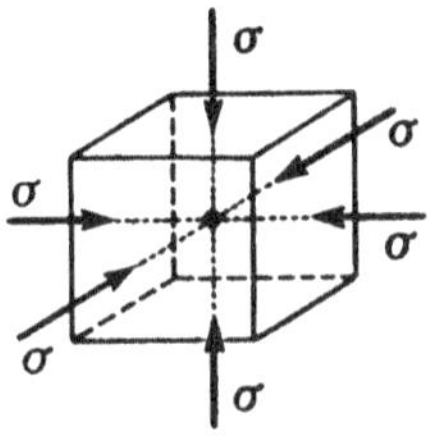

FIG. 5-14. The volume compressibility and Poisson's constant.

the cube is thereby changed to $a + \Delta a$, where Δa is the resultant of the decrease in length $-\dfrac{a\sigma}{Y}$, brought about by the longitudinal strain, and the increase in length $+\dfrac{2a\sigma\mu}{Y}$ resulting from the two transverse strains. Hence

$$\Delta a = -\frac{\sigma a}{Y} + \frac{2\sigma a}{Y}\mu = -\frac{\sigma a}{Y}(1 - 2\mu).$$

The new volume of the cube is

$$V + \Delta V = \left[a - \frac{\sigma a}{Y}(1 - 2\mu)\right]^3.$$

Keeping only the first power of the small quantity $\dfrac{\sigma}{Y}$, we have for ΔV,

$$\Delta V = -\frac{3a^3\sigma}{Y}(1 - 2\mu).$$

The fractional change in volume is therefore

$$\frac{\Delta V}{V} = -\frac{3\sigma}{Y}(1 - 2\mu). \tag{52-7}$$

The *isothermal compressibility* $\varkappa_T$ of a substance is defined as the ratio of the fractional decrease in volume, $\Delta V/V$, to the compressive stress σ, at constant temperature and under a uniform compressive stress. From Eqn. (52-7)

$$\varkappa_T = -\frac{1}{\sigma}\frac{\Delta V}{V} = \frac{3(1 - 2\mu)}{Y}.$$

Dimensions: $[\varkappa_T] = [M^{-1}LT^2]$

Units: cgs system: 1 cm^2/dyne
mks system: 1 m^2/newton
fps system: 1 ft^2/lbf.

In practice, *the units 1 in²/lbf and 1 atm⁻¹* are often used. *Example:* From tables of properties of matter, we find for aluminium

$$Y = 7.0 \times 10^5 \text{ kgf/cm}^2$$

$$G = 2.5 \times 10^5 \text{ kgf/cm}^2.$$

The constants μ and $\varkappa_T$ are therefore

$$\mu = \frac{Y - 2G}{2G} = 0.40,$$

$$\varkappa_T = \frac{3(1 - 2\mu)}{Y} = 8.6 \times 10^{-7} \text{ cm}^2/\text{kgf}.$$

For an incompressible substance, $\varkappa_T$: $= 0$ and therefore from Eqn. (52-7)

$$1 - 2\mu = 0,$$

$$\mu = \tfrac{1}{2}.$$

The value $\frac{1}{2}$ is therfore the upper limit of the Poisson ratio μ and is never reached by a real substance. That is

$$0 < \mu < 0.50.$$

If $\mu > 0.5$, the substance would *increase* in volume when pressure was exerted on it. The inverse of the compressibility, $\varkappa_T^{-1}$, is called the *volume* or *bulk modulus.*

From Eqn. (52-6), the quotient $\frac{Y - 2G}{2G}$ also lies between the limiting values of 0 and 0.5. If follows that

$$\frac{Y}{3} < G < \frac{Y}{2}.$$

In the subject of thermodynamics we often need the relations between the elastic constants and the compressibility. These are:

$$\varkappa_T = \frac{3(1 - 2\mu)}{Y} = \frac{3(3G - Y)}{GY}. \tag{57-8}$$

53. The Limits of Validity of Hooke's Law

If a rod is stretched or shortened by a tensile or compressive force acting along its axis, we have, from Hooke's law

$$\varepsilon = \frac{\sigma}{Y} \quad \text{or} \quad \Delta l = \frac{\sigma l}{Y},$$

and

$$\delta = \mu \frac{\sigma}{Y} \quad \text{or} \quad \Delta b = \frac{\mu \sigma b}{Y}.$$

Both of these equations are *empirical* and hence have a limited range of validity. The maximum value of the stress for which the longitudinal strain is proportional to the stress is called the *elastic limit* or *proportional limit* of the material. It can be determined from a *stress–strain diagram* for the material.

The most common form of the stress–strain diagram is shown in Fig. 5-15. It shows not only the relation between stress and strain in the elastic

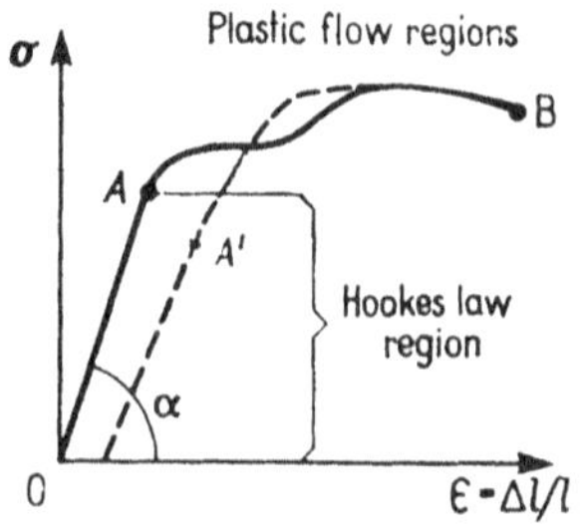

FIG. 5-15. A stress–strain diagram of a rod, showing the Hooke's Law region, plastic flow region, elastic limits, *A* and *A'*, and breaking point, *B*.

region, but also exhibits the properties of the substance in the irreversible region. As the stress is increased from zero, the strain increases in direct proportion until the elastic limit is reached at about point *A*. The range of stress in which the strain and stress are proportional is called the Hooke's Law region. From the definition of Young's modulus *Y*, the slope of this linear portion of the curve equals *Y*:

$$\tan\alpha = \frac{\sigma}{\varepsilon} = Y.$$

The elastic limit is not sharply defined, but in practice it can be located with sufficient accuracy.

When σ is increased beyond its value at the elastic limit, the strain increases much more rapidly than before. If the stress is decreased, after passing the elastic limit, the original curve is not re-traced. This phenomenon is called *elastic hysteresis*. For a given stress, the strain is greater than it was before the rod was stretched beyond its elastic limit. When the stress has been reduced to zero the rod has acquired a permanent strain and the value of Young's modulus is decreased. That is, if a stress is now applied, the stress–strain curve rises with a smaller slope. The elastic limit will also be found to have decreased. It follows that the elastic properties of a substance can only be considered constant if the substance is not stressed beyond its limit of proportionality.

In addition to the elastic modulus and the elastic limit, the stress–strain diagram provides information about other properties of a material. When the elastic limit has been passed, the material enters the region of *plastic flow*, in which permanent changes of shape take place. If this portion of the curve is very long, the material is said to be *ductile* (e.g.: wrought iron, copper). If it is short, the rod breaks soon after the elastic limit has been passed and the substance us called *brittle* (e.g.: glass, cast iron, bismuth, arsenic). The properties of ductility and brittleness are not related to the elastic properties of a substance, but refer to its properties in the region of plastic flow. Plastic flow is always accompanied by a noticeable rise in the temperature, since mechanical work is being converted into heat. The internal structure of the material changes, which is the reason for the changes in its elastic modulus and elastic limit. The origin of these phenomena in the irreversible region must be sought in the microcrystalline structure of a material. For large tensile or compressive stresses, slipping processes take place between the microcrystals in the planes of maximum shearing stress, and the material acquires a fibrous structure. The assumption that the body is isotropic is no longer satisfied and the relations derived in this chapter between stress and strain are no longer valid.

With still greater increase in stress, the rod eventually breaks. In the vicinity of the break, the cross-section of the rod decreases markedly as a result of plastic flow. As a result of this decrease in cross-sectional area, the tensile *stress* at the section increases more rapidly than the tensile *force*. The ratio of the tensile force to the cross-sectional area, just before breaking, s called the *breaking strength*.

54. Plane Bending of a Beam

If a straight beam of constant cross-sectional area A and length l is fixed at one end and acted on by its own weight and perhaps also by other external forces in a vertical plane, the beam becomes curved. In this section we shall neglect the weight of the beam and assume that only concentrated external forces such as $\boldsymbol{P}$ (Fig. 5-16) act on the beam. We shall also assume that all plane sections through the beam, parallel to the plane of the external forces (the x-y plane, as defined below), are in the same state of stress so that the stress is plane. Let us construct a coordinate system whose x-axis passes through the center of gravity of the cross-section A of the undistorted beam (Fig. 5-16). The z-axis points toward the reader and the y-axis verti-

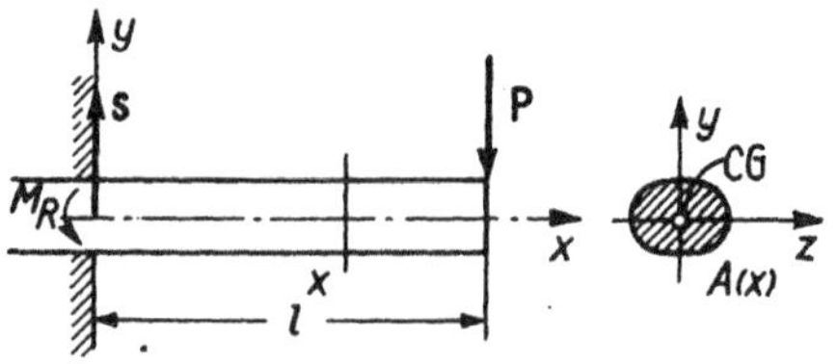

Fig. 5-16. The bending of a beam, showing the impressed force $\boldsymbol{P}$, the reaction force at the wall $\boldsymbol{S}$, and the reaction moment M_R.

cally upward. Since all forces are parallel to the y-axis, all cross-sections are at first subjected to a shearing stress only. This results in a bending of the beam, as a consequence of which new stresses are set up within it. Then the upper portion of the beam is in tension, the lower in compression. In the equilibrium state of the deformed beam, normal as well as shearing stresses exist at every cross-section. Since the stress is independent of z (because we have a state of plane stress) the shearing stresses at sections perpendicular to the axis of the beam are at first parallel to the y-axis. We neglect the longitudinal shear stress.

To investigate the equilibrium of the beam, we imagine it to be cut in two at an arbitrary position x by means of a cut perpendicular to its axis. We then assume that external forces are exerted on the portion to the left of the cut, so that the original stresses σ_{xx} and τ_{xy} are reproduced. As a result of these stresses, the left-hand portion of the beam remains in equilibrium.

Before proceeding with a discussion of the equilibrium of the beam, it will be advantageous to introduce two new quantities.

DEFINITION. The *transverse shearing force* $Q(x)$ at the section $A(x)$ is defined as the sum of all vertical shearing forces at the section, exerted on the portion to the left of the section by the portion to the right:

$$Q(x) = \sum \tau_{xy} \Delta A .$$

The normal stresses σ_{xx} at the section $A(x)$ produce a so-called *bending moment* about the z-axis.

DEFINITION. The *bending moment* $M_B(x)$ at the section $A(x)$ is defined as the moment of all the normal forces at the section with respect to the z-axis, exerted on the portion to the left of the section by the portion to the right.

$$M_B(x) = \sum \sigma_{xx} \Delta A \cdot y .$$

With the help of these concepts, the equilibrium conditions for the left-hand portion of the beam can be stated.

No external forces in the direction of the x-axis act on the portion of the beam considered. As a result of the deformation, however, tensile forces act on the upper portion of the section and compression forces act on the lower portion. For equilibrium, we must have

$$\sum_{\substack{\text{Left portion}\\ \text{of beam}}} F_x = \sum_{\substack{\text{cross-}\\ \text{section}}} \sigma_{xx} \Delta A = 0 . \tag{54-1}$$

The stress distribution over the section is such that there is a line in the plane of the section at which the normal stress is zero. It is called the *neutral axis* of the section, and is neither stretched nor compressed as a result of the beam's deformation. The totality of the neutral axes of all the sections forms the *neutral surface*, which is also undeformed under stress.

As a reaction to the external vertical forces exerted on the beam, a vertical force (see Fig. 5-16) acts on the beam at its fixed end. The magnitude of this force will be determined when we consider the equilibrium of the beam as a whole. A vertical shearing force given by $\sum \tau_{xy} \Delta A$ acts at the cross-section being considered. The equilibrium conditions for the y-direction therefore lead to the equation

$$\sum_{\substack{\text{Left portion}\\ \text{of beam}}} F_y = S + \sum_{\substack{\text{cross-}\\ \text{section}}} \tau_{xy} \Delta A = 0. \tag{54-2}$$

The moment equations, with respect to the three coordinate axes through the center of gravity of the section considered, are the following:

$$M_x = \sum \tau_{xy} \Delta A \cdot z = 0, \tag{54-3}$$

$$M_y = \sum \sigma_{xx} \Delta A \cdot z = 0, \tag{54-4}$$

$$M_z = - Sx + M_R + \sum \sigma_{xx} \Delta A \cdot y = 0, \tag{54-5}$$

or

$$M_B(x) = Sx - M_R, \tag{54-5'}$$

where M_R is the reaction moment exerted by the supports on the fixed end of the beam (see Fig. 5-16). From the equilibrium conditions for the beam as a whole, it follows that

$$\boldsymbol{S} = -\boldsymbol{P}, \quad \text{or} \quad S = P,$$

and

$$M_R - Pl = 0,$$

where $\boldsymbol{P}$ is the resultant external force acting on the beam, presumed to act on the free end. Hence

$$M_R = Pl = Sl.$$

Using the conditions of equilibrium, the transverse shearing force $Q(x)$ and the bending moment $M_B(x)$ can be computed for any given loading of the beam. From Eqn. (54-2), the shearing force $\boldsymbol{Q}(x)$ at any section, given by $\sum \tau_{xy} \Delta A$, is equal in magnitude and opposite in direction to the resultant vertical force acting on the left part of the beam:

$$\boldsymbol{Q}(x) = -\boldsymbol{S} = \boldsymbol{P}.$$

The bending moment at any section, $M_B(x) = \sum \sigma_{xx} \Delta A \cdot y$, is equal in magnitude and opposite in direction to the sum of all moments acting on the left part of the beam, where the moments are taken about the center of gravity of the section being considered:

$$M_B(x) = -(M_R - Sx) = -S(l - x).$$

However, the distribution of normal and shearing stresses over the section cannot be determined without further assumptions. To determine the stresses, and the deformations produced by them, it is necessary to make four assumptions, known as the *Navier–Bernoulli* assumptions. These are:

(i) *The shearing stress* τ_{xy} *is uniform over any cross-section of the beam*:

$$\tau_{xy} = \frac{Q(x)}{A} = \frac{P}{A}. \tag{I}$$

In the example we are considering, τ_{xy} is directed downward. Moment Eqn. (54-3) can be satisfied on the basis of assumption (i), since τ_{xy} can then be taken outside the summation sign and we have

$$\tau_{xy} \sum \Delta A \cdot z = 0.$$

Since $\tau_{xy} \neq 0$, assumption (i) requires that the y-axis pass through the center of gravity of the cross-section, for only then will $\sum \Delta A \cdot z$ vanish.

(ii) *A plane cross-section originally perpendicular to the axis of the beam remains plane when the beam is deformed.*†

This hypothesis can be satisfied only if the extension or contraction of each longitudinal fibre of the beam is proportional to its distance y from the origin. If this is the case, then from Hooke's law the normal stress σ_{xx} must be directly proportional to y (Fig. 5-17). That is, we must have

$$\sigma_{xx} = \alpha y \tag{II}$$

where α is a constant. Since the neutral surface experiences no normal stress, then according to the second assumption it must be located at $y = 0$.

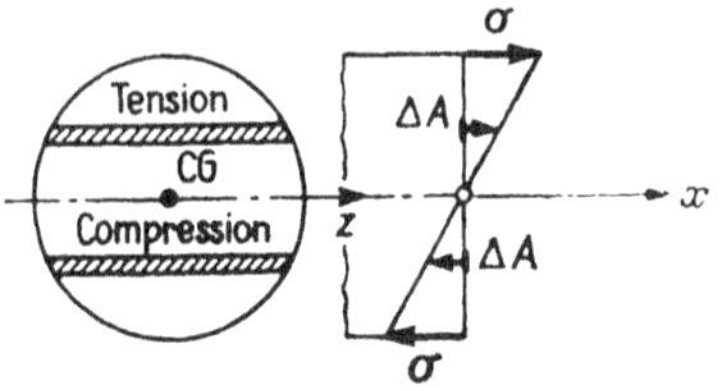

FIG. 5-17. Normal stresses at a typical cross-section, $A(x)$.

In the case of the cantilever beam under consideration, we then see that for positive values of y the stress is a tension; for negative values, a compression (Fig. 5-17). Making use of the second assumption, the equilibrium equation (1) takes the form

$$\sum \sigma_{xx} \Delta A = \alpha \sum y \cdot \Delta A = 0.$$

The condition that $\sum y \cdot \Delta A = 0$ requires that the x-axis must pass through the center of gravity of the section. Equation (54-4), together with assumption (ii), becomes

$$\sum \sigma_{xx} \Delta A \cdot z = \alpha \sum yz \cdot \Delta A = 0,$$

† An exact analysis shows that this hypothesis leads to an internal contradiction in the theory. See Sommerfeld, *Mechanics of Deformable Bodies*, Academic Press, 1952.

which is only possible if

$$\sum yz \cdot \Delta A = 0. \tag{III}$$

We shall return to this condition later. Finally, the definition of $M_B(x)$ and assumption (ii) give

$$M_B(x) = \sum \sigma_{xx}\Delta A \cdot y = \alpha \sum y^2 \Delta A,$$

or

$$M_B = \frac{\sigma_{xx}}{y} J_{zz}, \tag{54-6}$$

where

$$J_{zz} = \sum y^2 \Delta A.$$

The quantity J_{zz} is called the *second moment of the cross-sectional area* for bending about the z-axis. For bending about the y-axis, we would use the second moment of area J_{yy},

$$J_{yy} = \sum z^2 \Delta A.$$

The second moment of area has the dimensions

$$[J_{zz}] = [L^4].$$

It is expressed in cm^4, m^4, or ft^4. These quantities are purely geometrical properties of the cross-section and can be evaluated by integration. When they are known, Eqn. (54-6) gives the relation between the bending moment $M_B(x)$ and the normal stress:

$$\sigma_{xx} = \frac{M_B(x)}{J_{zz}} y. \tag{54-7}$$

Condition III above also imposes a purely geometrical condition on the cross-section. The quantity

$$J_{yz} = \sum yz\Delta A$$

is called the *product moment of area* of the cross-section (Fig. 5-18). According to condition III, the linear variation of the normal stress expressed by condition II is possible only if the product moment of area of the cross-section vanishes. Thus we come to the third Navier–Bernoulli assumption:

(iii) *The product moment of area of each cross-section must vanish.*

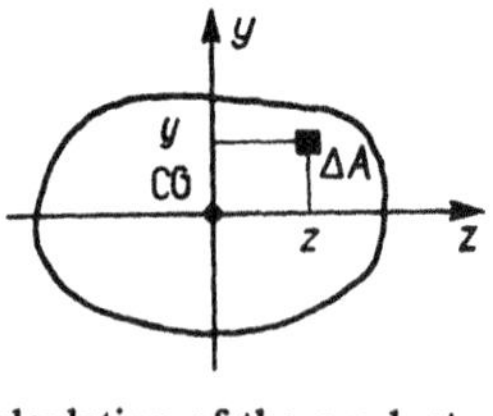

FIG. 5-18. The calculation of the product moment of area.

The fourth Navier–Bernoulli assumption will be introduced in connection with the evaluation of the deformation of the elastic curve (see Section 54-1).

In the calculation of bending moments it should be noted that the bending moment at any cross-section is equal and opposite to the moment of all external forces at the left of the section and is equal to the moment of all external forces at the right of the section. In every case, moments are computed about the center of gravity of the cross-section. The bending moment $M_B(x)$ varies smoothly without a break along every portion of a beam on which no concentrated loads are applied. At points where such loads act, the graph of $M_B(x)$ vs. x changes direction suddenly. Let us illustrate this by an example.

EXAMPLE. Figure 5-19 shows a beam of length l_0 resting on two supports at its ends and acted on by a concentrated load P at the point $x = p$. The weight of the beam itself is negligible. We wish to compute the bending moments at the left and at the right of the load.

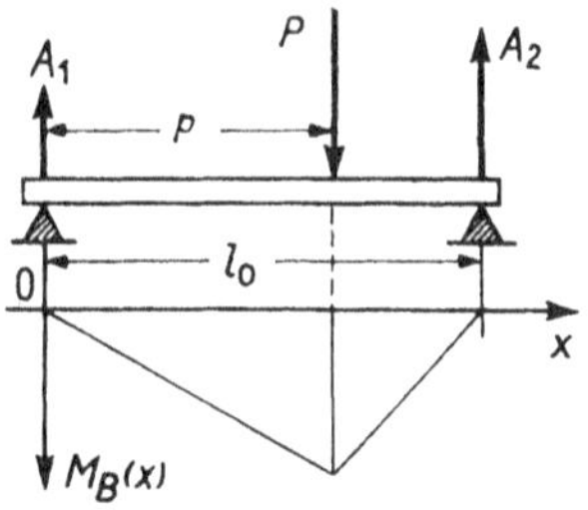

FIG. 5-19. The bending moments of a beam which rests on two supports and is acted on by a concentrated load, P.

At the left: $M_{B1}(x) = A_1 x$,
where

$$A_1 = \frac{P(l_0 - p)}{l_0}.$$

Therefore between $x = 0$ and $x = p$, we have

$$M_{B1}(x) = \frac{P(l_0 - p)}{l_0} x.$$

At the right: $M_{B2}(x) = A_2(l_0 - x)$,
where

$$A_2 = \frac{Pp}{l_0}.$$

Therefore in the region between $x = p$ and $x = l_0$,

$$M_{B2}(x) = \frac{Pp}{l_0}(l_0 - x).$$

Both moments are linear functions of x, and they are equal at the point $x = p$. The graph of M_B vs. x has a sharp break in slope at the point $x = p$.

As an illustration of the calculation of second moments of area, we consider two important special cases. Appendix II on the calculation of moments of inertia of variously shaped solids should be consulted for examples of more general cases.

(a) Rectangular cross-section (Fig. 5-20).

$$dJ_{zz} = y^2\, dA = ay^2\, dy,$$

$$J_{zz} = a\int_{-b/2}^{+b/2} y^2\, dy = \frac{ay^3}{3}\bigg|_{-b/2}^{+b/2} = \frac{1}{12}ab^3.$$

Similarly, we find that

$$J_{yy} = \frac{1}{12}a^3b.$$

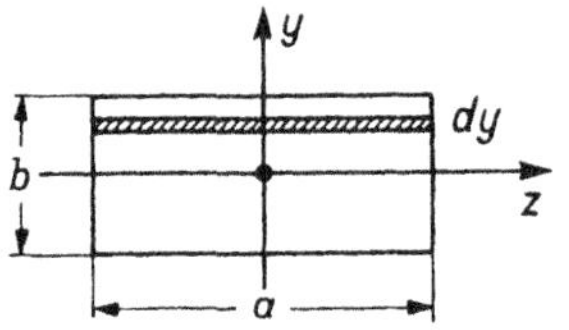

FIG. 5-20

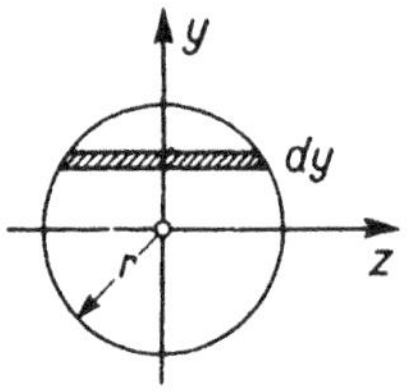

FIG. 5-21

FIG. 5-20. The calculation of the second moment of area of a rectangle.

FIG. 5-21. The second moment of area of a circle.

(b) Circular cross-section (Fig. 5-21). The equation of a circle is

$$z^2 + y^2 = r^2.$$

$$dJ_{zz} = 2zy^2\, dy = 2y^2 \sqrt{(r^2 - y^2)}\, dy,$$

$$J_{zz} = 2r\int_{-r}^{+r} y^2 \sqrt{[1 - (y/r)^2]}\, dy$$

Replacing y/r by $\sin\varphi$, we get

$$J_{zz} = 2r^4\int_{-\pi/2}^{+\pi/2} \sin^2\varphi \cos^2\varphi\, d\varphi = \frac{r^4}{2}\int_0^{\pi} \sin^2 2\varphi\, d(2\varphi),$$

$$J_{zz} = \frac{\pi}{4}r^4.$$

54-1 Stresses in a Bent Beam (plane stress)

Using the assumptions and relations developed above, the stress distribution in a bent beam can be obtained with the help of the Mohr circle. It is necessary to know the normal and shearing stresses at some one point of the beam, for two mutually perpendicular elements of area. We shall refer to these elements of area as *conjugate elements*. We shall take the first element in the plane of some arbitrary cross-section of the beam, and its conjugate element perpendicular to the cross-section and in the x-z plane. To correlate the stresses at the two elements, we must make one more assumption, namely, that any two layers of the beam which make contact in a horizontal plane exert only shearing stresses on each other. That is, such layers satisfy the condition $\sigma_{yy} = 0$. Figure 5-22 then shows the Mohr circles for a number of positions of two conjugate elements of area, relative to the neutral layer of the beam.

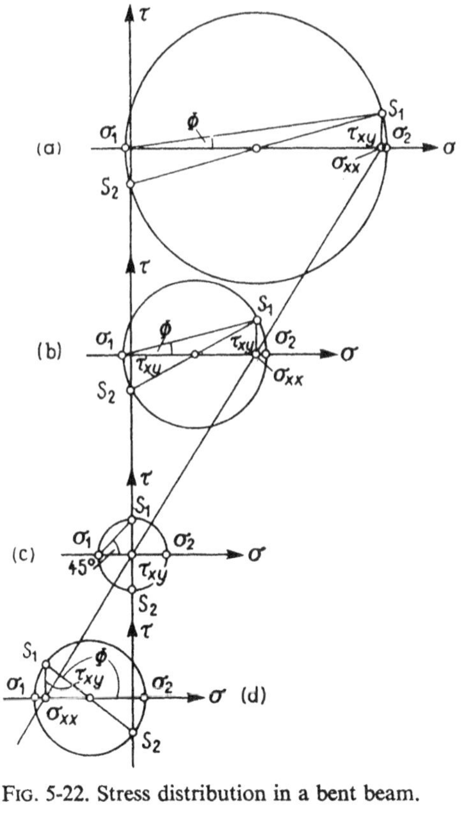

FIG. 5-22. Stress distribution in a bent beam.

(a) *Topmost layer*. Element of area I lies in a plane normal to the length of the beam and hence is subject to a very large normal stress σ_{xx} and a relatively small shearing stress τ_{xy}. For a given loading of the beam, both of these stresses can be computed. These two values together serve to locate the stress point S_1 on the Mohr circle. The conjugate element (II) has the same shearing stress and, according to the assumption above, no normal stress σ_{yy}. This locates the second stress point S_2 on the Mohr circle. The line joining S_1 and S_2 locates the center of this circle and hence the complete stress distribution in the topmost layer is determined.

The two elements of area at this layer at which the stress is purely normal (principal stress planes) make a small angle Φ with the conjugate elements. At one of these the stress is very large and is a tension (σ_2); at the other it is small and is a compression (σ_1).

(b) As the neutral surface is approached, the normal stress σ_{xx} at the element of area I decreases linearly. The shearing stresses do not change and at the conjugate element II the normal stress σ_{yy} is again zero. The Mohr circle is smaller, and the angle between the principal stress planes and the conjugate elements of area is larger. The compressive stress σ_1 in the second principal plane has increased.

(c) At the neutral surface, both normal stress components vanish, while the shearing stress retains the same value $\tau_{xy} = Q/A$. The neutral surface is therefore in a state of pure torsion. The principal planes of stress are inclined at an angle of 45° to the axis of the beam, and the normal stresses at these planes are equal to the shearing stress:

$$\sigma_1 = \sigma_2 = \tau_{xy} = \frac{Q}{A}.$$

(d) Below the neutral surface, the normal stress at element I is compressive and increases linearly with y and, since the shearing stress remains constant, the planes of principal stress are inclined at an increasingly large angle. The compressive stresses in these planes increase and the tensile stresses decrease. At the lowest layer, the planes of principal stress again coincide approximately with the conjugate elements of area.

55. The Elastic Curve of a Bent Beam

The *elastic curve* of a beam is the intersection of the plane of the external forces (generally the vertical x-y plane) with the plane of the neutral surface. To calculate the deformations of the elastic curve of the loaded beam, we require the fourth Navier–Bernoulli assumption:

(iv) *The cross-sections of a beam remain perpendicular to the elastic curve of the beam when the beam is deformed.*

This assumption is justified by the fact that, except in the innermost layers of the beam, the shearing stress is negligible compared with the normal stress. Thus we are ignoring the effect of the shearing stresses in changing the angle of the deformed cross-sections relative to the axis of the beam.

The mathematical expression of this condition can be written down at once from the diagram in Fig. 5-23. We have

$$l = R\Delta\Phi,$$

$$\Delta l = y_0 \Delta\Phi,$$

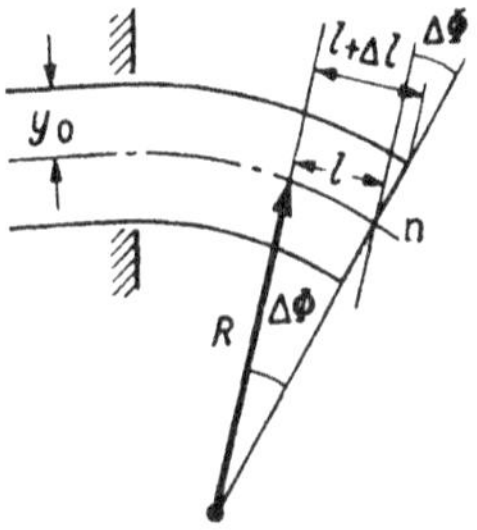

FIG. 5-23. Calculation of the elastic curve. The neutral surface is indicated by n.

where l is the unchanged length of a given section of the neutral axis and $l + \Delta l$ is the length of the corresponding topmost layer of the beam under stress. From this it follows that

$$\Delta l = \frac{y_0 l}{R}.$$

From Hooke's law,

$$\Delta l = \frac{y_0 l}{R} = \frac{\sigma_0 l}{Y},$$

where σ_0 is the normal stress in the topmost layer of the beam, and from Eqn. (55-7)

$$\frac{\sigma_0 l}{Y} = \frac{M_B y_0 l}{J_{zz} Y} = \frac{y_0 l}{R}.$$

Therefore

$$\frac{1}{R} = \frac{M_B}{J_{zz} Y},$$

where R is the radius of curvature of the elastic curve at the point x, and is given by†

$$R = \frac{\left[1 + \left(\frac{dy}{dx}\right)^2\right]^{3/2}}{d^2y/dx^2}$$

The Navier–Bernoulli assumptions are valid only for small distortions of a beam. In such a case, dy/dx is small compared with unity and the expression for R reduces to

$$\frac{1}{R} = \frac{d^2y}{dx^2}.$$

The differential equation for the elastic curve is then

$$\frac{d^2y}{dx^2} = \frac{M_B}{J_{zz} Y}.$$

† See any standard text on calculus for a derivation of this formula.

Hence, if M_B is known as a function of x, the equation of the elastic curve can be obtained by two integrations. It should be noted that $M_B(x)$ is integrable only within a region which is free from concentrated loads.

55-1 Examples of the Calculation of the Elastic Curve

1. A horizontal beam of length l (Fig. 5-24) is fixed at its left end, and a downward force $\boldsymbol{P}$ is exerted at its right end. The cross-section of the beam

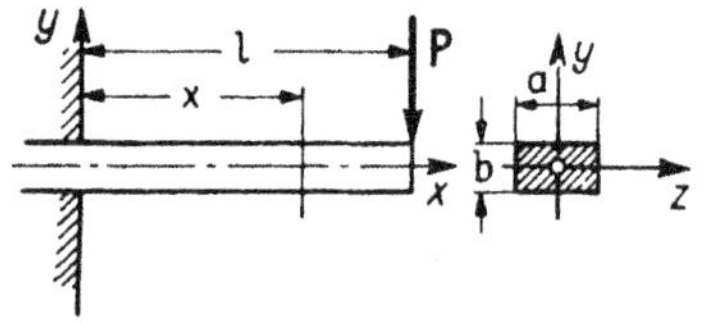

FIG. 5-24. Deflection of a beam supported only at one end (cantilever beam).

is a rectangle of width a and depth b. We wish to obtain the equation of the elastic curve and the deflection of the free end. We have:

$$M_B(x) = -P(l - x),$$

$$\frac{d^2y}{dx^2} = \frac{P(x - l)}{J_{xx}Y}.$$

Let

$$\frac{dy}{dx} = u, \quad \frac{d^2y}{dx^2} = \frac{du}{dx}.$$

Then

$$du = \frac{P}{J_{xx}Y}(x - l)\,dx,$$

$$u = \frac{P}{J_{xx}Y}\left(\frac{x^2}{2} - lx + C_1\right) = \frac{dy}{dx},$$

where C_1 is a constant of integration. One *boundary condition* imposed on the beam is

$$\frac{dy}{dx} = 0 \quad \text{when} \quad x = 0,$$

from which it follows that $C_1 = 0$. A second integration gives

$$y = \frac{P}{J_{xx}Y}\left(\frac{x^3}{6} - \frac{lx^2}{2} + C_2\right).$$

But $y = 0$ when $x = 0$, so $C_2 = 0$ also.

The equation of the elastic curve is therefore a cubic,

$$y(x) = \frac{P}{6J_{zz}Y}(x - 3l)\,x^2.$$

The displacement of the free end is obtained by setting $x = l$. We have

$$y(l) = \frac{-Pl^3}{3J_{zz}Y}.$$

For a rectangular cross-section

$$J_{zz} = \frac{ab^3}{12}$$

and

$$y(l) = \frac{-4Pl^3}{ab^3Y}.$$

2. A rod of length l_0 is supported near its ends as in Fig. 5-25, and bends under its own weight w. We wish to find the equation of the elastic curve and the deflection of the mid-point of the rod. The cross-section of the rod

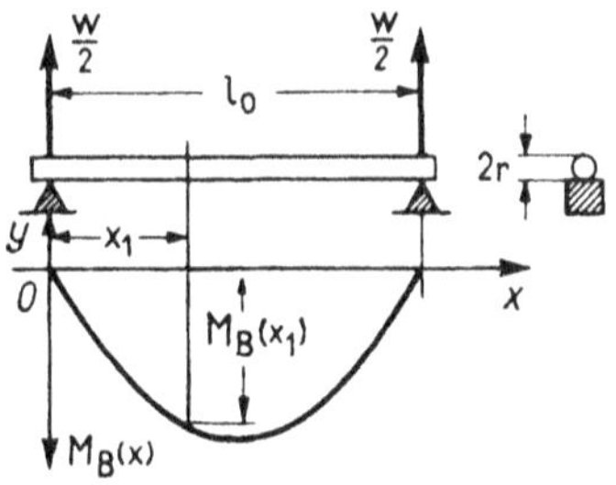

FIG. 5-25. Deflection of a beam supported at both ends (simple beam).

is a circle of radius r, area A, and the specific weight of the rod is s. Taking moments of the forces to the left about the neutral axis at x, we have

$$M_B(x) = -\left(-\frac{w}{2}x + \frac{As}{2}x^2\right) = \frac{As}{2}(l_0x - x^2),$$

which is the equation of a parabola. Then

$$\frac{d^2y}{dx^2} = \frac{As}{2J_{zz}Y}(l_0x - x^2),$$

$$\frac{dy}{dx} = \frac{As}{2J_{zz}Y}\left(\frac{l_0x^2}{2} - \frac{x^3}{3} + C_1\right).$$

When $x = l_0/2$, $dy/dx = 0$, and hence

$$C_1 = \frac{-l_0}{12}l_0^3.$$

Then

$$\frac{dy}{dx} = \frac{As}{24J_{zz}Y}(6l_0x^2 - 4x^3 - l_0^3)$$

and

$$y = \frac{As}{24J_{zz}Y}(2l_0x^3 - x^4 - l_0^3x + C_2).$$

When $x = 0$, $y = 0$ and hence $C_2 = 0$. It follows that y is also zero when $x = l_0$, as it should be.

In this case, the elastic curve is of fourth degree:

$$y(x) = \frac{-As}{24J_{zz}Y}(x^4 - 2l_0x^3 + l_0^3x).$$

The maximum deflection occurs at the point $x = l_0/2$, and is given by

$$y\left(\frac{l_0}{2}\right) = \frac{-5wl_0^3}{384J_{zz}Y}.$$

For a circular cross-section, $J_{zz} = \pi r^4/4$, and

$$y\left(\frac{l_0}{2}\right) = \frac{-5wl_0^3}{96\pi r^4 Y}.$$

56. Pure Torsion

Suppose that the left end of a rod of circular cross-section and of area A is rigidly fixed and that a torque of moment M_t is exerted on its right end, the torque vector lying along the axis of the rod. We say that a *torsion moment* acts on the rod. The x-axis of the coordinate system lies along the axis of the rod. The y-axis points upward and the z-axis outward toward the reader (Fig. 5-26). For small deformations, which is the only case we shall consider, it can be assumed that all cross-sections normal to the axis remain plane and perpendicular to the axis when the rod is twisted.

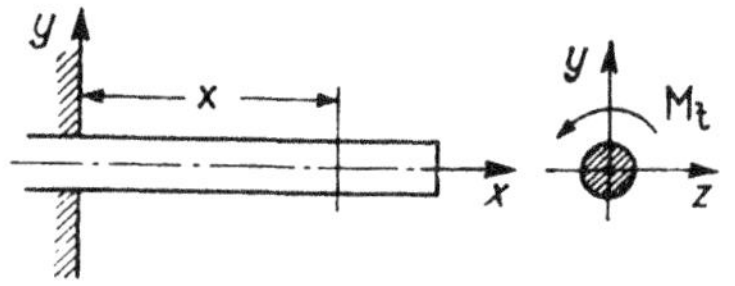

Fig. 5-26. Torsional stress on a rod.

If we consider a cross-section at the coordinate x, then since no external forces act parallel to the x-axis we must have from the conditions of equilibrium

$$\sum \sigma_{xx}\Delta A = 0.$$

The stress at an element of area in the plane of a cross-section is a pure shear.

To evaluate the deformation and the magnitudes of the stresses, certain simplifying assumptions are necessary. The simplest case is that of the purely torsional deformation of a thin rod or thin-walled hollow cylinder of circular cross-section.

We assume that the shearing force at every point of a cross-section is tangential to a circle through the point, having its center at the center of gravity of section, and also that the shearing stress τ is proportional to the

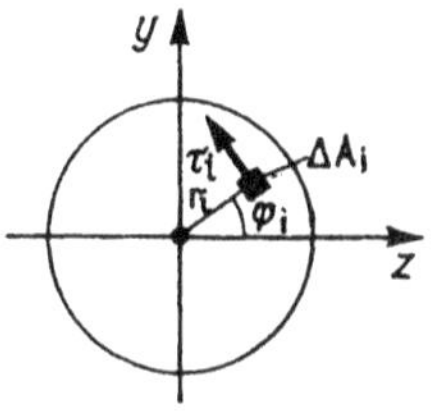

FIG. 5-27. Force in a rod under torsional stress.

distance r_i from the center of gravity. If this is the case, the z-component of the shearing force is (see Fig. 5-27)

$$\sum F_{zi} = 0 = -\sum \tau_i \Delta A_i \sin \varphi_i,$$

as we now prove.

If τ_1 is the shearing stress at unit distance from the center, the stress τ_i at a distance r_i is

$$\tau_i = r_i \tau_1$$

and

$$\tau_1 \sum r_i \sin \varphi_i \Delta A_i = \tau_1 \sum y_i \Delta A_i = 0. \tag{56-1}$$

Similarly, since $\sum F_{yi} = 0$, we get

$$\sum z_i \Delta A_i = 0. \tag{56-2}$$

Equations (56-1) and (56-2) merely state that the point at which $\tau_i = 0$ is the center of gravity of the cross-section.

The moment equation leads to the result:

$$M_t = \sum \tau_i \Delta A_i r_i = \tau_1 \sum \Delta A_i r_i^2 = \tau_1 J_p, \tag{56-3}$$

where

$$J_p = \sum r_i^2 \Delta A_i$$

and is called the *second polar moment of area* of the cross-section.

Dimensions: $[J_p] = [L^4]$

Units: 1 cm^4, 1 m^4, 1 ft^4.

Since $r_i^2 = z_i^2 + y_i^2$, then

$$J_p = \sum \Delta A_i (z_i^2 + y_i^2) = J_{yy} + J_{zz}.$$

That is, the second polar moment of area equals the sum of the second axial moments of area about two mutually perpendicular axes. For a circular cross-section of radius r (see p. 232),

$$J_p = \frac{\pi}{2} r^4.$$

If the torsion moment of the external forces about the axis of the rod is known, the shearing stress at a distance r from the axis can be found from Eqn. (56-3),

$$\tau_r = \tau_1 r = \frac{M_t r}{J_p}. \tag{56-4}$$

This equation corresponds to Eqn. (54-7) on page 230 which gives the normal stress in a bent beam, and is called the *torsion formula.*

A cylinder of length l and radius a is distorted by a torsional moment as shown in Fig. 5-28, in such a way that a line in its surface, originally parallel

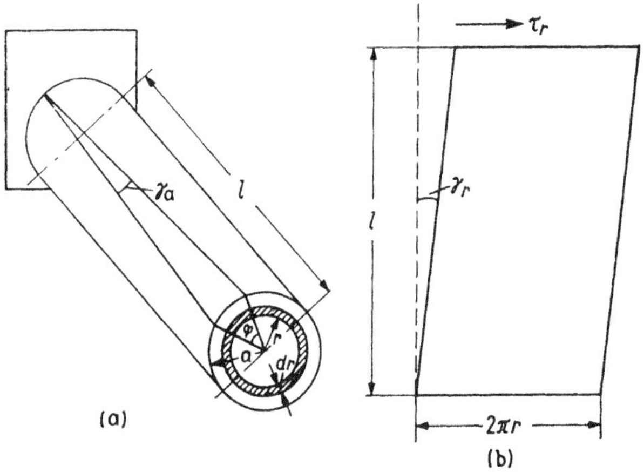

FIG. 5-28. Torsion in a cylinder.

to the axis, is twisted through an angle γ_a. Let us think of the cylinder as being made up of many coaxial cylindrical shells, a typical one having a radius r and thickness dr. Imagine that such a shell is slit down one side, parallel to the axis of the cylinder, and unrolled, giving a rectangular sheet. At each cylindrical shell there will be a shearing stress, given by Eqn. (56-4) above, and if this stress is applied to the rectangular sheet, the right angles will change by an amount $\pm\gamma_r$ (Fig. 5-28). For small angles, the value of γ_r is given by Eqn. (52-5) in Section 52-1,

$$\gamma_r = \frac{\tau_r}{G} = \frac{M_t r}{G J_p}. \tag{56-5}$$

If the torsion moment is the same for all cross-sections of a rod (which will always be the case when a cylindrical rod is twisted by torques applied at its ends) then the angle γ_r is the same at all points along the length of the rod at a given radius. A line on the surface, originally parallel to the axis, then becomes a helix of constant pitch. Every radius of the cross-section at the end of the rod rotates through a certain angle φ. From Fig. 5-28 the relation between the angles γ_r and φ is

$$l\gamma_r = r\varphi.$$

The angle of twist of the rod is therefore

$$\varphi = \frac{l\gamma_r}{r} = \frac{M_t l}{J_p G}. \tag{56-6}$$

This equation for the twisting of a rod of circular cross-section, when acted on by a torsion moment, has the same form as Hooke's law for extension or compression,

$$\Delta l = \frac{Fl}{AY}.$$

The torsion moment corresponds to the force, the second polar moment of area corresponds to the cross-sectional area, and the shear modulus to Young's modulus. The dependence of the torsion angle φ on the cross-section is contained in the quantity J_p.

It should be pointed out, however, that the torsion angle is determined by the second polar moment of area only for a rod of circular cross-section. For cross-sections of any other shape, the dependence of angle on cross-section is more complicated. However, the circular cross-section is the one most frequently encountered in practice.

EXAMPLE. (1) A steel wire ($G = 8 \times 10^{11}$ dynes/cm^2), 50 cm long and 1 mm in diameter, is twisted through an angle of 360°. What torsion moment is required?

$$M_t = \frac{\varphi J_p G}{l} = 9.87 \times 10^5 \text{ cm} \cdot \text{dyne}.$$

(2) A steel rod 1 cm in diameter and 1 m long is acted on by a torsion moment of 1 kgf · m ($G = 8 \times 10^5$ kgf/cm^2). Through what angle does one end of the rod turn, relative to the other end?

$$\varphi = \frac{M_t l}{J_p G} = 0.128 \text{ rad} = 7.3°.$$

57. THE SPIRAL SPRING

If a thin wire of circular cross-section, having a radius r, is wound in a helix on a cylindrical form of radius R, we obtain a helical spring, usually called a spiral spring. Let the ends of the wire each be bent through two right angles as in Fig. 5-29, the upper end fixed, and an axial force $\boldsymbol{P}$ exerted on the lower end. The weight of the spring itself will be neglected. To determine the effect of the force $\boldsymbol{P}$ on the spring, we first consider the effect produced by $\boldsymbol{P}$ at some arbitrary cross-section of the spring. Let us apply at the section A of the wire (Fig. 5-29) a force $\boldsymbol{P}'$, and also the force $\boldsymbol{P}''$, which is opposite in direction to $\boldsymbol{P}$ and which counteracts the effect of $\boldsymbol{P}'$. $\boldsymbol{P}'$ and $\boldsymbol{P}''$

both have the same magnitude as $\boldsymbol{P}$. Then the forces $\boldsymbol{P}$ and $\boldsymbol{P}''$ constitute a couple whose moment equals PR.

The effect of the force $\boldsymbol{P}$ at the section A is therefore equivalent to a single force of magnitude P and a torsion moment $M_t = PR$, whose axis is perpendicular to A. The stress at the section A can be considered as the superposition of a shearing stress produced by the force $\boldsymbol{P}'$, and a torsion stress produced by the torsion moment M_t. Strictly speaking, there is a small normal stress and a bending moment also, but these are negligible if the spring has a small pitch. The torsion moment plays the chief role in the distortion of the spring. In the following discussion, we shall assume that the spring has a small pitch and shall neglect both the bending and the shear.

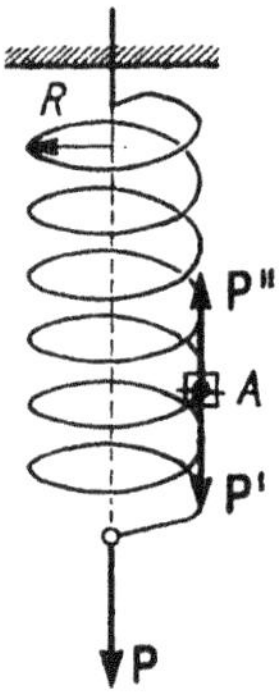

FIG. 5-29

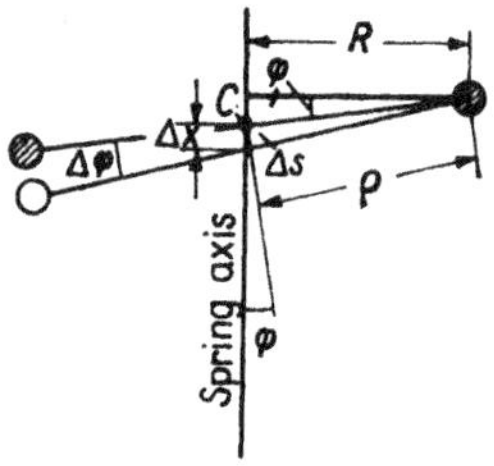

FIG. 5-30

FIG. 5-29. Spiral spring.

FIG. 5-30. Details of the stretching of a spiral spring.

To calculate the deformation of the spring, which according to the assumptions above is caused only by the torsion moment, we consider a short section Δl of one turn of a low pitch spring. As a result of the torsion moment M_t, the section is twisted through an angle $\Delta\varphi$ given by Eqn. (56-6):

$$\Delta\varphi = \frac{M_t \Delta l}{G J_p}. \qquad (57\text{-}1)$$

As a result of this twist, all of the spring below the element is slightly displaced downward. In particular, the single turn of which the element Δl is a part is displaced downward by an angle $\Delta\varphi$ from its original position and the mid-point C of the turn moves down through a distance

$$\Delta x = \Delta s \cos\varphi = \varrho \Delta\varphi \cos\varphi = R\Delta\varphi. \qquad (57\text{-}2)$$

At the same time, it is displaced slightly to one side of the axis of the spring. However, by reasons of symmetry the resultant of all such sidewise displacements will be zero and the net result of the torsion action is only an axial extension of the spring.

If we insert in Eqn. (57-1) the value RP for the torsion moment M_t, and make use of Eqn. (57-2), we get

$$\Delta x = \frac{R^2 P \Delta l}{GJ_p}.$$

The extension x of the entire spring is

$$x = \sum \Delta x = \frac{PR^2 l}{GJ_p},$$

where l is the total length of the wire composing the spring. If it contains n turns, then

$$l = 2\pi Rn.$$

For a circular cross-section,

$$J_p = \frac{\pi}{2} r^4$$

and the extension of the spring is

$$x = \frac{4nR^3}{Gr^4} P.$$

The extension of the spring is therefore proportional to the stretching force $\boldsymbol{P}$, a result which agrees with experiment and which we have already used. The *spring law* is usually written in the form

$$P = kx.$$

The spring constant k for a given spring is therefore

$$k = \frac{Gr^4}{4nR^3}.$$

To determine k experimentally, the spring is stretched by an axial force and the increase in its length is measured. The spring constant k is numerically equal to the force required to extend the spring a unit distance.

EXAMPLE. Consider a steel wire,

$$r = 0.25 \text{ mm}$$

$$n = 10$$

$$R = 1.5 \text{ cm}$$

$$G = 8 \times 10^5 \text{ kgf/cm}^2 \simeq 8 \times 10^{11} \text{ dynes/cm}^2$$

$$k = \frac{Gr^4}{4nR^3} = 2320 \text{ dynes/cm}.$$

Hence a force of 2320 dynes is needed to stretch the spring 1 cm.

CHAPTER 6

MECHANICS OF FLUIDS

THE subject of fluid mechanics, in a restricted sense, deals with the laws describing the behaviour of liquids. It can be subdivided into *statics of fluids*, the study of the equilibrium state of fluids at rest, and *fluid dynamics*, the study of the dynamic properties of fluids in motion.

In a wider sense, fluid mechanics also includes *aerostatics* and *aerodynamics*, the study of gases at rest and in motion.

In studying the mechanics of fluids it is helpful to begin by considering the laws of an *ideal liquid* and an *ideal gas*. An ideal liquid is *frictionless* and *incompressible*. The assumption that a liquid is frictionless is equivalent to the assertion that for such a liquid no shearing stresses act at the boundary surface between two layers of liquid moving with different velocities. The ideal gas is also frictionless but is compressible and obeys the *ideal gas law* $PV_m = RT$. In aerodynamics, gases can be considered incompressible provided velocities do not exceed 87 mi/hr† (as happens in the case of flow around a supersonic airplane or bullet). Real fluids depart more or less from these ideal laws, and we shall consider them in a special section.

58. Statics of Ideal Liquids

The three phenomena to be considered in this section are the following:

1. The state of stress in the interior of a liquid,
2. The variation of pressure with depth, and
3. Buoyancy.

58-1 The State of Stress in the Interior of an Ideal Liquid at Rest

The assumption of the absence of shearing stresses requires that the force on any arbitrary element of area, at any point in the interior of a liquid, must be perpendicular to the element and directed toward it. We then speak of the *pressure* P at the point, defining it as the ratio of the force dF to the area of the element, dA:

$$P = \frac{dF}{dA}.$$

† Because

$$\tfrac{1}{2}\varrho v^2 < 10^4 \text{ dyn/cm}^2 \sim 10^{-2} \text{ atm},$$
$$v \sim 40 \text{ m/sec} \sim 140 \text{ km/hr} \sim 87 \text{ mi/hr}.$$

Dimensions: $[P] = [ML^{-1}T^{-2}]$

Units: cgs system: 1 dyne/cm^2
mks system: 1 newton/m^2
fps system: 1 lbf/ft^2.

Many other units are in common use, and will be considered later.

Since pressure is the ratio of a force vector $\boldsymbol{dF}$ to an area dA, it is not a vector but is a stress in the sense in which this term is used in the theory of elasticity. It cannot be represented by an arrow. On the other hand, the force exerted on an element of area dA in a given orientation, at a point where the pressure is P, is a vector perpendicular to dA and of magnitude

$$dF = P\,dA.$$

We now prove the following statement. *The state of stress at a point in an ideal fluid at rest is such that all elements of area at the point, regardless of their orientation, experience the same normal stress P.*

Let us consider the equilibrium of an infinitesimal tetrahedron at any arbitrary point in the liquid (Fig. 6-1). As coordinate axes, we use the lines

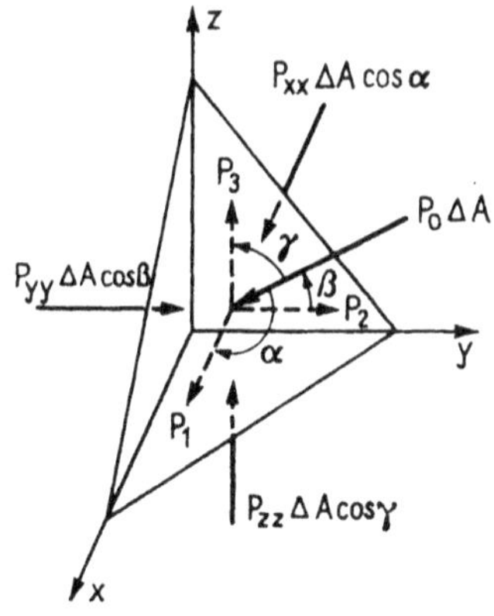

FIG. 6-1. Equilibrium of a tetrahedron of liquid.

of intersection of the three orthogonal surfaces of the tetrahedron. Since its dimensions are vanishingly small, the body force on the fluid in the tetrahedron (its weight) is negligible compared with the surface forces. The surfaces of the tetrahedron are acted on exclusively by the normal stresses P_{xx}, P_{yy}, P_{zz}, and P_0. There are no shearing stresses. The force on the inclined face of the tetrahedron is the normal force $P_0 \Delta A$. Its components are

$$P_1 \cdot \Delta A = P_0 \cdot \Delta A \cos\alpha,$$

$$P_2 \cdot \Delta A = P_0 \cdot \Delta A \cos\beta,$$

$$P_3 \cdot \Delta A = P_0 \cdot \Delta A \cos\gamma,$$

where α, β and γ are the angles between the surface normal and the coordinate axes.

The projections of the inclined face of the tetrahedron in the directions of the three coordinate axes are the three side faces of the tetrahedron:

$$\Delta A_x = \Delta A \cos\alpha,$$
$$\Delta A_y = \Delta A \cos\beta,$$
$$\Delta A_z = \Delta A \cos\gamma.$$

We therefore have the three equilibrium equations:

$$\sum F_x = 0 \quad \text{or} \quad P_{xx}\Delta A_x = P_1\Delta A = P_0\Delta A_x, \quad \text{i.e.,}$$
$$P_{xx} = P_0;$$
$$\sum F_y = 0 \quad \text{or} \quad P_{yy}\Delta A_y = P_2\Delta A = P_0\Delta A_y, \quad \text{i.e.,}$$
$$P_{yy} = P_0;$$
$$\sum F_z = 0 \quad \text{or} \quad P_{zz}\Delta A_z = P_3\Delta A = P_0\Delta A_z, \quad \text{i.e.,}$$
$$P_{zz} = P_0.$$

Therefore

$$P_{xx} = P_{yy} = P_{zz} = P_0,$$

which completes the proof that all elements of area at a given point experience the same pressure. This state of stress is called one of *hydrostatic pressure*. There are no shearing stresses.

58-2 Variation of Pressure with Depth

Now we take into consideration the fact that a volume element ΔV is acted on by a gravitational force $\varrho g\Delta V$, where ϱ is the density of the liquid and ϱg is its specific weight. We are assuming that ϱ is independent of the pressure in the liquid, which is true or almost true for most liquids. Using this fact, the pressure at any depth below the surface can be computed. We imagine a vertical right-circular cylinder to be cut out of a fluid at rest. Its upper and lower ends are horizontal and of area ΔA (Fig. 6-2).

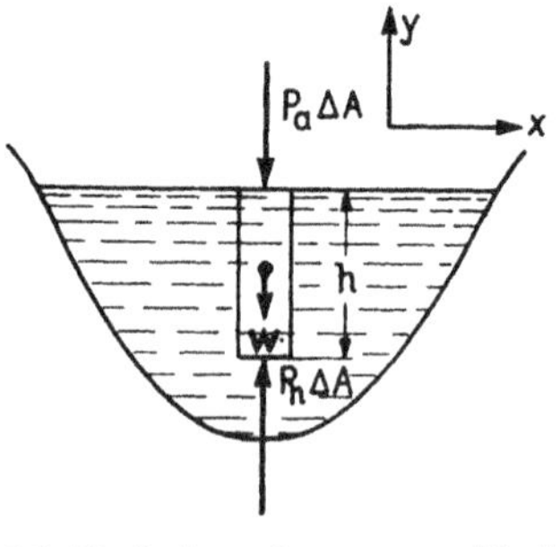

FIG. 6-2. Variation of pressure with depth.

The equilibrium conditions in the vertical direction lead to the equation

$$\sum F_y = -\varrho g h \Delta A + P_h \Delta A - P_a \Delta A = 0.$$

The pressure P_h at a depth h is therefore (P_a = atmospheric pressure)

$$P_h = P_a + \varrho g h.$$

Since the pressure on the vertical walls of the cylinder can give rise only to horizontal forces, the condition $\sum F_y = 0$ is not affected by forces on these walls. The difference $P_h - P_a = \Delta P$ between any two points separated by the height h is

$$\Delta P = \varrho g h.$$

The pressure on the walls of the cylinder increases linearly with the depth, since the pressure at a point is independent of the orientation of an element of area.

To investigate the equilibrium of a liquid, we need the expression for the change of pressure with depth, $\Delta P = \varrho g h$, together with a knowledge of the hydrostatic state of stress. It may be noted that the same atmospheric pressure P_a acts on all of the upper surfaces of liquids in open vessels, provided the difference in height of these surfaces is not too great. Many applications of hydrostatics will now be developed from these facts.

(a) *Communicating, open vessels.* In Fig. 6-3 the arms of two communicating vessels contain different liquids. The equilibrium state is readily

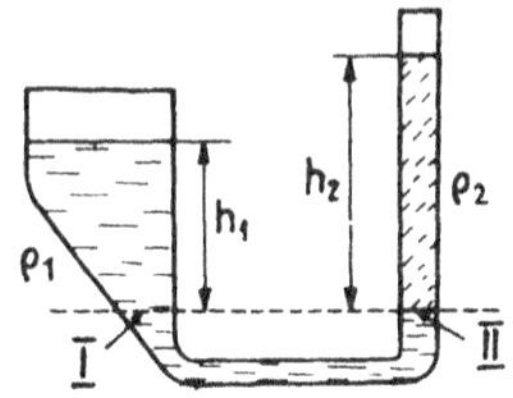

FIG. 6-3. Different liquids at equilibrium in communicating, open vessels.

determined by taking the reference level at the surface of separation between the liquids. The pressure is the same at all points in this plane (points I and II in Fig. 6-3). We have

$$P_I = P_a + \varrho_1 g h_1 = P_{II} = P_a + \varrho_2 g h_2,$$

and hence

$$\varrho_1 h_1 = \varrho_2 h_2.$$

The heights of liquid in the two arms, measured from the level of the surface of separation, are therefore inversely proportional to the densities, regardless of the value of g. In any number of open, communicating vessels containing the same liquid ($\varrho_1 = \varrho_2$), the liquid levels are all equal.

(b) *Pipettes and burettes* (Fig. 6-4). At points I and II the pressure is atmospheric and

$$P_a = P_1 + \varrho g h_1 = P_2 + \varrho g h_2 .$$

where P_1 and P_2 are the inside pressures above the liquids.

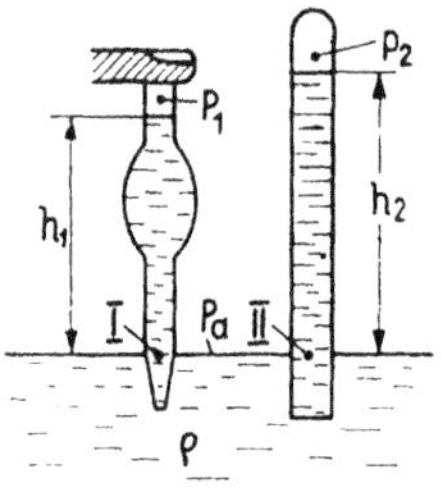

FIG. 6-4

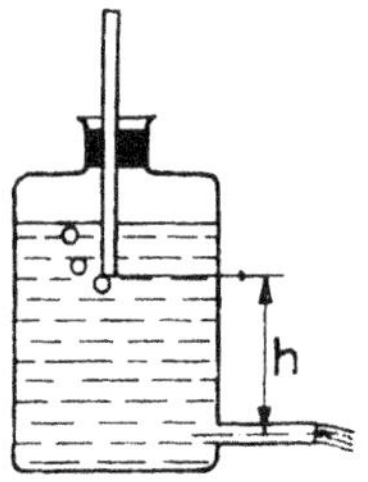

FIG. 6-5

FIG. 6-4. Liquid drawn up into a pipette or burette.

FIG. 6-5. A Mariotte bottle.

(c) *The Mariotte bottle.* When a liquid flows out of an opening in a bottle, the height of the liquid in the bottle decreases and the pressure due to the weight of the liquid decreases. The flow therefore takes place with a continually decreasing pressure. A constant pressure differential, within certain limits, can be obtained by using a *Mariotte bottle* (Fig. 6-5) which is provided with two openings. A long glass tube is tightly sealed into the upper opening. At the level of the lower end of this tube the pressure is atmospheric. The flow therefore takes place under a constant pressure

$$\Delta P = \varrho g h$$

until the liquid level sinks to a height h above the lower opening.

(d) *Pascal's principle and the hydraulic press.* We have shown above that the difference in pressure between two points depends only on the difference in height of the points and the density of the liquid. Thus, it follows that if we increase the pressure at one point, it must increase by an equal amount everywhere else in the liquid, provided the density does not change. This statement is usually known as *Pascal's principle of the transmissibility of pressure.* A good example of this principle is the hydraulic press, the essential parts of which are two communicating cylinders with tight-fitting pistons. The press enables one to produce a large force with the help of a relatively small pressure (Fig. 6-6). It is most commonly filled with oil, or with a mixture of water and glycerin. If the pistons are at the same elevation, then in equilibrium the pressure is the same at both. Since

$$P_1 = P_2 = P,$$

it follows that

$$F_1 = PA_1, \quad F_2 = PA_2,$$

that is,

$$F_2 = \frac{A_2}{A_1} F_1 .$$

EXAMPLE. If

$$\frac{A_1}{A_2} = \frac{1}{100}, \quad \text{then} \quad F_2 = 100F_1 .$$

For $A_2 = 100$ in^2, $P = 10$ lbf/in^2; we find $F_2 = 1000$ lbf.

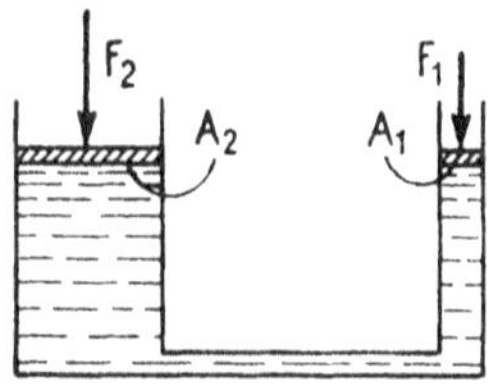

FIG. 6-6

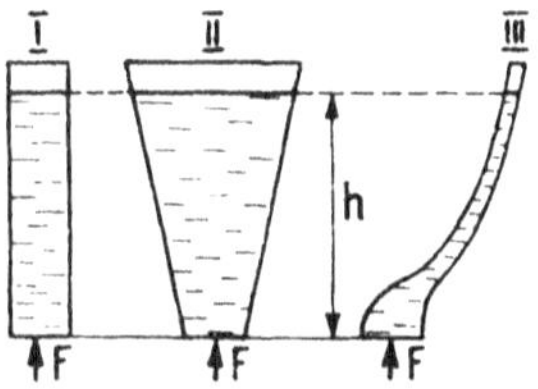

FIG. 6-7

FIG. 6-6. A hydraulic press.

FIG. 6-7. The hydrostatic paradox.

58-3 The Hydrostatic Paradox

The pressure $P = P_a + \varrho gh$ is independent of the shape and cross-section of the containing vessel (Fig. 6-7). The force F exerted on the bottom of the vessel depends only on the depth h of the liquid and the area A of the bottom.

$$F = (P - P_a)\, A = \varrho ghA .$$

At first sight this result appears paradoxical. It is surprising that the small quantity of liquid in vessel III should produce the same force on the bottom of the vessel as the much larger quantities in vessels I and II. The reason becomes clear when we realize that as a result of Newton's third law the sloping walls of a vessel exert on the liquid a force given by

$$\Delta F' = \Delta A(P_a + \varrho gh') .$$

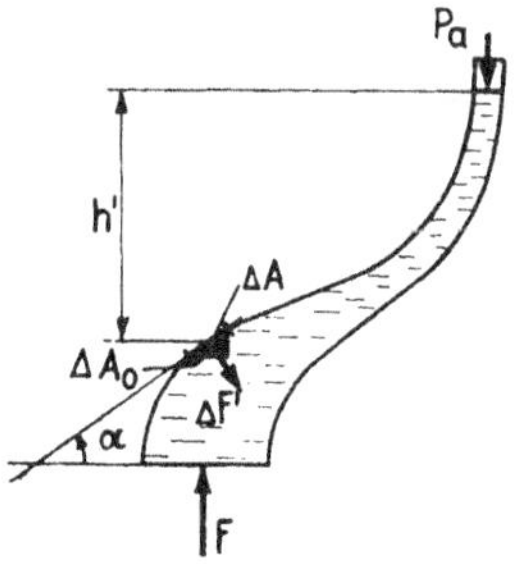

FIG. 6-8. Explanation of the hydrostatic paradox.

The vertical component of this reaction force (Fig. 6-8) is

$$\Delta F'_y = \Delta A(P_a + \varrho g h')\cos\alpha$$
$$= (P_a + \varrho g h')\,\Delta A_0,$$

where $\Delta A_0 = \Delta A \cos\alpha$ is the projection of ΔA on a horizontal plane. The vertical component $\Delta F'_y$ has the same magnitude as the force that would be produced by a column of liquid extending to a height h' above the area ΔA_0, increased by the force produced by the atmospheric pressure P_a.

58-4 Measurement of Pressure

(a) *The open-tube manometer.* Liquid manometers are practical applications of the principles of communicating vessels discussed above. They consist of a U-shaped tube, containing, according to the magnitude of the pressure to be measured, a heavy or light liquid of known density (usually water, mercury, or oil). One end of the tube is connected to the system in which the pressure is to be measured; the other is open (Fig. 6-9). At the level of point I we can measure the inside pressure P_i,

$$P_i = P_I = P_a + \varrho g h.$$

The height h is therefore proportional to the pressure difference

$$\Delta P = P_i - P_a = \varrho g h.$$

Open-tube manometers measure only the excess or deficiency of pressure above (or below) atmospheric pressure. This difference is known as the *gauge pressure.*

It is possible to make the reading of a manometer independent of atmospheric pressure. This is done by closing the top of the right-hand side of the U in Fig. 6-9, and evacuating the space above the liquid in this arm. In this case the pressure at the top of this arm is zero and at the point I

$$P_I = P_i = \varrho g h.$$

Such a manometer measures the *absolute pressure* P_i. A liquid such as mercury must be used, which has a low vapor pressure.

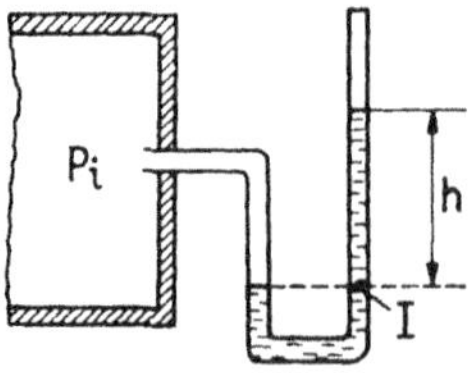

FIG. 6-9

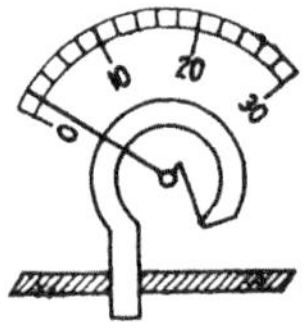

FIG. 6-10

FIG. 6-9. An open manometer.

FIG. 6-10. A Bourdon tube pressure gauge.

(b) *The Bourdon pressure gauge* (Fig. 6-10). A bent tube, closed at one end and of elliptical cross-section takes on a definite shape for a given pressure difference between its interior and its exterior. If the pressure within the tube is increased, the tube straightens somewhat, returning to its original shape when the original pressure is restored. The motion of the end of the tube is transmitted by a lever or a rack and pinion to a pointer. The scale of the instrument must be calibrated. It can be used for pressures either above or below atmospheric.

58-5 Buoyancy; Archimedes' Principle

The most familiar consequence of hydrostatic pressure is the so-called *buoyant force* exerted on a body in a liquid or a gas. A body completely immersed in a liquid is acted on by two external forces; its weight, and the resultant force exerted on its surface by the surrounding liquid. The latter force is called the *buoyant force, B.*

The buoyant force can be explained by the following considerations. The force exerted by the liquid on an element of surface of the immersed body depends only on the area of the element and its depth below the surface of the liquid. It does not depend on the material of the body. To determine these pressure forces, we can think of the body as composed of the same material as the liquid (Fig. 6-11). Then it need not be acted upon by any

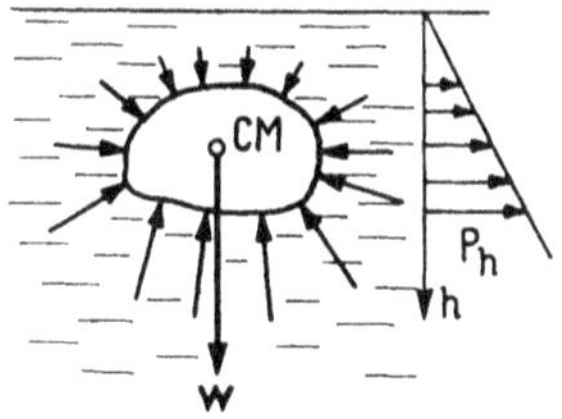

FIG. 6-11. Explanation of Archimedes' Principle.

external forces to remain in equilibrium, since every portion of a liquid at rest is in equilibrium. It follows that the resultant B of all pressure forces on the surface of the body is equal to the weight of the liquid displaced by the body, is vertically upward, passes through the centroid of the volume V of the body, and is of magnitude

$$B = \varrho g V,$$

where ϱ is the density of the liquid.

If a body is only partially submerged in a liquid, we can define a quantity called the *immersion factor* τ as the ratio of the volume submerged, V_s, to the total volume of the body, V:

$$\tau = \frac{V_s}{V} \leqq 1.$$

The bouyant force on a partially submerged body is

$$B = \varrho g \tau V.$$

From Newton's third law, the submerged body exerts a force on the liquid equal and opposite to the buoyant force. The level of the liquid is raised when the body is immersed in it, and this alters the pressure in the liquid and the force exerted on the base of the containing vessel.

The volume of a submerged body can be found from a measurement of the buoyant force exerted on it by a liquid of known density ϱ_l, since

$$V = \frac{B}{\varrho_l g}.$$

If the mass m of the body is known, its density (if it is homogeneous) can then be computed:

$$\varrho = \frac{m}{V} = \frac{m \varrho_l g}{B} = \varrho_l \frac{W}{B}.$$

This method of determining density can be used only with a liquid of known density.

58-6 Measurement of Liquid Densities

The density of a liquid can be measured by means of a *pyknometer* (Fig. 6-12). The pyknometer is a glass vessel with a ground glass stopper

FIG. 6-12. A pyknometer.

having a narrow hole along its axis. Three quantities must be known in order to make a density determination: the mass m_p of the empty pyknometer, the mass m_{p+l} of the pycnometer when filled with the liquid and the mass m_{p+l} of the pyknometer filled with distilled water. All three masses must be determined at the same temperature. Let V be the volume of liquid contained by the pyknometer. Then

$$V = \frac{m_w}{\varrho_w} = \frac{m_l}{\varrho_l}$$

$$= \frac{m_{p+w} - m_p}{\varrho_w} = \frac{m_{p+l} - m_p}{\varrho_l},$$

and

$$\varrho_l = \varrho_w \frac{m_{p+l} - m_p}{m_{p+w} - m_p}.$$

Density measurements can be made more rapidly, although less precisely, by means of either a *hydrometer* or a *Mohr–Westphal balance*. Both instruments make use of the fact that the buoyant force exerted on a body of constant volume depends only on the density of the liquid in which it is immersed. The depth to which a hydrometer (Fig. 6-13) sinks in a liquid is

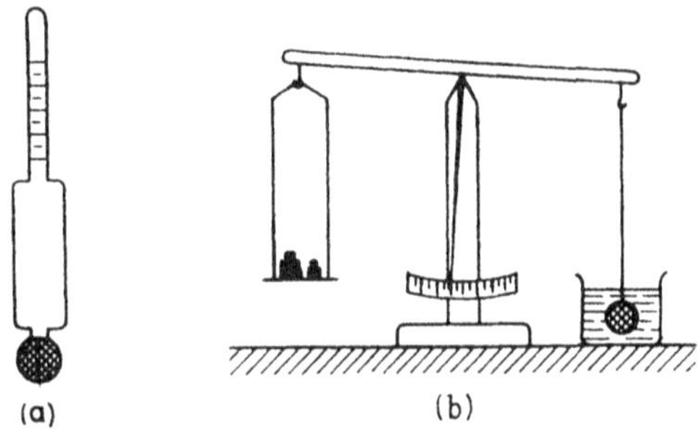

FIG. 6-13. (a) A hydrometer.
(b) A Mohr–Westphal balance.

a measure of the density of the liquid, and the latter can be read directly from a calibrated scale. In the Mohr–Westphal balance, a small body is suspended from one arm of a beam balance (Fig. 6-13). A balance is first obtained with the body in air. The body is then immersed in the liquid, and the balance is again brought to equilibrium by removing weights from the pan. Density measurements, accurate to four significant figures, can be made rapidly with this instrument.

58-7 *Equilibrium of Immersed Bodies*

A body immersed in a liquid is acted on by its weight and by the buoyant force exerted by the liquid. In general, these two forces are not equal in magnitude and do not have the same line of action, with the consequence that both a resultant force and a resultant moment act on the body and it is not in equilibrium (Fig. 6-14). Equilibrium is possible only if

$$\sum F = 0, \quad \text{i.e.,} \quad w = B,$$

and if

$$\sum M = 0.$$

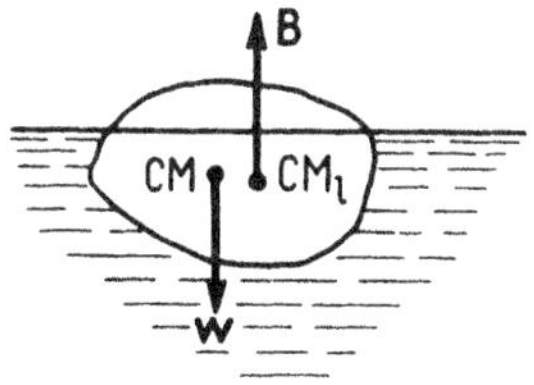

FIG. 6-14. Forces on a floating body.

The moment equation is satisfied if the buoyant force and the weight have the same line of action, which then passes through both the center of mass of the body and that of the displaced volume of fluid. The body will rotate until this condition is fulfilled.

Let us now consider homogeneous bodies. When completely submerged, such a body can be in equilibrium only if its density ϱ is equal to that of the liquid, ϱ_l. A homogeneous body will sink if $\varrho > \varrho_l$, and will remain in equilibrium at any depth if $\varrho = \varrho_l$. If $\varrho < \varrho_l$, then $\tau < 1$ and the body floats at the surface with enough of its volume submerged so that the buoyant force is equal in magnitude to its weight:

$$B = \tau V \varrho_l g = w = V \varrho g.$$

That is,

$$\tau = \frac{\varrho}{\varrho_l}.$$

58-8 Examples

1. What is the maximum height to which a suction pump can lift a column of water, if the atmospheric pressure is 700 mm Hg? The density of water is 1 gm/cm³ and that of mercury is 13.6 gm/cm³.

$$h_{\text{max}} = \frac{P_a}{\varrho_l g} = 952\text{ cm} = 31.3\text{ ft}.$$

(The vapor pressure of the water is neglected.)

2. A burette contains H_2SO_4 of density 1.450 gm/cm³. The acid stands in the burette at a height $h = 42.8$ cm. What is the pressure above the liquid column, if the barometer reads 720 mm Hg? The density of mercury is 13.6 gm/cm³.

$$P_i = P_a - \varrho_l g h = 918\text{ gf/cm}^2 = 675\text{ mm Hg}.$$

3. A cylindrical iron vessel of volume $V = 9.3$ ft³ has a weight w of 120 lbf. (a) What is its immersion factor in water? (b) What fraction f of its volume must be filled with oil of specific weight $\varrho_0 g = 50$ lbf/ft³ in order that the immersion factor shall be 95 per cent? The specific weight of water is $\varrho_w g = 62.4$ lbf/ft³.

(a)
$$V\tau\varrho_w g = w,$$

$$\tau = \frac{w}{V\varrho_w g} = 0.21.$$

(b)
$$V\tau\varrho_w g = w + Vf\varrho_0 g,$$

$$f = \frac{V\tau\varrho_w g - w}{V\varrho_0 g} = 93\%.$$

4. A wooden raft is to support a load L of 4 tons, and is to float with an immersion factor of 95 per cent. The specific weight of the wood is 42lbf/ft³. How many cubic feet of wood must be used?

$$V\tau\varrho_w g = L + V\varrho_{\text{wood}} g,$$

$$V = \frac{L}{\tau\varrho_w g - \varrho_{\text{wood}} g} = 460 \text{ ft}^3.$$

59. Aerostatics (the Study of Gases at Rest)

Gases differ from liquids chiefly in that they are readily compressed. On the other hand, they share with ideal fluids the property that they cannot support tensile and shearing stresses. It follows that the law we have already derived for the state of stress within a liquid can be applied to gases also. In addition, the expression for the buoyant force on a body immersed in a gas remains unchanged in most instances, since for small differences in elevation the density of a gas can be considered constant. The expression for the variation of pressure with elevation, however, no longer holds because the density of a gas increases with increasing pressure, or decreases with increasing elevation. In the subject of aerostatics we can therefore take over without change two of the laws of hydrostatics:

1. $P_{xx} = P_{yy} = P_{zz} = P_0$; the pressure at any point in a gas is independent of the orientation of the element of area under consideration. The force is always perpendicular to the element of area.

2. $B = V\varrho_g g$, insofar as ϱ_g can be considered constant throughout the volume V.

59-1 The Barometer

The *barometer* consists of a glass tube about 1 m long, closed at its upper end. It is completely filled with air-free mercury and placed with its open end beneath the surface of a vessel of mercury. The mercury level in the tube then sinks to a height h determined by the atmospheric pressure (Fig. 6-15). The vapor pressure of mercury at room temperature is only about 10^{-3} mm Hg, and hence the space above the mercury column is practically gas-free

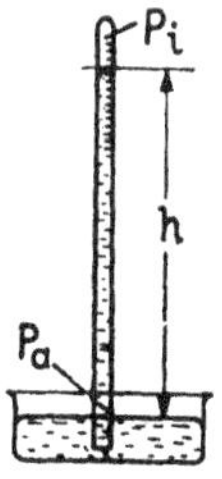

Fig. 6-15. A barometer or gas burette.

($P_t \cong 0$; Torricellian vacuum). The pressure at the level of the foot of the mercury column is therefore

$$P_a = \varrho_{Hg} g h .$$

The density ϱ_{Hg} of mercury varies with temperature. Therefore, at constant atmospheric pressure P_a, the height of the mercury column depends on the temperature. It is customary to reduce all barometric readings to 0°C by means of tables. The height h at 0°C is called the *reduced* barometer height. For a temperature of 20°C and a height of 700 mm, the correction is about 2 mm. For very accurate work it is also necessary to make a correction for the variation of the acceleration of gravity over the earth's surface.

The expression "mm Hg 0°C" is ungainly, and it is convenient to give it a new name. The pressure of a mercury column 1 mm high, at 0°C and at a point where g has its standard value of 980.665 cm/sec^2, is called 1 *torr*. The pressure of a mercury column 760 mm high at 0°C where g has the standard value is called 1 *atmosphere*. Barometric pressure at 0°C and sea level is about 1 atmosphere.

$$1 \text{ atm} = 1.0132 \times 10^6 \text{ dynes/cm}^2 = 14.68 \text{ lbf/in}^2 = 760 \text{ torr}.$$

In meteorology, the *bar* is used as a unit of pressure.

$$1 \text{ bar} = 10^6 \text{ dynes/cm}^2$$

$$1 \text{ millibar} = 10^{-3} \text{ bar}.$$

Table 6-1 lists the units of pressure commonly found in the literature, and the relations between them.

59-2 The Boyle–Mariotte Law

The density of a gas varies with the pressure. For gases which are far removed from their points of condensation, there is a simple relation between pressure and volume, provided *three* conditions are fulfilled. These are:

1. The mass m of the gas must be constant.
2. The process must take place at constant temperature T.
3. The gas must retain the same chemical constitution; that is, the molecular weight M of the constituent molecules must remain constant.

Some gases, of which NO_2 is an example, become polymerized with increasing pressure, as represented by the equation

$$2NO_2 \rightarrow N_2O_4 .$$

The Boyle–Mariotte law is not obeyed by such a gas, nor is it valid for a gas in which a chemical reaction takes place.

TABLE 6-1

	dyne/cm^2	Newton/m^2 = N/m^2	kgf/m^2	gf/cm^2	atm	torr = mm Hg (0°C)	bar
dyne/cm^2 =	1	0.1000	1.0197×10^{-2}	1.0197×10^{-3}	9.8692×10^{-7}	7.5006×10^{-4}	10^{-6}
1 N/m^2 =	10	1	0.10197	1.0197×10^{-2}	9.8692×10^{-6}	7.5006×10^{-3}	10^{-5}
1 kgf/m^2 =	98.0665	9.80665	1	0.1000	9.6784×10^{-5}	7.3556×10^{-2}	9.80665×10^{-5}
1 gf/cm^2 =	980.665	98.0665	10	1	9.6784×10^{-4}	0.73556	9.80665×10^{-4}
1 atm =	1.01325×10^{6}	1.01325×10^{5}	10332	1033.2	1	760	1.01325
1 torr =	1333.2	133.32	13.595	1.3595	1.3158×10^{-3}	1	1.3332×10^{-3}
1 bar =	10^{6}	10^{5}	10197	1019.7	0.98692	750.06	1

If the three conditions above are fulfilled,

$$m = \text{const},$$

$$T = \text{const},$$

$$M = \text{const},$$

then for such a gas

$$PV = \text{const}.$$

This law was discovered experimentally by Boyle in 1662 and by Mariotte in 1679, and is called the *Boyle–Mariotte law*, or more often, simply *Boyle's law*. It says that for different states of the gas, the product of the pressure and volume of a constant mass of any given gas is constant at constant temperature:

$$P_1 V_1 = P_2 V_2 = P_3 V_3, \quad \text{etc.}$$

An *ideal gas* is defined as one that obeys this law exactly.

The most precise measurements made for the purpose of verifying this law are those of Amagat (1880). Some of his results are given in the accompanying table. Amagat subjected an arbitrary initial volume of gas to an increasing pressure, and computed the product PV at each pressure from the measured pressure and volume. The following table gives the experimental results for a pressure range from 20 to 100 atm. The units of the product PV are atm · cm^3.

TABLE 6-2†

H_2		N_2		O_2		CO	
P atm	PV atm · cm³	P	PV	P	PV	P	PV
31.69	27,381	27.29	50,989	31.67	26,843	31.66	27,147
45.94	27,618	46.50	50,897	45.91	26,614	45.93	27,102
59.54	27,652	62.03	50,811	59.58	26,614	59.54	27,007
73.04	27,960	73.00	50,857	73.03	26,185	73.05	27,025
84.33	28,129	80.58	50,895	84.30	26,050	84.31	27,060
94.98	28,323	90.97	50,987	94.93	25,858	94.96	27,071
				110.77	25,745		
				133.50	25,639		
				176.15	25,671		
				233.66	25,891		

† E. H. Amagat, Mémoire sur la compressibilité des gaz à des pressions élevées, *Ann. de Chimie et de Physique,* N. 19, 1880, p. 345.

We see from these experimental results that the product PV is constant within a few per cent, for pressures between 20 and 100 atm. The Boyle–Mariotte law is therefore only an approximate relation, valid at low pressures. We shall return to this question later in the chapter on thermodynamics, in connection with the van der Waals theory of gases. Boyle's law makes it possible to compute the change in volume of a gas when the pressure is varied, and vice versa. The change of density resulting from a change in pressure can be determined also.

From the relation $P_1V_1 = P_2V_2$, it follows that

$$\varrho_1 = \frac{m}{V_1} = \frac{mP_1}{V_2P_2} = \varrho_2\frac{P_1}{P_2},$$

that is,

$$\frac{\varrho_1}{\varrho_2} = \frac{P_1}{P_2}.$$

The density of gas, under the conditions stated above, varies in the same way as the pressure.

59-3 Isothermal Compressibility of a Gas

The *isothermal compressibility* of a body is defined by the equation

$$\varkappa_T = -\frac{1}{V}\frac{\Delta V}{\Delta P}\bigg|_{T=\text{const.}}$$

For an ideal gas, this reduces to a very simple expression. From the Boyle–Mariotte law, we have

$$(P + \Delta P)(V + \Delta V) = \text{const.} = PV.$$

Neglecting the second order quantity $\Delta P \Delta V$, this reduces to

$$V\Delta P = -P\Delta V;$$

that is,

$$-\frac{\Delta V}{\Delta P} = \frac{V}{P}.$$

Therefore

$$\varkappa_T = -\frac{1}{V}\frac{\Delta V}{\Delta P} = \frac{1}{P}.$$

The isothermal compressibility of an ideal gas is equal to the reciprocal of the pressure.

59-4 Examples

(a) *The gas burette* (*Fig. 6-15*). This is a glass tube closed at one end and calibrated in cm³, similar to a barometer. The burette is initially filled with a liquid and a certain quantity of gas is introduced into it. The liquid in the burette then stands at a height h above the level of the liquid in the lower vessel. If the atmospheric pressure is P_a and the specific weight of the liquid is $\varrho_l g$, the pressure in the space above the liquid is

$$P_l = P_a - \varrho_l g h.$$

If the vapor pressure P_v of the liquid is negligible, the pressure of the gas is also P_l. If the vapor pressure cannot be neglected, it must be subtracted from P_l to obtain the pressure of the gas.

$$P_{\text{gas}} = P_l - P_v = P_a - \varrho_l g h - P_v.$$

(b) *The Magdeburg hemispheres.* (*Otto von Guericke, 1602–1686.*) Two hemispheres of radius r are placed in contact with an air-tight seal between them to form a hollow sphere (Fig. 6-16). The interior is partially evacuated so that the internal pressure P_i is smaller than the atmospheric pressure P_a. The force ΔF on a surface element of area ΔA is

$$\Delta F = (P_a - P_i)\,\Delta A = \Delta P\,\Delta A.$$

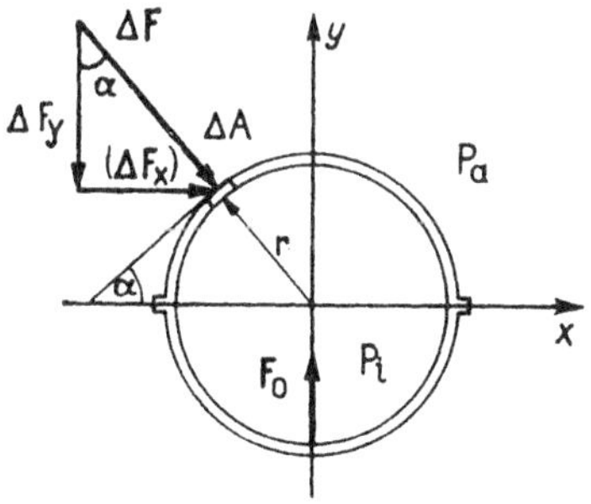

FIG. 6-16. The Magdeburg hemispheres.

Let us resolve ΔF into a vertical y-component and a horizontal x-component. Then if α is the angle between the element ΔA and the horizontal plane,

$$\Delta F_x = \Delta F \sin\alpha = \Delta P\,\Delta A \sin\alpha,$$
$$\Delta F_y = \Delta F \cos\alpha = \Delta P\,\Delta A \cos\alpha.$$

Neglecting its weight, the equilibrium conditions for the upper hemisphere are

$$\sum F_x = 0,$$
$$\sum F_y = F_0 - \Delta P\,\Delta A \cos\alpha = 0,$$

where F_0 is the upward force exerted on the upper hemisphere by the lower.

The first equation is satisfied by the reasons of symmetry. In the second equation, $\Delta A \cos\alpha$ is the projection of the surface element on the horizontal plane. Therefore

$$\sum \Delta A \cos\alpha = \pi r^2$$

and

$$F_0 = \pi r^2 \Delta P.$$

To separate the spheres, a force is required equal to the product of the pressure difference ΔP and the cross-sectional area of the sphere.

EXAMPLE.

$$r = 8 \text{ inches}$$
$$P_i = 5 \text{ mm Hg}$$
$$P_a = 720 \text{ mm Hg}.$$

Then

$$\Delta P = 715 \text{ mm Hg} = 13.8 \text{ lbf/in}^2,$$
$$F_0 = \pi r^2 \Delta P = 2{,}780 \text{ lbf}.$$

(c) *The air pump and the pump formula.* In Fig. 6-17(a) a vessel of volume V_0 contains a gas at an initial pressure P_0. The vessel is to be evacuated by

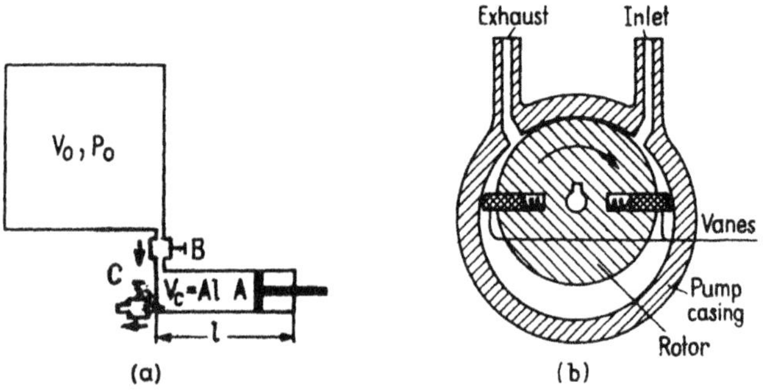

FIG. 6-17. (a) Derivation of the pump formula.
(b) A modern mechanical vacuum pump.

means of a pump whose cylinder has a volume $V_c = Al$. The so-called "dead space", the volume of the connecting tubes and of the valves, is assumed to be negligibly small. A valve B, located between the volume V_0 and the cylinder of the pump allows gas to flow only in the direction of the arrow. Valve C is a similar valve for the exhaust. The piston is initially at the left end of the cylinder. Let P_1 be the pressure at the end of the first exhaust stroke. If the temperature is constant and if there is no leakage, we have from the Boyle–Mariotte law

$$V_0 P_0 = (V_0 + V_c)\, P_1 .$$

After the gas in the cylinder has been expelled through the exhaust valve, a second exhaust stroke begins. For this stroke, we have

$$V_0 P_1 = (V_0 + V_c)\, P_2 ,$$

and for the nth stroke,

$$V_0 P_{n-1} = (V_0 + V_c)\, P_n .$$

That is, n exhaust strokes result in a pressure in the vessel of P_n, which, from the preceding equations, is

$$P_n = \left(\frac{V_0}{V_0 + V_c}\right)^n P_0 ,$$

a relation known as the pump formula. Replacing $\dfrac{V_0 + V_c}{V_0}$ by a, a pure number >1, we get

$$P_n = P_0 a^{-n}.$$

The pressure in the vessel therefore decreases exponentially. An exponential change in a quantity occurs only when the change is proportional to the magnitude of the quantity itself. In the preceding example the decrease in pressure in the vessel results from the fact that in each exhaust stroke a certain quantity of gas flows into the cylinder of the pump. This quantity of gas is proportional to the pressure in the space being evacuated, the proportionality factor depending only on other quantities which are constant. The pressure decrease therefore follows an exponential law. We shall encounter exponential variations in many other instances.

The pumps used today are for the most part rotary pumps, invented by Gaede. Figure 6-17(b) is a diagram of such a pump. Pumps of this sort can produce pressures as low as 10^{-2} to 10^{-3} mm of Hg. For still lower pressures down to 10^{-9} mm Hg, oil or mercury diffusion pumps are universally used (see Section 77).

High vacuum techniques, which are indispensable to present-day science and technology, have undergone significant development in recent years. Nature's "abhorrence of a vacuum" can now be ignored due to the development of efficient pumps and methods of detecting tiny leaks in vacuum equipment.

(d) *The McLeod gauge.* By the use of the proper measuring instruments pressures as low as 10^{-9} mm Hg can be measured accurately. A less sensitive instrument was developed by McLeod and makes use of the Boyle–Mariotte law in its operation. A bulb of known volume V (Fig. 6-18a) is connected to

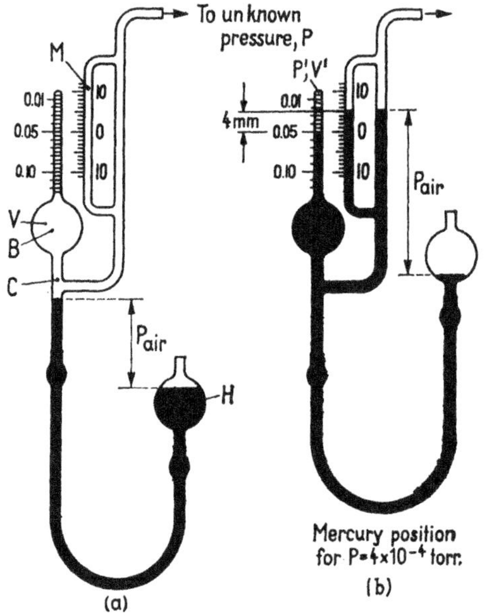

FIG. 6-18. A McLeod gauge measuring a pressure of 4×10^{-4} torr. $V = 500$ cm^3, $V' = 0.05$ cm^3, and $P' = 4$ torr.

the region in which the pressure P is to be measured. By raising the mercury container H, the mercury level rises and cuts off the volume V from the rest of the apparatus. The gas trapped in the volume V is then compressed to a smaller volume V' (Fig. 6-18b), and its pressure is increased to P'. From the Boyle–Mariotte law,

$$P = \frac{V'}{V} P'.$$

For a sufficiently great volume change, the pressure P' becomes large enough to be measured directly by a mercury manometer M.

Figure 6-18 illustrates the construction and operation of a McLeod gauge. The capillary tube connected to the top of the bulb B has a very small volume relative to the volume V of the bulb, where V includes the volume down to the cut-off point C. The ratio of the capillary volume to V is carefully calibrated. When the mercury bulb H is raised, the gas in bulb B is compressed and is finally all forced into the capillary. By means of the calibrated scale, that fraction of its original volume which it now occupies can be determined. The pressure P' of the compressed gas can be determined from the difference in levels of the mercury columns in the capillary containing the gas, and in the auxiliary capillary at its right. If both capillaries have the same diameter, any effect of capillary depression because of surface tension (see Section 60) is eliminated. The pressure P', multiplied by the known volume ratio V'/V, gives the desired pressure P. If the gas is completely free of condensable vapors such as water vapor, this manometer can be used for pressures as low as 5×10^{-6} mm Hg. The vapor pressure of mercury has no effect on the measurement.

60. Statics of Real Liquids

Real liquids differ from ideal liquids in two respects:

1. They are slightly compressible.
2. Small shearing stresses exist in their outer surfaces (surface tension).

60-1 Compressibility of Liquids

Definition. The *isothermal compressibility* $\varkappa_T$ of a liquid is defined as the fractional decrease in volume per unit increase in pressure, at constant temperature.

$$\varkappa_T = -\frac{1}{V}\frac{\Delta V}{\Delta P}\bigg|_{T=\text{const.}}$$

Dimensions: $[\varkappa_T] = [M^{-1}LT^2]$

Units: cgs system: 1 cm²/dyne
mks system: 1 m²/newton
fps system: 1 ft²/lbf

The unit atm^{-1} is also frequently used.

The compressibility is constant only within a limited range of pressures, and in general decreases with increasing pressure and increases with increasing temperature, except in the case of water, for which $\varkappa_T$ decreases with increasing temperature. For small changes in pressure at constant temperature, the change in volume can be computed from the defining equation, considering $\varkappa_T$ to be constant:

$$\Delta V = -\varkappa_T V \Delta P.$$

The change of density with pressure can also be expressed in terms of the compressibility.

$$\Delta\varrho = \varrho_1 - \varrho_0 = -\frac{m(V_1 - V_0)}{V_0 V_1}$$

$$= -\frac{m}{V_0}\frac{\Delta V}{V} = \varrho_0 \varkappa_T \Delta P.$$

EXAMPLE. Find the decrease in volume and the increase of density of 1000 cm^3 of water at 20°C, when the pressure is increased by 15 atm, if $\varkappa_T = 49.1 \times 10^{-6}$ atm^{-1}.

$$\Delta V = -\varkappa_T V \Delta P = -0.736 \text{ cm}^3,$$

$$\Delta \varrho = \varkappa_T \varrho_0 \Delta P = 7.35 \times 10^{-4} \text{ gm/cm}^3.$$

60-2 Surface Tension

Every real liquid at rest can support small shearing stresses. This is easily demonstrated by placing on the surface of a liquid a small body whose density is greater than that of the liquid. The liquid supports the body in a depressed region on its upper surface (Fig. 6-19). The depression disappears immediately when the body is removed. The outer surface of a

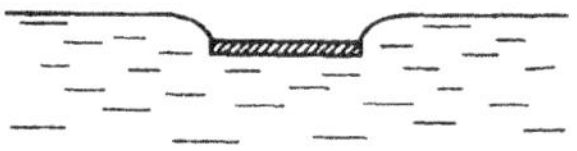

FIG. 6-19. Surface of a liquid supporting a body whose density is greater than that of the liquid.

liquid behaves in some respects like an elastic membrane in which there are forces tangent to the surface. However, it differs significantly from a stretched sheet of rubber, for example, in that the forces in the surface of a liquid are independent of the magnitude of the surface area. The existence of these forces is not apparent at the free surface of a large body of liquid since they mutually cancel one another. If the outer surface has a boundary line, however, the surface of the liquid exerts a force on the body at the boundary. This force lies in a plane tangent to the surface (Fig. 6-20). According to Newton's third law, the body exerts an equal reaction force on the surface of the liquid. The concept of *surface tension* is related to these forces at the boundary of the surface of a liquid.

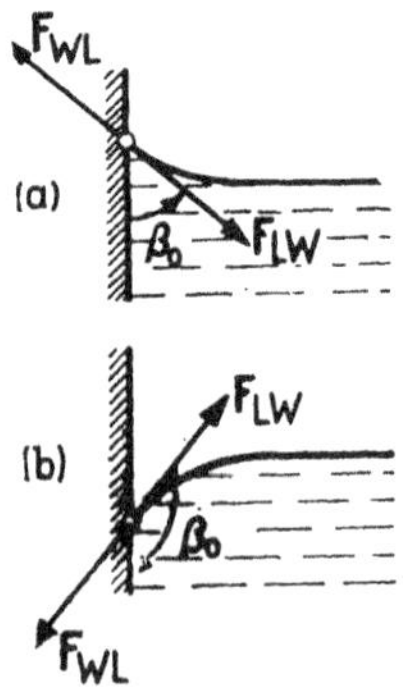

FIG. 6-20. Surface forces at the boundary line of a liquid.

DEFINITION. The surface tension σ is defined as the force per unit length of the boundary line of a surface.

Dimensions: $[\sigma] = [MT^{-2}]$

Units: cgs system: 1 dyne/cm,
mks system: 1 newton/m,
fps system: 1 lbf/ft.

Figure 6-21 represents a soap film formed on a rectangular U-shaped wire frame provided with a sliding wire cross-piece. As a result of surface

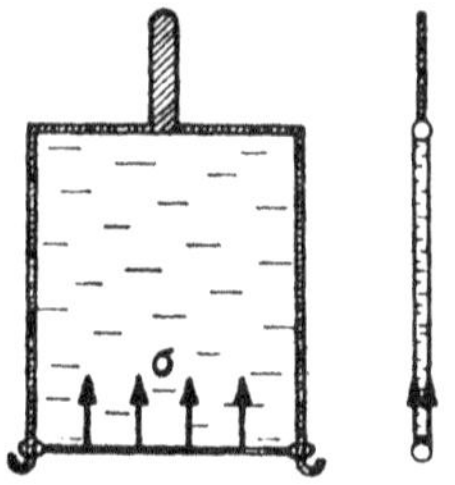

FIG. 6-21. Surface tension acting on a soap film in a wire frame.

tension, the cross-piece is drawn up to the top of the frame provided its weight is less than the surface tension force. To pull the cross-piece down again, a force F is required, given by

$$F = 2b\sigma,$$

where b is the length of the moving wire. The factor 2 arises from the fact that the soap film has two lines of contact with the cross-piece. To move the cross-piece a distance l, an amount of work W is required, given by

$$W = 2bl\sigma = \sigma \Delta A. \quad (60\text{-}1)$$

The product $2bl = \Delta A$ equals the new surface area created in the process (the film has two sides). From this equation, the surface tension can be considered as the work required to create a unit area of new surface. This aspect of surface tension is also evident from its units, since

$$1\,\frac{\text{dyne}}{\text{cm}} = 1\,\frac{\text{erg}}{\text{cm}^2},$$

$$1\,\frac{\text{newton}}{\text{m}} = 1\,\frac{\text{joule}}{\text{m}^2}.$$

As a result of this stored work or potential energy, surface forces are called into play. The surface forces act in such a way as to make this potential energy a minimum. If no external forces whatever act on the surface of a liquid of given volume, it will take on that shape for which its surface is a minimum, namely, a sphere. Thus, closed soap bubbles have a spherical shape.

If a drop of oil is placed on a water surface, three boundary surfaces must be considered (Fig. 6-22): water-oil (1,2), oil-air (2,3), and air-water (1,3). If the surface (1,2) is increased by ΔA, the surface (2,3) also increases by

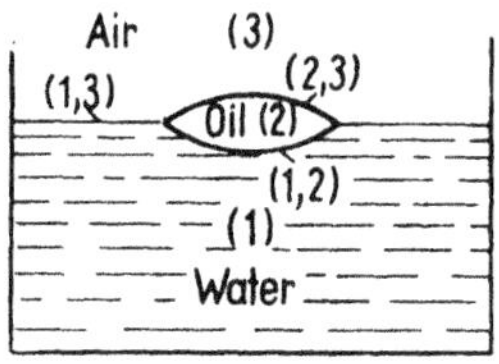

FIG. 6-22. Surfaces in an air, water, and oil system.

ΔA while the surface (1,3) decreases by ΔA. The work ΔW required in the process is

$$\Delta W = (\sigma_{12} + \sigma_{23} - \sigma_{31})\,\Delta A.$$

It is known that $\sigma_{12} + \sigma_{23} - \sigma_{31} \leqq 0$, and hence the work is negative; that is, the potential energy of the system decreases and the process will take place spontaneously. The oil spreads out over the water surface until an oil layer of the maximum possible area, namely, a *monomolecular* layer, has been formed.

The surface tension of a liquid can be measured by measuring the force required to pull a ring of fine platinum wire out of the surface. Typical values of surface tension are:

Water, 20°C:	$\sigma = 73$ dynes/cm,
Soap solution:	$\sigma \cong 25$ dynes/cm,
Mercury:	$\sigma \cong 500$ dynes/cm.

We emphasize again that the surface tension is not a true tension, such as is found in elastic media, but is a fiction which allows us to handle easily certain phenomena in a quantitative manner. The true explanation of these phenomena lies in the intermolecular forces between the molecules of the liquid (see Section 75-3).

A liquid comes in contact with a solid with a characteristic *contact angle* β_0 along the line of contact (Fig. 6-20). The angle depends both on the liquid and on the material at its boundary. If the contact angle is zero, the liquid is said to *wet* the surface completely. This occurs when the molecules of the liquid are attracted more by the solid than they are by each other. A contact angle $0° < \beta_0 \leqq 90°$ corresponds to a partially wetted surface (Fig. 6-20a). If the contact angle is greater than 90°, as in Fig. 6-20b, we say that the liquid does not wet the surface. The contact angle of water and glass is about 5°, and of mercury and glass is about 128°. For soap solutions and most surfaces, the contact angle is nearly zero.

The behavior of a liquid in contact with a boundary is completely determined by the contact angle and the surface tension of the liquid.

(a) *Soap films and their properties.* Surface tension forces are small compared with gravitational forces, and hence to demonstrate surface tension effects the weights of the objects concerned must be kept as small as possible. We must therefore reject liquids like mercury with large surface tensions because they are too heavy, and make use of those which will form thin films, so that the weight is negligible. Soap solutions are liquids of this sort. As a result of linkages between the exceptionally long chain molecules of oleic acid, they form exceedingly thin films.

If a soap film is formed on a horizontal plane wire ring, the film will be very nearly plane also, because of the surface tension. The weight of the film produces only a small distortion. If a loop of silk thread is placed on the film, it retains its original shape since the surface forces on both sides of the thread are of the same magnitude. However, if the film inside the loop is punctured, the thread is pulled out into a circle by the surface tension forces on its periphery (Fig. 6-23).

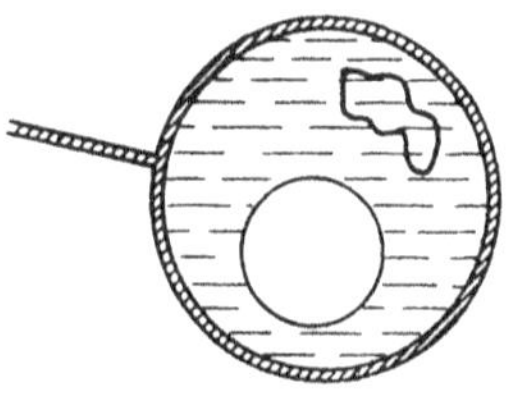

FIG. 6-23. A loop of silk thread stretched into a circle by the surface tension of a soap film. Above, an unstretched loop.

The pressure in the interior of a soap bubble is slightly greater than the external pressure. The magnitude of the excess pressure ΔP, for a spherical bubble, can be found by considering the equilibrium of a segment of the sphere (Fig. 6-24). Let us think of the segment as separated from the remainder of the sphere. Its weight is negligible. Its inner surface is acted on

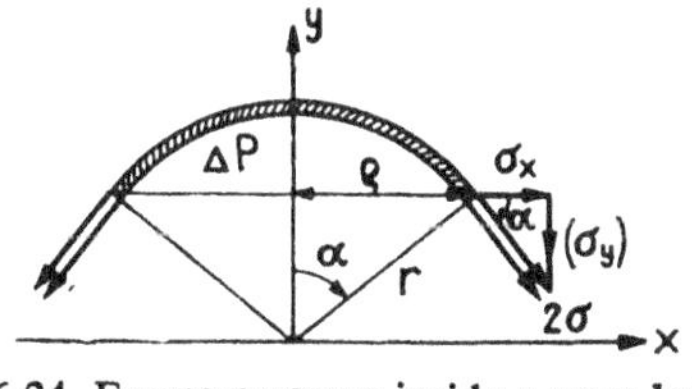

FIG. 6-24. Excess pressure inside a soap bubble.

by the excess pressure ΔP. The surface tension forces exert a force of magnitude σ per unit length along the two boundary lines, so that in the coordinate system indicated, the equilibrium conditions for the y-direction (compare the Magdeburg hemisphere, p. 259) are

$$\sum F_y = 0 = -2 \cdot 2\pi\varrho\sigma \sin\alpha + \pi\varrho^2 \Delta P .$$

The excess pressure is therefore

$$\Delta P = \frac{4\sigma \sin\alpha}{\varrho},$$

or, since $\varrho = r \sin\alpha$,

$$\Delta P = \frac{4\sigma}{r}.$$

The excess pressure is inversely proportional to the radius, as can easily be demonstrated by a micromanometer or by blowing two bubbles of different radii at opposite ends of a tube.

EXAMPLE. Radius of sphere; $r = 5$ mm; $\sigma = 25$ dynes/cm.

$$\Delta P = \frac{4\sigma}{r} = 200 \text{ dynes/cm}^2.$$

The pressure in small bubbles is greater than in large bubbles. This explains why the common surface of a large and a small bubble bends into the larger bubble as in Fig. 6-25a, and also why the surface of separation of two bubbles of the same size is flat.

Whenever three soap films come together, they always make angles of 120° with one another. This is simply a consequence of the principles of statics, since three equal forces in equilibrium and acting at the same point must form an equilateral trangle (Fig. 6-25b).

Soap films that are not closed surfaces do not support an excess pressure. They form so-called *minimal* surfaces, that is, surfaces which have a minimum area for a given boundary. Such surfaces are of the type called *saddle* surfaces, and have equal and opposite radii of curvature at the lines of inter-

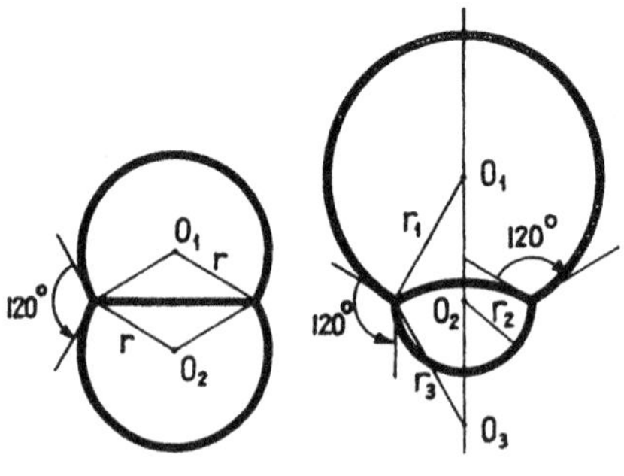

FIG. 6-25a. Twin bubbles of various sizes.

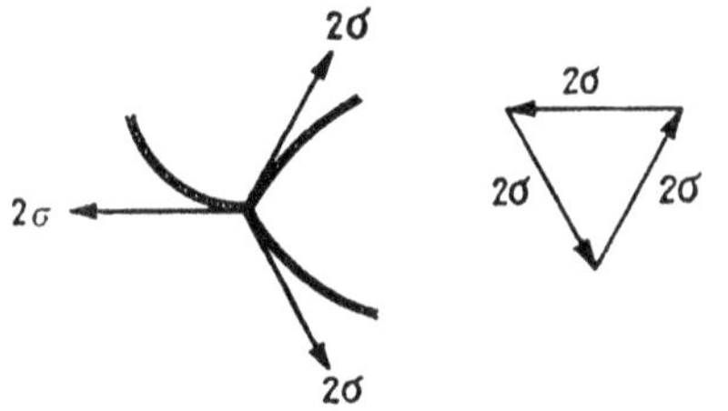

FIG. 6-25b. Common meeting point of three soap film surfaces.

section of the surface with two mutually perpendicular planes (Fig. 6-26). This follows from a consideration of the equilibrium of a small element of surface having sides of length Δs_1 and Δs_2. The forces acting at the

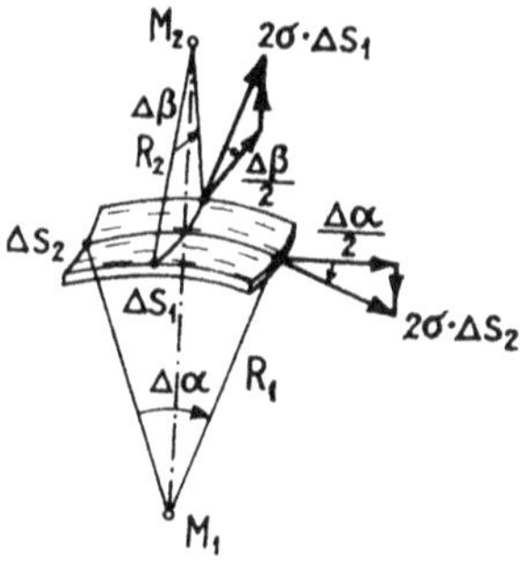

FIG. 6-26. Equilibrium of an open surface soap film (a minimal surface).

four edges must be in equilibrium. Let the surface tension on the liquid be σ, the radii of curvature be R_1 and R_2 and the centers of curvature M_1 and M_2. The lengths of the sides can be expressed in terms of the small central angles $\Delta\alpha$ and $\Delta\beta$,

$$\Delta s_1 = R_1 \Delta\alpha,$$
$$\Delta s_2 = R_2 \Delta\beta.$$

The force on a side of length Δs_1 is $F_1 = 2\sigma \Delta s_1$. Then, since $\sin \Delta\beta/2 \approx \Delta\beta/2$, the equilibrium condition in the direction $M_1 M_2$ gives:

$$4\sigma \Delta s_1 \, \Delta\beta/2 - 4\sigma \Delta s_2 \, \Delta\alpha/2 = 0.$$

Therefore

$$R_1 \Delta\alpha \, \Delta\beta = R_2 \Delta\alpha \, \Delta\beta$$

and

$$R_1 = R_2.$$

The interior of a liquid having a curved outer surface is under an excess pressure because of the action of surface forces. It should be noted that in this case there is only one surface to be considered, in contrast to the soap film, where there are two. If R_1 and R_2 are the radii of curvature of the outer surface in two mutually perpendicular planes, the excess pressure ΔP can be found from equilibrium considerations similar to those applying to minimal surfaces. The equilibrium conditions for the y-direction (Fig. 6-27) lead to the equation

$$\sum F_y = 0 = \Delta P\,\Delta s_1\,\Delta s_2 - 2\sigma\,\Delta s_1\,\frac{\Delta\beta}{2} - 2\sigma\,\Delta s_2\,\frac{\Delta\alpha}{2}.$$

Since $\Delta s_1 = R_1\Delta\alpha$ and $\Delta s_2 = R_2\Delta\beta$, this becomes

$$\Delta P\,\Delta s_1\,\Delta s_2 = \sigma\Delta s_1\,\Delta s_2\left(\frac{1}{R_1} + \frac{1}{R_2}\right)$$

and hence

$$\Delta P = \sigma\left(\frac{1}{R_1} + \frac{1}{R_2}\right).$$

For a spherical film ($R_1 = R_2 = R$) having two surfaces (such as a spherical soap bubble) there is a pressure change of the above magnitude across each surface. Hence in agreement with the result on page 268 there is an excess pressure $\Delta P = 4\sigma/R$ in the interior of the bubble.

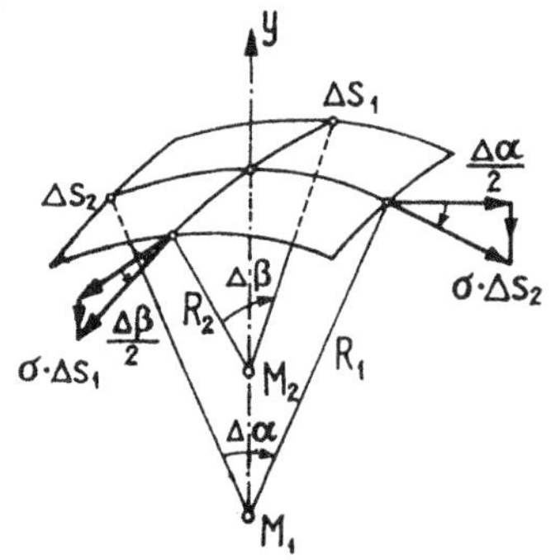

FIG. 6-27. Calculation of the excess pressure produced in the interior of a fluid by the surface forces of a curved surface.

This excess pressure arising from the curvature of its surface adds to any hydrostatic pressure within a liquid. If the surface is convex outward, the pressure difference is positive. There is no pressure difference across a plane surface, and for concave surfaces the pressure difference is negative. This fact must be taken into account in precise measurements of pressure (see the following discussion of capillarity).

(b) *Capillary phenomena.* When a liquid makes contact with a solid body, there exists a certain line of contact and an angle of contact. The surface exerts a force of attraction on the solid body and the body attracts the liquid with an equal force per unit length of contact line. Both forces lie in a plane tangent to the liquid surface. If the contact angle β_0 is less than 90°, the level of the liquid is pulled up by the vertical component of the force exerted on it by the body. The body, on the other hand, is pulled down by the liquid (Fig. 6-20). If the contact angle is greater than 90°, the liquid is pulled down and the solid is pulled up.

FIG. 6-32. Streamline patterns of laminar flow around bodies of various shapes. The apparatus for showing those phenomena was first constructed by Pohl. The streamlines are made visible by coloring parts of the flow pattern with ink.

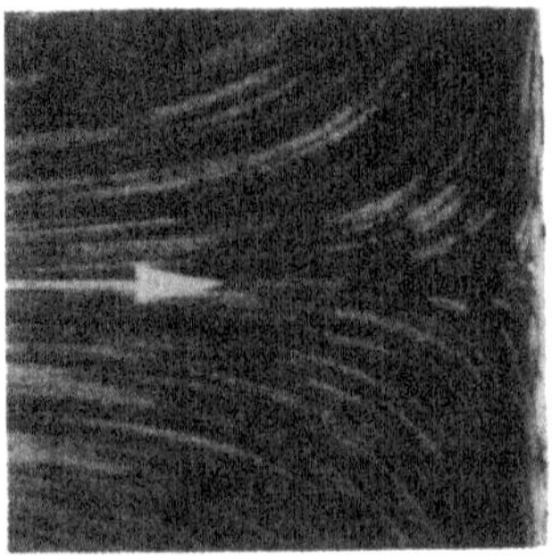

FIG. 6-33. Stagnation point in a flow pattern. The streamlines were made visible by particles of aluminium powder suspended in the liquid (Prof. J. Ackeret).

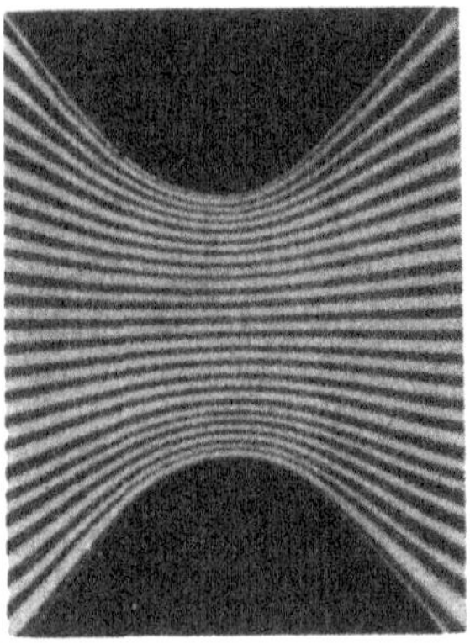

FIG. 6-34. Streamline pattern in a constricted channel.

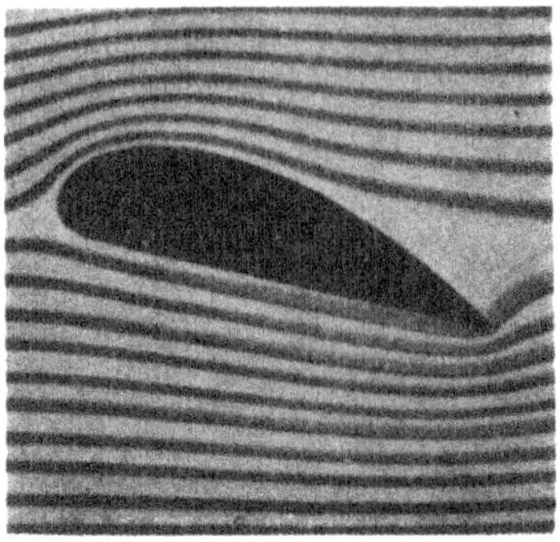

FIG. 6-35. Streamline pattern around a wing section.

Plate II

FIG. 6-37. Daniel Bernoulli (1700–1782), from a medal in the possession of Prof. O. Spiess, Basel.

Bernoulli studied medicine and wrote his doctor's dissertation at the age of 21 on the subject, "Concerning the Amount of Air Which Enters the Lungs Upon Inhaling". After a long residence outside his homeland, he was appointed to the chair of Anatomy and Botany at Basel in 1732 and in 1750 to the Professorship of Physics, which suited him better. His major work was the Hydrodynamica, which appeared in 1738, but his scientific investigations resulted in some 70 papers, among which were many prizewinning works. At the age of 25, he won a large prize from the Paris Academy with the paper "Sur la perfection des clepsydres ou des sabliers sur mer". At one time or another he worked on the theory of bent beams and the kinetic theory of gases, formulated the principle of superposition for an oscillating system, and determined the mechanical power of the human heart.

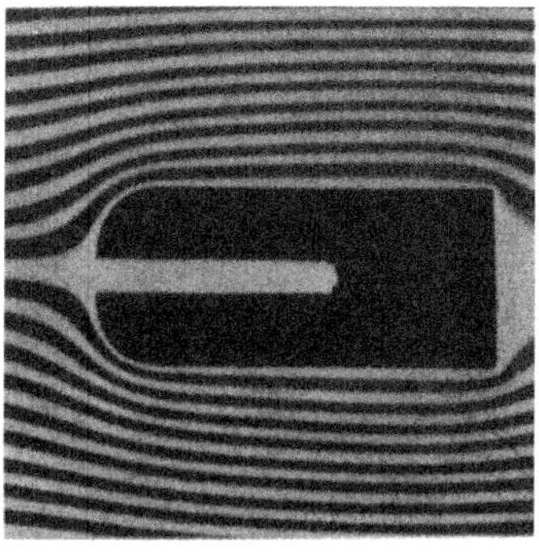

FIG. 6-41. Flow pattern around a pitot tube.

Cylindrical capillary. If a cylindrical capillary of radius r dips into a liquid which wets it ($\beta_0 < 90°$) the liquid rises in the capillary as a result of the surface tension forces distributed along the line of contact. Equilibrium results when the weight of the liquid in the capillary equals the vertical component of the surface forces (Fig. 6-28). Disregarding the small volume segment at the meniscus, the height h of the capillary rise is given by the equation

$$2\pi r\sigma \cos \beta_0 = \pi r^2 h\varrho g,$$

from which

$$h = \frac{2\sigma \cos \beta_0}{r\varrho g}.$$

The capillary rise h is inversely proportional to the capillary radius. For liquids which do not wet the capillary ($\beta_0 > 90°$), h is negative and the liquid is depressed.

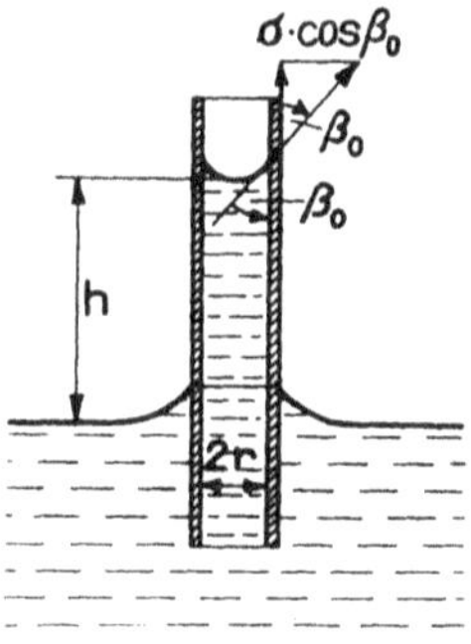

FIG. 6-28

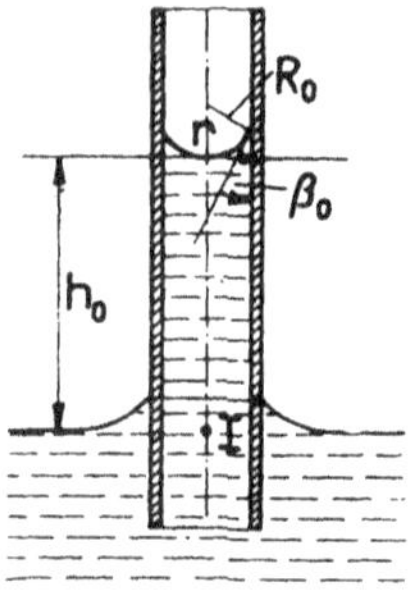

FIG. 6-29

FIG. 6-28. Liquid rising in a capillary, calculation method 1.

FIG. 6-29. Liquid rising in a capillary, calculation method 2.

EXAMPLE. For a capillary 1 mm in diameter, in water at 20°C ($\sigma = 73$ dynes/cm, $\cos \beta_0 \approx 1$) the capillary rise is $h = 2.92$ cm.

The capillary rise can also be computed by realizing that because of the concave surface of the meniscus (radius of curvature $R_0 = r/\cos \beta_0$) the liquid is under a reduced pressure (Fig. 6-29) of

$$\Delta P = \frac{2\sigma}{R_0}.$$

We have assumed here that the meniscus surface is spherical, which is true only for very small tubes. Since the pressure at point I in Fig. 6-29 is atmospheric, then

$$P_a = P_a - \frac{2\sigma}{R_0} + \varrho g h_0,$$

and

$$h_0 = \frac{2\sigma}{\varrho g R_0} = \frac{2\sigma \cos \beta_0}{r\varrho g}.$$

Flattening of the meniscus. If a capillary tube is connected to a wide funnel, as in Fig. 6-30, the form of the meniscus depends on the difference in level h between the meniscus and the liquid surface in the funnel. As the

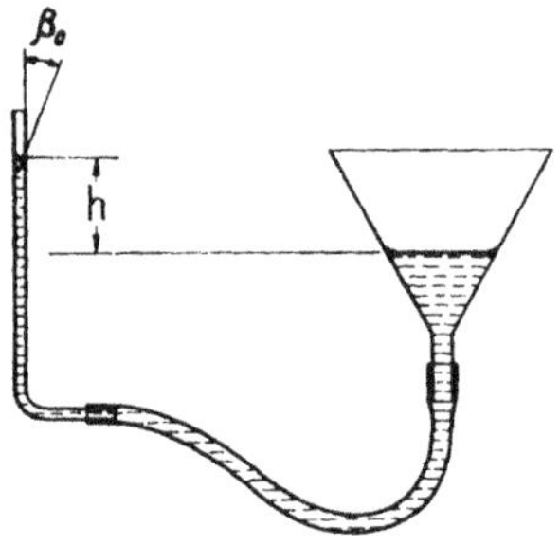

FIG. 6-30. Apparatus for investigating the meniscus surface.

funnel is raised, the meniscus rises in the capillary tube by the same amount and the liquid in the capillary has its normal contact angle, β_0 (Fig. 6-29). When the liquid reaches the top of the capillary, and the funnel is raised still further, the contact angle and the radius of curvature of the meniscus both increase (Fig. 6-31 a). For equilibrium, in this case,

$$P_a = P_a - \frac{2\sigma}{R_1} + \varrho g h_1$$

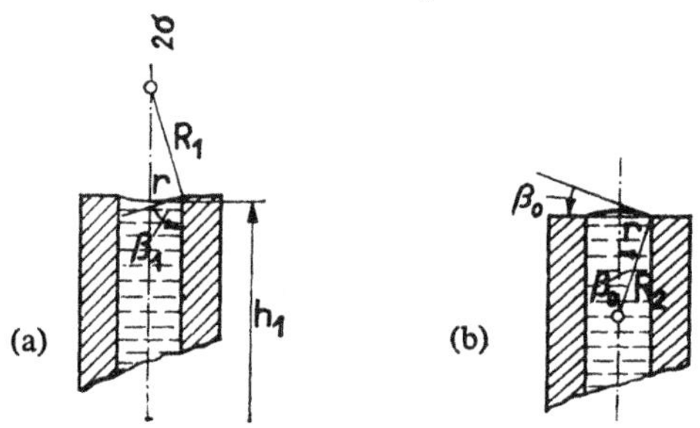

FIG. 6-31 a, b. Change of the meniscus shape with increasing pressure difference.

and the radius of curvature of the meniscus is

$$R_1 = \frac{2\sigma}{\varrho g h_1}.$$

Also, since $R_1 \cos \beta_1 = r$, the contact angle is given by

$$\cos \beta_1 = \frac{r}{R_1} = \frac{r \varrho g h_1}{2\sigma}.$$

When $h = 0$, the radius of curvature of the meniscus becomes infinite. The liquid level in the funnel is then at the same height as the end of the capillary and the liquid surface in the capillary is plane. The contact angle is 90°.

The limiting case for which liquid starts to flow out of the capillary is reached when the surface of the liquid in the capillary makes its normal angle of contact with the horizontal upper surface of the capillary tube (Fig. 6-31 b). We now have

$$P_a + \frac{2\sigma}{R_2} = P_a + \varrho g h_2,$$

where h_2 is the height of the liquid surface in the funnel above the end of the capillary tube. Since $R_2 \sin \beta_0 = r$, we have

$$h_2 = \frac{2\sigma \sin \beta_0}{\varrho g r}.$$

Liquid will not flow out of the capillary until the height of the liquid in the funnel exceeds that in the capillary by h_2.

It will be realized from this discussion why the natural rise of a liquid in a capillary cannot be utilized in the construction of a perpetual motion machine. If a fine capillary tube whose length is less than the natural capillary rise h_0 is dipped in a liquid, no liquid flows out of the top of the capillary; the liquid surface at the upper end of the capillary simply flattens out to form a surface of larger angle of contact.

61. Hydrodynamics of Ideal Liquids

The subject of *hydrodynamics* deals with liquids in motion. To simplify the problem, we first consider an idealized liquid. This is characterized by two properties:

1. No shearing stresses exist in the liquid, i.e. it is frictionless.
2. The ideal liquid is incompressible.

The motion of liquids of this sort, and the pressure distribution within them, is governed by relatively simple laws. The forces acting on the liquid result from pressure and from the weight of the liquid. The fact that no shearing forces exist makes it possible for two adjacent layers of the liquid to move relative to one another without friction. Although real liquids do not satisfy these conditions, they may nevertheless be considered ideal provided the relative velocities of adjacent layers are small enough so that the frictional forces are negligible compared with the other forces. The assumption of incompressibility is closely fulfilled for all real liquids. However, we find from experience that even for ideal liquids only so-called *stationary flow* can be treated by elementary methods.

Every case of fluid flow originally passes through a state in which the flow is not stationary. In many cases, however, flow paths eventually become established which do not change with time. If the paths change only slowly, the flow is called *quasi-stationary*. We shall concern ourselves for the present only with stationary flow.

61-1 Stationary Flow

A flow is called *stationary* if the velocity at a given point of space is independent of the time. In this stationary state every element of the liquid describes the same path as the preceding element. The path followed by an element of the liquid is called its *line of flow*. A *streamline* is defined as a curve whose tangent at every point is in the same direction as the velocity of the liquid. The streamlines describe the direction of the velocity at every point of space. In stationary flow the lines of flow are identical with the streamlines. If one constructs all the streamlines passing through the boundary of a surface element in the fluid, these enclose a tubelike structure, a so-called *stream tube*. No particles of liquid can cross the walls of a stream tube. The liquid within the stream tube is called a *flow filament*. In the stationary state, filaments of flow can never mix.

The simplest type of flow is *homogeneous* flow. In flow of this nature the flow filaments are all parallel to one another and the velocity in all filaments is the same. Other examples of stationary flow, around obstacles of various forms, are illustrated in Fig. 6-32. These stream photographs were taken in a so-called *streamline apparatus*. The flow is made visible by introducing alternate filaments of clear and colored liquid. They show the following typical behavior:

1. One of the streamlines encloses the entire obstacle, splitting into two parts at the *forward* or *upstream stagnation point* (Figs. 6-32 and 6-33, Plate I) and rejoining at the rear of the obstacle (the rear or downstream stagnation point). The stagnation points are characterized by the fact that the velocity at these points is zero.

2. In regions where the channel narrows, the cross-sections of all the flow filaments diminish. By introducing small particles in the flow, it can be seen that the velocity of the flow increases markedly. In spite of this, no mixing of the filaments of flow takes place. If the channel becomes wider, the cross-sections of the flow filaments increase (Fig. 6-34, Plate I).

3. The flow around a symmetrical obstacle is symmetrical.

4. The flow lines retain their identity even around sharp edges of an obstacle (aircraft wing profile, Fig. 6-35, Plate I).

The following two simple laws govern the stationary flow of ideal fluids.

61-2 The Continuity Equation

This equation is a consequence of the incompressibility of the fluid and states that the product of the cross-section A and the velocity v remains constant along a flow filament in the stationary state. Thus we have the *continuity equation*,

$$A_0 v_0 = A_1 v_1 = \text{const}.$$

The product $A_0 v_0$ has the dimension $[L^3 T^{-1}]$ and is the volume of fluid entering the tube of flow per unit time across the area A_0 (Fig. 6-36). In stationary flow, the flow filaments retain their form and the density of the

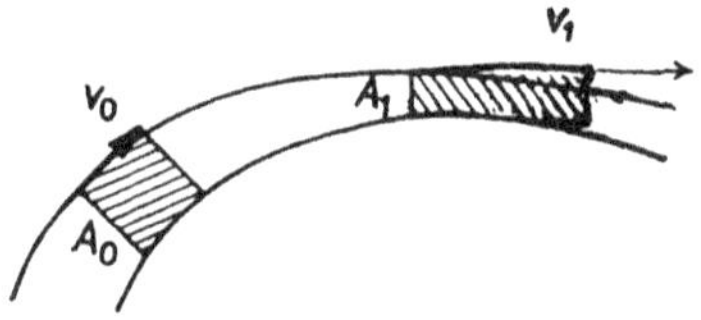

FIG. 6-36. Illustration of the continuity equation.

fluid remains constant (incompressibility). Therefore the same volume of fluid must cross A_1 per unit time, as crosses A_0. It follows from the continuity equation that at points where the cross-section of the flow filaments decreases the velocity increases.

EXAMPLE. A stream of water flows vertically downward from an opening of area A_1 in the bottom of a tank of water. The exit velocity is v_1. The liquid falls with the acceleration of gravity and, since its velocity increases, the cross-section of the stream decreases (see Fig. 6-45).

The continuity equation is valid even for a non-stationary flow, since only the condition of incompressibility is used in its derivation. In this case, however, it is meaningless to speak of individual filaments of flow. The continuity equation asserts here that in an arbitrary given volume the quantity of fluid flowing in per unit time equals that flowing out.

61-3 Bernoulli's Equation

(Daniel Bernoulli, 1700–1782, Fig. 6-37, Plate II)

This equation is the form taken by the work-energy equation for frictionless stationary flow. Consider the volume element *ABCD* (Fig. 6-38) lying within a tube of flow. In a time Δt it is displaced to $A'B'C'D'$. In this process, work is done on the element by the pressure forces exerted on its end faces *AB* and *CD*. The work-energy theorem states that the increase in energy, ΔE, of the element equals the work, ΔW, done on it by the pressure force. The energy E of the element of liquid is part kinetic and part potential. The change in energy is given by the energy change of the shaded volumes

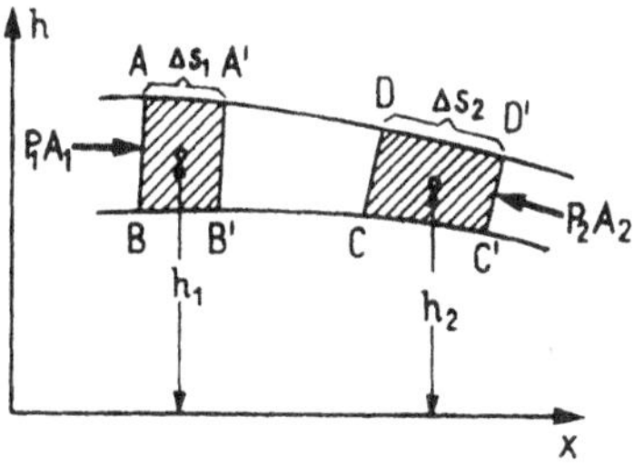

FIG. 6-38. Illustration of Bernoulli's equation.

only, since the volume element $ABCD$ is displaced to $A'B'C'D'$ in the time Δt:

$$\Delta E = E_2 - E_1 = \left(\frac{m_2 v_2^2}{2} + m_2 g h_2\right) - \left(\frac{m_1 v_1^2}{2} + m_1 g h_1\right).$$

According to the equation of continuity the masses m_1 and m_2 are equal:

$$m_1 = m_2 = \varrho \Delta V = \varrho A_1 \Delta s_1 = \varrho A_2 \Delta s_2.$$

Hence we obtain the expression

$$\Delta E = \Delta V\left(\frac{\varrho v_2^2}{2} + \varrho g h_2\right) - \Delta V\left(\frac{\varrho v_1^2}{2} + \varrho g h_1\right).$$

The work done on the element is given by

$$\Delta W = P_1 A_1 \Delta s_1 - P_2 A_2 \Delta s_2 = (P_1 - P_2)\Delta V,$$

where P_1 and P_2 are the static pressures at points 1 and 2. Since $\Delta E = \Delta W$, it follows that

$$\Delta V\left(\frac{\varrho v_2^2}{2} + \varrho g h_2\right) - \Delta V\left(\frac{\varrho v_1^2}{2} + \varrho g h_1\right) = (P_1 - P_2)\Delta V,$$

or

$$P_2 + \frac{\varrho v_2^2}{2} + \varrho g h_2 = P_1 + \frac{\varrho v_1^2}{2} + \varrho g h_1.$$

That is, along a flow filament the sum

$$P + \frac{\varrho v^2}{2} + \varrho g h = \text{const.}$$

Note that the constant is not necessarily the same for different filaments.

This is the famous equation of *Daniel Bernoulli*, which appeared in 1783 (but not in the above general formulation) in his *Hydrodynamica*.† Bernoulli's equation is the most important principle of hydrodynamics. The first term P represents the *static pressure*. This is the pressure that would be measured with a manometer by an observer moving with the fluid. At any one point of the flow the static pressure is independent of the orientation of the surface element against which the pressure is exerted. If the streamlines are curved, the pressure is increased as a result of centripetal forces.

The second term of the equation, $\varrho v^2/2$, is the kinetic energy per unit volume. It has the dimensions of a pressure and is known as the *dynamic pressure*. (In older books it is called the *stagnation pressure*.)

The last term represents the potential energy per unit volume and is called the *gravity pressure*.

† This work also contained the first account of the kinetic theory of gas pressure and the theory of the bending of a beam.

The sum $P + \varrho v^2/2 = P_{st}$ is called the *stagnation pressure* or occasionally the *total pressure*. This latter designation is only justified, however, if the flow is horizontal. In this case the Bernoulli equation reduces to

$$P + \frac{\varrho v^2}{2} = \text{const} = P_{st}.$$

As soon as differences in elevation occur, the stagnation pressure $P_{st} = P + \varrho v^2/2$ varies along a flow filament in the same way as does the hydrostatic pressure in a fluid at rest.

Bernoulli's equation applied to two neighbouring flow filaments leads to the same sum of pressures, i.e. the same constant, only if both filaments have the same initial energy.

61-4 Measurement of Various Pressure Quantities

Each of the pressure terms in Bernoulli's equation is measured with a different apparatus. Measuring instruments, however, measure only the static pressure at the measuring point. It is necessary to construct each instrument in such a way that the proper quantity is measured, always being careful that the flow is disturbed as little as possible.

(a) *Measurement of static pressure, P.* The static pressure in a closed tube, in which the flow is laminar, is measured by making an opening in the tube wall perpendicular to the direction of flow and connecting a manometer to it (Fig. 6-39). Such an opening is called a *piezometric orifice* and a manometer used in this manner is known as a *piezometer* when the fluid in the tube is also used as the manometer fluid.

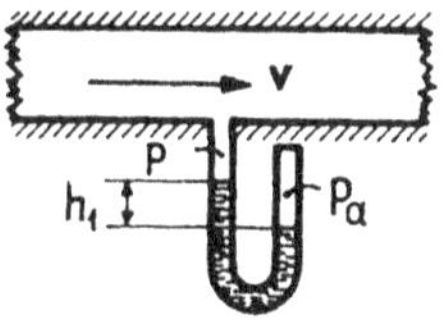

FIG. 6-39

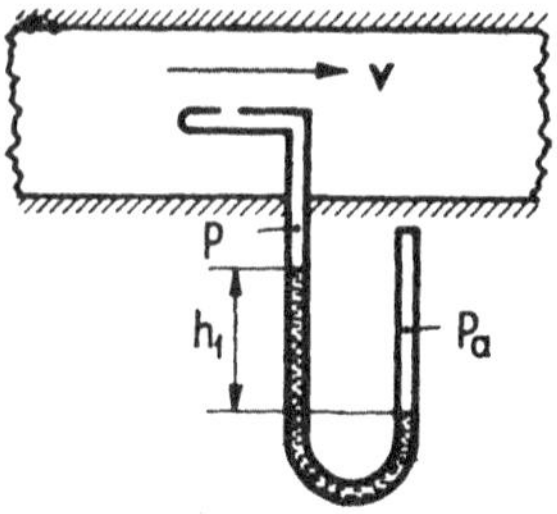

FIG. 6-40

FIG. 6-39. Measurement of the static pressure with a piezometer.

FIG. 6-40. A pressure probe.

To measure the static pressure in the interior of a liquid, a pressure probe is used. It consists of a closed, well rounded tube having an aperture in one side (Fig. 6-40). The instrument is parallel to the flow within the flow channel and is connected to a manometer.

(b) *Measurement of the stagnation pressure, P_{st}.* This is measured with a *Pitot tube*, which measures the pressure at the stagnation point on the end of a rounded off tube (Fig. 6-41, Plate II). At this point $v = 0$, and hence the static pressure, relative to that at a point where the velocity is v, is greater by an amount $\varrho v^2/2$. The instrument must be set parallel to the lines of flow and its transverse dimensions must be small relative to the cross-section of the flowing liquid (Fig. 6-42). From the Bernoulli equation,

$$P_1 = P_2 + \frac{\varrho v_2^2}{2} = P_{st}.$$

A manometer measures the stagnation pressure P_{st} as a static pressure at the stagnation point. If an open tube manometer is used,

$$P_{st} = P_a + \varrho_m g h_1$$

where ϱ_m is the density of the manometer liquid.

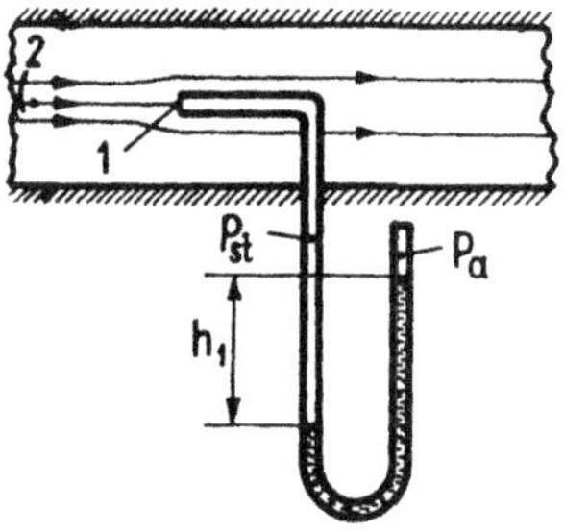

FIG. 6-42

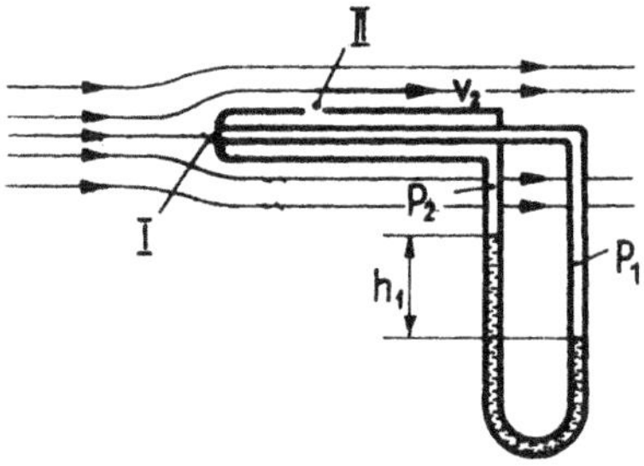

FIG. 6-43

FIG. 6-42. Measuring the stagnation pressure with a pitot tube.

FIG. 6-43. A pitot-static tube for measuring the dynamic pressure.

(c) *Measurement of the dynamic pressure.* This is measured with a *pitot-static tube*. Figure 6-43 is a sectional view of such an apparatus, which is a combination of a pitot tube and a pressure probe. It is important here also that the cross-section of the tube be small relative to that of the main stream, so that the velocity v_2 will not be appreciably different from that in the undisturbed flow. The difference in elevation between points I and II is negligible. At point I a stagnation point is formed at which $v_1 = 0$. For points I and II, in the same tube of flow, the Bernoulli equation gives:

$$P_1 = P_2 + \tfrac{1}{2}\varrho v_2^2$$

and hence

$$\tfrac{1}{2}\varrho v_2^2 = P_1 - P_2 = \varrho_m g h_1.$$

The manometer connected between the two pressure measuring points measures the dynamic pressure directly. It also measures the speed of flow, if we neglect the change in density (i.e. compressibility) since $v_2 = \sqrt{\left(\frac{2\varrho_m g h_1}{\varrho}\right)}$. Provided the velocity is small compared with the velocity of sound, gases can in a first approximation be considered as incompressible.

EXAMPLE. The dynamic pressure in an airstream is measured as

$$h_1 = 4.20 \text{ cm } H_2O$$

at 18°C. Taking $\varrho_{H_2O} = 0.999$ gm/cm³, we get

$$\Delta P = 4116 \text{ dynes/cm}^2.$$

The velocity of the air stream ($\varrho_{air} = 1.20 \times 10^{-3}$ gm/cm³) is therefore

$$v = \sqrt{\left(\frac{2\Delta P}{\varrho_{air}}\right)} = 2619 \text{ cm/sec.}$$

61-5 Applications of Bernoulli's Equation

In spite of the restrictive condition of incompressibility, the continuity equation and Bernoulli's equation can often be applied to the flow of gases. The changes in density of a gas arising from differences in elevation, or from its motion, are often so small as to be negligible. For air at 15°C, for example, the influence of its compressibility on the static pressure, at a velocity of 200 ft/sec, is only about 1 per cent. Differences in height up to about 250 ft alter the magnitude of the pressure by less than 1 per cent due to the effect of compressibility.

(a) *Flow of gas out of a large container.* A large cylinder (Fig. 6-44) has a small opening at one end, out of which flows a stream of gas under an excess pressure ΔP. Provided the excess pressure does not exceed about

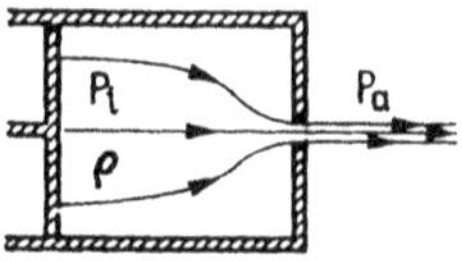

FIG. 6-44. Gas streaming from a container.

10^4 dynes/cm² ≅ 10.2 cm of water, the gas can be considered incompressible and Bernoulli's equation can be applied. The outflowing stream of gas, together with the container, forms a tube of flow. If the amount of gas flowing out is small, so that the velocity of the moving piston which keeps the pressure head ΔP constant is neglibibly small, then neglecting differences in elevation, we have

$$P_i = P_a + \tfrac{1}{2}\varrho v^2,$$

$$v = \sqrt{\left[\frac{2(P_i - P_a)}{\varrho}\right]} = \sqrt{\left(\frac{2\Delta P}{\varrho}\right)}.$$

EXAMPLE. Consider illuminating gas for which

$$\Delta P = 9.5 \text{ gf/cm}^2, \quad \varrho = 7 \times 10^{-4} \text{ gm/cm}^3.$$

Then

$$v = \sqrt{\left(\frac{2 \times 9.5 \times 983}{7 \times 10^{-4}}\right)} = 5.1 \times 10^3 \text{ cm/sec.}$$

The velocity of the escaping gas is very large in spite of the small excess pressure, because of its small density.

(b) *Efflux of liquids.* A reservoir of large cross-section has an opening of area A_1 in its base. The area A_1 is sufficiently small so that the velocity of descent of the liquid surface in the reservoir can be neglected (Fig. 6-45). At the upper surface of the liquid, and at the opening in its base, the pressure is atmospheric. Let h be the height of the liquid surface in the tank above the opening. If we consider the tank and the outflowing stream of liquid as a tube of flow, then from Bernoulli's equation

$$P_a + \varrho g h = P_a + \tfrac{1}{2}\varrho v_1^2,$$

and

$$v_1 = \sqrt{(2gh)}.$$

This relation was obtained by Torricelli, about 100 years before the publication of Bernoulli's *Hydrodynamica.* The equation is obviously correct, since it states that the potential energy of the water is completely converted to kinetic energy. Since we have assumed the liquid to be frictionless, this must be the case. The velocity of efflux is the same as if the water had fallen freely from the level of the free surface in the tank to the opening.

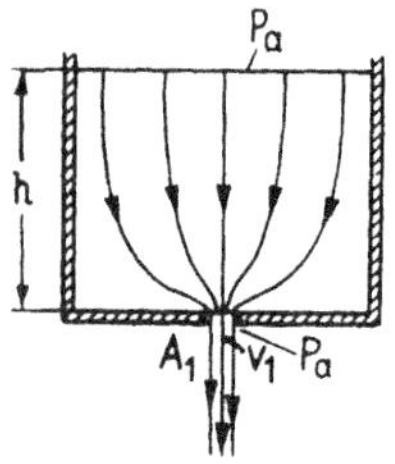

FIG. 6-45

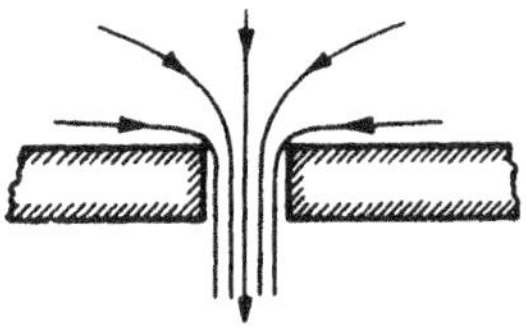

FIG. 6-46

FIG. 6-45. Efflux of a liquid.

FIG. 6-46. Contraction of a stream of liquid.

Discharge rate and jet contraction factor. When a liquid flows out of an orifice with sharp edges, as in Fig. 6-46, the effect of its inertia becomes noticeable. The flow filaments cannot bend suddenly around the sharp corners, since a radius of curvature of zero would require an infinite pressure. It follows that the flow tubes leave the walls of the opening and have a finite radius of curvature. As a consequence, the geometrical cross-section of the opening is not identical with the cross-section of the outflowing stream. This phenomenon can be observed at every sharp-edged opening through which a liquid is flowing. The stream of fluid makes contact with the opening only at its inner edge (Fig. 6-46) and we can write

$$A_{\text{stream}} = \alpha A_{\text{geom}},$$

where A_{geom} is the actual area of the opening and α is called the *contraction factor.* The rate of flow at the opening may be considerably reduced as a result of the contraction factor. However, by proper design of the exit passage the contraction factor can be made very nearly unity.

(c) *The aspirator pump.* A stream of water, flowing in the entrance tube I in Fig. 6-47 where the velocity is v_1, the pressure P_1, and the cross-section A_1, emerges from the narrow opening II into the low-pressure region V.

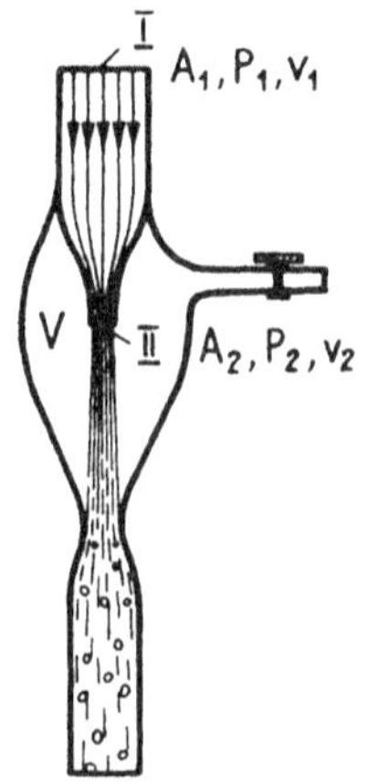

FIG. 6-47. The aspirator pump.

As a result of the increase in its velocity, the static pressure P_2 at point II is much less than the pressure P_1. The stream of water therefore absorbs air and carries it out of the volume V until the pressure is reduced to the static pressure P_2. From Bernoulli's equation, neglecting differences in elevation, we have:

$$P_1 + \tfrac{1}{2}\varrho v_1^2 = P_2 + \tfrac{1}{2}\varrho v_2^2.$$

From the continuity equation,

$$A_1 v_1 = A_2 v_2,$$

and hence

$$P_2 = P_1 - \frac{1}{2}\varrho(v_2^2 - v_1^2)$$

$$= P_1 - \frac{1}{2}\varrho v_1^2\left(\frac{A_1^2}{A_2^2} - 1\right).$$

The pressure difference is

$$P_1 - P_2 \cong \frac{1}{2}\varrho v_1^2 \frac{A_1^2}{A_2^2}.$$

In practice, the pressure in the volume V can be reduced only to the vapor pressure of water, which is a function of temperature and at 10°C is about 9 mm Hg. One can obtain a better vacuum by using a liquid with a lower vapor pressure. For instance, the mercury vapor pump will produce pressures as low as 10^{-3} mm of Hg.

61-6 Application of Newton's Second Law to Fluid Flow

Newton's second law, the most important principle in mechanics, can also be applied to problems in stationary flow, even though nothing can be said regarding the details of the flow process. Figure 6-48 represents a portion of a rigid tube through which a liquid is flowing with stationary flow.

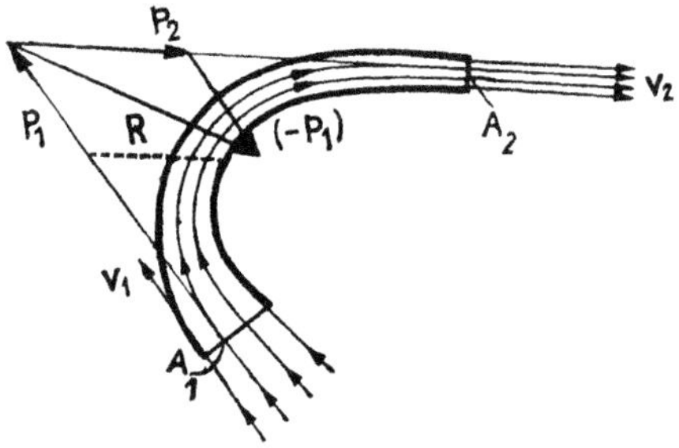

FIG. 6-48. Applying Newton's Second Law to fluid flow.

The liquid exerts forces on the tube because of the pressure distribution at the tube walls. External forces keep the tube in equilibrium. The resultant $\boldsymbol{R}$ of these forces must be equal in magnitude and opposite in direction to the vector sum of the pressure forces. The latter are not easy to calculate, but with the help of Newton's second law we can draw some conclusions about them, for the case of an ideal liquid. Let us apply this law to the liquid in the tube at any instant. Let $\boldsymbol{v}_1$ be the velocity of the liquid at the entrance of the tube, and $\dot{m}$ the mass of liquid flowing into the tube in one unit of time. The momentum flowing in per unit time, is therefore

$$\boldsymbol{p}_1 = \dot{m}\boldsymbol{v}_1 .$$

If A_1 is the cross-sectional area at the entrance, then

$$\dot{m} = \varrho A_1 v_1 .$$

Similarly, the momentum flowing out of the tube per unit time is

$$\boldsymbol{p}_2 = \dot{m}\boldsymbol{v}_2 .$$

There is no change in the momentum of the tube itself. Hence the change in the momentum of the system per unit time is

$$\frac{d\boldsymbol{p}}{dt} = \dot{m}(\boldsymbol{v}_2 - \boldsymbol{v}_1).$$

This change in momentum is produced by forces exerted on the liquid by the walls of the tube. Their resultant is

$$\boldsymbol{R} = \dot{m}(\boldsymbol{v}_2 - \boldsymbol{v}_1) = \varrho A_1 v_1(\boldsymbol{v}_2 - \boldsymbol{v}_1).$$

The force $\boldsymbol{R}$ is exerted on the liquid by the walls of the tube. The reaction to this force, $\boldsymbol{R}_1 = -\boldsymbol{R}$, is exerted on the tube by the liquid. To keep the tube in equilibrium, an external force equal in magnitude and opposite in direction to $\boldsymbol{R}_1$, i.e. equal to $\boldsymbol{R}$, must be exerted on it.

EXAMPLE. 1. Figure 6-49 shows a tube having a right-angled bend. The tube has a constant cross-section A of 7.0 cm^2 and a liquid of density $\varrho = 1$ gm/cm^3 flows through it with a velocity $v = 40$ cm/sec.

$$\boldsymbol{R} = Av\varrho(\boldsymbol{v}_2 - \boldsymbol{v}_1),$$

$$R_1 = R = A\varrho v^2 \sqrt{2} = 1.58 \times 10^4 \text{ dynes.}$$

The line of action of the resultant passes through the point of intersection of the axes of the horizontal and vertical portions of the tube.

2. Reaction of a liquid flowing out of a vessel (Fig. 6-50). The stream of liquid flowing out of the vessel with a velocity $v_2 = \sqrt{2gh}$ gives rise to a sidewise reaction force F_x in the vessel, where

$$F_x = 2\varrho gh A_2 .$$

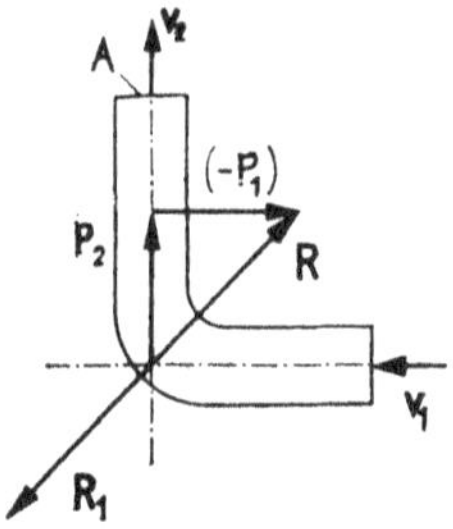

FIG. 6-49

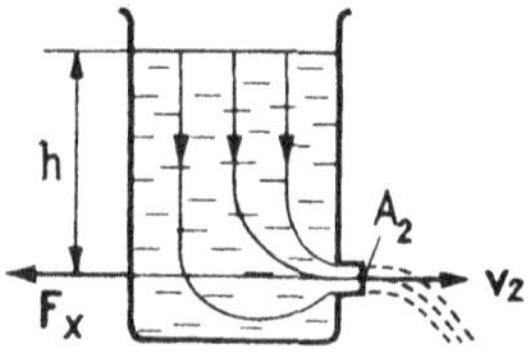

FIG. 6-50

FIG. 6-49. Force exerted on a pipe elbow by the liquid flowing through it.

FIG. 6-50. Reaction force of a liquid flowing out of a vessel.

62. Hydrodynamics of Real Liquids

Real liquids differ from ideal liquids in two respects; they are *compressible*, and *frictional* forces act between two adjacent flow filaments having different velocities. Since the compressibilities of liquids are so small, this factor can usually be neglected. The shearing stresses arising from the friction forces, however, are of great importance. They can result in two very different types of flow, known as *laminar* flow and *turbulent* flow. In laminar flow, adjacent flow filaments remain side by side and a so-called *boundary layer* adheres to the surfaces of the bodies around which the liquid flows. In turbulent flow, on the other hand, neighbouring flow filaments separate from one another and a violent mixing and vortex motion takes place.

62-1 Viscosity

The frictional forces which two adjacent layers of a real fluid exert on each other can be understood through the concept of *viscosity*. Let us consider a layer of liquid (glycerin, for example) between two plane parallel plates as in Fig. 6-51. The lower plate is stationary, the upper plate moves

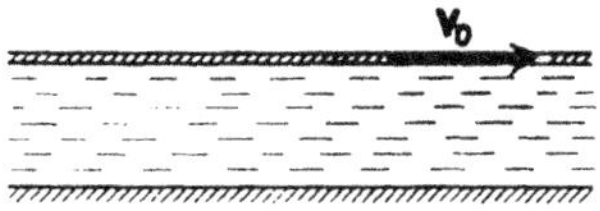

FIG. 6-51. Production of plane, viscous flow.

to the right with a constant velocity v_0. The moving upper plate sets the liquid adjacent to it in motion. The layer of liquid in direct contact with the upper plate is dragged along with the velocity v_0. The neighbouring layers are acted on by frictional shearing forces and move with a velocity which is smaller, the greater the distance from the upper plate.

The shearing stress τ at the surface between adjacent layers of the liquid, in both stationary and non-stationary flow, is given by Newton's law of viscosity,

$$\tau = \eta \frac{dv}{dy}.$$

The derivative dv/dy is the rate of change of velocity in the y-direction and is called the *velocity gradient*. Newton's law of viscosity then states that *the shearing stress τ is proportional to the velocity gradient dv/dy*. In the absence of external forces, the effect of the shearing stress is to equalize the velocity difference between adjacent layers.

The coefficient η in the equation above is a phenomenological constant and is called the *coefficient of viscosity*, or simply the *viscosity* of the liquid.

Dimensions: $[\eta] = [ML^{-1}T^{-1}]$

Units: cgs system: 1 gm/cm · sec = 1 poise
= 100 centipoise (after Poiseuille)
mks system: 1 kgm/m · sec = 10 poise
fps system: 1 lbm/ft · sec

The velocity distribution over the cross-section of a stationary flow can now be found from the viscosity law given above. As an example, consider the case of plane, viscous flow (Fig. 6-52). A viscous liquid flows in a channel, the velocity of the bottom layer being zero and of the top layer v_0. In stationary flow, every element of liquid between two streamlines must be in equilibrium. If the velocity gradient at a height y is $(dv/dy)_y$, the retarding force at the lower face of the element is

$$F_x(y) = \tau(y)\,\Delta A = \eta \left(\frac{dv}{dy}\right)_y \Delta A.$$

The liquid above the element exerts a forward force on the upper surface of the element, given by

$$F_x(y + dy) = \eta \left(\frac{dv}{dy}\right)_{y+dy} \Delta A.$$

If the flow is stationary (zero acceleration) the forces on the element must be in equilibrium. Since no other forces act on it in a horizontal direction.

$$\sum F_x = \eta \Delta A \left[\left(\frac{dv}{dy}\right)_{y+dy} - \left(\frac{dv}{dy}\right)_y\right] = 0.$$

Therefore

$$\left(\frac{dv}{dy}\right)_{y+dy} = \left(\frac{dv}{dy}\right)_y,$$

and the change of velocity per unit distance has the same value at all elevations. That is, the velocity gradient in this type of flow is constant and the velocity increases linearly with y (Fig. 6-53).

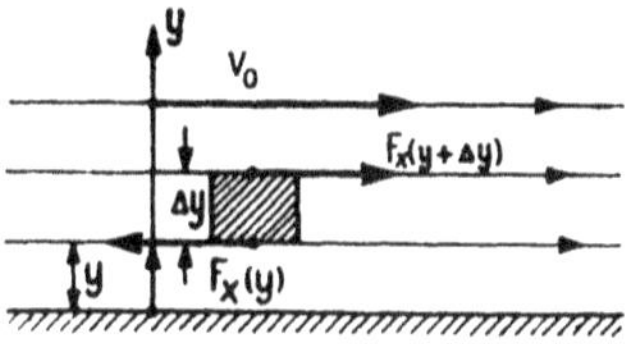

FIG. 6-52

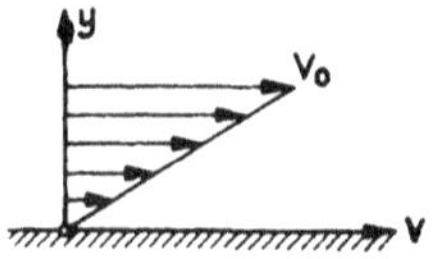

FIG. 6-53

FIG. 6-52. Forces in plane, viscous flow.

FIG. 6-53. Velocity distribution in plane, viscous flow.

62-2 *Effects of Viscosity*

The viscosity of a liquid is the decisive factor which determines the nature of a flow. There are two limiting special cases. In one, the predominating forces are purely viscous forces. In the other, which occurs at sufficiently high velocities, the viscous forces cause the formation of vortices which give rise to an additional resistance to the flow.

(a) *Flow determined by viscous forces only.* Let us consider the flow of a liquid in a smooth-walled cylindrical tube, under the influence of purely viscous forces. This special case is of great practical importance.

In Fig. 6-54, a liquid flows through a tube of radius R. The pressure difference between two cross-sections separated by a distance l is ΔP. Now

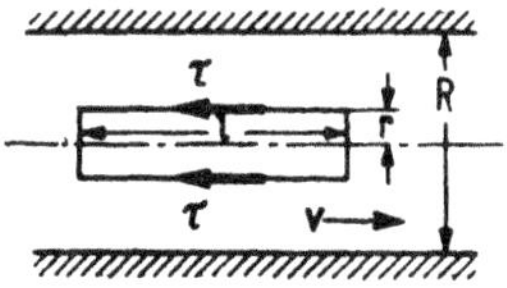

FIG. 6-54. Derivation of Poiseuille's Law.

consider the flow of a cylindrical element of the fluid of length l and radius r whose axis coincides with the axis of the tube. As a result of the pressure gradient, a force $\pi r^2 \Delta P$ acts on the element in the direction of decreasing pressure. The surrounding liquid exerts on the cylindrical walls of the element a viscous force given by $2\pi r l \tau$, where τ is the shearing stress at the surface of the element. The magnitude of τ depends on the radial velocity gradient at the surface, that is,

$$\tau = \eta \frac{dv}{dr}.$$

If the flow is stationary (zero acceleration) the velocity gradient must be such that the pressure force and the viscous force have equal magnitudes and are oppositely directed. Therefore,

$$\pi r^2 \Delta P + 2\pi r l \eta \frac{dv}{dr} = 0.$$

From this equation, the velocity gradient at the surface of the element is

$$\frac{dv}{dr} = -\frac{\Delta P}{2 l \eta} r.$$

The velocity gradient increases from the center of the tube linearly with the radius. Integration of this equation, taking into account the fact that the liquid is stationary at the walls of the tube ($v = 0$ when $r = R$), leads to the result:

$$v = \frac{\Delta P}{4 l \eta}(R^2 - r^2).$$

The velocity distribution across a section of the tube is therefore parabolic [Figs. 6-55 and 6-56 (Plate III)].

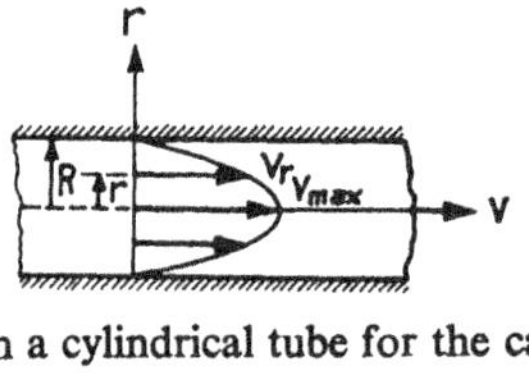

FIG. 6-55. Velocity in a cylindrical tube for the case of laminar flow.

The volume of liquid ΔQ which flows in time Δt across an annular zone of radii r and $r + \Delta r$ is

$$\Delta Q = 2\pi r \Delta r \cdot v \Delta t.$$

The total flow in time Δt across the entire cross-section is found by integration of this equation using the above expression for v. The result is:

$$Q = \frac{\pi R^4 \Delta P}{8 l \eta} \Delta t.$$

(a)

(b)

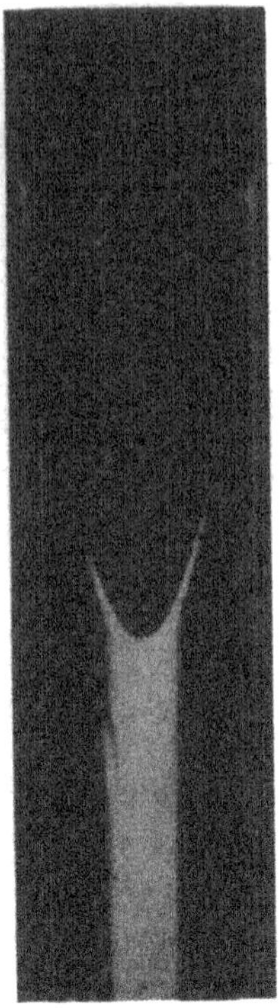

FIG. 6-56. Parabolic deformation due to laminar flow of a plane interface between regions of colored and colorless glycerin.

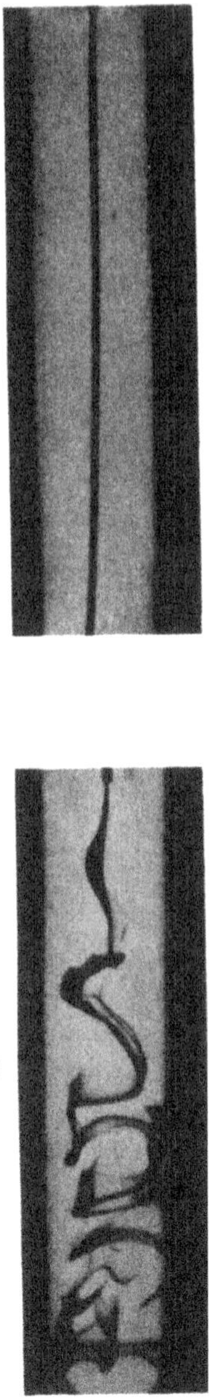

FIG. 6-58. Laminar (a) and turbulent (b) flow in a cylindrical tube. In order to make these phenomena visible a fine stream of ink has been introduced into the water through a capillary tube which is just visible at the top of each of the pictures.

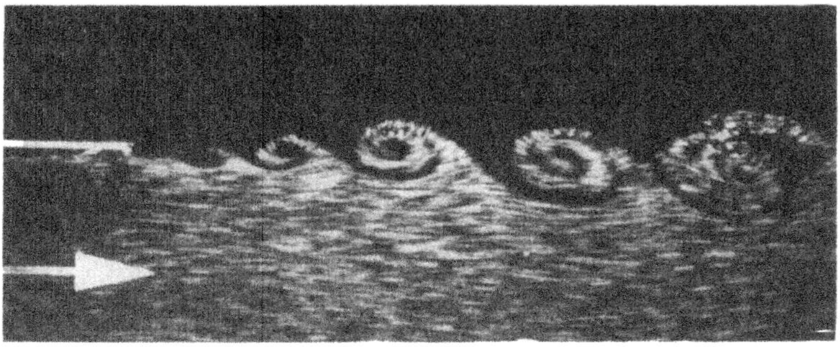

Fig. 6-60. The formation of vortices when a moving stream of water flows into a region where the water is at rest. The flow is made visible with the help of aluminium powder suspended in the moving stream.
(Prof. J. Ackeret.)

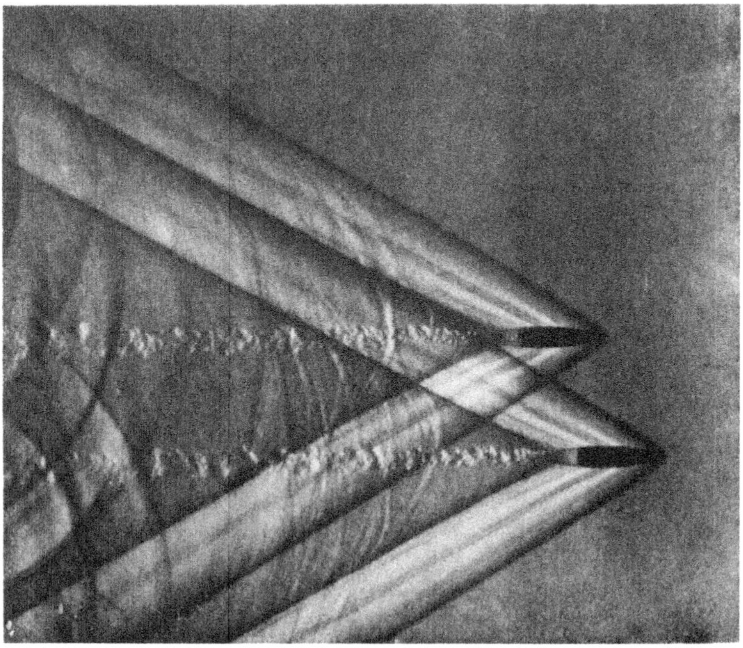

Fig. 6-64. The Mach cones of supersonic bullets. Note the turbulent vortex stream behind each bullet. (Prof. H. Schardin.)

This law was discovered by Hagen and Poiseuille and is named after them. It shows the strong influence of the radius R on the quantity of liquid flowing through the tube. Increasing the radius of a tube is very effective in increasing the quantity of flow.

Another example of laminar flow occurs when very small spherical drops of liquid fall through a gas at rest, or when small spheres fall in a highly viscous liquid. The exact expression for the friction force $\boldsymbol{F}$ in this case was derived by Stokes in 1850. To a first approximation

$$\boldsymbol{F} = 6\pi\eta r\boldsymbol{v},$$

where $\boldsymbol{v}$ is the velocity of the sphere and r is its radius.

The fact that the resisting force is proportional to the velocity leads to the result that a small sphere, after falling for a short time, ceases to accelerate and moves with constant velocity. The sphere initially accelerates under the influence of an accelerating force equal to its own weight minus the buoyant force. As its velocity increases, the resisting force increases also and the velocity becomes constant when the resisting force equals the accelerating force. From the conditions of equilibrium,

$$6\pi\eta rv = \frac{4\pi}{3} gr^3(\varrho_b - \varrho_{fl}).$$

The constant *terminal velovity* v_{ter} is

$$v_{\text{ter}} = \frac{2}{9}\frac{\varrho_b - \varrho_{\text{fl}}}{\eta} gr^2,$$

where ϱ_b is the density of the body and ϱ_{fl} that of the surrounding fluid.

For fog droplets of radius $r = 2 \times 10^{-4}$ cm, in air of viscosity $\eta = 1.60 \times 10^{-4}$ poise, the terminal velocity is

$$v = \frac{2}{9}\frac{\varrho_{\text{H}_2\text{O}} - \varrho_{\text{air}}}{\eta} gr^2 = 5.44 \times 10^{-2} \text{ cm/sec}.$$

TABLE 6-3

Material	Temp. °C	η (poise)
Water	0	1.79×10^{-2}
Water	20	1.00×10^{-2}
Water	50	0.55×10^{-2}
Air	0	1.71×10^{-4}
Hydrogen	0	8.35×10^{-5}
Glycerin	20	14.90
Mercury	20	1.55×10^{-2}

The Hagen–Poiseuille law and Stokes law can both be used, with the proper experimental apparatus, to measure the coefficient of viscosity η. A few numerical values are given in Table 6-3. Note that the viscosity varies markedly with temperature.

These numerical values show that for the fluids most commonly encountered in practice, namely, water and air, the viscosity is relatively small. Hence, if the velocity gradients are not unduly large we may neglect the viscous forces and consider the fluid as ideal. However, the velocity gradient in the layer of fluid adjacent to a body (the boundary layer) is always large and is greater, the smaller the viscosity. This was first recognized by Prandtl in 1904, and it follows as a consequence that frictional forces in the boundary layer must always be taken into consideration.

(b) *Turbulent flow.* At high velocities, viscous flow in a tube is transformed to a new type of flow for which the Hagen–Poiseuille law does not apply. Fluid flow, which takes place smoothly in the laminar case, here becomes an eddying, irregular, mixing motion. This eddying motion is called *turbulence.* In a cylindrical tube it results in an almost rectangular or uniform velocity distribution which decreases to zero in a narrow region adjacent to the tube walls (Fig. 6-57). In this region the very high velocity gradient results in a large shearing force which gives rise to the formation of vortices. The character of the flow is fundamentally altered and new expressions must be derived to describe it.

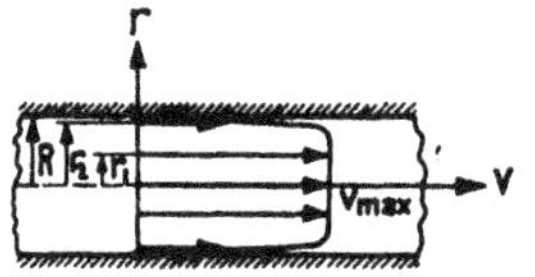

FIG. 6-57. Velocity distribution in the case of turbulent flow in a cylindrical tube.

Both laminar and turbulent flow can easily be demonstrated in a glass tube through which water is flowing, by introducing into the flow a fine stream made visible by ink or dye. In laminar flow, a well-defined filament of dye is formed (Fig. 6-58 a, Plate III). Neighboring filaments flow smoothly beside it. In turbulent flow, on the other hand, the flow filaments are completely destroyed as a result of the mixing motion (Fig. 6-58 b, Plate III).

The phenomenon of turbulence occurs not only in tubes, but also in the case of flow around obstructing bodies when the flow velocity is sufficiently large. In flow of this sort, certain characteristic processes take place just at the surface of the bodies, where a layer of liquid at rest clings to the walls. Very close to the walls a layer is formed, the so-called *boundary layer*, in which there is a large velocity gradient and within which the frictional or viscous effects are chiefly developed. The flow in the boundary layer is greatly slowed down by the resulting viscous forces. Despite this friction, however, the fluid in the boundary layer on the upstream side of an obstacle is driven forward by the pressure gradient which results from the contraction of the streamlines. On the downstream side of the obstacle, however, the flow takes place *against* a pressure gradient, and because of the previous friction losses it has insufficient kinetic energy to surmount this pressure rise. It therefore comes to rest and then reverses its direction of motion as a result of the pressure distribution due to the adjacent liquid layers. The mass of this fluid with reversed motion increases and eventually, as a result of energy acquired from that nearby fluid which still has a forward motion, it becomes rolled up in the form of a vortex. When the vortex has grown to a sufficient size it separates from the surface and moves away from it. As a result of this vortex formation a mixing process takes place in the liquid which destroys the laminar flow. This vortex formation takes place on the downstream side of the body. Around a cylinder, for example, the vortices become detached alternately from the upper and lower sides and form a so-called *vortex stream.* This periodic detachment of vortices may occur at sufficiently high frequency to result in an audible tone (a sound like that of wind blowing through telephone wires, noise of the wind in the trees).

The energy required for the formation of vortices can only be supplied by the liquid itself; that is, it has to overcome a resistance. As a result of vortex formation on the downstream side of an obstacle the pressure distribution is so altered (complete stagnation pressure upstream, reduced pressure downstream) that the pressure forces give rise to a resultant force on the body. The component of this force in the direction of flow is called the *pressure drag*. For turbulent flow, this resistance is added to any purely frictional drag due to skin friction. Since the body is in equilibrium, it exerts an equal and opposite force on the fluid, and the work done against this force appears as kinetic energy of the vortices. It has been found experimentally that the pressure drag D_p of a body is proportional to the stagnation pressure $\varrho v^2/2$ and proportional to the surface A of the body at right angles to the direction of flow,†

$$D_p = c_{Dp}\frac{\varrho v^2}{2}A.$$

The sum of the pressure drag and the frictional drag is the *total drag D*, which, in analogy to the expression for the pressure drag, is often written as

$$D = c_D\frac{\varrho v^2}{2}A.$$

The coefficient c_D is called the *coefficient of drag* of the body and is a pure number. We shall return to the discussion of the coefficient c_D in the next section. It should be noted that both pressure drag and friction drag are effects due to viscosity.

† For an aircraft wing, one uses the product of wing span and chord instead of the projected area in the direction of motion.

This resisting force, which increases proportionally to v^2, is the most important force opposing the motion of rapidly moving vehicles such as automobiles and aircraft. It is important that the coefficient of drag should be kept small by proper design. Since vortex formation occurs only on the rear or downstream side, the form of this portion of the vehicle has the greatest influence on the drag. One of the problems in the design of autos and aircraft is to select a shape for which the vortex development will be small in spite of the rise in pressure. It may be remarked in this connection that a turbulent boundary layer, because of the violent mixing that goes on within it, is more forcibly dragged along by the surrounding fluid and is less effective in producing vortices. Hence the drag coefficient is smaller for a turbulent than for a laminar boundary layer.

When a body moves on the surface of a liquid, such as a ship on the surface of the ocean, surface waves are developed which represent a transport of kinetic energy away from the body. Work must be done by the moving body to produce these waves and a resisting force known as the *wave resistance* must be added to the forces of viscosity and drag.

In the preceding discussion, it was assumed that viscosity was the cause of vortex formation. Vortices can also be produced in a nonviscous medium, if two adjacent layers of fluid are in motion relative to one another (Fig. 6-59).

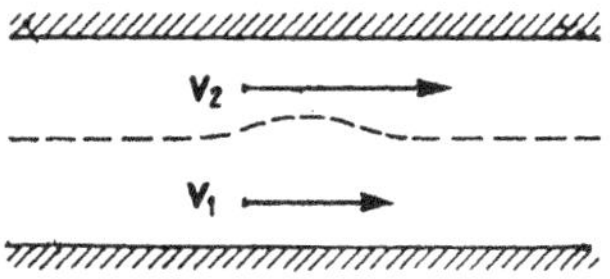

FIG. 6-59. The start of the formation of a vortex in nonviscous flow.

The boundary surface between layers is an unstable region. Suppose that at some point this boundary becomes perturbed in some way, for example, by a slight upward displacement. As a result of this, the stream tubes on the lower side become wider and those on the upper side, narrower. Acording to the continuity equation, therefore, the velocity decreases below the boundary and increases above it. From Bernoulli's equation, the static pressure below the boundary increases and that above decreases. This pressure distribution increases the perturbation still further. If the original disturbance had been downward, the resulting pressure changes would have been reversed. This situation, in which the original disturbance always increases, leads eventually to the production of vortices. The growth of an initial disturbance and the resulting formation of vortices is well demonstrated in the photograph of Fig. 6-60 (Plate IV). The same phenomenon is chiefly responsible for the frequently observed "mackerel sky".

62-3 *Laws of Similarity. Reynolds Number*

In this section we shall investigate the question of the *dynamic similarity* of two different flows having geometrically similar boundary surfaces. It was the English physicist Osborne Reynolds who in 1883 first solved this problem and thereby discovered an important new principle of hydrodynamics.

Similarity laws are of the greatest interest in physics, and we therefore take this opportunity to discuss the concept briefly. In contrast to a purely geometrical similarity, such as that of two triangles whose sides are proportional to one another, here we deal with a *physical similarity* which applies to all physical quantities involved.

DEFINITION. Two processes are physically similar if corresponding physical quantities at similarly located points are proportional to one another.

In mechanics it is possible to express all quantities in terms of the three fundamental quantities length, time, and mass. *Two processes are therefore physically similar if corresponding lengths, times, and masses at similar points are proportional to one another.*

We shall now apply this requirement to two geometrically similar flows. To fix the conditions, let us think of two geometrically similar bodies in a liquid (Fig. 6-61). The bodies can be described in terms of their *characteristic*

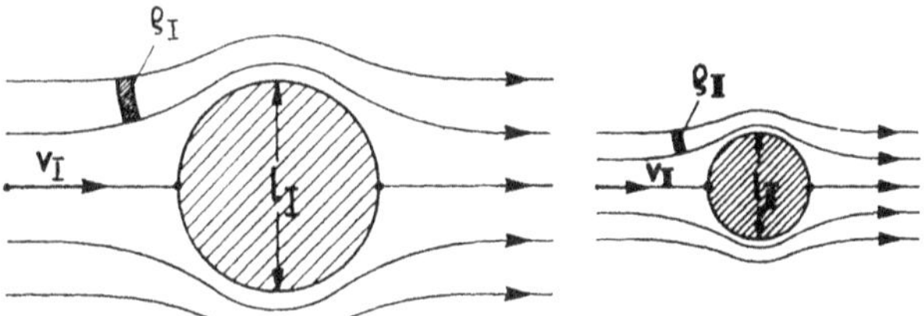

FIG. 6-61. Geometrically similar flows.

lengths. For spheres, for example, these lengths would be the diameters l_I and l_{II} respectively. In similarity investigations, it is convenient to use as the unit of length not 1 cm or 1 m but some characteristic length which enters directly in the problem such as the diameters l_I or l_{II}. Then all lengths and length differences are expressed in units of l_I or l_{II}. Similar lengths, areas, and volumes in the two flows then have the same numerical values when expressed in these units. Suitable units of time and mass are chosen in the same way. The unit of time can be fixed by the selection of a characteristic velocity (for example, the upstream velocity v_I or v_{II}). Then $t_I = l_I/v_I$, $t_{II} = l_{II}/v_{II}$. Finally, the unit of mass is determined by the density of the liquid, ϱ_I or ϱ_{II}. All quantities of interest in the two flows can then be expressed in terms of these fundamental units, namely, l_I, v_I (instead of t_I), ϱ_I (instead of m_I), and l_{II}, v_{II}, and ϱ_{II}.

In what follows, we shall assume that geometrically similar flows are completely determined by pressure forces and viscous forces. Other forces such as gravity will be ignored. The accelerations occurring in the flow are then determined by the pressure and viscous forces only. Instead of dealing with accelerations we can introduce the D'Alembert inertial forces and consider the flow as a system in equilibrium. At every point the pressure force, the viscous force, and the inertial force must be in equilibrium. One of the three forces can then be determined from the conditions of equilibrium. We select as the forces determining the flow pattern of the fluid the viscous force V and the inertial force I. It follows from this that if the ratio of the inertial force I to the viscous force V is the same at similarly located points, the flows are dynamically similar (*similarity condition*). This requirement is the basis of the *Reynolds similarity law*. Specifically, this becomes

$$\frac{I_I}{I_{II}} = \frac{V_I}{V_{II}},$$

that is

$$\left(\frac{I_I}{V_I}\right)_p = \left(\frac{I_{II}}{V_{II}}\right)_{p'}, \tag{62-1}$$

where p and p' are similarly located points in the two flows (Fig. 6-62).

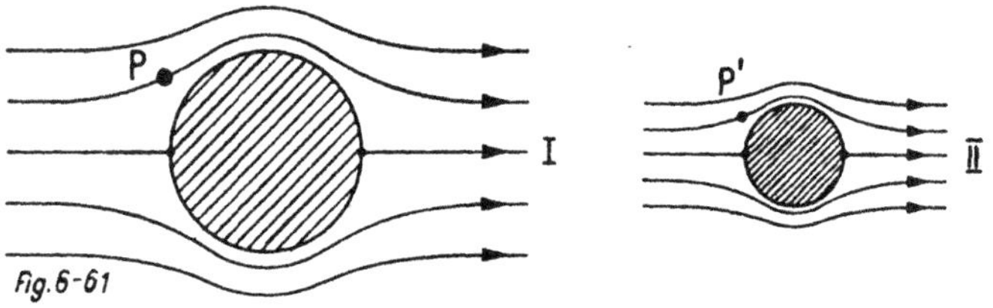

FIG. 6-62. Diagram for the derivation of the Reynolds similarity law. P and P' are similarly located points.

From p. 284 and p. 85,

$$V = \eta \frac{dv}{dy} \Delta A$$

and

$$I = m \frac{dv}{dt}.$$

Then from Eqn. (62-1)

$$\frac{\left(m_I \frac{dv_I}{dt_I}\right)_p}{\left(\eta_I \frac{dv_I}{dy_I} \Delta A_I\right)_p} = \frac{\left(m_{II} \frac{dv_{II}}{dt_{II}}\right)_{p'}}{\left(\eta_{II} \frac{dv_{II}}{dy_{II}} \Delta A_{II}\right)_{p'}}. \tag{62-2}$$

This equation can now be expressed in terms of the special fundamental units l_I, v_I, ϱ_I and l_{II}, v_{II}, and ϱ_{II}. For physically similar flow we have at the similar points:

$$\frac{m_I}{m_{II}} = \frac{\varrho_I l_I^3}{\varrho_{II} l_{II}^3}; \quad \frac{dv_I/dt_I}{dv_{II}/dt_{II}} = \frac{v_I^2 l_I^{-1}}{v_{II}^2 l_{II}^{-1}};$$

$$\frac{\left(\frac{dv}{dy}\right)_I}{\left(\frac{dv}{dy}\right)_{II}} = \frac{v_I l_I^{-1}}{v_{II} l_{II}^{-1}}; \quad \frac{\Delta A_I}{\Delta A_{II}} = \frac{l_I^2}{l_{II}^2}.$$

Inserting these expressions in Eqn. (62-2), we get

$$\frac{\varrho_I l_I^3 v_I^2 l_I^{-1}}{\eta_I v_I l_I^{-1} l_I^2} = \frac{\varrho_{II} l_{II}^3 v_{II}^2 l_{II}^{-1}}{\eta_{II} v_{II} l_{II}^{-1} l_{II}^2},$$

or, after simplification,

$$\frac{\varrho_I v_I l_I}{\eta_I} = \frac{\varrho_{II} v_{II} l_{II}}{\eta_{II}}.$$

It follows that for dynamical similarity the quantity $\varrho vl/\eta$ must have the same value in both flows. Since this quantity is the ratio of two forces (V and I), it is a pure number, and is called the *Reynolds number R*. The quantity η/ϱ = viscosity/density is called the *kinematic viscosity* and is designated by ν. Two geometrically similar flows will thus be physically similar if the Reynolds number is the same in both flows:

$$R_I = \frac{v_I l_I}{\nu_I} = R_{II} = \frac{v_{II} l_{II}}{\nu_{II}},$$

which is Reynolds' similarity law.

Instead of using the ratio I/V it would also be possible, from the conditions of d'Alembert equilibrium to introduce the ratio of inertial force I to pressure force F as characteristic of the flow. The similarity condition then leads to the requirement that at similarly located points

$$\left(\frac{I_I}{F_I}\right)_p = \left(\frac{I_{II}}{F_{II}}\right)_{p'}. \tag{62-3}$$

Since the ratio of the pressure forces at p and p' is

$$\frac{F_I}{F_{II}} = \frac{P_I l_I^2}{P_{II} l_{II}^2},$$

then from eqn. (62-3)

$$\frac{m_I v_I^2 l_I^{-1}}{P_I l_I^2} = \frac{m_{II} v_{II}^2 l_{II}^{-1}}{P_{II} l_{II}^2}.$$

In this formulation, the similarity law becomes (since $\varrho = m/l^3$),

$$\frac{\varrho_I v_I^2}{P_I} = \frac{\varrho_{II} v_{II}^2}{P_{II}}.$$

That is, the quantity $\varrho v^2/P$ must have the same value at corresponding points in similar flows. This condition, however, adds nothing new to our previous formulation since the three forces I, V, and F are related through the conditions of equilibrium. The second requirement can only be satisfied if the first is satisfied also.

The great practical significance of the Reynolds similarity law is that it makes possible experiments with models. In place of the actual object, smaller models can be investigated at correspondingly higher velocities. It is only necessary that the Reynolds number shall be the same for the actual system and for the model. In many instances, however, the ratio vl/ν, in the model experiment, cannot be increased sufficiently. This happens, for instance, if the same fluid is used in the model tests as in the actual situation and if the actual flow past the object already has a large velocity, so that strictly speaking the two flows are not similar. Such model experiments are nevertheless useful because the properties of the flow change only slightly even with large changes in velocity.

The Reynolds similarity law in the form developed above is applicable only when the significant forces are those of friction and inertia. For a ship moving on the upper surface of a body of water, the significant force (in addition to the inertial force) is not a pressure but the force of gravity. Physically similar flows are to be expected in these cases if the ratio

$$\frac{\text{inertial force}}{\text{gravity force}} = \frac{m\dfrac{dv}{dt}}{mg} = \text{const.}$$

Using the same considerations as those used in developing the Reynolds number, this leads to the requirement that for similarity of two flows the dimensionless quantity v^2/lg must have the same value in both,

$$\frac{v^2}{lg} = \text{const.}$$

This similarity law is known as *Froude's law*. The pure number v^2/lg is called the *Froude number*. Similar wave systems can therefore be expected for a real ship and its model if the velocities are proportional to the square roots of the respective lengths.

Every flow of a viscous fluid can be characterized by a corresponding Reynolds number:

$$R = \frac{lv}{\nu}.$$

Since R is the ratio of inertial force to viscous force, a large Reynolds number means that the inertial forces outweigh the viscous forces and the flow becomes turbulent. If, on the other hand, R is small, the flow is laminar. The designations "large" and "small" have a meaning only with respect to a critical value R_{crit} at which the transition from laminar to turbulent flow occurs. In cylindrical tubes the critical Reynolds number R_{crit} is very strongly dependent on the entrance conditions and the degree of roughness of the tube walls. With smooth walls and undisturbed entrance flow, the critical Reynolds number may be as great as 20,000. The lower limit is about 1200, below which disturbances are damped out. Falling spheres obey *Stokes' law* (see p. 287) provided the Reynolds number is small compared with unity. Extremely precise measurements show that Stokes' law holds for all velocities that occur in practice.

EXAMPLES. 1. Water at 20°C ($\nu_{20°} = 1.00 \times 10^{-2}$ cm²/sec) flows in a smooth tube of radius $r = 1$ cm. The smallest critical velocity ($R_{\text{crit}} = 1200$) is then

$$v_{\text{crit}} = \frac{\nu}{d} R_{\text{crit}} = 6 \text{ cm/sec}.$$

If air at 0° and 1 atm pressure ($\nu = 0.133$ cm²/sec) flows in the same tube, then

$$v_{\text{crit}} = 80 \text{ cm/sec}.$$

2. If the flow of water ($\nu = 10^{-2}$ cm²/sec) around an obstacle is laminar, then air ($\nu = 0.133$ cm²/sec) will flow around an obstacle whose linear dimensions are 100 times greater in a similar but enlarged streamline pattern, provided the velocity is correspondingly decreased. Similar flows will result if the Reynolds numbers are the same,

$$\left(\frac{vl}{\nu}\right)_{\text{water}} = \left(\frac{vl}{\nu}\right)_{\text{air}}.$$

It follows from this that

$$\frac{v_{\text{water}}}{v_{\text{air}}} = \frac{\nu_{\text{water}}\, l_{\text{air}}}{\nu_{\text{air}}\, l_{\text{water}}} = 7.5.$$

We shall conclude this section with two remarks.

1. On page 289 the following expression for the total drag resistance was given:

$$D = c_D \frac{\varrho v^2}{2} A.$$

This simple formula can only be written, however, by incorporating in the factor c_D a complicated dependence of the resistance on ϱ, v, η, and A. The pure number c_D is in fact not a constant but rather is a function of the Reynolds number:

$$c_D = f(R).$$

As long as the Reynolds number is constant, so is the drag coefficient c_D. In experiments with models, correct values of resistance can be determined only if one works with equal Reynolds numbers. In practice, this is not always possible. Fortunately, the drag coefficient c_D is very nearly constant over a wide range of Reynolds numbers.

The viscous resistance given by Stokes' law can be put in the form

$$D = 6\pi r \eta v = c_{\text{st}} \frac{\varrho v^2}{2} A.$$

In this case, the area A is the projected area of the sphere, πr^2. The *Stokes drag coefficient* is therefore

$$c_{st} = \frac{12\pi r\eta v}{\varrho v^2 \pi r^2} = \frac{12}{rv\varrho/\eta} = \frac{12}{R},$$

and thus is a function only of the Reynolds number.

2. If a body moving through the air acquires a velocity v which approaches or exceeds the velocity of sound c, the drag coefficient c_D is no longer a function of R alone. In place of the Reynolds number, the nature of the flow essentially depends on the ratio v/c. This important quantity is called the *Mach number*. In the region of Mach number 1 (the transonic region), the value of c_D increases rapidly as a result of the production of sound energy in the form of shock waves (analogous to the production of wave energy by a ship).

If a body moves with a *supersonic* velocity v, greater than the sound velocity c, then with respect to a reference system moving with the body no propagation of compressional waves in the direction of motion is possible. At a time t_1 the pressure wave originating at time $t = 0$ occupies a spherical surface of radius $r = ct_1$. In the same time interval the body covers a distance $x = vt_1$. The waves originating in the interval between $t = 0$ and $t = t_1$ are spherical surfaces whose envelope is a circular cone called the *Mach cone*. Its apex coincides with the position of the body (Fig. 6-63) and its half-angle (the *Mach angle*) is given by

$$\sin\varphi = \frac{c}{v}.$$

These relations are beautifully illustrated in the photographs of bullets in flight (Fig. 6-64, Plate IV).

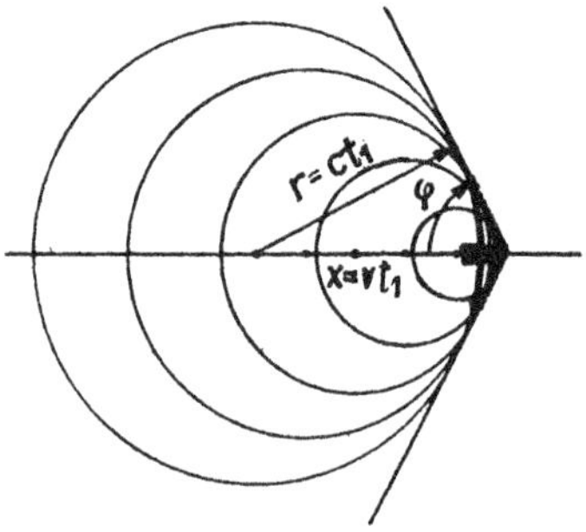

FIG. 6-63. The Mach cone of a supersonic body.

62-4 Dynamic Lift

A balloon rises because of the buoyant force or lift resulting from the air it displaces, a so-called *static lift*. The lifting force exerted on the wings of an aircraft is called a *dynamic lift*. Its origin is entirely different from that of a static lift. It can, however, be explained in a relatively simple way. The hydrodynamics of non-viscous fluids is adequate for the explanation.

A body about which there is symmetrical fluid flow experiences no force, since by symmetry the resultant of all static pressure forces on the body is zero. If this flow is modified in such a way that the velocity above the body is increased, while that below it is decreased, then according to Bernoulli's principle the static pressure on the lower side of the body increases, and that on the upper side decreases. This modified pressure distribution results in a force perpendicular to the direction of flow which is called the *dynamic lift*.

An asymmetry in the flow pattern can be thought of as a superposition of a symmetrical flow and a so-called *circulatory flow*. The latter is a flow which circulates around the object and in which the streamlines are closed curves. It is characterized by the so-called *circulation* Γ. The circulation of a flow is defined as the sum of the scalar product of the flow velocity $\boldsymbol{v}$ and the element of path length $\Delta \boldsymbol{s}$, along a closed curve surrounding the body,

$$\Gamma_{\circ} = \sum \boldsymbol{v} \cdot \Delta \boldsymbol{s} = \sum_{\circ} v \Delta s \cos \alpha$$

where α is the angle between the line element $\Delta \boldsymbol{s}$ and the flow velocity $\boldsymbol{v}$ at that element (Fig. 6-65).

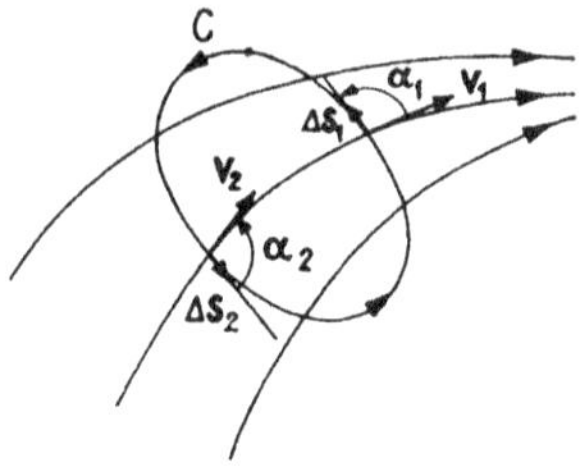

FIG. 6-65. Calculation of the circulation Γ.

A circulation exists, for example, when a fluid rotates about an axis in such a way that the velocity is inversely proportional to the radius: $v = a/r$. The circulation is then the same for any closed curve enclosing the axis. To show this we start with

$$\Delta \Gamma = v \Delta s \cos \alpha,$$

but

$$\Delta s \cos \alpha = r \Delta \varphi,$$

so

$$\Delta \Gamma = a \Delta \varphi$$

and

$$\Gamma_{\circ} = 2\pi a = \text{const.}$$

Such a circulation exists in the vortex that is often seen when a bowl of water is emptied through an opening at the bottom of the bowl.

If we superpose a circulatory flow on the purely translational and hence symmetrical flow about a cylinder (Fig. 6-66), the velocities of the two flows add vectorially at every point (Fig. 6-67). Above the cylinder the two velocities are in the same direction while below it they are in opposite directions.

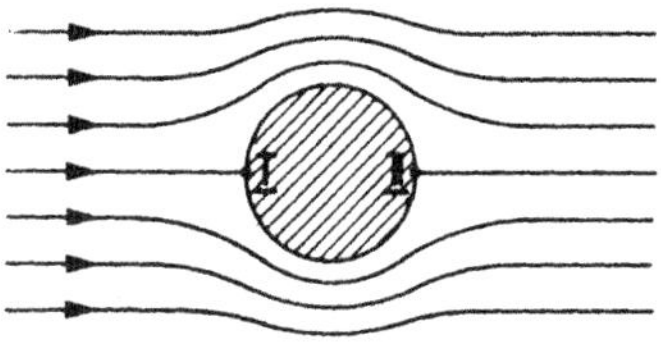

FIG. 6-66

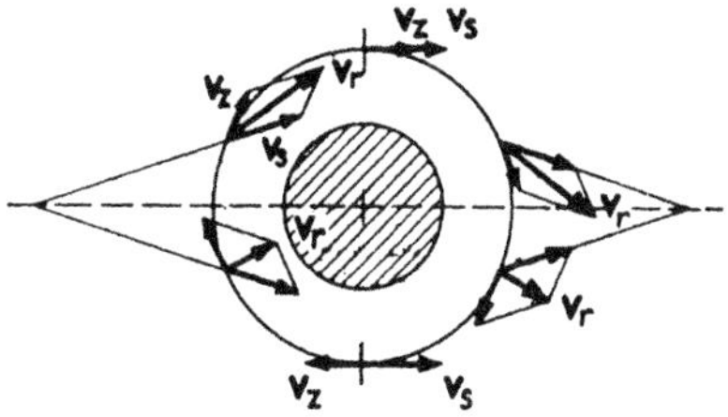

FIG. 6-67

FIG. 6-66. Symmetric flow around a cylinder.

FIG. 6-67. Vectorial addition of a circulatory flow to a symmetric flow around a cylinder.

The result is that the velocity above the cylinder is increased and that below it is decreased. The two stagnation points are displaced toward the lower side of the body (Fig. 6-68). A deformation of the streamlines results, which depends on the magnitude of the circulation. As a result of this deformation, higher velocities and hence lower pressures exist on the upper surface of the cylinder. On the lower side of the cylinder, as a result of the decreased velocity, the pressure is increased. (Increases and decreases in pressure are relative to the prevailing pressure without the circulation.) As a result of the

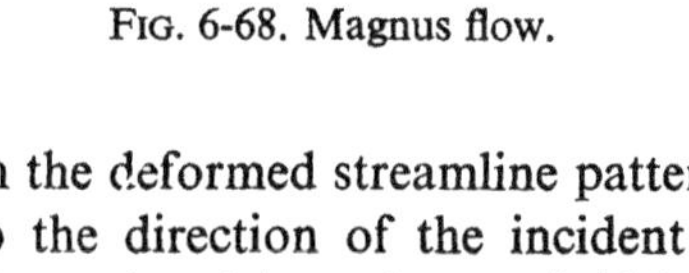

FIG. 6-68. Magnus flow.

pressure distribution in the deformed streamline pattern there is an upward force perpendicular to the direction of the incident flow, that is, a *lift*. When the circulation is produced in a viscous fluid by rapidly rotating the cylinder, this phenomenon is called the *Magnus effect*. The resulting lift L has been shown by Kutta and Joukowski to be

$$L = \varrho v_\infty \Gamma l,$$

where ϱ is the density of the fluid, v_∞ the flow velocity very far from the object, and l the length of the cylinder. There is an exact analogue of this equation in Electrodynamics (see Vol. II). A conductor of length l carrying a current I in a magnetic field H experiences a force F perpendicular to both I and H and given by

$$F = \mu_0 H I l.$$

The circulation Γ corresponds to the current I and the flow velocity v_∞ to the magnetic field H.

A circulation can be produced by the rotation of a cylinder. The air adjacent to the cylinder is set in motion by viscous forces. A rotor ship designed by Flettner made use of the Magnus effect, but for economic reasons it could not compete with propeller driven vessels.

An asymmetric flow can also be set up in a stream of fluid moving with translatory motion by the proper choice of profile of a body in the stream. An aircraft wing or airfoil produces this result. The flow around an airfoil (Fig. 6-69c), in which the fluid flows smoothly away from the trailing edge, can be represented by the superposition of a so-called *potential flow* (Fig. 6-69a) and a circulatory flow (Fig. 6-69b). The potential flow exhibits

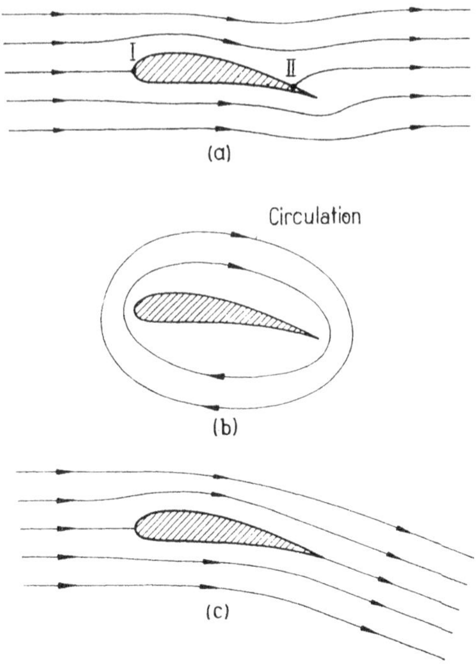

FIG. 6-69. The occurrence of lift on an airfoil.
(a) Potential flow (no circulation) around an airfoil.
(b) Circulatory flow around an airfoil.
(c) Actual flow, which can be considered to be a superposition of the potential flow and the circulatory flow shown above.

the same pattern of streamlines as is obtained in a streamline apparatus with a fluid of high viscosity (Fig. 6-35). The superposed circulatory flow is essential if there is to be a dynamic lift. It can be seen without a numerical calculation that the effect of the circulation is to increase the velocity above the airfoil and decrease the velocity below it. From Bernoulli's equation, this results in a decrease of pressure on the upper surface of the airfoil and an increase of pressure on the lower surface. As with the Magnus effect, the resulting lift is proportional to the circulation

When an aircraft begins its motion, a circulation is set up around its wings. Since the flow as a whole can be considered an isolated system whose angular momentum is conserved, the development of a circulation must be accompanied by an equal and oppositely directed angular momentum. This appears in the form of vortices, which remain behind the aircraft at the starting point.

The lift L on an airfoil is markedly dependent on the form of its profile. The lift, like the drag, can be expressed by an equation of the form

$$L = c_L \frac{\varrho v_\infty^2}{2} A,$$

where v_∞ is the velocity of the undisturbed air stream, A is the product of the wing span and the chord, and c_L is the *coefficient of lift* of the profile.

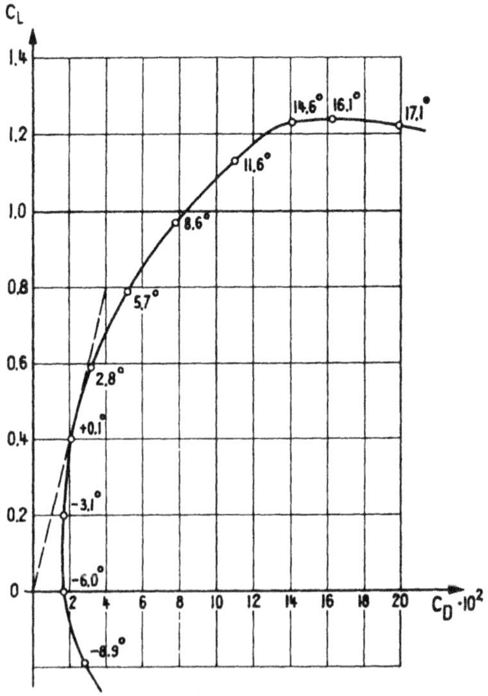

FIG. 6-70. The polar diagram of an airfoil.

The coefficient of lift is very nearly directly proportional to the *angle of attack* α, that is, the angle between the chord of the airfoil and the direction of the undistorted flow.

The coefficient of drag of an airfoil is also dependent on the angle of attack α, but to a first approximation it increases quadratically with α. It follows that a graph in which c_L is plotted vertically and c_D horizontally is roughly parabolic in shape. Such a diagram is called the *polar diagram* of the airfoil (Fig. 6-70). For the foil to which Fig. 6-70 applies, the lift is a maximum at an angle of attack of about 16° and it decreases rapidly as the angle is increased. This results from the fact that with increasing α the air flow does not follow the profile of the airfoil. Vortices form at the upper surface of the foil and result in a marked decrease in the circulation as a consequence of the law of conservation of angular momentum. This causes the aircraft to stall.

The polar diagram shows that a noticeable lift exists even for negative angles of attack. The ratio of lift to drag, for a given airfoil, is a maximum for the angle of attack at which a line from the origin is tangent to the polar diagram.

EXAMPLE. An aircraft whose wing span is 48 ft and chord is 5.0 ft flies at an angle of attack of 8.6°. Its velocity relative to the air is $v = 150$ mi/hr $= 220$ ft/sec. From the polar diagram of Fig. 6-70, which we will assume refers to this aircraft, we find that at an angle of attack of 8.6° the coefficients of lift and drag are respectively $c_L = 0.98$ and $c_D = 0.079$. The density of the air is 0.0807 lbm/ft³. Find the lift and the drag.

$$L = c_L \frac{\varrho v^2}{2} A = 46{,}000 \text{ lbf}$$

$$D = c_D \frac{\varrho v^2}{2} A = 3700 \text{ lbf.}$$

CHAPTER 7

BASIC CONCEPTS OF THERMODYNAMICS

63. Introduction

The science of mechanics is based on the three laws of Newton, which form the basis for the calculation of the motions of single particles and of systems of particles. We postulate that these laws are unrestricted in their validity for non-atomic phenomena, in contrast to empirical laws such as Hooke's law and the laws of friction. Newton's laws are characterized by the absence of any empirical constants, and by the fact that they make no assumptions regarding the structure of matter.

In an exactly analogous way, the science of *thermodynamics* is based on certain general laws, known as the *three laws of thermodynamics*.† Like Newton's laws, the laws of thermodynamics contain no empirical constants and involve no assumptions regarding the structure of matter. Hence, they may be universally applied to all physical phenomena without the results being dependent upon the properties of the substances being studied. Just as Newton's laws are used to derive relations between the quantities characterizing the mechanical properties of matter (mass and momentum, stress and strain, etc.), so the laws of thermodynamics are used to obtain relations between the quantities describing the thermodynamic state of a system. In general, empirical constants appear in these relations. Thus, this aspect of thermodynamics is also similar to the application of Newton's laws in mechanics.

The molecular structure of matter provides a basis for understanding these empirical constants and their interrelations. The study of those properties of matter that depend on some of the simpler consequences of the assumption that the constituent molecules are in constant motion is known as *kinetic theory*. The name derives from the fact that some properties of matter (e.g.: pressure, viscosity, temperature) can be explained as manifestations of the kinetic energy of the constituent molecules. In this sense, kinetic theory gives a deeper understanding of thermodynamic processes and concepts.

The macroscopic objects with which thermodynamics deals are composed of exceedingly large numbers of individual particles (molecules, atoms,

† The third law, the so-called *Nernst heat theorem*, will not be discussed.

ions). It is hopeless to attempt to apply the laws of mechanics to such a large collection of objects, simply because of the computational labor involved. Kinetic theory deals with this problem by postulating the *principle of molecular disorder* or *molecular chaos*. This principle states that the motion of individual particles in a macroscopic object is random. Once this assumption is made, the observable properties can then be shown to be related to the *average* value of velocity or energy.

In making molecular chaos a fundamental principle, we abandon the idea of following the motion of each individual particle in the microscopic object. It then becomes necessary to make some assumption about the distribution of molecular velocities on order to calculate the average values. The branch of physics that deals with this topic is known as statistical mechanics and is beyond the scope of this book. However, the distribution law can be related to the theory of probability and we discuss this aspect of the problem in Chapter 8.

64. The Concept of Temperature

64-1 The Fundamental Thermal Process

Careful experiments carried on over many years have led to the following conclusion: If two bodies of the same material and having the same surface properties, one of which feels warm to our sense of touch and the other cold, are brought in contact, their thermal states become the same after a finite time; that is, they feel *equally warm* to our sense of touch. (For two bodies of different material, our sense of touch may not verify the attainment of the same thermal state.) On the basis of this fact we are led to postulate that the following statement is universally valid:

> Fundamental Postulate of Thermometry. Two isolated bodies of the same or different material, which make contact with each other and whose outer surfaces have the same properties, reach the same thermal state after a sufficiently long time of contact. We say that the bodies are then in *thermal equilibrium*.

The attainment of this state of thermal equilibrium was formerly believed to result from the transfer of a material substance, a quantity of heat Q, from the warmer to the cooler body. This explanation, however, has proven to be incorrect and we now know that *heat* is not a substance but rather a *form of energy*.

64-2 Temperature

The first step in a quantitative analysis of heat transfer is to give a precise meaning to the terms "hot" and "cold". The two concepts can be replaced by one, since hot and cold are purely subjective perceptions and in particular are merely relative sensations which can provide only qualitative information regarding the intensity of a given thermal state. A quantitative, if not entirely satisfactory, determination of the intensity concept is possible if some property of a substance can be found which changes with the intensity of the thermal state in a unique and reproducible way. One such property is the volume of a body. With its help, it is possible to measure the intensity of the thermal state, known as the *temperature*.

The formulation of rules for the measurement of temperature has caused considerable difficulty. These rules can only be formulated in a manner free from physical objections with the help of the second law of thermodynamics. A preliminary and rather crude formulation follows, based on the phenomenon of volume expansion.

If a thin walled glass bulb, to which a capillary tube is attached, is filled with mercury (Fig. 7-1) and the apparatus is inserted in melting ice when

FIG. 7-1. Fixed points on a fluid thermometer.

the external pressure is 760 mm Hg, the top of the mercury column in the capillary stands at a certain reproducible position indicated by position I in the diagram. The mercury then obviously has the same thermal state (that is, the same temperature) as the melting ice. Similarly, when the instrument is placed in steam at a pressure of 760 mm Hg, the mercury always stands at position II. The mercury then has the same temperature as the steam.

This simple instrument, a glass bulb and a capillary of uniform cross section containing mercury, is called a *thermometer*. The *fundamental interval* of temperature measurement means the temperature difference between these *fixed points*, determined by the temperature of *melting ice* at 760 mm Hg and the temperature of *boiling water* at 760 mm Hg.

DEFINITION. One centigrade degree (also called Celsius degree, 1 C°) is defined as one one-hundredth of the temperature difference between the fixed points, and is the unit of temperature.†

Dimensions: [Degree]
Units: 1 centigrade degree = 1 C°.

One Fahrenheit degree (1 F°) is 1/180th of the temperature difference between the fixed points.

† One centigrade degree (1C°) is a temperature *interval* (Δt) of one unit measured on the centigrade scale. One degree centigrade (1°C) is a specific reading on that scale of a temperature just above the freezing point of water.

The arbitrary origin or zero point of temperature on the centigrade scale is the melting point of ice at 760 mm Hg. The zero point on the Fahrenheit scale lies 32 F° below the ice point (Fig. 7-2).

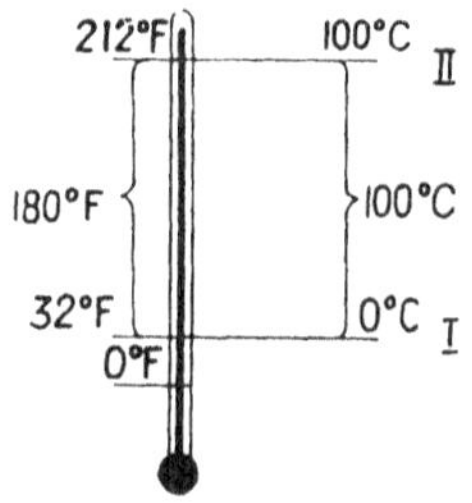

FIG. 7-2. The two temperature scales and their zero points.

The fundamental interval and the size of the degree are therefore uniquely defined. The problem of measuring a temperature which does not coincide with either of the fixed points is another matter. For the present, we proceed arbitrarily. Using mercury as the indicating liquid, we divide the fundamental interval on the thermometer stem into 100 equal parts, and continue the numbering above and below the fixed points. This procedure is to some extent arbitrary. It does, however, provide a practical method of measuring temperature with a mercury thermometer. This uniform subdivision of a thermometer stem cannot be done for other liquids, if the readings of thermometers containing these liquids are to agree with those obtained with a mercury thermometer. If we were to use equal subdivisions, it would be necessary to specify the thermometric liquid in every statement of a temperature and there would be as many different values for the same temperature as there were thermometer liquids. It is impossible to identify the physically meaningful concept of temperature with any of these readings, thus indicating that it is unsatisfactory to define temperature in terms of the properties of a particular material. We can get somewhat better and more consistent results with a gas thermometer. However, only the second law of thermodynamics is able to provide a completely satisfactory solution to the problem.

If for the moment we disregard these difficulties, we can assume that a temperature scale has been determined and calculations of temperature are then simply problems in proportionality.

EXAMPLE. (1) 20°C = ?°F.

$$\frac{20}{100} = \frac{y}{180}, \quad y = 36.0°\text{F above the ice point.}$$

20°C therefore corresponds to 68.0°F.

(2) 100°F = x°C.

$$\frac{100-32}{180} = \frac{x}{100}, \quad x = 37.78°\text{C}.$$

65. The Principles of Thermal Expansion

Having defined the temperature scale, it is now possible to make precise studies of the thermal expansion of a body. The following discussion will show that matter is subject to many different types of thermal processes.

65-1 Linear Expansion of a Solid at Constant Pressure

We shall restrict the discussion here to a given isotropic substance which has no crystallographic transition points in the temperature interval considered. A substance is said to be isotropic if its properties are independent of direction. If the external pressure is constant, we find experimentally that for a small temperature change Δt, the change in length Δl of the body can be expressed as

$$\Delta l = \lambda l_0 \Delta t$$

where l_0 is the original length of the body. The quantity λ is called the *isobaric coefficient of linear expansion* of the material.

The isobaric coefficient of linear expansion can be described as the increase in length of a rod, per unit of original length and per unit change in temperature, when the pressure is constant:

$$\lambda = \left[\frac{1}{l_0}\frac{dl}{dt}\right]_{P=\text{const.}}$$

Units: $1\ (\text{C}^\circ)^{-1}$.

The expression for Δl is valid in the above form only within a limited range of temperature, because λ varies with temperature. For large temperature intervals, l can be expressed as a power series in t. The coefficients must be determined experimentally.

$$l_t = l_0(1 + \lambda_1 t + \lambda_2 t^2 + \lambda_3 t^3 + \cdots).$$

Table 7-1

Nickel-iron alloy (% nickel)	$\lambda\ (\text{C}^\circ)^{-1}$
10	1.30×10^{-5}
20	1.95
30	1.20
36	0.09
40	0.60
50	0.97
80	1.25
100	1.28

There are certain alloys which have exceedingly small coefficients of linear expansion at room temperature. One such is the nickel-iron alloy, Invar, containing 36 per cent nickel. The above table gives the expansion coefficients of a number of nickel-iron alloys and shows the sharp decrease of λ at the critical composition of 36 per cent nickel.

Fused quartz also has an exceptionally small coefficient of linear expansion, considerably smaller than that of crystalline quartz.

65-2 Volume Expansion at Constant Pressure

The *volume expansion* of a body can be computed from the expression for linear expansion. Consider a rectangular parallelepiped having sides of length a, b, c. In a small temperature interval we have for the expansion of the three sides

$$a_t = a_0(1 + \lambda\Delta t),$$
$$b_t = b_0(1 + \lambda\Delta t),$$
$$c_t = c_0(1 + \lambda\Delta t),$$

from which

$$V_t = (abc)_t = V_0(1 + \lambda\Delta t)^3.$$

Neglecting the quantities $(\lambda\Delta t)^2$ and $(\lambda\Delta t)^3$, which are small compared with $\lambda\Delta t$, this becomes

$$V_t = V_0(1 + 3\lambda\Delta t),$$

that is,

$$\Delta V = V_t - V_0 = 3\lambda V_0\Delta t = \alpha_p V_0 \Delta t,$$

where

$$\alpha_p = 3\lambda.$$

The quantity α_p is called the *isobaric coefficient of volume expansion.* It can be described as the change in volume of a body per unit of original volume and per unit change in temperature:

$$\alpha_p = \frac{1}{V_0}\frac{dV}{dt}\bigg|_{P = \text{const.}}$$

Units: $(1\ \mathrm{C}^\circ)^{-1}$.

The expression given above for computing ΔV is restricted to small temperature intervals Δt. For large ranges of temperature a power series is necessary, whose coefficients can be computed from the linear coefficients λ_1, λ_2, etc.:

$$V_t = V_0(1 + \alpha_1 t + \alpha_2 t^2 + \alpha_3 t^3 + \cdots).$$

The *temperature dependence* of the coefficient of volume expansion can easily be computed from this equation. Since the quantity α_p will frequently play a role in later thermodynamic calculations, such formulas are of considerable interest. The volume coefficient of expansion of all substances approaches zero at very low temperatures.

EXAMPLE. For mercury,

$$V_t = V_0(1 + 181.538 \times 10^{-6}t + 0.7548 \times 10^{-8}t^2 + 1.533 \times 10^{-11}t^3 + 5.36 \times 10^{-14}t^4)$$

in the interval between 0°C and 300°C.

From the equation

$$\alpha_p = \left(\frac{1}{V_0}\frac{dV}{dt}\right)_{P=\text{const}},$$

we find

$$\alpha_p = 181.538 \times 10^{-6} + 1.5096 \times 10^{-8}t + 4.599 \times 10^{-11}t^2 + 21.44 \times 10^{-14}t^3.$$

Thus at the temperatures of 0°C and 20°C,

$$\alpha_{0^\circ C} = 181.538 \times 10^{-6}\,(C^\circ)^{-1},$$

$$\alpha_{20^\circ C} = 181.858 \times 10^{-6}\,(C^\circ)^{-1}.$$

Water exhibits an unusual behavior in that it contracts in the temperature range from 0° to 4°C and then expands as the temperature is increased further. Since the density of water is an important physical quantity and is used in the preparation of many tabulated values of density, the internationally accepted table of the absolute density of water is given in Appendix III.

The reason for this anomalous behavior of water is not fully understood. It appears that the liquid formed by the melting of ice retains a certain ordered state, which gradually disappears as the temperature is increased. According to recent investigations,† liquid water consists of a mixture of three modifications (states of ordering) which differ in their closeness of packing and have different degrees of stability. A state of maximum density exists because at a temperature of 0°C one of the modifications of lower density predominates, but is gradually replaced by more dense modifications as the temperature increases. It tends toward the most dense state of closely packed spheres (see Section 80) until the effects of thermal agitation become strong enough to decrease the density again.

65-3 Variation of Density with Temperature

It follows from the law of expansion and the definition of density, that the density in general decreases with increasing temperature:

$$\varrho_t = \frac{m}{V_t} = \frac{m}{V_0(1 + \alpha_p \Delta t)} = \frac{\varrho_0}{(1 + \alpha_p \Delta t)},$$

and to a first approximation

$$\varrho_t = \varrho_0(1 - \alpha_p \Delta t),$$

that is

$$\varrho_t - \varrho_0 = \Delta\varrho = -\varrho_0 \alpha_p \Delta t.$$

† J. D. Bernal and R. Fowler, *J. Chem. Phys.* 1933; J. Morgan and E. Warren, *J. Chem. Phys.* 1938.

Because of its variation with temperature, tabulated values of density are frequently expressed, not in the absolute units of gm/cm^3, but as relative values which can be determined more precisely. Measurements are usually made relative to water.

DEFINITION. The *relative density* $d_{t°/4°C}$ of a substance is defined as the ratio of the density ϱ of the substance to the density of water at 4°C:

$$d_{t°/4°C} = \frac{\varrho_{t°C}}{\varrho_{\text{water at } 4°C}}.$$

This pure number is often called the *specific gravity* of the substance, obviously an improper designation.

The absolute density is obtained by multiplying the relative density by the density of water at 4°C.

65-4 Temperature Measuring Instruments and their Ranges

The instrument most commonly used in the measurement of temperature is the *mercury-in-glass thermometer* It can be used over a temperature range from −30 to 300°C. (The freezing point of mercury at 760 mm Hg is −38.9°C, its boiling point is 356.9°C.) A mercury thermometer containing nitrogen under high pressure, and using a high melting-point borosilicate glass, can be used up to a temperature of 550°C.

Higher temperatures are measured with *gas thermometers* (see p. 314), with *bimetallic thermometers*, which utilize the unequal expansions of strips of two different metals, with *thermocouples*, and with *resistance thermometers* (see Vol. II). Temperatures above 1500°C are measured with an *optical pyrometer*, whose principle of operation cannot be explained without an understanding of heat radiation (see Vol II.)

Temperatures down to −90°C can be measured with *alcohol* and *toluene* thermometers. The freezing point of alcohol is −114.6°C, that of toluene is −95°C. *Pentane* thermometers have a range from 20° to −190°C. Lower temperatures can be measured with thermocouples, resistance thermometers, and gas thermometers.

All of these thermometers must be calibrated in terms of the gas thermometer scale (see Section 66). Even the mercury thermometer does not agree with this scale to within a hundredth of a degree, if the length of the thermometer stem is divided uniformly. The following table shows the differences in the readings of a number of different thermometers when their stems are divided uniformly, for temperatures between 0° and −100°C.

The table shows that the scales of toluene, alcohol, and pentane thermometers cannot be divided uniformly. Corrections must be made in experiments of even moderate precision. This is also true of the various thermometers used to measure high temperatures.

TABLE 7-2

$t_{\text{Gas thermometer}}$	t_{Toluene}	t_{Alcohol}	t_{Pentane}
0°	0.00°	0.00°	0.00°
−10	−8.54	−9.31	−11.79
−20	−16.90	−18.45	−23.08
−30	−25.10	−27.44	−33.91
−40	−33.15	−36.30	−44.31
−50	−41.08	−45.05	−55.34
−60	−48.90	−53.71	−64.01
−70	−56.63	−62.31	−73.39
−100			−100.00

65-5 Gay-Lussac's Law (1802)

Gases, like solids and liquids, increase in volume when their temperature is increased at constant pressure. In contrast to solids and liquids, all ideal gases follow the *same* law of expansion, *Gay-Lussac's law*. The validity of this law, like that of the Boyle–Mariotte law, is subject to three conditions: constant pressure, constant mass, and unvarying molecular composition. When these conditions are fulfilled,

$$V_{t^\circ\mathrm{C}} = V_{0^\circ\mathrm{C}} \cdot (1 + \alpha_p t),$$

or

$$\Delta V = V_t - V_0 = V_{0^\circ\mathrm{C}} \alpha_p t.$$

The quantity α_p is the *isobaric coefficient of volume expansion*,

$$\alpha_p = \left(\frac{1}{V_0} \cdot \frac{dV}{dt}\right)_{P = \mathrm{const}}.$$

The magnitude of α_p is

$$\alpha_p = 0.003661 (\mathrm{C}^\circ)^{-1} = \frac{1}{273.15} (\mathrm{C}^\circ)^{-1}.$$

That is, all ideal gases expand by (about) 1/273 of their volume at 0°C, per centigrade degree increase in temperature, at constant pressure.

In contrast to solids and liquids, the expansion of a gas is always referred to its volume at 0°C. Only then does α_p have the value given above.

66. ABSOLUTE TEMPERATURE AND THE IDEAL GAS LAW

66-1 Absolute Temperature

According to the previous section a simple relation exists between the increase in volume of an ideal gas and its temperature (Fig. 7-3). The change in volume, $\Delta V = V_0 \alpha_p t$, is proportional to the centigrade temperature t and to the original volume V_0. For ideal gases with different initial volumes at 0°C, say V_0 and V_0', we have

$$\frac{V_t'}{V_o'} = \frac{V_t}{V_0} = 1 + \alpha_p t = \frac{273.15 + t}{273.15}.$$

According to this equation, the straight lines representing the volume of an ideal gas as a function of centigrade temperature t, intersect the axis of t at a common point, whatever the volume at 0°C. This point corresponds to a temperature of −273.15°C. Because of this fact, it is convenient to measure

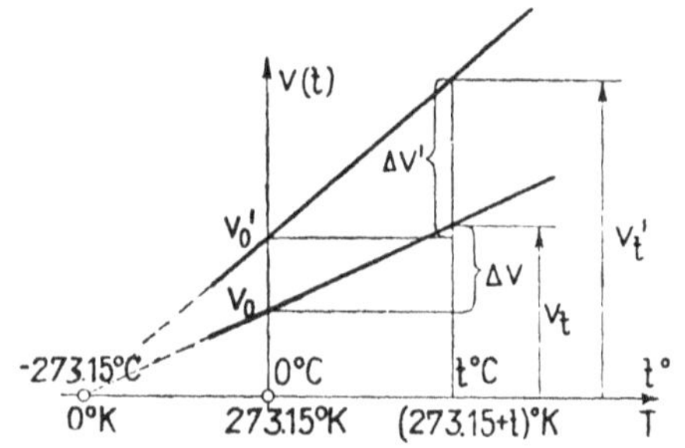

FIG. 7-3. Derivation of the absolute temperature scale.

temperature from the common point of intersection. The volumes at arbitrary temperatures are then directly proportional to the temperatures measured from this new origin. The new origin is called *absolute zero* and lies at −273.15°C. The justification for the term "absolute" will be given later. Temperatures measured from absolute zero are called *absolute temperatures* or *Kelvin temperatures* and are expressed in degrees Kelvin, °K. We shall represent the absolute temperature by the symbol T, to distinguish it from the centigrade temperature t. A temperature interval of 1 K° is equal in magnitude to a temperature interval of 1 C°. The two scales differ only in the location of their zero points.†

$$T(°\text{K}) = t(°\text{C}) + 273.15.$$

The Kelvin temperature scale was introduced in 1848 by Sir William Thomson (Lord Kelvin).

In terms of the absolute temperature scale, the Gay-Lussac law simplifies to

$$\frac{V_2}{V_1} = \frac{T_2}{T_1}.$$

† The relative position of the two zero points was most recently fixed by the International General Conference on Weights and Measures, Paris and Sevres, 1954. They decided: "to define the thermodynamic scale of temperature by means of the triple point of water as the fixed fundamental point, by assigning to it the temperature 273.16°K, exactly." The triple point of water has the centigrade temperature of 0.01°C, and hence the ice point or 0.00°C is equivalent to 273.15°K.

Gay-Lussac's law is not valid down to the temperature of absolute zero. Real gases condense at low temperatures and the equation above does not then describe their behavior.

In engineering work in the United States and England, the *Rankine* temperature scale is used for absolute temperatures corresponding to the Fahrenheit scale The temperature interval on this scale is the same as on the Fahrenheit scale, that is

$$1 \text{ Rankine degree} = 1\ \text{R}^\circ = 1\ \text{F}^\circ.$$

The zero point of the Rankine scale lies at $-459.67°$F. Hence

$$T(°\text{R}) = t(°\text{F}) + 459.67.$$

66-2 The Ideal Gas Law

The Boyle–Mariotte law and the Gay-Lussac law are special cases of a more general relation, which we now derive.

(a) Consider a certain mass of gas of volume V_0 at a pressure P_0 and a temperature T_0. Let the pressure be increased to P_1, with the temperature, mass, and molecular composition remaining constant. Then from the Boyle–Mariotte law:

$$P_0 V_0 = P_1 V_1^*. \tag{66-1}$$

The new volume is represented by V_1^* since we are to make further changes in the volume.

(b) The temperature of the gas is now increased from T_0 to T_1, keeping the pressure, mass, and molecular composition constant. Then from the Gay-Lussac law,

$$\frac{V_1}{V_1^*} = \frac{T_1}{T_0}.$$

Using Eqn. (66-1), we can replace V_1^* by $P_0 V_0 / P_1$, and hence

$$\frac{P_1 V_1}{P_0 V_0} = \frac{T_1}{T_0},$$

or

$$\frac{P_1 V_1}{T_1} = \frac{P_0 V_0}{T_0} = \text{const}.$$

This relation is known as the *ideal gas law*. It applies to an ideal gas whose mass and molecular composition remain constant. The constant in the equation has the dimensions [energy/degree].

Real gases obey this law only approximately. Furthermore, for equal masses of gas, the constant has different values for different gases. This makes the law inconvenient to apply. In Section 68 we shall derive the equation of state of an ideal gas, which will remove this difficulty.

To represent the ideal gas law graphically, it is necessary either to draw a network of parametrically indexed curves, or to construct a three-dimensional P-V-T diagram. Figure 7-4 is an isometric drawing of a portion of the three-dimensional P-V-T surface and its three projections, and on it we see the following:

(a) Planes of constant temperature (T = const) intersect the surface in *equilateral hyperbolas*, PV = const. (Boyle–Mariotte law.)

(b) Planes of constant pressure (P = const) intersect the surface in straight lines, whose slope in the T-V plane increases with increasing pressure.

$$\frac{V}{T} = \text{const (Gay-Lussac law).}$$

The straight lines all intersect the pressure axis at right angles.

(c) Planes of constant volume (V = const) also intersect the surface in straight lines whose slope in the P-T plane is inversely proportional to V.

$$\frac{P}{T} = \text{const}.$$

These lines all intersect the V-axis at right angles.

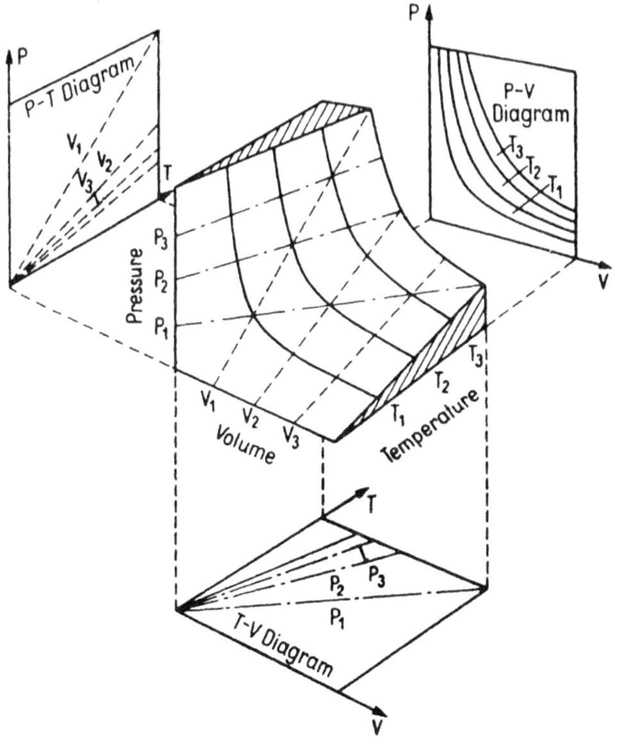

FIG. 7-4. Three-dimensional surface representing the state of an ideal gas. The projections give the P-V, P-T, and T-V diagrams.

EXAMPLE, A rigid balloon of volume 1200 m^3 is filled with gas at the surface of the earth where the pressure $P_0 = 760$ mm Hg and the temperature $t_0 = 18$°C. It rises to a point where the pressure $P_1 = 460$ mm Hg and the temperature is 2.3°C. What volume of gas, reckoned at a pressure of 460 mm Hg and a temperature of 2.3°C, has escaped from the balloon?

$$V_1 = \frac{P_0 V_0 T_1}{T_0 P_1} = 1777 \text{ m}^3.$$

The volume expelled is therefore 577 m^3.

66-3 The Jolly Gas Thermometer

The ideal gas law makes it possible to measure temperatures by measuring the pressure of a gas. Measurements of T are particularly simple if the volume is kept constant, since then

$$\frac{T_1}{T_0} = \frac{P_1}{P_0},$$

that is, the temperatures are directly proportional to the corresponding pressures. The *Jolly gas thermometer* (Fig. 7-5) makes use of this fact. It

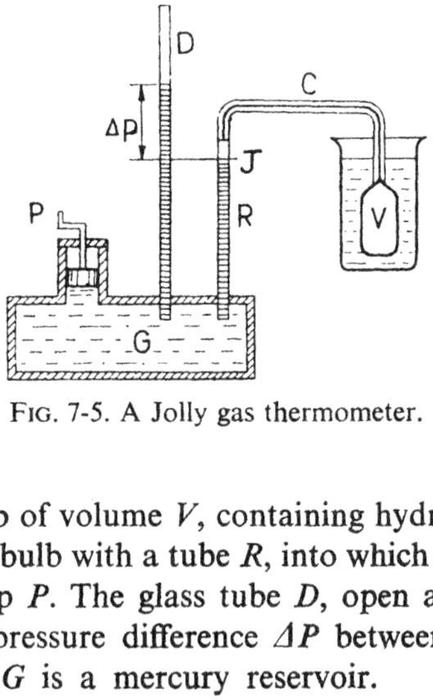

FIG. 7-5. A Jolly gas thermometer.

consists of a glass bulb of volume V, containing hydrogen or helium. A capillary C connects the bulb with a tube R, into which mercury can be forced by means of the pump P. The glass tube D, open at the top, permits the measurement of the pressure difference ΔP between the gas in the bulb and the atmosphere. G is a mercury reservoir.

To measure an unknown temperature T_1, the bulb containing the gas is brought to this temperature by a heat bath and mercury is forced into or withdrawn from the tube R until the mercury level stands at a reference mark J. The pressure P_1 of the gas is then measured. (Small corrections must be made for the expansion or contraction of the bulb and for the fact that the gas in the capillary and in the tube R is not at the proper temperature.) In the same way the pressure P_0 is measured at the fixed point of 0°C. The temperature T_1 can then be computed from the known values of T_0, P_0, and P_1:

$$T_1 = \frac{P_1}{P_0} T_0.$$

However, the law

$$\frac{P_1}{P_0} = \frac{T_1}{T_0}$$

applies to ideal gases only. For real gases, it must be replaced by

$$P_t = P_{0°C}(1 + \beta_v t),$$

where β_v is the *isochoric coefficient of pressure change* of the gas. It has nearly the same value for all quasi-ideal gases and is approximately equal to 1/273, if the initial pressure is that at 0°C.

DEFINITION. The *isochoric coefficient of pressure change* of a real gas is defined as the ratio of increase in pressure to the increase in temperature of a gas, when the gas is heated at constant volume, divided by the pressure at 0°C:

$$\beta_v = \left(\frac{1}{P_{0°C}} \frac{dP}{dT}\right)_{V = \text{const}}.$$

For ideal gases,

$$\beta_v = \alpha_p = 1/273.15 \text{ K}° = 3.66099 \times 10^{-3} (\text{K}°)^{-1}.$$

The difference between β_v and α_p is not large for quasi-ideal gases, as is shown by the following table.

TABLE 7-3

Gas	α_p	βv
Ideal gas	$3.66099 \cdot 10^{-3}$/K°	$3.66099 \cdot 10^{-3}$/K°
H_2	3.6603	3.6627
He	3.6591	3.6613
N_2	3.671	3.672
CO_2	3.725	3.711

66-4 The Gas Temperature Scale

The measurement of temperature with the help of an ideal gas constitutes a significant advance in the science of thermometry. For real gases (such as H_2, He, and N_2) at a sufficiently low pressure, these so-called *gas temperatures* are very nearly independent of the thermometric material, a property that is not shared by liquid-in-glass thermometers using different liquids. A liquid-in-glass thermometer must be calibrated by means of a gas thermometer and its readings brought into agreement with the more satisfactory gas scale (Fig. 7-6). However, gas thermometer scales do not agree *exactly* with one another. Precise measurements made with, say, hydrogen and helium gas thermometers show small differences at one and the same temperature. In precise measurements, it is still necessary to state what thermometric substance was used. A definitive solution of the problem of temperature measurement can be obtained only with the help of the second law of thermodynamics.

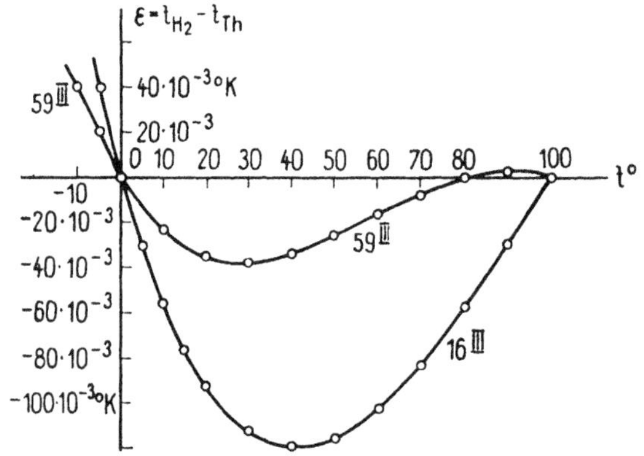

FIG. 7-6. Deviation of the temperature measured by Hg thermometers (made of Jena glass no. 16^{III} and Hard glass no. 59^{III}) from the standard gas temperature scale as determined with a H_2 gas thermometer.

67. THE MASS UNITS OF THERMODYNAMICS

The common units of mass used in mechanics, that is, the gram, kilogram, pound-mass, and slug, are also often used in thermodynamics. In many chemical problems, however, other units are not only more convenient, but from the viewpoint of kinetic theory are the only ones justified. One of these units is the *gram molecular mass* M^* or *mole*. A gram molecular mass of any substance is an amount of that substance whose mass, expressed in grams, is equal numerically to the molecular mass M (often incorrectly called the molecular weight) of the substance. Similar definitions hold for the mass units in the *mks* and *fps* systems, but these units are rarely used. The asterisk serves to distinguish the symbol for *molar* mass from the un-starred letter representing its magnitude, the so-called molecular weight M.

The concept of the *mole* originated in the science of chemistry, and is based on that of molecular weight. The exact definition of this concept can be given from two different points of view, and since one of these still enjoys great popularity, while the other is more up-to-date, both points of view will be considered in more detail.

67-1 The Traditional Viewpoint of Chemistry

The traditionally minded chemist defines the relative molecular mass M of a compound as the sum of the relative atomic masses A_i of the elements which constitute the molecule according to its formula. The relative chemical atomic mass A of an element is defined as the equivalent mass of that element multiplied by its chemical valence. As a standard of comparison, it was decided at recent international conferences† that the relative atomic mass $A_{C^{12}}$ of the carbon isotope C^{12} would be defined to be exactly equal to 12:

$$A_{C^{12}} = 12.00000.$$

The equivalent mass is a pure number, equal to the number of grams of the substance which will combine with 7.99795 grams of the oxygen isotope O^{16}. The relative atomic mass, relative molecular mass, and equivalent mass are all pure numbers.

Before this new definition of relative atomic mass, it was necessary to consider the difference between the *chemical* and *physical* atomic masses. This difference occurred because the *chemical* relative atomic mass was based on the natural mixture of oxygen isotopes (the relative atomic mass of natural oxygen was defined as $A_{O_{nat}} = 16.00000$), while the *physical* relative atomic masses depended on the isotope O^{16} (the relative atomic mass of O^{16} was defined as $A_{O^{16}} = 16.00000$). Having two different definitions had the result that the physical atomic masses were about 0.028 per cent larger than the chemical ones. To this somewhat unhappy situation was recently added the discovery that the relative proportions of oxygen isotopes in naturally occurring oxygen varied widely with the place of origin of the gas. This gave the necessary impulse for establishing a more universal definition of atomic mass. With this new definition, the relative atomic masses differ only slightly from their previous values and we have the advantage of having exactly the same atomic masses in physics and chemistry. The conversion factors between the new and the old scales are:

$$A \text{ (new } C^{12} \text{ scale)} = 0.999957 \text{ A (natural oxygen scale)}$$
$$= 0.999682 \text{ A } (O^{16} \text{ scale}).$$

The mole can now be defined as follows:

DEFINITION. A *mole* M^* of a substance is the amount of the substance having a mass in grams, numerically equal to the relative molecular mass M of the substance.

In a similar way we define the following:

(a) A *gram-atom* A^* of a substance is the amount of the substance having a mass in grams, numerically equal to the atomic mass A.

(b) A *gram-equivalent* of a substance is the amount of the substance having a mass in grams, numerically equal to the equivalent mass of the substance.

† Decision of the International Union of Physics (IUPAP, Ottawa, 1960) and the International Union of Chemistry (IUPAC, Montreal, 1961).

It is evident that when the masses of chemical compounds and of elements are expressed in moles or gram-atoms, there are as many units of mass as there are compounds and elements. To a non-chemist, this procedure appears at first to be peculiar, but it is justified chemically by the laws of stoichiometry.

In physics there is a more fundamental reason for the adoption of these chemical units of mass, which arises from the *atomic structure* of matter. The meaning of the concept of the mole is obtained from *Avogadro's law* together with the concept of the *molar volume* of an ideal gas.

AVOGADRO'S LAW (1811). Equal volumes of an ideal gas, at the same pressure and temperature, contain equal numbers of molecules.

This law was deduced by Avogadro from experimental measurements of the volume changes in gaseous reactions, on the basis of an atomic hypothesis. In the light of our present-day knowledge, we know that it is not strictly true for real gases, but this is not of importance. It is correct to a very good approximation at low pressures and high temperatures.

MOLAR VOLUME. The volume of a mole of an ideal gas, the so-called *molar volume* V_M, is the same for all gases at the same temperature and pressure. At 0°C and 760 mm Hg (normal temperature and pressure) the molar volume is 22,420.7 cm^3. The molar volume of a real gas is similarly defined and differs only slightly in magnitude.

It follows from these facts that, independent of the chemical nature of the gas, one mole of every gas contains the same number of molecules. The number of molecules in one mole is called *Avogadro's number*, N_0. The unit of N_0 is $(\text{mole})^{-1}$. Its value was first obtained in 1865 by *Loschmidt*, using kinetic methods. The best experimental value at present is

$$N_0 = 6.025 \times 10^{23} \text{ molecules/mole}.$$

If a gas is condensed to a liquid, and if no polymerization occurs, that is, if the mass of a molecule does not change, the number of molecules per mole in the liquid state is also N_0. However, even today our knowledge of the liquid state is only scanty. We speak of a mole of water, meaning 18 gm of water, although we know that in fact the water is polymerized. It must be realized that in the liquid state the concept of a molecule loses the precise significance which it has in the gaseous state. It represents only a statistical average value. The molecule is a typical concept of the chemistry of gases. In the solid state we often make use of the molecular concept although we know that only in the exceptional cases of molecular crystals does the concept of a molecule retain its meaning.

The more fundamental *atomistic* justification for the use of the mole units rests on the fact that we almost always refer to the same number of molecules in all equations. It might then be expected, as is confirmed by experiment, that the "molecular" formulation of laws brings out the true relations between the properties of substances much better than does the "specific" formulation (that is, per gram or per pound) (see Section 68-1).

67-2 The Modern, Atomistic Viewpoint

The *atomic hypothesis* of the 19th century has long since outgrown its designation as a "hypothesis," and is today one of the most firmly established concepts of physics. No one who is acquainted with the results of contemporary experimental physics can doubt the reality of atoms. A purely atomistic basis for the definition of a mole is therefore not only simpler, but has a solid experimental foundation. The mole can be defined in the following way:

DEFINITION. One mole M^* is defined as the amount of a substance in the gaseous state containing N_0 molecules, where N_0 is Avogadro's number. The numerical value of M^* is the mass in grams of N_0 molecules.

The mass of a mole M^* and the mass m_M of a molecule are related by the equation

$$M^* = N_0 m_M .$$

We can define the gram-atom in an analogous way to that above for the mole, and a corresponding equation,

$$A^* = N_0 m_A$$

connects the gram-atomic mass A^* and the atomic mass m_A. These equations form the connecting link between the macroscopic, measurable molar masses and the masses of molecules and atoms.

EXAMPLE. One mole of hydrogen equals 2.01626 gm. The mass m_{H_2} of the hydrogen molecule is

$$m_{H_2} = \frac{M^*}{N_0} = \frac{2.01626}{6.025 \times 10^{23}} = 3.346 \times 10^{-24} \text{ gm}.$$

The mass m_H of a hydrogen atom is

$$m_H = \frac{A^*}{N_0} = \frac{1.00813}{6.025 \times 10^{23}} = 1.673 \times 10^{-24} \text{ gm}.$$

The molar concept can also be made use of in the liquid and solid states, even when the molecular formula is unknown, by simply using the formula for the gaseous state. That is, we use the *gram formula weight*, which has a mass in grams numerically equal to the sum of the atomic weights of the atoms appearing in the formula.

The *magnitude* of the mass of the mole is, unfortunately, usually called the *molecular weight*, and that of the gram-atom is called the *atomic weight*. Since they are magnitudes, these so-called "weights," are in reality pure numbers.

68. The Equation of State of an Ideal Gas

68-1 Chemically Pure Gas

The constant in the ideal gas law,

$$\frac{PV}{T} = \text{const},$$

is different for different gases and hence the above equation is inconvenient to use when numerical calculations are to be made. The reason for this is that equal masses of different gases contain different numbers of molecules. The difficulty disappears if the equation is transformed to molar form and we work with the molar mass M^*. The volume occupied by the gas is then called the *molar volume* V_M.

To begin with, we express the following quantities in terms of M^* and V_M.

1. The density ϱ is defined as the mass per unit volume. It follows that

$$\varrho = \frac{M^*}{V_M}.$$

2. The *specific volume* v is the volume per unit mass,

$$v = \frac{V_M}{M^*} = \frac{1}{\varrho}.$$

The specific volume equals the reciprocal of the density.

The *molar* form of the gas equation can now be obtained by writing the ideal gas equation for a mass of 1 mole. The volume V is replaced by the molar volume V_M. The quantity PV_M/T, which formerly was different for different gases, now has the same value for all ideal gases and is a universal constant, since all ideal gases have the same molar volume at the same pressure and temperature. This constant is called the *universal gas constant* and is represented by R. The *universal equation of state for an ideal gas* is therefore

$$PV_M = RT.$$

This equation applies to all ideal gases. The magnitude of R can be determined from the known value of $V_M = 22{,}420.7\ \text{cm}^3$ at a temperature of 0°C (273.15°K) and a pressure of 760 mm Hg (1.01325×10^6 dynes/cm^2). In cgs units,

$$R = \frac{P_0 V_M}{T_0} = 8.31696 \times 10^7 \frac{\text{ergs}}{\text{mole} \cdot \text{K}^\circ}.$$

Other combinations of units are often used. For example, if the pressure is expressed in atmospheres and the volume in liters,

$$R = \frac{1 \times 22.4207}{273.15} = 0.082082 \frac{\text{liter} \cdot \text{atm}}{\text{mole} \cdot \text{K}^\circ}.$$

The universal ideal gas equation can easily be transformed into a form which is applicable to any arbitrary mass m of a gas. The number of moles n of the gas is

$$n = \frac{m}{M^*}$$

and the volume V is

$$V = nV_M.$$

The equation of the state of an ideal gas then becomes

$$PV = nRT.$$

The ideal gas equation is now in a convenient form for all quantitative calculations. We shall illustrate its application by two examples.

EXAMPLE 1. A mass of 200 gm of oxygen (O_2) is at a temperature of 60°C and a pressure of 3.2 atm. We wish to calculate the quantities ϱ, v, V, and y, the number of molecules per cubic centimeter.

It will be helpful to first calculate the molar volume. This is

$$V_M = \frac{RT}{P} = 8.542 \times 10^3 \frac{\text{cm}^3}{\text{mole}}.$$

It follows that

$$\varrho = \frac{M^*}{V_M} = 3.746 \times 10^{-3} \frac{\text{gm}}{\text{cm}^3},$$

$$v = \frac{1}{\varrho} = 267 \frac{\text{cm}^3}{\text{gm}}.$$

The number of moles is

$$n = \frac{m}{M^*} = 6.25 \text{ moles}.$$

The total volume V is therefore

$$V = nV_M = 5.34 \times 10^4 \text{ cm}^3.$$

The number of molecules per cm^3 is

$$y = \frac{N_0}{V_M} = 7.053 \times 10^{19} \frac{\text{molecules}}{\text{cm}^3}.$$

EXAMPLE 2. 176 cm^3 of chloroform ($CHCl_3$, $M = 119.4$) are contained in a vessel at a pressure $P = 804$ mm Hg. The temperature $t = 82$°C. (a) Find the values of m and ϱ. (b) What volume V_1 is required to contain a mass $m_1 = 12.6$ gm at this temperature and pressure?

$$V_M = \frac{RT}{P} = 27556 \text{ cm}^3/\text{mole},$$

$$\varrho = \frac{M^*}{V_M} = 4.333 \times 10^{-3} \text{ gm/cm}^3,$$

$$v = \frac{1}{\varrho} = 230.8 \text{ cm}^3/\text{gm},$$

$$m = \varrho V = 0.7626 \text{ gm},$$

$$V_1 = vm_1 = 2908 \text{ cm}^3.$$

68-2 Mixtures of Ideal Gases

Mixtures of gases are frequently encountered (e.g., air, initial and final states of a gaseous reaction, etc.). Suppose that the mixture consists of $n_1, n_2, n_3, \cdots$ moles of pure gases having molecular masses ("weights") $M_1, M_2, M_3, \ldots$ etc. The total volume of the mixture is V, the total pressure is P_t, and the temperature is T. The thermodynamic treatment of the mixture is the same as for a single ideal gas. It is helpful, however, to define some new concepts.

1. *The partial pressure* P_i. The *partial pressure* P_i of the ith component is defined as the pressure which this gas would exert if it alone occupied the entire volume at the same temperature.

2. *The partial volume* V_i. We imagine the components of the gas to be completely separated from one another by semi-permeable membranes, without any changes in their pressure and temperature. The ith component then occupies the *partial volume* V_i. The partial volume V_i of the ith component is the volume which this component would occupy at a pressure and temperature equal to the pressure and temperature of the mixture.

It follows from the Boyle–Mariotte law that

$$P_i V_t = P_t V_i.$$

Hence the ratio of the partial pressure to the total pressure equals the ratio of the partial volume to the total volume,

$$\frac{P_i}{P_t} = \frac{V_i}{V_t}.$$

If the gases are ideal, their molecules exert no forces on one another (see Section 74) and the total volume equals the sum of the partial volumes,

$$V_t = \sum V_i.$$

Summing the preceding ratio equation over all values of i, we get

$$\frac{\sum P_i}{P_t} = \frac{\sum V_i}{V_t} = 1;$$

that is,

$$P_t = \sum P_i.$$

This is *Dalton's law* (1801): In a mixture of ideal gases, the sum of the partial pressures is equal to the total pressure. These simple relations are not satisfied exactly by real gases.

3. *The partial volume concentration* q_i. The partial volume concentration of the ith component of a mixture of gases is defined as the ratio of the partial volume of this component to the total volume.

The derivation of Dalton's law shows that for ideal gases, q_i is also equal to the ratio of the partial pressure of this component to the total pressure:

$$q_i = \frac{V_i}{V_t} = \frac{P_i}{P_t}.$$

Since the molar volume is the same for all ideal gases at the same temperature and pressure, it follows from this equation that

$$q_i = \frac{V_i}{V_t} = \frac{n_i V_M}{n_t V_M} = \frac{n_i}{n_t},$$

where n_i is the number of moles of the ith component and n_t is the total number of moles in the mixture. The ratio n_i/n_t is called the *mole-fraction* of the ith component, and is represented by x_i,

$$x_i = \frac{n_i}{n_t}.$$

For ideal gases, $x_i = q_i$. It is evident that

$$\sum q_i = 1,$$

since

$$\sum V_i = V_t, \quad \sum P_i = P_t, \quad \sum n_i = n_t.$$

EXAMPLE. The air in a room, at a total pressure of 720 mm Hg, has the following composition: 76% N_2, 17.5% O_2, 4.2% CO_2, and 2.3% H_2O. At these low pressures, CO_2 and H_2O can be considered ideal gases. Find the partial pressures of the components.

$$P_{N_2} = q_{N_2} \cdot P_t = 547.2 \text{ mm Hg}$$
$$P_{O_2} = q_{O_2} \cdot P_t = 126.0 \text{ mm Hg}$$
$$P_{CO_2} = q_{CO_2} \cdot P_t = 30.2 \text{ mm Hg}$$
$$P_{H_2O} = q_{H_2O} \cdot P_t = 16.6 \text{ mm Hg}$$
$$P_t = 720.0 \text{ mm Hg}$$

4. *The mean molecular weight $\overline{M}$ of a mixture of gases.* All of the components will be assumed ideal. The molar volume of the mixture is

$$V_M = \frac{RT}{P_t}.$$

DEFINITION. The mass of a mole of a mixture of gases is defined as the mass of a molar volume of the mixture. Its magnitude is the *mean molecular mass.*

To compute $\overline{M}$, we express the total mass m_t of the mixture in two different ways, namely,

$$m_t = \sum n_i M_i^*, \quad \text{and} \quad m_t = n_t \overline{M}^*.$$

It follows that

$$\overline{M}^* = \sum M_i^* \frac{n_i}{n_t} = \sum M_i\, q_i,$$

and the mean molecular weight is then

$$\bar{M} = \sum M_i q_i .$$

The mean molecular weight of a mixture of ideal gases is the sum of the products of the molecular weights of its components and their partial volume concentrations. Calculations of this quantity are often necessary in the study of gas reactions.

EXAMPLE. Find the mean molecular weight of air, from the data in the example above.

$$M_{N_2} q_{v_{N_2}} = 21.25$$

$$M_{O_2} q_{v_{O_2}} = 5.60$$

$$M_{CO_2} q_{v_{CO_2}} = 1.85$$

$$M_{H_2O} q_{v_{H_2O}} = 0.41$$

$$\bar{M} = 29.11$$

69. The Concept of Quantity of Heat

69-1 The Fundamental Equation of Calorimetry

In the development of the science of thermodynamics, it has proved as difficult to attain a rigorous understanding of the concept of *quantity of heat* as of the concept of temperature. For a discussion of this problem, the corresponding chapter in Mach's "Principles of Thermodynamics" is highly recommended. The results of many years of painstaking effort can be summarized by the equation:

$$\Delta Q = mc\Delta t .$$

This equation states: When the temperature of a *homogeneous* body of mass m is increased by Δt, a quantity of heat ΔQ must be supplied to the body. The quantity of heat depends on the rise in temperature, on the mass, and on a property of the material represented by the factor c.

69-2 Units of Heat. Specific Heat Capacity

Every temperature change Δt of a mass m is associated with the transfer of a quantity of heat

$$\Delta Q = mc\Delta t .$$

The empirical constant c is called the *specific heat capacity*, or often just the *specific heat*, of the material of which the body is composed. However, it must be specified more precisely. Depending on the nature of the process in which the temperature is increased, we distinguish between the *specific heat capacity at constant volume*, c_v, and the *specific heat capacity at constant pressure*, c_p.

A definition of the unit of heat is possible only when some standard substance has been agreed on.

DEFINITION. The *unit of heat* is defined as the quantity of heat which increases the temperature of a unit mass of water at constant pressure by a specified one-degree interval of temperature.

In the cgs system, the unit of mass is the gram and the specified temperature interval is from 14.5 to 15.5°C. The corresponding unit of heat is called 1 *calorie*. In the fps system, the unit of mass is the pound-mass and the temperature interval is from 59.5 to 60.5°F. The corresponding unit of heat is called the *British thermal unit*, abbreviated Btu.

For the present, the calorie and the Btu will be considered as new fundamental units. When the first law of thermodynamics is discussed, we shall see that they are actually units of energy.

The temperature intervals between 14.5 and 15.5°C, and between 59.5 and 60.5°F are specified for the following reason. The calorie was originally defined as the quantity of heat required to raise the temperature of one gram of water at constant pressure by one centigrade degree. Later, more precise measurements showed that this quantity depended on the temperature of the water. The heat absorbed in the interval from 14.5 to 15.5°C was found to agree most closely with the accepted magnitude of the calorie. Similar considerations determined the selection of the temperature interval on the Fahrenheit scale.

The specific heat capacities c_p and c_v are now defined as follows.

DEFINITION. The *specific heat capacity at constant pressure* c_p of a material is defined as the quantity of heat ΔQ flowing into a mass m of the material at constant pressure, divided by the product of the mass m and the increase in temperature Δt:

$$c_p = \left(\frac{1}{m}\frac{\Delta Q}{\Delta t}\right)_{P=\text{const}}.$$

Hence in a process at constant pressure, (an isobaric process)

$$\Delta Q_p = mc_p\Delta t.$$

DEFINITION. The *specific heat capacity at constant volume* c_v of a material is defined as the quantity of heat ΔQ flowing into a mass m of the material at constant volume, divided by the product of the mass m and the increase in temperature Δt:

$$c_v = \left(\frac{1}{m}\frac{\Delta Q}{\Delta t}\right)_{V=\text{const}}.$$

Hence in a process at constant volume

$$\Delta Q_v = mc_v\Delta t.$$

Units of specific heat capacity:

cgs system: 1 cal/gm · C°
mks system: 1 Kcal/kgm · C°
fps system: 1 Btu/lbm · F°.

The numerical values of the specific heat capacity of a given material are the same in all three systems.

The range of validity of the equations above is restricted to temperature intervals of the order of 10–20 centigrade degrees, since specific heat capacities vary appreciably with the temperature.

It is often more convenient to use the *heat capacity* of a system, represented by C^s.

DEFINITION. The *heat capacity* C^s of a *system* is defined as the product of its mass and its specific heat capacity.

Thus the heat capacities of a system, at constant pressure and at constant volume, are respectively

$$C_p^s = mc_p; \quad C_v^s = mc_v.$$

Units: cgs system: 1 cal/C°
mks system: 1 Kcal/C°
fps system: 1 Btu/F°.

We see that *specific heat capacity* can be described as the *heat capacity per unit mass.*

The *molar heat capacity* is defined by the equations

$$C_p = M^*c_p; \quad C_v = M^*c_v.$$

Units: cgs system: 1 cal/mole · C°.

Specific heat capacities at constant volume are difficult to measure directly. From the second law of thermodynamics, however, we can derive a relation by which C_v can be computed from C_p, namely,

$$C_p - C_v = \frac{\alpha_p^2 V_M T}{\varkappa_T},$$

where α_p is the isobaric coefficient of volume expansion and $\varkappa_T$ is the isothermal compressibility (see Sections 92 and 93-2).

70. Measurement of Heat. Calorimetry

Quantities of heat are measured by means of *calorimeters.* A calorimeter is a system which cannot exchange heat with its surroundings. If this condition is fulfilled exactly, the total heat content of the calorimeter can change only if some process takes place within it which liberates or absorbs heat.

Such processes can take place under various external conditions. There are four different processes which are of importance in thermodynamics. They are distinguished by their differing external conditions, which at least partially determine the course of the process.

(a) *Isochoric* or *isovolumic processes*, which take place at constant volume. These processes are particularly simple in their theoretical relations, but rarely occur in practice. Chemical reactions taking place in an autoclave are isovolumic.

(b) *Isobaric processes*, which take place at constant pressure. These are also theoretically simple and they do occur frequently in practice. Many processes take place under constant atmospheric pressure and are therefore isobaric.

(c) *Isothermal processes.* These take place at constant temperature. Their theoretical analysis is somewhat more difficult than for the two processes above. They are, however, of great theoretical and practical importance.

(d) *Adiabatic processes.* An adiabatic process is one in which no heat flows into or out of a system.

Most calorimetric measurements are isobaric processes. In general, the process taking place in a calorimeter consists of an interaction at constant atmospheric pressure between a number of different masses m_i having different specific heat capacities c_{pi} and different initial temperatures t_i. We can distinguish two important special cases.

(a) The process taking place in the calorimeter proceeds *without evolution of heat.* In this case, the calorimeter equation is

$$\sum m_i c_{pi} \Delta t_i = 0.$$

This equation states that in the process of reaching thermal equilibrium the quantity of heat given up by the warmer bodies equals the quantity of heat absorbed by the cooler bodies.

Such isobaric calorimetric measurements serve primarily as experimental methods for measuring the specific heat capacity c_p. It is necessary that, as far as possible, the calorimeter should be thermally insulated against any interchange of heat with its surroundings. For this purpose the calorimeter is often separated from its surroundings by a layer of still air (Fig. 7-7).

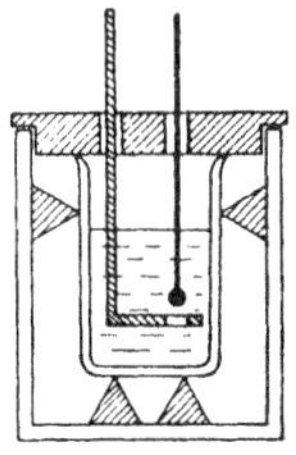

FIG. 7-7. A Bunsen calorimeter.

The vessel in which measurements are made is insulated from the outer container and closed by a cover. A thermometer is inserted for measurement of the temperature. This simple apparatus is called the *Bunsen calorimeter.* The accuracy can be increased by replacing the layer of air by a high vacuum, which is a better heat insulator. Heat transfer by radiation is minimized by silvering the outer surface of the calorimeter. It should be noted that this instrument measures only the *mean* specific heat capacity for the temperature range between the initial and final temperatures of the substance being measured.

Solid bodies in air exchange heat with their surroundings very slowly. It is therefore customary to immerse the bodies which exchange heat with one another in a liquid of known heat capacity. Water is usually used. Naturally, care must be taken that no chemical reactions take place between the calorimeter fluid and the bodies immersed in the calorimeter, and that these are not soluble in the fluid.

EXAMPLE. A copper calorimeter of mass $m_1 = 190$ gm contains a mass $m_2 = 150$ gm of water. The initial temperature $t_i = 18.21$°C. A mass $m_3 = 450$ gm of copper, at a temperature $t_0 = 98.32$°C, is introduced in the calorimeter and the final temperature is $t_f = 33.75$°C. We wish to find the mean specific heat capacity c_p of copper. The heat capacity of the thermometer is $C_{\text{th}} = 4$ cal/C°. The specific heat capacity of water can be considered constant and equal to 1 cal/gm · C°. From the calorimeter equation, we have

$$m_3(t_0 - t_f)\, c_{pCu} = (m_1 c_{pCu} + m_2 c_{pH_2O} + C_{\text{th}})\,(t_f - t_i),$$

$$c_{pCu} = \frac{(m_2 c_{pH_2O} + C_{\text{th}})\,(t_f - t_i)}{m_3(t_0 - t_f) - m_1(t_f - t_i)} = 0.0917\,\frac{\text{cal}}{\text{gm}\cdot\text{C}^\circ}.$$

(b) The reactions in the calorimeter take place with the evolution of a quantity of heat called the heat of reaction, Q. The calorimeter equation is then

$$\sum m_i c_{pi} \Delta t_i = Q.$$

The term *heat of reaction* means the quantity of heat liberated or absorbed in a process. The reaction is said to be *exothermic* if heat is given up by the reacting system, and is said to be *endothermic* if heat is absorbed. The usual sign convention is to consider heat given up as negative, heat absorbed as positive.

Examples of heats of reaction are heats of fusion and solidification, heats of vaporization and condensation, heats of sublimation and crystallization; heat of solution, of dilution, and of hydration, as well as the heats of all chemical reactions.

71. Specific Heat Capacities of Gases, Liquids and Solids

The most important thermal property of a substance is its specific heat capacity. The internal energy of a substance, its entropy, enthalpy, and free energy, i.e., just those quantities which determine the course of all chemical reactions, all depend on the specific heat capacity (see Section 93). A knowledge of specific heat capacities is therefore of great importance. An even more significant quantity than the specific heat capacity (per gram), is the *molar* heat capacity (per mole). From a knowledge of this quantity, we can obtain information about the properties of individual molecules since the molar heat capacity always refers to the same number of molecules.

71-1 The Molar Heat Capacities of Gases

(a) *Ideal Gases.* Gases such as He, Ne, A, H_2, N_2, O_2, CO, and NO behave approximately like ideal gases. It is found that the molar heat capacities of all monatomic gases are nearly equal to one another, and that those of diatomic gases, while they differ from those of monatomic gases, are also nearly equal to one another.

1. *Monatomic gases,* such as He, Ne, A. All of these have the same molar heat capacities:

$$C_p = 4.97 \text{ cal/mole} \cdot \text{C}^\circ \cong 5 \text{ cal/mole} \cdot \text{C}^\circ,$$

$$C_v = 2.98 \text{ cal/mole} \cdot \text{C}^\circ \cong 3 \text{ cal/mole} \cdot \text{C}^\circ.$$

The ratio of the heat capacities is

$$\gamma = \frac{C_p}{C_v} \cong \frac{5}{3} \cong 1.67.$$

2. *Diatomic gases* such as H_2, N_2, O_2, CO, NO. The molar heat capacities for an ideal diatomic gas are:

$$C_p = 6.96 \text{ cal/mole} \cdot \text{C}^\circ \cong 7 \text{ cal/mole} \cdot \text{C}^\circ,$$

$$C_v = 4{\cdot}97 \text{ cal/mole} \cdot \text{C}^\circ \cong 5 \text{ cal/mole} \cdot \text{C}^\circ.$$

The ratio of heat capacities is

$$\gamma = \frac{C_p}{C_v} \cong \frac{7}{5} \cong 1.40.$$

Polyatomic gases such as H_2O, SO_2, CO_2, NO_2, NH_3, CH_4, etc., usually deviate considerably from the behavior of ideal gases (see the section on real gases below).

The kinetic theory of gases provides an explanation for the simple experimental relations given above. We shall return to this question in Section 74-3.

(b) *Real gases.* Even diatomic gases begin to show marked departure from the ideal gas state. As a result, their specific heat capacities vary with temperature. Precise experimental measurements are available only for C_p. The results can be stated with reasonable accuracy by expressing C_p^0 as a power series in the absolute temperature T, where C_p^0 is the heat capacity as the gas pressure approaches zero. The correction for most gases at moderate pressures is small. Some representative values for the temperature range from 300 to 2000°K are given in the table below.

TABLE 7-4†

Gas	C_p^0(cal/mole · K°)
Monatomic gases	4.97
H_2	$6.52 + 0.78 \times 10^{-3}T + 0.12 \times 10^5 T^{-2}$
O_2	$7.16 + 1.00 \times 10^{-3}T - 0.40 \times 10^5 T^{-2}$
N_2	$6.83 + 0.90 \times 10^{-3}T - 0.12 \times 10^5 T^{-2}$
F_2	$8.26 + 0.60 \times 10^{-3}T - 0.84 \times 10^5 T^{-2}$
Cl_2	$8.85 + 0.16 \times 10^{-3}T - 0.68 \times 10^5 T^{-2}$
CO	$6.79 + 0.98 \times 10^{-3}T - 0.11 \times 10^5 T^{-2}$
CO_2	$10.57 + 2.10 \times 10^{-3}T - 2.06 \times 10^5 T^{-2}$
H_2O	$7.30 + 2.46 \times 10^{-3}T$
CH_4	$5.65 + 11.44 \times 10^{-3}T - 0.46 \times 10^5 T^{-2}$

† G. N. Lewis and M. Randall, *Thermodynamics*, 2nd Edition, revised by K. S. Pitzer and L. Brewer, p. 66 (McGraw-Hill, New York, 1961).

The specific heat capacities of real gases vary with the pressure as well as with the temperature. For an ideal gas, the specific heat is independent of the pressure. The reason for this dependence on pressure will also be discussed in the next chapter on kinetic theory.

The values given above for the molar heat capacities of gases apply only in the temperature range above 300°K. Investigations by A. Eucken of the specific heat of H_2 at low temperatures showed that the molar heat capacity of diatomic hydrogen decreases as the temperature is decreased below 300°K, approaching a constant value in the neighborhood of 60°K (see Fig. 7-8). This limiting value is equal to the molar heat capacity of a *monatomic* gas, that is, 4.97 cal/mole · C°. This is an interesting and important result for the kinetic theory of gases. Experiments on the specific heats of other gases show that they also decrease with decreasing temperature.

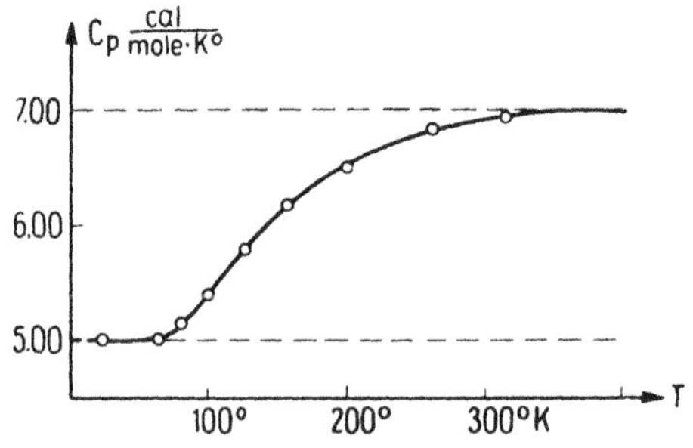

FIG. 7-8. The decrease in the molar specific heat C_p of H_2 at low temperature.

However, most other gases liquefy before becoming so cool that their specific heat has reached a constant value, independent of temperature.

The fact that the heat capacity of most substances varies with temperature cannot be understood in terms of non-atomic physics. Instead, a description of matter on the microscopic level, called quantum theory, must be employed. The variations of the heat capacities are explained by quantum theory.

71-2 The Specific Heat Capacities of Liquids

The liquid state is intermediate between the complete molecular disorder of the gaseous state and the complete order of the solid state (see Chapter 9). It is not surprising, therefore, that the specific heat capacities of liquids do not display the regularities observed in solids and gases. Roughly speaking, the specific heat capacities of non-associated liquids lie between 0.3 cal/gm · C° and 0.6 cal/gm · C°. All highly associated liquids have much larger specific heat capacities, which often increase rapidly with the temperature.

Of special interest is the specific heat capacity of water and its variation with temperature. The value of c_p for water is about 1 cal/gm · C°. In contrast to other associated liquids, however, the specific heat capacity of water varies only slightly with temperature. Since water is so often used as a calorimetric substance, some values of c_p are given in the accompanying table. The specific heat has a minimum at 35°C and has its "normal" value at 15°C and at about 65°C.

TABLE 7-5

t°C	c_p (cal/gm · C°)
0	1.00764
10	1.00153
15	1.00000
20	0.99907
35	0.99818
50	0.99878
65	1.00024
80	1.00253
100	1.00721

Up to the present time, no theory has been developed for the temperature dependence of c_p for liquids. We must rely wholly on experimental measurements. The results can usually be represented by smooth curves. Departures from smooth curves occur only when sudden changes in structure take place. Thus, liquid sulfur (the bright yellow, very fluid modification) transforms quite suddenly at 159°C to a viscous brown modification. This transition shows up in the graph of c_p vs. T as a sudden rise of c_p from 0.25

to 0.45 cal/gm · C°, within a range of a few degrees (see Fig. 7-10). After the transition, c_p decreases again to a value of 0.26 cal/gm · C°. The temperature at which the cusp appears in the graph of c_p vs. T is called a *lambda point* from the shape of the curve.

71-3 The Molar Heat Capacities of Solids

The molar heat capacities of solids at constant pressure show no regularity of behavior. For C_v, however, there exists both a limiting value at high temperatures and a law governing the dependence on temperature.

The limiting value was discovered by Dulong and Petit (the *Law of Dulong and Petit*) and it applies for temperatures in the neighborhood of the melting point of the material. The law states:

The molar heat capacity C_v, except for the strongly electropositive elements Li, Na, K, Ca, and Mg, has the same value for all elements at high temperatures: $C_v = 5.97$ cal/mole · C°. For the strongly electropositive elements, C_v is greater than the preceding value.

The Dulong–Petit law is an approximation, and it implies a temperature dependence of C_v. In precise calculations reference must be made to the experimentally determined values, which are computed from C_p by the equation given at the end of Section 69-2.

(a) *Temperature dependence of C_v*. The extensive experimental investigations of Nernst and Lindemann, in 1911, which were extended to temperatures near absolute zero, showed a strong dependence of C_v on temperature. The results of these experiments can be summarized as follows: The molar heat capacity C_v decreases with temperature and, for all substances, approaches zero as T approaches zero (Fig. 7-9). As the temperature is increased from zero, C_v increases slowly at first, then shows a rapid rate of rise, and at high temperatures approaches asymptotically the value $C_v = 5.97$ cal/mole · C° (except for the strongly electropositive elements). The temperature range in which the rapid rise occurs varies from element

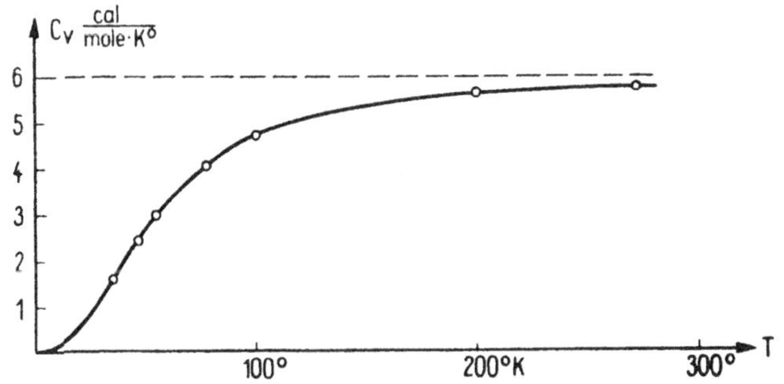

FIG. 7-9. The fall of the molar heat capacity C_v of Ag with decreasing temperature.

to element. Not only do the elements obey this law, but also all cubic crystalline compounds provided a limiting value of 5.97 cal/gm-atom · C° is assumed for each atomic constituent. The law does not apply if radicals such as CO_3^{--} or NO_3^- exist in the solid. For example, NaCl has a limiting value of $C_v = 2 \times 5.97$ cal/gm-mole · C° and behaves normally. $NaNO_3$, on the other hand, is not normal because of the occurrence of the radical NO_3^- in the crystal lattice.

P. Debye, in 1912, derived the law of the temperature dependence of C_v, for all elements and radical-free compounds which exist as cubic crystals using a combination of quantum theory and the theory of elasticity. The Debye law has been extensively verified and it shows that at very low temperatures C_v increases as T^3.

(b) *The temperature dependence of* C_p. Values of the molar heat capacity at constant pressure are more useful in the thermodynamics of chemical reactions than are those at constant volume, because nearly all experimental studies of reactions are made at constant pressure. For this purpose, the values of C_p for gases were expressed in the form of power series in the temperature T. Similar relations can be written for liquids and solids. A large number of such equations, based on experimental values, will be found in Handbooks and Tables.†

72. Transition Points. Latent Heats

72-1 Polymorphic Solids and Liquids

In solids as well as in liquids, the variation of specific heat capacity with temperature is a sensitive indicator of changes in intermolecular structure. If a body is cooled through small temperature intervals, and its specific heat capacity is determined for each interval, certain substances show sudden variations in specific heat at definite temperatures. Usually this behavior is associated with an alteration of the lattice structure of the solid. In liquids, changes in the mode in which the molecules pack together or changes in the structure of the molecules themselves give rise to the same effect (Fig. 7-10).

In the solid state, the most important factor in uniquely determining a modification is the degree of symmetry in the structure of the unit crystallographic cell. In general, a change in the form of the unit cell occurs suddenly, at a definite transition temperature. To every temperature there corresponds a definite modification. Although other modifications may exist at a given temperature, they change to the stable form more or less rapidly. This transformation process may require days or years, depending on the material and on the circumstances. The diamond-graphite transformation is a

† J. H. Perry, *Chemical Engineers Handbook*, McGraw-Hill, 1950.

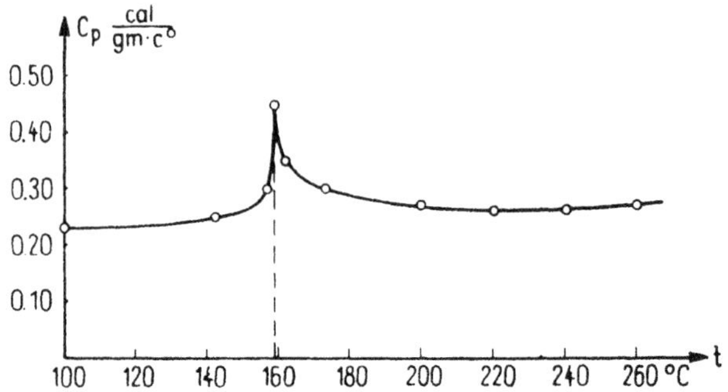

FIG. 7-10. The specific heat C_p of liquid sulfur as a function of temperature in the region of the $S_8 \rightarrow S_6$ transformation point at 159°C.

striking example. At atmospheric pressure and room temperature, graphite is the stable modification of carbon, while diamond is the stable form at pressures above about 8000 atm. Nevertheless, diamonds can exist for thousands of years at atmospheric pressure, without transforming into graphite.

In the liquid state, the relations are much more complicated. On the one hand, the structure of the molecule plays an important role in determining the nature of a modification, while on the other, the degree of polymerization, which changes continuously with the temperature, is also a significant factor. Furthermore, it has been made clear by recent investigations that a certain amount of ordering exists even in the liquid state. This ordering differs from the ordered state of a solid, which extends to regions of macroscopic size, in that it is confined to extremely small regions. The result is that different liquid modifications of one and the same substance can exist side by side at the same temperature. They form a system in equilibrium, the relative proportions of each changing with the temperature.

A chemically homogeneous substance is said to be *polymorphic*,

(a) if in the solid state it can exist in different crystallographic structures corresponding to different temperatures,

(b) if in the liquid state it can have different molecular structures (e.g. S_6 and S_8) as well as different forms of packing which occur in different ranges of temperature. The term allotropic is often used to mean the same thing for an elemental substance; e.g. diamond and graphite are allotropes of carbon.

The different lattice types are called *polymorphic modifications* of the substance. Very sharply defined transition points often exist for the solid state between different modifications.

The following table lists some polymorphic modifications of elements and compounds. The third column gives their regions of stability, as far as these are known. Certain modifications transform in either direction at a well-defined transition temperature, while others show a gradual transformation process within certain temperature ranges. Still other transitions can be caused to proceed in one direction but not in the other.

TABLE 7-6

Substance	Type of lattice†	Region of stability
$\alpha - C$	A_4 (diamond)	Low temperature
$\beta - C$	A_9 (graphite)	High temperature
$\alpha - S$	A_{16}, orthorhombic	Below 95.5°C
$\beta - S$	Monoclinic	Between 95.5–119°C
$\alpha - Fe$	A_2, cubic, ferromagnetic	Below 761°C
$\beta - Fe$	A_2, cubic, paramagnetic	Between 761–906°C
$\gamma - Fe$	A_1, cubic	Between 906–1400 °C
$\delta - Fe$	A_2, cubic	Above 1400°C
$\alpha - Sn$	A_4, cubic (gray)	Below 13.2°C
$\beta - Sn$	A_5, tetragonal (white)	Between 18–203°C (Rossini)
$\gamma - Sn$	Orthorhombic	Between 170–232°C
ZnS	Hexagonal (wurtzite)	Above 1020°C
ZnS	Cubic (zincblende)	Below 1020°C

† The designation of lattice types is explained in Chapter 9.

72-2 The Melting Point

Solid bodies exhibit a particularly sharp transition when the effect of the ordering forces has been sufficiently diminished by an elevation of temperature. The previously existing ordered state, extending through regions of macroscopic size, gives way to the more or less disordered liquid state. The transition, for a chemically homogeneous crystalline substance, takes place at a definite temperature. A satisfactory explanation of the phenomenon is still lacking. An amorphous substance, such as glass, has no definite transition temperature, but only becomes less viscous as the temperature increases.

The *melting point* of a crystalline substance is the temperature at which it changes from the solid to the liquid state under a pressure of one atmosphere. The *heat of fusion*, L_f, is the quantity of heat per unit mass that is required to change a substance at its melting point from the solid to the liquid state. The quantity of heat Q necessary to melt a mass m of material is therefore

$$Q = mL_f.$$

Since the temperature of the substance is not changed by the absorption of this heat, it is often spoken of as *latent* heat.

If a chemically homogeneous substance is crystalline in its solid form, then in the liquid state it always freezes or solidifies at the same temperature, the *freezing point*, and gives up the same quantity of heat that it absorbed on melting. The *heat of solidification* is equal to the *heat of fusion*.

The most important heat of fusion is that of water. At a temperature of 0°C its magnitude is

$$L_f = 1436 \frac{\text{cal}}{\text{mole}} = 79.71 \frac{\text{cal}}{\text{gm}}.$$

The heat of fusion of other materials varies from substance to substance in a quite unpredictable way.

Although a solid cannot be heated above its melting point, a liquid can easily be cooled below its freezing point. It is then said to be *supercooled*. To supercool a liquid it is necessary that it contain no crystals of the solid to serve as nucleation centers. The instant the tiniest crystalline particle is introduced, solidification (in part or in whole) takes place with liberation of heat. The process is so strictly reproducible that it is always used in the precise measurement of melting points. That is, melting points are not determined by warming the solid, but by supercooling the liquid and then inducing freezing.

72-3 Dependence of the Melting Point on Pressure

The melting point of a pure substance is slightly pressure dependent. With the help of the two laws of thermodynamics a quantitative expression can be obtained for this dependence (see Section 95-1). For the present, we merely state the following experimental facts:

(a) If the volume of a substance decreases on melting (and hence if crystals of the solid float in the liquid) the melting point of the substance decreases with increasing pressure. (Examples: water, bismuth.)

(b) If the volume of a substance increases on melting (and hence if crystals of the solid sink in the liquid) the melting point increases with increasing pressure. (Examples: paraffin, lead, phosphorus, sulfur.)

72-4 The Sublimation Point

Some solids transform directly from the solid to the gaseous state at atmospheric pressure, and crystallize directly from the gas state as the temperature is lowered. Both processes take place at the same temperature, the *sublimation point*. That is, the sublimation point of a substance is the temperature at which the substance changes from the solid to the gaseous state. The *heat of sublimation* is the quantity of heat, per unit mass, which is necessary to change the substance at its sublimation point from the solid to the gaseous state.

The phenomenon of sublimation has not been investigated as thoroughly as that of melting. The processes are, however, exactly analogous. Thus the heat of sublimation is equal to the heat of crystallization. The sublimation point is pressure dependent, and more strongly so than the melting point. An increase of pressure always increases the temperature of the sublimation point.

A few sublimation points, at a pressure of 1 atm, are given below.

TABLE 7-7

Substance	Sublimation point (1 atm)
As	615°C
As_2O_3	218°C
NH_4Cl	335°C
CO_2	−78.5°C

72-5 The Boiling Point

A liquid in an open vessel is always observed to evaporate. If the vessel is closed, the process of evaporation continues until an equilibrium state is reached which is independent of any other gases that may simultaneously occupy the vessel. The *partial pressure* of the vapor (gasified liquid) in the equilibrium state is called the *vapor pressure* of the liquid. It is independent of the volume occupied by the vapor and depends only on the nature of the liquid and on the temperature. At 20°C the vapor pressure of water is 17.54 mm Hg.

It is important to note that for a vapor in contact with its liquid, the vapor pressure is independent of the volume and depends only on the temperature of the liquid. If the volume is increased (keeping the temperature constant), enough liquid evaporates to keep the pressure constant. Conversely, if the volume is decreased (at constant temperature), some of the vapor condenses. A vapor in contact with its liquid (at a constant temperature) constitutes a self-regulating pressure reservoir.

When heat is supplied to a liquid in an open vessel, the temperature of the liquid, and its vapor pressure, both increase. As long as the vapor pressure is less than the external pressure, the liquid evaporates relatively slowly and the chief result of the heat supplied is to raise the temperature of the liquid. As the addition of heat continues, a temperature is eventually reached at which the entire heat supply is used to evaporate the liquid. This occurs when the vapor pressure becomes equal to the external pressure. This temperature, known as the *boiling point*, is characterized by the fact that the temperature remains constant even though heat is continually being supplied to the liquid. The boiling point at an external pressure of 760 mm Hg is called the *normal boiling point*.

The *heat of vaporization* of a liquid, L_v, is the heat per unit mass necessary to transform the liquid at its normal boiling point from the liquid to the gaseous state. Heats of vaporization, like heats of fusion and sublimation, mass are spoken of as *latent* heats.

If a quantity of heat Q is supplied to a liquid at its boiling point, the mass of liquid which evaporates is

$$m = \frac{Q}{L_v}.$$

The same mass of liquid condenses when a quantity of heat Q is withdrawn from a vapor in contact with its liquid.

Heats of vaporization of liquids are of the order of magnitude of 10 times greater than heats of fusion. As is the case for heats of fusion, heats of vaporization do not change in any regular way from one substance to another.

In contrast to the phenomena observed at the melting point, both superheating and supercooling can occur at the boiling point. With careful heating, the temperature of a liquid can be increased several degrees above its boiling point before an explosive transformation into vapor takes place. Small sharp-edged bits of glass or metal are effective in preventing superheating of a liquid.

When a vapor is cooled below the boiling point temperature, it does not ordinarily condense to a liquid without some help. Condensation is greatly aided by the presence of so-called *condensation nuclei*, which may be small droplets of the liquid, particles of dust or smoke, or ions.

Since the vapor pressure varies rapidly with the temperature, the boiling point, in contrast to the melting point, is strongly pressure dependent. The table below can be considered to represent either the vapor pressure of water as a function on its temperature, or the boiling point as a function of external pressure.

TABLE 7-8

Temperature or boiling point	Vapor pressure or external pressure
0°C	4.58 mm Hg
10	9.21
20	17.54
30	31.8
40	55.3
50	92.5
60	149.4
70	234
80	355
90	526
100	760
200	11660
300	64432

The boiling points of solutions do not follow any simple laws, except when the solutions are highly dilute. Extremely small amounts of dissolved substances produce noticeable changes in the boiling point, which can be both increased and decreased.

This introductory section has presented the elementary foundations of the theory of heat. It has dealt almost exclusively with a clarification of basic concepts, not with the physical significance of the phenomena described. In the next chapter we shall be concerned with an interesting and well-founded theory by means of which these phenomena can be understood.

CHAPTER 8

KINETIC THEORY

ONE goal of kinetic theory is to provide a deeper understanding of at least some of the thermodynamic properties of matter. The basic postulates of kinetic theory are the molecular structure of matter and the principle of molecular chaos. In order to treat the problem quantitatively, it is necessary to make various assumptions, most of which have arisen from some of the early concepts of the structure of matter, or idealizations of that structure. In this sense, kinetic theory is quite inferior to thermodynamics, which is based on only a very few general principles. However, kinetic theory offers us something which thermodynamics lacks, due to the very nature of its formulation, and that is an insight into the behavior of matter on a microscopic level.

To explain the behavior of gases at low pressures, it is sufficient to ascribe to their molecules only a few very general properties. The more closely the molecules are packed, the more specific must our assumptions be. Eventually, in order to explain the properties of liquids and solids, we shall have to take into account the structure of molecules and atoms themselves.

These considerations point the way in which our study of kinetic theory should proceed. We will consider first the kinetic theory of an ideal gas. From the discrepancies with experiment, the theory of real gases will follow naturally, and this will lead us to an understanding of the liquid and solid states

73. Thermal Motion

The concepts of temperature and quantity of heat were introduced earlier in a purely formal way. Both of these quantities have a physical interpretation from the molecular point of view. Temperature and quantity of heat can be clearly explained in terms of other concepts with which we are already familiar.

In terms of the molecular picture of matter, solids, liquids, and gases are all composed of *atoms* and *molecules*. In a crystalline *solid*, these basic elements are arranged in a regular space lattice, each element of the lattice having a definite equilibrium position. In non-crystalline solids, the atoms are fixed in position, but there is no regularity to the positions. In *liquids* and *gases*, on the other hand, the molecules and atoms are not fixed in position. Liquids and gases differ chiefly in the closeness of packing of their molecules. Liquids are only very slightly compressible, which indicates that their molecules are very close together. The large compressibility of gases, on the other hand, shows that their molecules are widely separated from one another.

When this crude picture of the structure of matter is examined more closely, we find that it is far from complete. Molecules and atoms are not at rest but are in a state of continuous motion called *thermal motion*. In general, this motion cannot be directly observed. However, a true representation of it exists in *Brownian motion*, which was discovered in 1827 by the botanist *Brown* in a microscopic study of a water suspension of plant pollen. Brown observed, under the microscope, that the pollen spores were in a state of continuous agitation, an observation which was one of the most significant in the history of natural science. He showed that this *Brownian motion* is shared by all microscopically visible particles in both liquids and gases.

Using a dark-field microscope and appropriate particles, such as those in a suspension of rutile (crystalline TiO_2), Brownian motion can easily be demonstrated in a most striking way. The particles execute a completely irregular zigzag motion without any apparent cause. Brownian motion gives us a rather direct encounter with molecules and atoms, for their thermal motion is the cause of this strange particle agitation. The phenomenon can be understood with the help of a simple model. Imagine a vessel containing a number of small steel balls which are moving about with random velocities. By means of a lens, a slightly out-of-focus image of the vessel is formed on a screen, so that the moving balls can scarcely be distinguished. If a somewhat larger wooden ball is now dropped in the vessel, we then have a model of Brownian motion. Driven to and fro by the invisible but nonetheless real moving steel balls, the wooden ball executes a completely irregular motion which *is* visible to us. The motion is irregular because the instantaneous force exerted on the wooden ball by the steel balls in their collisions with it is not zero and the direction of this resultant force is purely random.

Brownian motion provides us with our first and most direct evidence of molecular agitation. It appears impossible, at first glance, that this complex and very disordered motion can be predicted. However, if we assume that the molecules obey Newton's laws of motion, then we know that in principle it is possible to calculate their motions. In practice, it is impossible to calculate the motion of even one given molecule over a long period of time. In order to calculate the path of one molecule, one must calculate the motions of them all because of the very many collisions that take place. This would require the use of an electronic computer much larger than any yet conceived to handle even a small body of gas. What we can do is calculate the *average behavior* of a large number of molecules on the basis of a few simplifying assumptions. As we shall show in the following sections, this average behavior follows definite laws.

74. Kinetic Theory of an Ideal Gas

The gaseous state is the simplest to understand from the kinetic viewpoint. Daniel Bernoulli, more than two hundred years ago, concerned himself with the properties of gases. His chief work, *Hydrodynamica,* contains the earliest treatment of kinetic gas theory. Later work by Krönig (1856), Clausius (1857), Maxwell (1859), and above all Ludwig Boltzmann (1895) helped to put this theory on a firm foundation and bring it to maturity.

In beginning our discussion of the gaseous state, it is convenient to idealize certain properties of gases, thus arriving at the concept of an ideal gas. Ideal gases do not actually exist. However, all gases that are not readily condensed, such as He, H_2, N_2, O_2, etc., behave approximately like ideal gases at room temperature. We have already given the experimental criterion for an ideal gas, namely, that the quantity PV/T shall be constant for a constant amount of gas.

In order to construct our model of an ideal gas, we consider a vessel containing a very large number of identical molecules. The following conditions must be fulfilled by these molecules in order that the quantity PV/T shall remain constant.

1. The volume of a molecule is vanishingly small.

2. The molecules exert no forces of attraction or repulsion on one another, except at the instant of a collision.

3. A collision between two molecules, or between a molecule and the wall of the container, is perfectly elastic.

4. The principle of *molecular chaos* applies. The first three conditions characterize an ideal gas and make possible a simple calculation of its mechanical properties. The fourth requirement, the principle of molecular chaos, goes beyond and is unnecessary in the field of mechanics, but it is essential in thermodynamics. It is also an inherent feature of the kinetic theory of gases and we shall now consider it in more detail.

74-1 The Principle of Molecular Chaos

The irregular Browniai motion which is observed with particles of microscopic size leads to the conclusion that the motion of the molecules of a gas is also completely random. It is this complete disorder, as well as the consideration of very large numbers of the molecules, which simplifies the problem of calculating average values of the properties of a system to such an extent that a calculation is possible. It is therefore of great benefit to postulate a *principle* which states that the thermal motion in a state of thermal equilibrium is completely irregular. In particular, this principle asserts that at any instant of time there exist in the gas all possible velocities, with respect to both magnitude and direction. This is not to say that all magnitudes are equally probable; they are not (see Section 74-9). But, if the total number of molecules is large enough, there will be found some molecules in every interval of velocity magnitude. If the velocity vectors of all of the molecules of a gas are drawn from a common origin, then in a state of molecular chaos these vectors form a symmetrical "vector porcupine" (Fig. 8-1). The principle also states that the distribution of the molecules

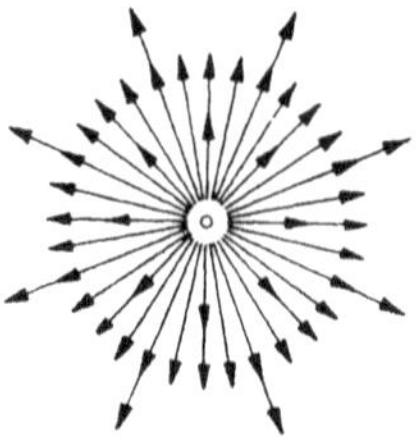

FIG. 8-1. The velocity vectors of the gas molecules in the case of molecular chaos.

in space is uniform (i.e. homogeneous density) and that there is no correlation between position and velocity.

The principle of molecular chaos requires not only that velocities of all *magnitudes* are possible, but also that if each velocity vector $\boldsymbol{v}$ is resolved into its three components v_x, v_y, and v_z, then there are exactly as many positive as negative components of a given magnitude. The *algebraic mean values* $\bar{v}_x$, $\bar{v}_y$, and $\bar{v}_z$, represented by a bar over the symbol, are therefore zero. The *mean values of the squares* of the components, however, are not zero and are the same for all three directions:

$$\overline{v_x^2} = \overline{v_y^2} = \overline{v_z^2},$$

where

$$\overline{v_x^2} = \frac{1}{n} \sum n_i v_{xi}^2, \text{ etc.},$$

n_i = number of molecules with x-velocity component v_{xi},
n = total number of molecules.

The principle of molecular chaos therefore implies a *uniform distribution* of molecular kinetic energy in the directions of the three axes, and hence in all directions in space. Since

$$v_x^2 + v_y^2 + v_z^2 = v^2,$$

it follows that

$$\overline{v_x^2} = \overline{v_y^2} = \overline{v_z^2} = \frac{1}{3}\overline{v^2}.$$

It is now easy to derive the properties of an ideal gas, making use of the above four conditions.

74-2 The Pressure Exerted by a Gas

Consider a gas in an enclosure whose walls are at rest. The gas exerts a pressure on the walls, which we shall refer to as the *gas pressure*. It is caused by the *thermal motion* of the molecules. A gas molecule striking the wall of the container rebounds elastically and thus exerts a force on the wall. The gas pressure P is the force per unit area exerted on the wall by the reflected molecules.

To calculate P, we must know the molecular velocities. Consider an arbitrary molecule i having the velocity $\boldsymbol{v}_i$, in a direction making an angle α_i with the normal to the wall (Fig. 8-2). The molecule approaches the wall

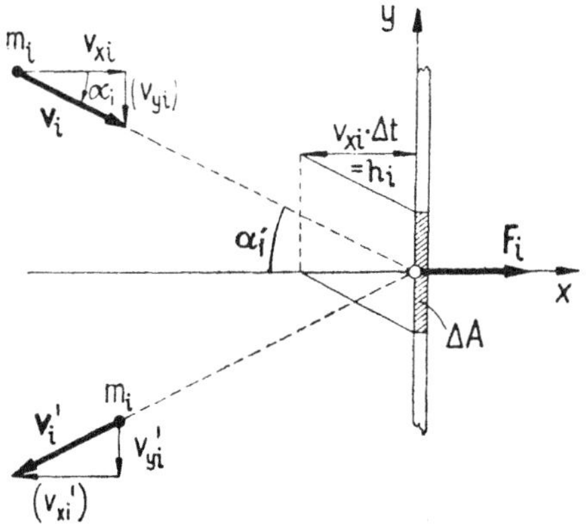

FIG. 8-2. Diagram for the derivation of the gas pressure on a wall.

with the velocity $v_{xi} = v_i \cos \alpha_i$, and is elastically reflected. It leaves the wall with a velocity v_i' whose x-component is $v'_{xi} = -v_{xi}$ and whose y-component is unaltered; thus the magnitude of $\boldsymbol{v}_i'$ is equal to that of $\boldsymbol{v}_i$. We now consider those molecular impacts with the wall which take place in an arbitrary time interval Δt. All molecules which strike the element of area ΔA at an angle α_i in the time interval Δt lie in a slant cylinder of base ΔA and of height $h_i = v_{xi}\Delta t$. Let z_i be the number of molecules in the cylinder having a velocity of magnitude v_i which makes an angle α_i with the normal to the wall. From the principle of molecular chaos, half of these molecules will strike the wall in the time interval Δt. (The other half are moving away from the wall.) Then from the momentum law, we have for the impulse $F\Delta t$ of the force exerted on the wall (m is the mass of a molecule)

$$\Delta p_i = F_i \Delta t = 2mv_{xi}\frac{z_i}{2}.$$

The force F_i is therefore

$$F_i = \frac{mv_{xi}z_i}{\Delta t}.$$

Let n_i be the number of molecules per unit volume whose velocity of magnitude v_i makes an angle α_i with the normal to the wall. Then

$$z_i = n_i \Delta A v_{xi} \Delta t$$

and

$$F_i = m \Delta A v_{xi}^2 n_i.$$

Summing over all molecules i and their corresponding velocities, we have

$$F = \sum F_i = m \Delta A \sum v_{xi}^2 n_i = m \Delta A n \overline{v_x^2},$$

where n is the total number of molecules per unit volume. But from the principle of molecular chaos

$$\overline{v_x^2} = \frac{1}{3} \overline{v^2}.$$

The pressure P is therefore

$$P = \frac{F}{\Delta A} = mn\overline{v_x^2} = \frac{1}{3} nm\overline{v^2} = \frac{2}{3} n \frac{m\overline{v^2}}{2},$$

that is,

$$P = \frac{2}{3} n \frac{m\overline{v^2}}{2}. \tag{74-1}$$

Kinetic theory therefore leads to the conclusion that *the gas pressure P equals 2/3 of the mean molecular kinetic energy per unit volume.* This relation is the starting point of all further kinetic considerations.

Since the different types of molecules in a mixture of ideal gases act independently of one another, the total pressure P_t is equal to the sum of the partial pressures P_i:

$$P_t = \sum P_i = \frac{2}{3} \left\{ n_1 \frac{m_1 \overline{v_1^2}}{2} + n_2 \frac{m_2 \overline{v_2^2}}{2} + \ldots \right\}$$

$$= P_1 + P_2 + \ldots \text{ (Dalton's law).}$$

In Eqn. (74-1) for the gas pressure, the number of molecules per unit volume, n, can be replaced by the quotient of Avogadro's number N_0 and the molar volume V_M:

$$n = \frac{N_0}{V_M}.$$

Multiplying Eqn. (74-1) by V_M, and writing u^2 for the mean square velocity, we get

$$PV_M = \frac{2}{3} N_0 \frac{mu^2}{2}.$$

This *equation of state*, derived from kinetic theory, recalls the ideal gas equation of state, which is an expression for the product PV_M based on experiment:

$$PV_M = RT.$$

If we now equate the right sides of these equations, we obtain an expression for the temperature, a quantity which was introduced earlier on purely phenomenological grounds but which now acquires physical significance. We have

$$RT = \frac{2}{3} N_0 \frac{mu^2}{2},$$

an equation which relates *temperature* to *average molecular kinetic energy*, $mu^2/2$. The concept of temperature thus acquires physical meaning in terms of mechanical quantities. Its significance can also be expressed by the equation

$$\frac{mu^2}{2} = \frac{3}{2} \frac{R}{N_0} T = \frac{3}{2} kT.$$

The *universal constant* R/N_0 in this equation is called the *Boltzmann constant k*, in honor of Boltzmann. Its magnitude is

$$k = \frac{R}{N_0} = 1.38044 \times 10^{-16} \text{ erg/K}^\circ.$$

The relation

$$\frac{mu^2}{2} = \frac{3}{2} kT$$

leads to the interesting conclusion that the molecules of all ideal gases, regardless of their chemical composition, have the same average kinetic energy at the same temperature. (This equality of the mean kinetic energies is one facet of the *principle of equipartition of energy*. This aspect of the principle can be thought of as a corollary of a more general statement of the principle of equipartition, which is given at the end of this section.) For example, if the gases H_2, O_2 and CH_4 are at a pressure low enough for them to be considered ideal gases, then at the same temperature

$$\left(\frac{mu^2}{2}\right)_{H_2} = \left(\frac{mu^2}{2}\right)_{O_2} = \left(\frac{mu^2}{2}\right)_{CH_4} = \frac{3}{2} kT. \tag{74-2}$$

From Eqn. (74-2) we can compute the root mean square velocity u for each gas. Replacing m by M^*/N_0, we have

$$u = \sqrt{\left(\frac{3N_0kT}{M^*}\right)} = \sqrt{\left(\frac{3RT}{M^*}\right)}.$$

At a given temperature, the root mean square velocity u is greater, the smaller the molecular weight. For O_2 and H_2, at 17°C (290°K), the velocities are:

$$u_{O_2} = 475 \text{ m/sec},$$
$$u_{H_2} = 1894 \text{ m/sec}.$$

The mean kinetic energy of both types of molecule is

$$\frac{mu^2}{2} = \frac{3}{2}kT = 6.005 \times 10^{-14} \text{ erg}.$$

The total random molecular kinetic energy of one mole, at this temperature, is

$$E_k = N_0\frac{mu^2}{2} = \frac{3}{2}RT = 3.618 \times 10^{10} \text{ ergs} = 3618 \text{ joules}.$$

This is a sizeable amount of energy. However, this energy cannot be utilized directly because the molecular motions are completely random. It is not of the same economic value as if it were ordered.

An additional interesting conclusion is that the kinetic viewpoint makes the existence of an absolute zero of temperature understandable. When $T = 0°K$, then $mu^2/2 = 0$. This is certainly the smallest possible value the mean kinetic energy could have. This is the way an *absolute zero* of temperature is defined for an ideal gas.

The molecules of an ideal gas, which behave like the particles or mass points of mechanics, can have three independent motions in space, corresponding to the three directions in space. We say that they have three *degrees of freedom* with respect to their translatory motion. From the principle of molecular chaos, we found that the same amount of energy is associated with each degree of freedom, namely $\frac{1}{2}kT$. This distribution of energy also applies to the more complicated motions of molecules which are not mass points and which have more than three degrees of freedom. This uniform energy distribution among all the degrees of freedom of a molecule is also part of the *principle of equipartition*. A more general statement of the principle of equipartition of energy is that the energy of a system is distributed uniformly among all the degrees of freedom of the system.† Many experimental results can be understood in the light of this principle and take on a new significance.

74-3 The Molar Heat Capacity at Constant Volume C_v

When the temperature of n moles of an ideal gas is increased by ΔT at constant volume, a quantity of heat $\Delta Q = nC_v\Delta T$ must be supplied to the gas. Since the volume is constant, no expansion work is done by the gas. Furthermore, no potential energy can be accumulated in an ideal gas since its molecules exert no forces on one another. All of the heat supplied to

† As is shown in texts on kinetic theory, the uniform distribution of energy occurs only when the energy of the system is quadratically dependent on position and velocity. We shall only consider systems with this type of energy dependence.

the gas is therefore utilized to increase the kinetic energy of its molecules, and C_v is equal to the increase in molecular kinetic energy per mole when the temperature increases by 1 K°.

(a) *Monatomic gases like He, Ne, A.* The molecules of a monatomic gas have three translational degrees of freedom. Hence

$$\Delta Q = nC_v\Delta T = nN_0\frac{3}{2}k\Delta T,$$

$$C_v = \frac{3}{2}N_0k = \frac{3}{2}R.$$

We find by experiment, for monatomic gases,

$$C_v = 2.98 \text{ cal/mole} \cdot \text{K}^\circ,$$

from which it follows that

$$R = \frac{2}{3}C_v = 1.99 \text{ cal/mole} \cdot \text{K}^\circ.$$

On the other hand, the value of R can be obtained from the gas equation:

$$R = \frac{P_0V_0}{T_0} = 8.31696 \text{ joules/mole} \cdot \text{K}^\circ.$$

Equating these two values of R gives the so-called *mechanical equivalent of heat*, that is, the value of a calorie in mechanical energy units:

$$1 \text{ calorie} = 4.185 \text{ joules}.$$

(b) *Diatomic gases such as H_2, N_2, O_2, CO, NO.* For so-called *dumbbell* molecules such as these, whose atoms are tightly bound to each other, there exists the possibility of rotation about the center of mass in addition to the translatory motion of the center of mass with its corresponding three degrees of freedom. It is to be expected that molecular collisions would produce such rotation. The principle of equipartition requires that the average energy associated with each rotational degree of freedom is also $\frac{1}{2}kT$. In view of the rod-like structure of these molecules, it is evident that in a collision only two of the principal axes of inertia can actually acquire energy, since the moment of inertia about the axis of the dumbbell is effectively zero. Figure 8-3 is a schematic diagram of a dumbbell molecule. Every such molecule has five degrees of freedom and its mean energy is $5kT/2$.

The amount of energy necessary to raise the temperature of n moles of diatomic gas by ΔT is

$$\Delta Q = nC_v\Delta T = nN_0\frac{5}{2}k\Delta T,$$

and hence for diatomic ideal gases,

$$C_v = \frac{5}{2}N_0k = \frac{5}{2}R = 4.97 \text{ cal/mole} \cdot \text{K}^\circ.$$

(Note that strictly speaking, a diatomic gas cannot be ideal, since the molecules are not mass points if they can rotate. However, if the gas is very rarified so that the molecular volume is negligible, we can still consider a diatomic gas to be ideal.) Experimentally, we find for rarified diatomic gases

$$C_v \cong 4.97 \text{ cal/mole} \cdot \text{K}^\circ,$$

in agreement with the theory.

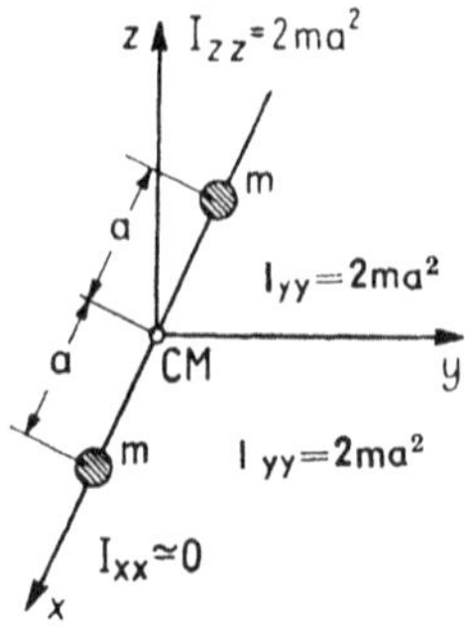

FIG. 8-3. Moments of inertia of a dumbell-shaped gas molecule.

74-4 The Molar Heat Capacity at Constant Pressure C_p

When the temperature of an ideal gas is raised at constant pressure, the energy supplied to the gas can be divided into two parts:

1. The mean kinetic energy of the molecules of the gas must be increased. This requires a quantity of heat equal to $nC_v\Delta T$.

2. If the pressure of the gas is kept constant, it expands and does work against the external pressure. A lowering of the temperature during the expansion, corresponding to a decrease in the molecular kinetic energy could account for the work expended, but since the temperature actually increases, during the expansion, an additional amount of heat must be supplied to take care of the work done. For gases, an increase of temperature at constant pressure is always accompanied by an increase in volume, and the molar heat capacity at constant pressure C_p is always greater than that

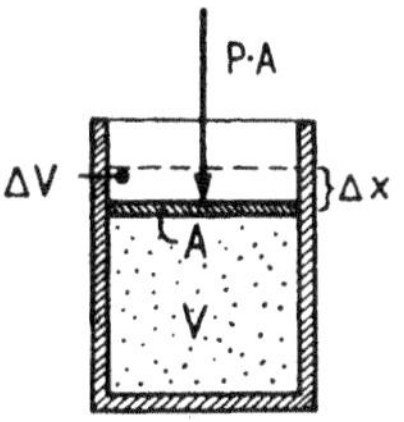

FIG. 8-4. Diagram for the calculation of $C_p - C_v$.

at constant volume C_v. The difference between C_p and C_v is easily computed for an ideal gas, by calculating the work done (Fig. 8-4),

$$W = PA\Delta x = P\Delta V,$$

where ΔV is the increase in volume for a temperature rise ΔT of 1°C. From the ideal gas equation of state,

$$\Delta V = \frac{nR}{P}\Delta T,$$

$$W = P\Delta V = nR\Delta T.$$

Therefore

$$nC_p\Delta T = nC_v\Delta T + W = nC_v\Delta T + nR\Delta T,$$

or,

$$C_p - C_v = R.$$

The physician Julius Robert Mayer, in 1842, was the first to calculate the mechanical equivalent of heat, making use of this equation. It has been verified experimentally for many monatomic and diatomic gases.

Monatomic cases: $C_v = 2.98$ cal/mole K°
$C_p = 4.97$ cal/mole K°

Diatomic gases: $C_v = 4{\cdot}97$ cal/mole K°
$C_p = 6.96$ cal/mole K°

The values of molar heat capacity calculated from kinetic theory are limiting values approached by real gases at high temperatures. With decreasing temperature, the heat capacities of real gases decrease. This decrease is already evident at room temperature for H_2, O_2, N_2, and CO, as the following data show:

Ideal gases (assuming there could be a diatomic ideal gas):

$$C_p = 6.96 \text{ cal/mole K°}$$

Experimental value at 288°K for

H_2: $C_p = 6.83$ cal/mole K°
O_2: $C_p = 6.96$ cal/mole K°
N_2: $C_p = 6.94$ cal/mole K°
CO: $C_p = 6.94$ cal/mole K°

Readily condensible gases of the dumbbell type have heat capacities in excess of the theoretical value. The measured values at 288°K for HCl and Cl_2 are

HCl: $C_p = 7.07$ cal/mole K°
Cl_2: $C_p = 8.16$ cal/mole K°

Kinetic theory provides an explanation of these facts. For diatomic and polyatomic molecules, a motion of oscillation is possible in addition to translation and rotation. This increases the number of degrees of freedom and, according to the principle of equipartition, increases the mean energy. The experimental values of C_v for H_2, O_2, N_2 and CO show that at room temperature these oscillations are not yet excited. On the other hand, the molecules HCl and Cl_2 have an appreciable amount of vibrational energy at 288°K. The binding forces in these molecules must therefore be weaker than for the molecules H_2, N_2, and O_2. The principle of equipartition, which is only a limiting principle, can provide no information regarding the extent to which the vibratory motion is excited. The problem lies in the field of quantum mechanics.

The measurements made by Eucken of the heat capacity of hydrogen at very low temperatures show that as the temperature decreases not only the vibrational, but also the rotational degrees of freedom gradually become inactive. The hydrogen molecule, H_2, first begins to acquire an appreciable rotational kinetic energy at about 60°K, and the energy becomes fully developed at about 300°K. Below 60°K the heat capacity is constant and equal to that of a monatomic gas (Fig. 7-8).

The molar heat capacities of complex molecules do not follow any simple laws. This is to be expected from kinetic theory, since the possible number of degrees of freedom increases rapidly with the number of atoms in the molecule. The table below summarizes the number of degrees of freedom f of various molecules (n represents the number of atoms in the molecule).

TABLE 8-1

	f (translation)	f (rotation)	f (vibration)	f (total)
Monatomic gases	3	0	0	3
Linear molecules	3	2	$3n - 5$	$3n$
Nonlinear molecules	3	3	$3n - 6$	$3n$

The total number of degrees of freedom is therefore always equal to $3n$. In general, however, not all of these are fully excited and the simple theory above can give only an upper limit for C_v. An average energy of $\frac{1}{2}kT$ is assigned to each *translational* and *rotational* degree of freedom. The mean energy of a *vibrational* degree of freedom, however, is kT, since in vibrational motion the molecule has potential as well as kinetic energy.

74-5 The Molar Heat Capacity C_v of Solids

The molecules or atoms (ions) of a solid oscillate about fixed equilibrium positions called lattice points. They have three vibrational degrees of freedom. To raise the temperature of $N = nN_0$ lattice points of a solid body by an amount ΔT, assuming all degrees of freedom are fully excited, requires a quantity of heat given by

$$\Delta Q = nC_v\Delta T = nN_0 \cdot 3k\Delta T = 3nR\Delta T$$

from which

$$C_v = 3R.$$

Kinetic theory therefore provides an explanation of the Dulong–Petit law (p. 332). The energy $3R$ corresponds to the molar heat capacity C_v, since only when the volume is constant is all of the added energy available for increasing the thermal motion of the lattice points. At constant pressure, work is done against the external pressure and against the forces of attraction between the lattice points as the distance between them increases (thus storing potential energy in the lattice). Both of these quantities of work are different for different materials, so that the difference $C_p - C_v$ for solids does not obey any simple law [see Eqn. (93-15) on p. 484]. The decrease of the molar heat capacity C_v of solids with decreasing temperature cannot be accounted for by the classical kinetic theory; its explanation must also be sought on quantum-mechanical grounds (Debye's T^3 law).

74-6 The Isothermal Atmosphere

The composition of air at the earth's surface is determined by the partial pressures P_i of its components. For an idealized atmosphere composed of ideal gases at uniform temperature, the decrease of pressure with elevation is easily computed. For this purpose we consider a column of air of constant cross-section (Fig. 8-5). In the equilibrium state, the pressure

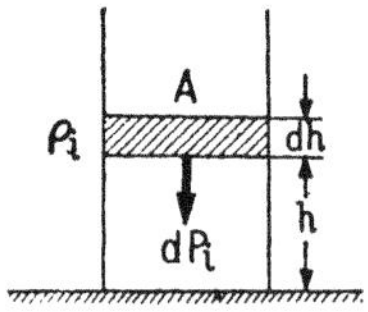

FIG. 8-5. Derivation of the pressure in an isothermal atmosphere.

at any elevation is produced solely by the weight of the gas above it. If the variation of g with elevation is neglected, we can then make the following assertions:

The weight of the ith component in a volume element of height dh increases the force on the lower cross-section by

$$dw_i = Adh\varrho_i g.$$

The pressure at an elevation h is greater than that at an elevation $h + dh$, that is, an *increase* of height dh results in a *decrease* of pressure dP_i, and hence

$$dP_i = -\frac{dw_i}{A} = -\varrho_i g dh.$$

The density ϱ_i of the ith component is given by the ideal gas equation (M_i is the molar mass of the ith component)

$$\varrho_i = \frac{M_i}{V_M} = \frac{M_i P_i}{RT}.$$

Hence ϱ_i is not constant, but is proportional to the pressure P_i. Since gases are readily compressed, it follows that the pressure change dP_i, for a given change in height dh, becomes smaller as the pressure P_i decreases. That is, the pressure does not decrease linearly with the height as it does in an incompressible liquid.

From the equations above, we have

$$dP_i = -\frac{M_i P_i}{RT} g dh,$$

and

$$\frac{dP_i}{P_i} = -\frac{M_i g}{RT} dh.$$

Integration of this equation gives for the pressure of the ith component as a function of height,

$$P_i = P_{i,0} \exp\left[-\frac{M_i g}{RT} h\right],$$

where $P_{i,0}$ is the pressure of this component at the earth's surface. In an isothermal atmosphere, then, the partial pressure decreases exponentially with height.

From Dalton's law, the total gas pressure P_t equals the sum of the partial pressures P_i:

$$P_t = \sum P_i.$$

The total pressure in an isothermal atmosphere, at a height h, is therefore the sum of a number of partial pressures each of which decreases exponentially with h. The exponent also contains the molar mass M_i as a factor. The greater the value of M_i, the more rapidly does the partial pressure decrease with increasing height. As a consequence, the composition of the atmosphere changes with elevation and only the lighter gases are found at high altitudes. There is, however, no sharp upper boundary to the atmosphere; the density of each component decreases asymptotically to zero with increasing height.

In this discussion of the atmosphere kinetic theory was only indirectly involved. However, the thermal motion of the molecules is a determining factor in the origin of our atmosphere. If it were not for this thermal motion, the molecules of the air would fall to the earth's surface under the influence of their weight and would form a layer of air only about 30 feet thick. In the absence of gravitational forces, on the other hand, the molecules would fly off into empty space as a result of their thermal motion, since the atmosphere is not enclosed. They are prevented from doing this only by their weight. The stability of the atmosphere is the result of these two opposing influences.

74-7 Determination of Avogadro's Number from Sedimentation Equilibrium (Perrin, 1912)

In an isothermal atmosphere (T = const) all molecular species have the same average kinetic energy. As a consequence, the decrease of partial pressure with height depends only on the quantity $M_i gh$:

$$P_i = P_{i,0} e^{-\frac{M_i gh}{RT}}.$$

Since $E_k = m_i u_i^2/2 = 3kT/2$, it follows that

$$\frac{M_i gh}{RT} = \frac{N_0 m_i gh}{N_0 kT} = \frac{m_i gh}{kT} = \frac{E_p}{\frac{2}{3}E_k}.$$

Thus the exponent in the above equation can be considered as the ratio of the potential energy to the kinetic energy, for molecules of type i. Because of their greater potential energy, the partial pressure of heavy particles decreases with increasing altitude more rapidly than that of light particles, a feature that has already been pointed out in the preceding section. The formulation above, however, expresses this behaviour in a more graphic manner.

At constant temperature, and hence at constant kinetic energy, the partial pressure P_i depends only on the number of molecules per unit volume, n_i. From the discussion on p. 345, we see that

$$P_i = \tfrac{2}{3} n_i E_k.$$

The ratio of the partial pressure at a height h to that at the earth's surface is equal to the corresponding ratio of the number of molecules per unit volume, and from the equations above

$$\frac{P_i}{P_{i,0}} = \frac{n_i}{n_{i,0}} = e^{-\frac{N_0 m_i gh}{RT}}.$$

This same equation also applies to the distribution with height of particles suspended in a liquid, if the particles are of uniform mass and radius and the concentration of particles is not too great. By measuring the number of suspended particles per unit volume as a function of height in the liquid, and applying this equation, J. Perrin made one of the earliest determinations of Avogadro's number N_0 in 1912.† He obtained particles of uniform size

† For further details of this interesting experiment see *Textbook of Physical Chemistry*, Samuel Glasstone, 2nd ed., p. 257, D. Van Nostrand Co., Inc., New York.

by the process of fractional centrifugation. It was necessary to measure the mass of the particles very precisely, and to count the numbers at different heights by observing them with a microscope. Table 8-2 gives the typical results of two series of measurements of the number of particles per unit volume at different heights.

TABLE 8-2

h	$5\,\mu$	$35\,\mu$	$65\,\mu$	$95\,\mu$	
n_1	100	47	22.6	12	Particles per unit volume (o)
n_2	100	48	23	11.2	Particles per unit volume (x)

$(1\,\mu = 10^{-4}\text{ cm} = 10^{-6}\text{ m})$

With the exception of the value $n_1 = 12$, the measured values show an exponential dependence on h (Fig. 8-6). Perrin's measurements of N_0 are given in Table 8-3.

TABLE 8-3

Material	Radius of particle	Mass of particle	N_0
Gamboge (gum resin) in water	$0.5\,\mu$	6.06×10^{-13} gm	8.0×10^{23}
Mastic in water	$5.50\,\mu$	7.5×10^{-13} gm	7.8×10^{23}
Mastic in water	$0.52\,\mu$	6.5×10^{-13} gm	7.25×10^{23}
Gamboge in glycerin	$0.385\,\mu$	2.9×10^{-13} gm	6.4×10^{23}
Particularly homogeneous particles in water	$0.367\,\mu$	2.46×10^{-13} gm	6.88×10^{23}

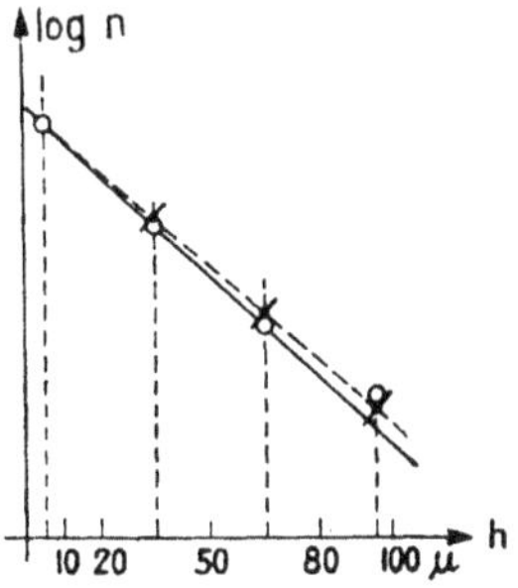

FIG. 8-6. A semilogarithmic plot showing the exponential density-distribution of sedimentary particles.

Particles of mass 10^{-13} gm correspond to a gas of molar mass $M = 10^{10}$ gm/mol. The experiments have been repeated in recent times by Westgren, using colloidal suspensions of gold and selenium, with the considerably more accurate result of $N_0 = 6.05 \times 10^{23}$ molecules per mole. The best value, based on the results of all methods of measurement, is 6.023×10^{23}.

74-8 Temperature Change of an Ideal Gas in Expansion and Compression

When a gas expands against a moving piston, its temperature decreases. When the gas is compressed, its temperature rises. To understand the molecular nature of these processes, let us consider an ideal gas in a cylinder provided with a movable piston. We shall assume that the walls are nonconductors of heat, and are of negligible heat capacity. Let P_0 and T_0 represent the initial pressure and temperature. If the piston is now moved inward, the molecules of the gas make collisions with a moving wall. This type of collision process was studied in Section 40-2, with the following results: Take the x-axis in the direction of motion of the piston. When a molecule collides with the piston it experiences an increase in the x-component of its velocity and hence in its kinetic energy. These facts are expressed by the following equations:

$$u'_x = u \cos \alpha + 2v,$$

$$\frac{1}{2} mu'^2 = \frac{1}{2} mu^2 \left[1 + \frac{4v(u \cos \alpha + v)}{u^2}\right],$$

where u is the velocity of the molecule before collision, u' is its velocity after collision, and v is the velocity of the piston. The increase in energy is provided by the external work done on the moving piston. The reflected molecules carry their excess energy into the interior of the gas, increase the mean translational kinetic energy of the other molecules by collisions, and hence raise the temperature of the gas. If the molecules are polyatomic, their rotational and vibrational energies also increase. For a given compression, the temperature rise in a monatomic gas is greater than in one whose molecules are diatomic or polyatomic, by an amount depending on the ratio of the number of rotational and vibrational degrees of freedom to the number of translational degrees of freedom. The temperature rise is easily computed for monatomic and diatomic gases.

(a) Monatomic gases:

$$P\Delta V = nN_0 \tfrac{3}{2} k\Delta T_1,$$

$$\Delta T_1 = \frac{2}{3} \frac{P\Delta V}{nR}.$$

(b) Diatomic gases:

$$P\Delta V = nN_0 \tfrac{5}{2} k\Delta T_2,$$

$$\Delta T_2 = \frac{2}{5} \frac{P\Delta V}{nR}.$$

It is easy to see that in an expansion process the kinetic energies of the molecules are reduced and the temperature of the gas decreases.

One further point should be noted. Immediately in front of the piston, the state of complete molecular chaos is disturbed as a result of the collisions with the piston. In a compression process, the molecules coming from the interior of the gas and striking the piston all have a smaller velocity than those returning to the gas after colliding with the piston. The velocities of the molecules coming from the gas are uniformly distributed in direction. The velocities of the reflected molecules, on the other hand, are not randomly distributed because of the added x-velocity component, $2v$, which makes the velocities more parallel (Fig. 8-7). In this zone of the gas, the concepts of gas pressure and temperature lose their thermodynamic meaning. The average velocity, at a given point, is different in different directions. The same is true of the pressure and temperature. This state of partial molecular order, and the pressure and temperature variations associated with it, propagates through the gas with the velocity of sound and becomes dispersed through collisions with the molecules of the undisturbed gas. The relationships between pressure, volume, and temperature in the body of the gas will remain simple only if the piston velocity is much smaller than the velocity of sound. In particular, a measurement of the temperature of this truly thermo-"*dynamic*" system has a meaning only if the piston is moved so slowly that the pressure and temperature waves, which travel outward from the moving piston, have time to become completely dispersed throughout the gas before the piston has traveled a large fraction of the length of the cylinder.

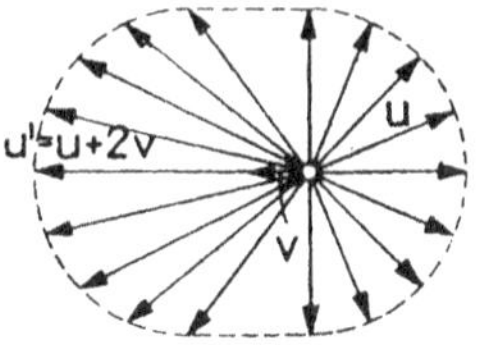

FIG. 8-7. Deviation from molecular chaos of the velocity distribution in the boundary zone in front of the moving piston.

74-9 The Maxwell Velocity Distribution Law

In the preceding discussion we have assigned to each molecule a *root mean square velocity u*, of such magnitude that in spite of our neglect of the distribution of velocities over all possible values, the total kinetic energy of the gas was unaltered. The quantity $\frac{1}{2}mu^2$ corresponded to the mean kinetic energy per molecule, and the velocity u was a very useful quantity in simplifying our considerations. The actual velocity distribution of the molecules of a gas at a given temperature is a much more complicated matter. The magnitudes of the molecular velocities range from zero to very large values. The *Maxwell velocity distribution law* gives the frequency of occurrence of a velocity of any specified magnitude. It is closely related to the *Gaussian law of errors*. This latter law describes the way in which a large number of measured values of a given quantity are distributed, when these values are affected by purely random disturbing influences. (In fact, the definition of "purely random disturbances" is that the measured values do obey the Gauss law of errors.) As a result of such a disturbing influence, the measured values differ from one another and are grouped around a certain *mean value*. The fractional number of values which differ appreciably from the mean is smaller, the larger the number of individual measurements. For the average values of molecular quantities, the number of individuals is so large (10^{20} or more) that the overwhelming majority have values very close to the mean. Large deviations, however, can always occur.

(a) *Probability density and the Gauss error function.* To illustrate the concept of *probability density*, let us perform the following imaginary experiment.

Suppose that we measure very precisely the density ϱ of a gas in a number of very small volume elements, all of the same size. As a result of thermal motions, we would find different numbers of molecules in each of these volume elements and the density would fluctuate from one element to another. The majority of the measured values, however, would be grouped close to the mean value $\bar{\varrho}$, which is the value that would be obtained in a macroscopic measurement. The probability density W, which states the frequency with which the individual values of ϱ differ from the mean value $\bar{\varrho}$ by an amount $z = \varrho - \bar{\varrho}$ is given by the Gauss error function:

$$W(z) = \frac{1}{p\sqrt{(2\pi)}} e^{-z^2/2p^2},$$

where p is the root-mean-square deviation. That is

$$p^2 = \overline{z^2},$$

(see p. 361) and p is inversely proportional to the square root of the average number of molecules q in a volume element. When q is large the mean deviation is small, and the curve is steep. Figure 8-8 shows the shape of the curve $W(z)$ for $m = p/\bar{\varrho} = 1$ per cent (that is, $q = 10{,}000$ molecules in a volume element) and $m = 2$ per cent (2500 molecules in each volume element). The abscissa in Fig. 8-8 is $x = z/\bar{\varrho}$ and the ordinate is $y = W\bar{\varrho}$.

To understand fully the significance of this law, the concept of *probability* must be explained more precisely. The probability of an event is defined as the ratio of the number of instances in which the event actually occurs to the number of all instances in which it might possibly occur. This is often stated as follows: *The probability is the ratio of the number of actual events to the number of possible events.* In our example of density measurements we ask, what is the probability that a measured value of ϱ will differ from the mean value $\bar{\varrho}$ by an amount between z ($= \varrho - \bar{\varrho}$) and $z + dz$? That is, dz is an infinitesimal interval of the deviation being considered. *The probability of the event is proportional to this interval.* Let the number of cases falling

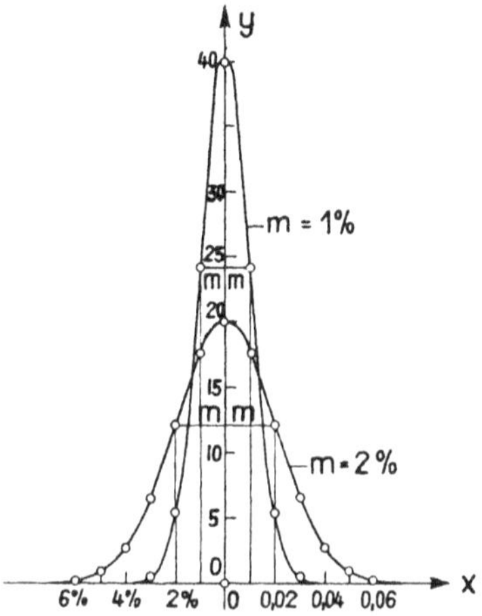

FIG. 8-8. Gaussian error curves.

within this range be dn (the number of actual events) and let n be the number of all possible events. The probability $W(z)\,dz$ that the deviation lies between z and $z + dz$ is then

$$W(z)dz = \frac{dn}{n}. \tag{74-1}$$

The quantity $W(z)$ itself is called the *probability density* and equals the ratio of the number of actual events *per unit interval* to the total number of possible events:

$$W(z) = \frac{dn/n}{dz} = \frac{1}{n}\frac{dn}{dz}.$$

This probability density, as stated above, is given by the *Gauss error function*:

$$W(z) = y(z) = \frac{1}{p\sqrt{(2\pi)}}e^{-z^2/2p^2}.$$

The number of actual events dn in an interval dz is, from Eqn. (74-1),

$$dn = nW(z)dz = \frac{n}{\sqrt{(2\pi)}p}e^{-z^2/2p^2}dz.$$

In the example of the distribution of values of the gas density, we obtain on setting $z = \varrho - \bar{\varrho}$ and $dz = d\varrho$,

$$dn = \frac{n}{p\sqrt{(2\pi)}}e^{-(\varrho-\bar{\varrho})^2/2p^2}\,d\varrho$$

and

$$W(\varrho - \bar{\varrho}) = \frac{1}{p\sqrt{(2\pi)}}e^{-(\varrho-\bar{\varrho})^2/2p^2}$$

We note that the integral of $W(\varrho - \bar{\varrho})\, d\varrho$ over all possible values of $\varrho (0 \leqq \varrho \leqq \infty)$ is unity; i.e. this simply is an alternative expression for $n^{-1} \int dn = 1$. We say that W is normalized to unity, corresponding to the fact that the maximum probability for the occurrence of any event is unity, while the total probability for the occurrence of all events is also unity. The probability density can be used to determine the mean value of a function of ϱ as follows:

$$\bar{f} = \int d\varrho f(\varrho)\, W(\varrho - \bar{\varrho}),$$

where $\bar{f}$ is the mean value of $f(\varrho)$.

In Fig. 8-9, the probability density $W(\varrho - \bar{\varrho}) = 1/n(dn/d\varrho)$ is plotted as a function of ϱ. If we consider a number of intervals all of the same breadth $d\varrho$, the ordinate of the curve gives the fractional frequency of occurrence, dn/n, of the corresponding values of density.

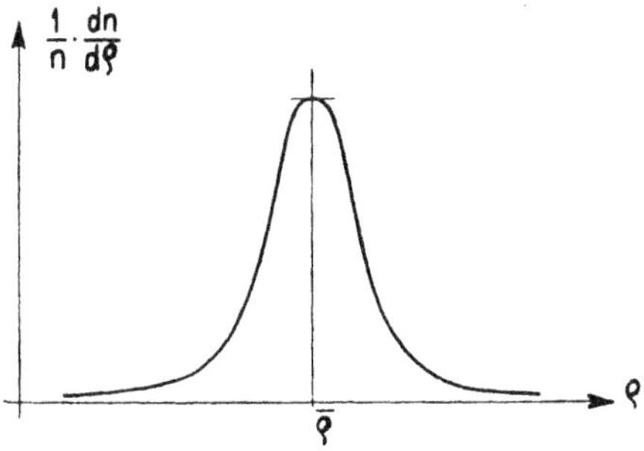

FIG. 8-9. A curve giving the frequency of various measured values of the density, ϱ.

(b) *Derivation of the Maxwell velocity distribution law.* This derivation is based on the assumption that the *magnitudes* of the three velocity components, v_x, v_y, and v_z are distributed according to the Gauss error function. From the principle of molecular chaos, to every positive value of a velocity component there corresponds an equal negative value. The average values $\overline{v_x}$, $\overline{v_y}$, and $\overline{v_z}$ are therefore all zero. The components are grouped around this mean value of zero according to the Gauss error curve:

$$W(v_x)\, dv_x = \frac{1}{p_x \sqrt{(2\pi)}} e^{-v_x^2/2p_x^2}\, dv_x,$$

$$W(v_y)\, dv_y = \frac{1}{p_y \sqrt{(2\pi)}} e^{-v_y^2/2p_y^2}\, dv_y,$$

$$W(v_z)\, dv_z = \frac{1}{p_z \sqrt{(2\pi)}} e^{-v_z^2/2p_z^2}\, dv_z.$$

To determine the as yet unknown quantity p_x, we compute the mean value of v_x^2, that is $\overline{v_x^2}$. This will not equal zero, since the squares of the velocities are all positive. It can be computed from the probability distribution,

$$\overline{v_x^2} = \frac{1}{p_x \sqrt{(2\pi)}} \int_{-\infty}^{\infty} v_x^2 e^{-v_x^2/2p_x^2}\, dv_x.$$

We find from a table of integrals that

$$\int_{-\infty}^{\infty} x^2 e^{-\alpha^2 x^2}\, dx = \frac{\sqrt{\pi}}{2\alpha^3}.$$

Therefore (letting $\alpha^2 = 1/(2p_x^2)$)

$$\overline{v_x^2} = \frac{1}{p_x \sqrt{(2\pi)}} \frac{\sqrt{\pi}\,(p_x \sqrt{2})^3}{2} = p_x^2.$$

The constant p_x is therefore the square root of the mean square value of the velocity component v_x:

$$p_x = \sqrt{(\overline{v_x^2})}.$$

In the same way, for the y- and z-components,

$$p_y = \sqrt{(\overline{v_y^2})}, \quad p_z = \sqrt{(\overline{v_z^2})}.$$

But as we showed in Section 74-1, $\overline{v_x^2} = \overline{v_y^2} = \overline{v_z^2}$, and hence $p_x = p_y = p_z = p$.

If μ represents the molecular mass, the principle of equipartition of energy requires that

$$\frac{\mu \overline{v_x^2}}{2} = \frac{\mu \overline{v_y^2}}{2} = \frac{\mu \overline{v_z^2}}{2} = \frac{1}{2} kT.$$

It follows that

$$p = \sqrt{\left(\frac{kT}{\mu}\right)}.$$

The distribution function of the velocity components is therefore completely determined:

$$W(v_x)\, dv_x = \frac{1}{\sqrt{2\pi}} \sqrt{\left(\frac{\mu}{kT}\right)} e^{-\mu v_x^2/2kT}\, dv_x,$$

with corresponding expressions for $W(v_y)$ and $W(v_z)$.

To obtain the Maxwell distribution law, we must find the distribution function for the magnitude of the velocity. According to probability theory, the probability of the simultaneous occurrence of the components v_x, v_y, and v_z equals the product of the individual probabilities:

$$W(v_x)\, W(v_y)\, W(v_z)\, dv_x dv_y dv_z =$$

$$\left(\frac{\mu}{2\pi kT}\right)^{3/2} e^{-\mu(v_x^2 + v_y^2 + v_z^2)/2kT}\, dv_x dv_y dv_z.$$

To obtain the probability that the magnitude of the velocity lies between v and $v + dv$, regardless of the direction of $\boldsymbol{v}$, this equation must be integrated, with a fixed value of $v = \sqrt{(v_x^2 + v_y^2 + v_z^2)}$, over all possible directions of $\boldsymbol{v}$. For this purpose it is more convenient to introduce in velocity space the polar coordinates v, θ, and φ. In terms of these coordinates, an infinitesimal volume element is $vd\theta \cdot v \sin\theta d\varphi \cdot dv$, and we then obtain, remembering that $v = \text{const}$,

$$W(v)\,dv = \left(\frac{\mu}{2\pi kT}\right)^{3/2} e^{-\mu v^2/2kT}\, v^2 dv \int_0^{2\pi} d\varphi \int_0^{\pi} \sin\theta d\theta .$$

Since

$$\int_0^{2\pi} d\varphi = 2\pi$$

and

$$\int_0^{\pi} \sin\theta d\theta = 2$$

we have

$$W(v)\,dv = \sqrt{\left(\frac{2\mu^3}{\pi k^3 T^3}\right)}\, v^2 e^{-\mu v^2/2kT} dv .$$

This is the *Maxwell velocity distribution law* (Fig. 8-10) and the new distribution function $W(v)$ is called the Maxwell velocity distribution function. It has a very different character from the Gauss error function. The quantity $W(v)\,dv$ equals the number molecules dn having velocities whose magnitudes lie between v and $v + dv$, divided by the total number of molecules n,

$$W(v)\,dv = \frac{dn}{n} .$$

The distribution function $W(v)$ vanishes at $v = 0$ and $v = \infty$, and passes through a maximum between these limits. The maximum of the distribution curve corresponds to the velocity that occurs most frequently, or the *most*

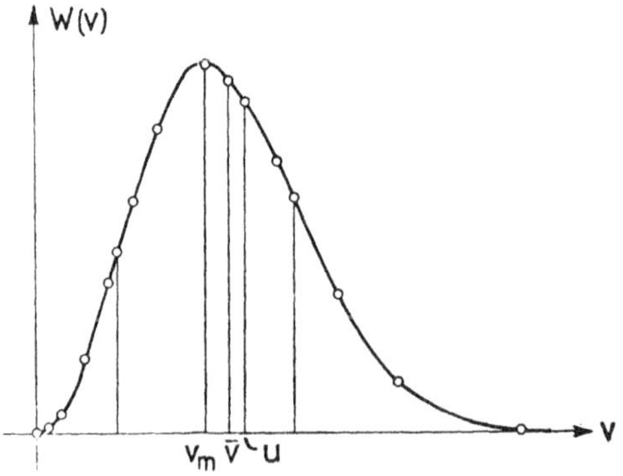

FIG. 8-10. The Maxwell velocity distribution.

probable velocity v_m. This is the velocity that results most frequently from the interplay of molecular collisions. Its magnitude is

$$v_m = \sqrt{\left(\frac{2kT}{\mu}\right)} = \sqrt{\left(\frac{2RT}{M}\right)}.$$

The graph of $W(v)$ falls off quite rapidly on both sides of the maximum. This belatedly justifies our simplifying assumption noted at the beginning of this sub-section that all molecules have the same root mean square velocity. The *root-mean-square* velocity $u = \sqrt{(\overline{v^2})}$ is, however, not equal to the most probable velocity v_m. From the equation

$$u^2 = \overline{v^2} = \overline{v_x^2} + \overline{v_y^2} + \overline{v_z^2} = \frac{3kT}{\mu},$$

we have as the relation between u and v_m:

$$u = v_m \sqrt{\left(\frac{3}{2}\right)} = 1.225\, v_m.$$

The *arithmetic mean* velocity $\bar{v}$ *is*

$$\bar{v} = \int_0^\infty v \frac{dn}{n} = \frac{2}{\sqrt{\pi}} v_m = 1.128\, v_m = \sqrt{\left(\frac{8kT}{\pi\mu}\right)} = \sqrt{\left(\frac{8RT}{\pi M}\right)}.$$

The arithmetic mean velocity is somewhat larger than the most probable velocity v_m, but is smaller than the root-mean-square velocity u.

$$v_m < \bar{v} < u.$$

The differences, however, are not large.

75. The van der Waals Theory of a Gas

75-1 Van der Waals Gases and Real Gases

Real gases do not obey exactly the equation of state of an ideal gas. The ideal gas is a concept which is only approximated by real gases. This is best demonstrated by examining the behaviour of the product PV_M as the pressure is increased at constant temperature. For an ideal gas, the product PV_M is equal to RT and therefore at constant temperature it has the same value at all pressures. A graph of PV_M vs. P would be a horizontal straight line. All real gases, however, show marked deviations from this condition as shown by the graphs of Figs. 8-11 and 8-12. These deviations must result from the fact that the assumptions which were made regarding the molecules of an ideal gas are only partially satisfied by the molecules of real gases. In 1873, van der Waals replaced the earlier assumptions by others which correspond more closely to reality. This new approach leads to an equation of state which can no longer be described as *universal*, since it contains two

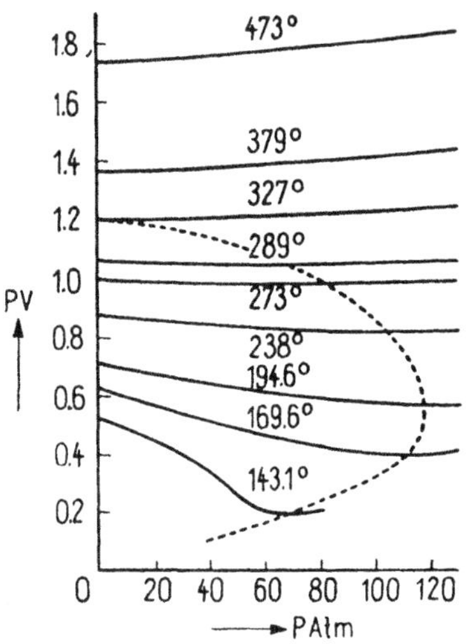

FIG. 8-11

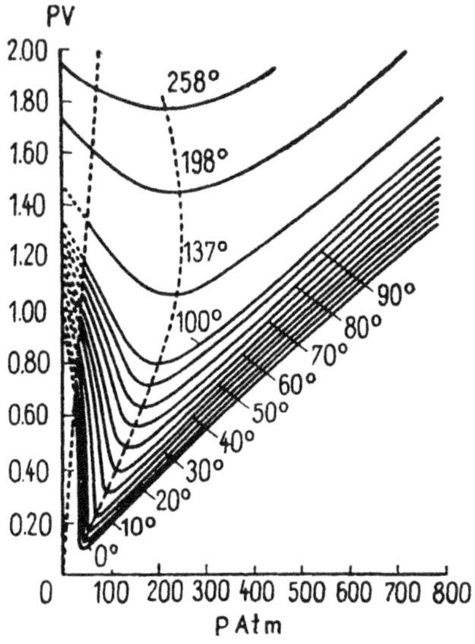

FIG. 8-12

FIG. 8-11. PV vs. P diagram for air.

FIG. 8-12. PV vs. P diagram for carbonic acid.

empirical constants which must be determined experimentally for every gas. It does, however, provide a theory of the condensation of a gas and enables us to understand qualitatively the nature of the liquid state. The van der Waals equation also does not describe the properties of real gases exactly, but it does furnish a more realistic picture of them. The values of the state variables P, V_M, and T, as computed from the van der Waals equation, agree only approximately with the experimental values for real gases. The agreement is best at low pressures and high temperatures. The reason for this is that the two empirical "constants" introduced by van der Waals turn out, on closer examination, not to be constants but to depend on both pressure and temperature. We should therefore distinguish between

(a) *van der Waals gases*, that is, gases which obey the van der Waals equation exactly and which do not exist in nature, and

(b) *real gases*, which behave like a van der Waals gas only to a first approximation.

75-2 The Covolume

The first refinement made by van der Waals to the theory of ideal gases was to recognize that the molecules of a gas occupy a *finite volume*. Molecules are of many different forms: spherical, linear, angled, plane, tetrahedral, pyramidal, etc. When we speak of the "volume" of such a molecule, we use the term only in a dynamical sense, not in a static, geometrical one. We may think of the molecules of a gas as being in a state of rapid rotation about their center of mass. They can then be considered to occupy a certain volume of spherical shape, independent of their actual geometric configuration at rest. The radius of this dynamically equivalent sphere will be represented by r_μ. The molecular volume V_μ is then given by

$$V_\mu = \frac{4\pi}{3} r_\mu^3.$$

This volume was considered by van der Waals as a constant for a given gas. This is true only as a first approximation. It is found experimentally that for real gases the radius r_μ and the volume V_μ decrease with increasing pressure and temperature.

The assumption that the molecules of a gas occupy a certain volume leads to the following modification of the ideal gas equation. When two molecules collide, their centers of mass can approach each other no closer than a distance $2r_\mu$. This means that the volume available for the motion of the molecules of a gas is smaller than the geometrical volume of the gas container. The obvious result is that the number of collisions per unit time, at a given temperature, is greater than would be the case if the molecules were geometrical points. The correction that must be made to the volume can be derived as follows:†

Consider a collision of two spherical molecules. In contrast to the "point" molecules of an ideal gas, the molecules we are now considering occupy space. Their centers are separated by at least a distance equal to twice the molecular radius. The center of one of the molecules is excluded from a spherical volume of radius $2r_\mu$, that is, a volume equal to eight times the molecular volume V_μ. However, when one takes a proper average over all the molecules, each molecule in a large body of gas is excluded from a volume of only half as much, or four times the molecular volume. The above considerations apply to *binary* collisions, i.e., collisions between only two molecules. These are by far the most frequent, and neglecting higher order collisions there results a reduction in volume, relative to that for point molecules, of $\frac{1}{2}\frac{4\pi}{3}(2r_\mu)^3$. The total *volume of exclusion* b, for one mole of a gas, is therefore

$$b = \frac{N_0}{2}\frac{4\pi}{3}(2r_\mu)^3 = 4N_0\frac{4\pi}{3}r_\mu^3.$$

This equals four times the proper volume of all the molecules in one mole and is called the *covolume*. With this correction, the ideal gas equation

† A more rigorous derivation will be found in Max Born, *Atomic Physics*, G. E. Stechert & Co., Sixth Edition, 1957, App. III, or A. Sommerfeld, *Thermodynamics and Statistical Mechanics*, Academic Press, 1956, Sec. 26A.

becomes

$$P(V_M - b) = RT.$$

Figure 8-13 shows an isotherm T = const for a gas obeying this equation. The existence of the covolume results in a pressure which is greater than that of an ideal gas.

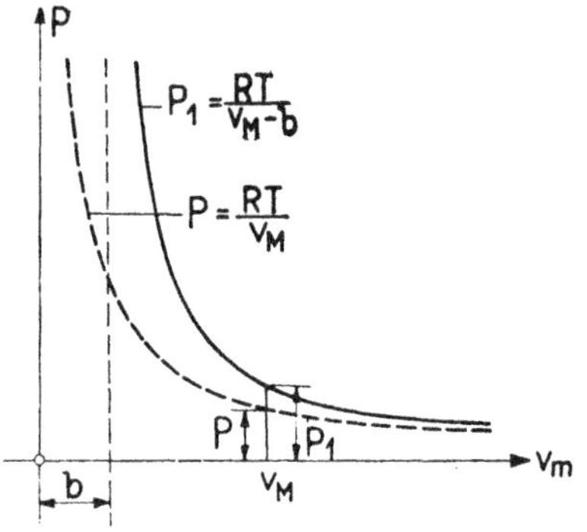

FIG. 8-13. Isotherms of a gas with and without the consideration of the covolume.

75-3 Internal Pressure. The van der Waals Equation of State

Real gases differ from ideal gases in another respect. The molecules of a real gas exert forces of attraction on one another, forces which decrease rapidly with increasing distance and are only appreciable at extremely small separations (short-range forces). A so-called *sphere of molecular attraction*, of radius r_a, can be assigned to each molecule, within which the forces are effective. These forces, called the *van der Waals cohesive forces*, depend on the structure of the molecule and require the introduction of a second empirical constant in the equation of state.

The introduction of the van der Waals force has the following important consequence: namely, that the molecules now have *potential* as well as kinetic energy. When the gas expands, work must be done not only against the external pressure but also against the mutual forces of attraction between the molecules. The potential energy of a pair of molecules increases with increasing distance between them, but since the cohesive force decreases very rapidly with increasing distance the increase of potential energy is significant only within the sphere of molecular attraction.

The great contribution of van der Waals was to discover a way by which these molecular cohesive forces could be taken into account in the equation of state. This can be done as follows. From the principle of molecular chaos, every molecule in the interior of a gas is surrounded uniformly by other molecules. The attractive forces exerted on a given molecule by its neighbors will, on the average, cancel one another, so that no resultant force is exerted on the molecule by the other molecules which surround it. For a molecule at the outer surface of the gas, however, the situation is very different. Let us think of the gas as bounded by a wall, which exerts no forces on the molecules. A molecule in the surface layer is surrounded by neighbors on one side only. Those within its sphere of molecular attraction exert on it an inward force, which by symmetry considerations is perpendicular to the boundary surface (Fig. 8-14). Hence there exists at the outer surface of the

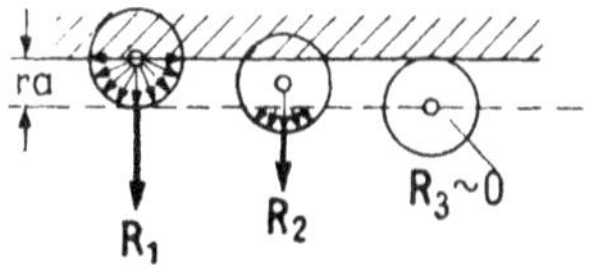

FIG. 8-14. The resultant force **R** acting on a surface molecule.

gas a *boundary layer* of thickness r_a, within which all molecules experience a resultant inward force. A molecule of kinetic energy E_k, approaching this zone from the interior of the gas, loses kinetic energy and gains an equal amount of potential energy when it enters the boundary layer. If it is brought to rest, the same forces accelerate it again toward the interior and it returns with its original kinetic energy. It follows that at a sufficiently low temperature every van der Waals gas (and every real gas also) has a boundary layer which is practically impenetrable to its molecules. This is particularly noticeable for very dense gases, that is, for liquids, and it is the reason for the existance of surface tension.

The force on a molecule in the boundary layer is obviously greater, the greater the number of molecules within its sphere of action. That is, the force is proportional to the density ϱ. These forces give rise to a pressure on the gas which can never be externally ascertained, and which is called the van der Waals *internal pressure* P_i. Since it is a force per unit area of the surface layer, produced by the forces of cohesion, it is greater the greater the number of molecules per unit area of the boundary layer, a number which is also proportional to the density. Therefore the pressure P_i is proportional to ϱ^2, or,

$$P_i = a_0 \varrho^2.$$

The proportionality factor a_0 depends on the structure of the molecules of the gas and hence is an empirical constant which is different for different gases. It is a measure of the strength of the cohesive forces. Since

$$\varrho = \frac{M}{V_M},$$

it follows that

$$P_i = \frac{a_0 M^2}{V_M^2} = \frac{a}{V_M^2},$$

where

$$a = a_0 M^2.$$

For a real gas the pressure P_i acts on the gas in addition to any external pressure P. The actual pressure on the gas is therefore $P + P_i$ and the gas equation takes the form

$$(P + P_i)(V_M - b) = RT.$$

Replacing P_i by a/V_M^2, we have finally

$$\left(P + \frac{a}{V_M^2}\right)(V_M - b) = RT.$$

This is the *van der Waals equation of state*, which can be used to describe the behavior of a real gas over much wider ranges of P, V, and T, than can the equation of state of an ideal gas. For very large molar volumes it reduces to the ideal gas equation. The covolume b and the internal pressure P_i then become negligibly small.

The empirical constants a and b are called the *van der Waals constants* of the gas. Their values can be determined from the differences between the values of the product PV_M for a real gas and for an ideal gas. As a first approximation, they can also be found from the experimentally determined critical constants of the gas (see p. 371). Since a/V_M^2 must represent a pressure, the unit of a in the cgs system is 1 dyne · $cm^4/mole^2$ and in the mks system it is 1 newton · $m^4/mole^2$. If pressures are expressed in atmospheres and volumes in cubic centimeters, the unit of a is 1 atm · $cm^6/mole^2$.

The constant b is a volume per mole. The unit of b in the cgs system is 1 cm^3/mole and in the mks system it is 1 m^3/mole.

TABLE 8-4

VAN DER WAALS CONSTANTS AND CRITICAL DATA†

Gas	$a\left(\frac{\text{atm} \cdot \text{cm}^6}{\text{mole}^2}\right)$	$b\left(\frac{\text{cm}^3}{\text{mole}}\right)$	P_c (atm)	T_c (°K)	$\varrho_c\left(\frac{\text{gm}}{\text{cm}^3}\right)$
He	0.3412×10^5	23.71	2.26	5.2	0.0693
Ne	2.107	17.10	25.86	44.4	0.4835
A	13.46	32.20	48.0	150.7	0.5308
H_2	2.444	26.62	12.80	33.2	0.0310
N_2	13.91	39.15	33.54	126.2	0.3110
O_2	13.61	31.88	50.14	154.8	0.41
CO	14.87	39.87	34.5	133.0	0.3010
NO	13.40	27.90	64	180	0.52
CO_2	35.96	42.69	72.85	304.2	0.468
SO_2	67.14	56.38	77.79	430.7	0.524
Cl_2	64.98	56.25	76.1	417.2	0.573
HCl	36.67	40.83	81.5	324.6	0.423
H_2S	44.34	42.89	88.9	373.6	0.349
H_2O	54.64	30.52	218.3	647.4	0.326

† Critical data from *American Institute of Physics Handbook*, McGraw-Hill, 1957.

(a) *Discussion of the van der Waals equation.* The equation

$$\left(P + \frac{a}{V_M^2}\right)(V_M - b) = RT$$

defines a *surface* in a coordinate system in which P, V_M, and T are plotted along the axes. To investigate its form we can intersect it by planes T = const, P = const, V_M = const.

The most important curves of intersection are the *isotherms*, T = const. They satisfy the equation

$$P = \frac{RT}{V_M - b} - \frac{a}{V_M^2}.$$

These are third order curves, for which only the first term contains the temperature T as a parameter. Figure 8-15, plotted in arbitrary units, shows schematically the variation of P with V_M at six different temperatures. From the diagram we can see that at large molar volumes and sufficiently high temperatures a van der Waals gas behaves approximately like an ideal gas. As the temperature is decreased the hyperbolas begin to develop a bulge in the region marked A, and as the temperature is further decreased this develops into an "oscillation". In this region of the diagram liquefaction takes place. If a gas is compressed isothermally at a temperature T_1, for which the isothermal shows this oscillation, the curve would require that the pressure should at first increase but that when a certain point is reached (point S_2 in Fig. 8-15) it should decrease in spite of a decrease in volume. Since the temperature is constant and the pressure P is proportional to $\frac{1}{3}$ nmu^2, this means that the number of molecules per unit volume decreases. This is what actually happens in that some of the gas molecules separate out by condensation. Actually, however, condensation sets in before point S_2 is reached. Figure 8-16 shows some experimentally determined isotherms for CO_2. The experimental points, indicated by crosses, show that the computed isotherms (full lines) agree very closely with the experimental values.

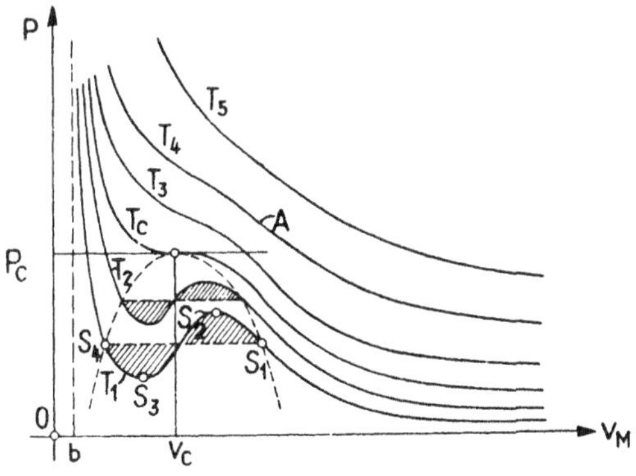

FIG. 8-15. The isotherms of a van der Waals gas.

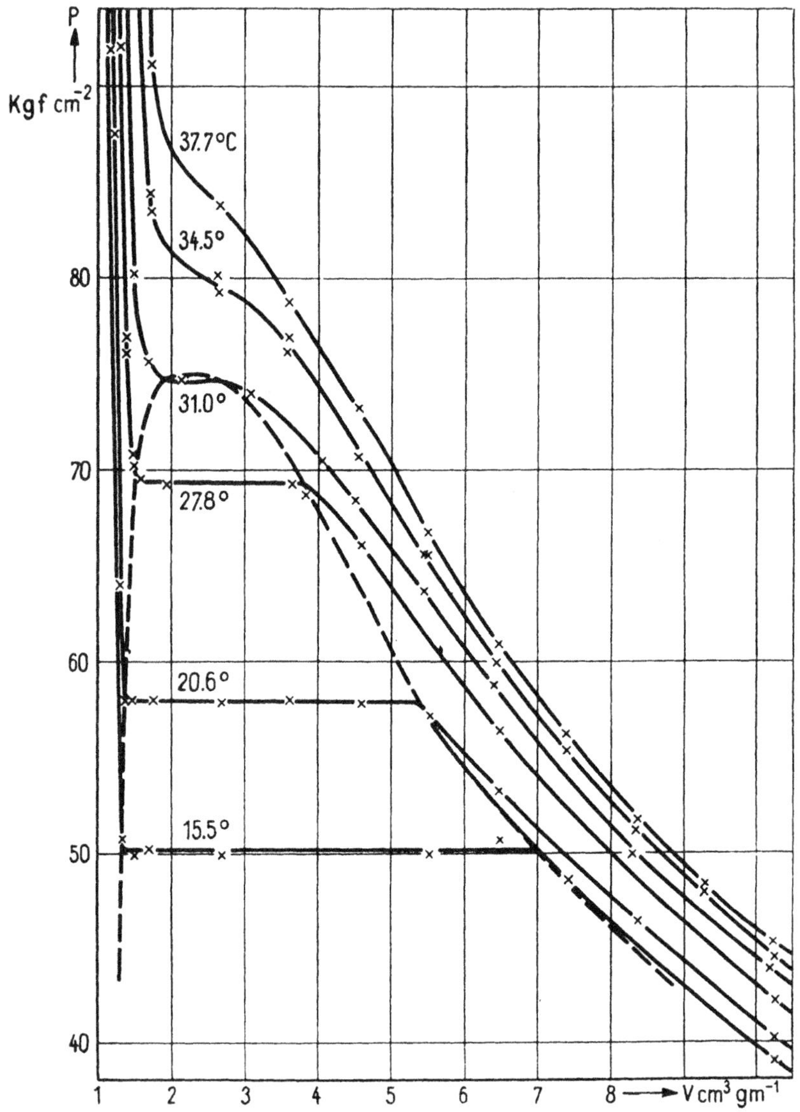

FIG. 8-16. Experimentally determined CO_2 isotherms for various temperatures.

The condensation of a gas to a liquid is related to the appearance of oscillations in the isothermal curves. These oscillations vanish when the temperature is raised to such a value that the isotherm has a point of inflection at which the tangent is horizontal. The temperature of this isotherm is called the *critical temperature*, T_c. At temperatures above T_c the gas can of course be compressed until its density equals that of the liquid. However, in this state it is impossible to make any distinction between the gas and liquid states. If the pressure is decreased at constant temperature, the gas molecules can again move freely throughout the entire volume. At temperatures below the critical temperature this is not possible. At the lower temperature the kinetic energies of the molecules are so small that only a few of them are able to penetrate through the boundary layer. A liquid-vapor *meniscus* or interface is formed, which prevents the molecules from leaving the liquid phase. That is, when the external pressure is decreased the molecules remain imprisoned in the liquid phase by the boundary layer, under a high van der Waals internal pressure (about 1200 atm ($T\alpha_p/\varkappa_T$), for water at 18°C) which cannot be observed externally. It is therefore the van der Waals forces of cohesion which actually cause a gas to condense. An increase of external pressure only creates the possibility by approaching the molecules. The boundary layers can be penetrated only by those few molecules which, as a results of the Maxwell velocity distribution, have the requisite amount of energy. It is these molecules that give rise to the phenomenon of evaporation.

The oscillations shown in Fig. 8-15 can be only partially verified experimentally. In the absence of condensation nuclei, a vapor can be compressed isothermally to a point in the neighborhood of the turning point S_2 of the curve. Ordinarily, however, condensation sets in suddenly and the pressure remains constant until all of the vapor is condensed. We shall show later (Section 90) with the help of the second law of thermodynamics that the horizontal line P = const must lie at such a pressure that the shaded areas in Fig. 8-15 are equal.

The *isochoric curves*, V_M = const, are straight lines, i.e. linear functions of T:

$$P = \frac{R}{V_M - b}T - \frac{a}{V_M^2}.$$

The *isobars*, P = const, are again third-order curves:

$$\frac{RT}{V_M - b} - \frac{a}{V_M^2} = \text{const}.$$

The isothermal curves are helpful in providing an insight into the process of condensation. The *critical isotherm* is of particular significance. Geometrically, it is identified by the property that it has a horizontal tangent at its point of inflection. This point of inflection is called the *critical point* of the substance, and the corresponding pressure, volume, and temperature are called the *critical constants*, P_c, V_c, and T_c. They can be computed in terms of a and b by setting both the first and second derivatives of the equation for the isotherm equal to zero:

$$\left.\frac{dP}{dV_M}\right|_T = -\frac{RT}{(V_M - b)^2} + \frac{2a}{V_M^3} = 0,$$

$$\left.\frac{d^2P}{dV_M^2}\right|_T = \frac{2RT}{(V_M - b)^3} - \frac{6a}{V_M^4} = 0.$$

From these equations the critical volume is

$$V_c = 3b.$$

The critical volume is therefore three times the covolume. Inserting this expression for V_c in either of the preceding equations, we get

$$T_c = \frac{8a}{27bR}.$$

The critical pressure can now be found by inserting V_c and T_c in the equation of the isotherm. This gives

$$P_c = \frac{a}{27b^2}.$$

The critical constants are therefore completely determined by the van der Waals constants a and b, and a and b can be calculated from any pair of critical constants. For example, we find in terms of V_c and P_c,

$$b = \frac{V_c}{3},$$

$$a = 27b^2 P_c = 3P_c V_c^2.$$

(b) *The law of corresponding states.* The van der Waals equation takes an interesting form when the empirical constants a and b, and the gas constant R, are expressed in terms of the critical constants. Introducing in the equation of state the values

$$b = \frac{V_c}{3}, \quad a = 3P_c V_c^2, \quad R = \frac{8P_c b}{T_c} = \frac{8P_c V_c}{3T_c},$$

we get

$$\left(P + \frac{3P_c V_c^2}{V_M^2}\right)\left(V_M - \frac{V_c}{3}\right) = \frac{8P_c V_c}{3T_c} T.$$

Dividing this equation by $P_c V_c/3$, and introducing as new variables the *reduced* pressure, volume, and temperature, defined as

$$P_r = \frac{P}{P_c}, \quad V_r = \frac{V_M}{V_c}, \quad T_r = \frac{T}{T_c},$$

we get the van der Waals equation in terms of the reduced variables,

$$\left(P_r + \frac{3}{V_r^2}\right)(3V_r - 1) = 8T_r.$$

Expressed in this form, all van der Waals gases obey the same equation of state, since the equation contains no empirical constants. If, for two different gases, the quantities P and V_M are the same multiples of their respective critical values, the temperature T is also the same multiple of the critical temperature. The gases are said to be in *corresponding states* (law of corresponding states). Since most cases obey the law of corresponding states fairly well, the critical constants of a gas become important for use in this application when there is little other data available about the gas.

(c) *The Boyle point of a van der Waals gas.* There exists for every van der Waals gas a certain temperature T_B at which, to a first approximation, the gas is ideal for pressures ranging from zero up to several atmospheres. At this temperature, the product PV_M, as a function of P, remains constant. The temperature T_B is called the *Boyle point* and it can be computed from the equation of state,

$$P = \frac{RT}{V_M - b} - \frac{a}{V_M^2}.$$

Let us write this in the form

$$\frac{PV_M}{RT} = \frac{V_M}{V_M - b} - \frac{a}{V_M RT}.$$

The first term on the right can now be expanded in a series by the binomial theorem, to give

$$\begin{aligned}\frac{PV_M}{RT} &= 1 + \frac{b}{V_M} + \frac{b^2}{V_M^2} + \cdots - \frac{a}{V_M RT} \\ &= 1 + \frac{1}{V_M}\left(b - \frac{a}{RT}\right) + \frac{b^2}{V_M^2} + \cdots\end{aligned}$$

At low pressure $V_M \gg b$, and hence, neglecting the small quantity b^2/V_M^2 and higher powers, a van der Waals gas will behave like an ideal gas when

$$b - \frac{a}{RT} = 0.$$

The *Boyle point* T_B is therefore

$$T_B = \frac{a}{Rb} = \frac{27}{8} T_c.$$

The following table gives the theoretical and some experimental values of T_B for a few gases:

TABLE 8-5

Gas	T_c	T_B (theoretical)	T_B (experimental)
He	5.2°K	17.6°K	
H_2	33.2	112.1	106°K
N_2	126.2	426	
O_2	154.8	522	423°K
CO_2	304.2	1027	
H_2O	647.4	2185	

75-4 Calculation of the Internal Pressure

To obtain a better idea of the magnitude of the internal pressure P_i, let us consider a few numerical examples. In those instances in which the internal pressure is less than 0.1 atm and the temperature of the gas is greater than 300°K, the molar volume that can be computed from the equation of state of an ideal gas pressure is 10 atm or more.

The internal pressure of a gas is of the order of one atmosphere, when

$$\frac{a}{V_M^2} \cong 1 \text{ atm}.$$

Since for most gases

$$a \cong 10^6 \frac{\text{atm} \cdot \text{cm}^6}{\text{mole}^2},$$

this condition is fulfilled when

$$V_m \cong 1000 \text{ cm}^3.$$

EXAMPLE. For oxygen at 20°C and 1 atm external pressure, the internal pressure is

$$P_i = \frac{a}{V_M^2} = 2.36 \times 10^{-3} \text{ atm} = 1.79 \text{ mm Hg}.$$

One application of the van der Waals equation is in calculating small corrections to quantities derived from experiment. In the Victor Meyer method of determining molecular weights a weighed quantity of material is vaporized in a closed vessel. The volume of air displaced by the vapor is measured in a gas burette attached to the vessel, where it has cooled from the temperature of the vaporization vessel to room temperature. The molar volume V_M of the vapor is usually computed from the ideal gas equation, since the chemist is often interested only in distinguishing between a simple or multiple state of molecular aggregation. The results obtained by this procedure are always about 2 per cent too high.

If the van der Waals internal pressure of the vapor is taken into account, the accuracy of the results is significantly increased. The van der Waals correction term is generally small enough that to a first approximation the molar volume computed from the ideal gas law can be used to calculate the internal pressure P_i. This pressure can then be imagined to become zero and correspondingly the gas expands a small amount according to the Boyle–Mariotte law. That is, if V is the actual measured volume displaced in the burette, and V^* the volume that would be occupied if the internal pressure were zero, then

$$V(P + P_i) = V^*P.$$

The volume V^* is in general about 2 per cent greater than V, and is the value which should be used for the most accurate results.

EXAMPLE. A mass $m = 0.3855$ gm of CCl_4 is vaporized at 100°C and causes a displacement in the measuring burette of a volume $V = 65.4 \text{ cm}^3$ at a pressure $P = 690$ mm Hg and at a temperature $t = 22°C$. The value of the van der Waals constant a for CCl_4 is 19.5×10^6 atm · cm^6/mole/mol^2, and the pressure in the vaporization vessel is $P_v = 720$ mm Hg.

At a temperature of 100°C and a pressure of 720 mm Hg, the molar volume of an ideal gas is 32,300 cm^3. The internal pressure is therefore

$$P_i = \frac{a}{V_M^2} = 1.87 \times 10^{-2} \text{ atm} = 14.2 \text{ mm Hg}.$$

The corrected volume V^* is

$$V^* = V\frac{P_v + P_i}{P_v} = 66.6 \text{ cm}^3.$$

The molar volume at 690 mm Hg and 22°C is $V_{M'} = 26{,}600$ cm^3. The molecular weight M^* is therefore

$$M^* = \frac{mV_{M'}}{V^*} = 154.0 \text{ gm/mole}.$$

Using the ideal gas equation, we would have found $M^* = 156.9$ gm/mole. It is evident from this correction that even in the Victor Meyer method of determining molecular weights the temperature of the evaporating chamber influences the result.

For substances in the liquid and solid states, the internal pressure cannot be computed from the equation $P_i = a/V_M^2$ since the van der Waals equation is not applicable to these states. The pressure can be found, however, from the second law of thermodynamics. We shall show in Section 93-2 that the internal pressure for the case of a homogeneous phase is given to a good approximation by the equation

$$P_i \cong \frac{T\alpha_p}{\varkappa_T},$$

where α_p is the isobaric coefficient of volume expansion and $\varkappa_T$ is the isothermal compressibility. This equation can be used to compute P_i for both the solid and the liquid states. The following example serves to show the great difference between this equation and the van der Waals value of P_i.

Example. (a) For water at 18°C,

$$V_M = 18 \text{ cm}^3$$

$$\alpha_p = 2.07 \times 10^{-4} (\text{C}°)^{-1}$$

$$\varkappa_T = 4.65 \times 10^{-5} \text{ atm}^{-1}.$$

The value of P_i computed from the second law is

$$P_i = \frac{291 \times 2.07 \times 10^{-4}}{4.9 \times 10^{-5}} = 1070 \text{ atm}.$$

The van der Waals equation gives

$$P_i = \frac{a}{V_M^2} = \frac{5.46 \times 10^6}{324} = 16{,}900 \text{ atm}.$$

(b) For mercury at 18°C

$$\alpha_p = 1.82 \times 10^{-4} \, (\mathrm{C}^\circ)^{-1}$$

$$\varkappa_T = 3.9 \times 10^{-6} \, \mathrm{atm}^{-1},$$

and

$$P_i \cong 13{,}600 \, \mathrm{atm}.$$

75-5 The Condensation of Gases

The condensation of a gas to form a liquid is brought about only indirectly by the van der Waals cohesive forces. Even a large mutual force of attraction cannot cause two molecules to stick together when they collide. The van der Waals forces only serve to increase the kinetic energies of the two particles as they approach each other. The binding energy decreases by the same amount. The collision of the molecules, traveling with this larger kinetic energy, is still completely elastic† and the molecules separate again after the collision in spite of the force of cohesion between them. Condensation can take place only when two additional conditions are simultaneously satisfied:

1. The average kinetic energy must be made sufficiently small by reducing the temperature below a critical value;
2. The density must be increased to a sufficiently large value.

Both of these conditions can be understood from the kinetic viewpoint. When the density is low, the collisions are still perfectly elastic even when the temperature is less than the critical temperature. In the attenuated vapor state, practically all collisions are binary. At higher densities, however, a new feature enters, namely that many of the molecules are linked to several neighbors by strong van der Waals forces. In the collision of a single molecule with such a group of molecules, the energy of the colliding molecule is transferred to all members of the group and shared among them. It is no longer available to the single molecule so that it can separate from the group again. The incident molecule can then remain bound to another molecule by van der Waals forces. In terms of this picture, we can also understand the familiar fact that condensation takes place only when *condensation nuclei* such as dust particles, or electrically charged atoms or molecules, are present in the gas.

The first condition is also easy to understand. If the temperature is above a certain critical value, the colliding particle has enough kinetic energy to break the molecular bond, even though some of its original energy is partitioned among a group of molecules. Only below a certain temperature is condensation possible.

75-6 Potential Energy of a van der Waals Gas

When a van der Waals gas expands, work must be done against the forces of attraction between its molecules. This work is stored up in the gas in the

† Provided that no excited states, in the quantum theoretical sense, are produced.

form of potential energy and increases the internal energy of the gas. In addition, work is done on the surroundings of the gas against the external pressure. The total work ΔW done when the volume of a van der Waals gas increases by ΔV_M is

$$\Delta W = -P\Delta V_M = -\frac{RT}{V_M - b}\Delta V_M + \frac{a}{V_M^2}\Delta V_M.$$

The first term on the right side of this equation is the work done on the surroundings of the gas, against the external pressure. Because of the existence of the covolume it is somewhat larger than for an ideal gas. The second term is positive and represents the increase in energy of the gas, as a result of the work done in separating its molecules against the forces of attraction between them (i.e., work done against the internal pressure). This term corresponds to the increase in potential energy of the gas.

75-7 Origin of the van der Waals Cohesive Force

Van der Waals' assumption that molecules attract one another can be explained qualitatively from the fact that all molecules are composed of positively and negatively charged particles. The results of atomic physics leave no doubt that this is true. The fact that the positive and negative charges in a molecule do not merge and neutralize each other, but are separated in space, suffices to explain the existence of van der Waals forces. At a sufficiently great distance from such a charge distribution the molecule appears neutral, since its net charge is zero.

In the immediate neighborhood of the molecule, on the other hand, the electric field, whose lines of force are from the positive to the negative charges, is quite strong. It was first shown by P. Debye how to take account of these electric fields in order to explain the cohesive force.

Although the net charge of every molecule is zero, the centers of gravity of the positive and negative charges do not necessarily coincide. At large distances this charge distribution can be considered in a first approximation to be that of a *dipole*, whose field is shown in Fig. 8-17. A dipole may result

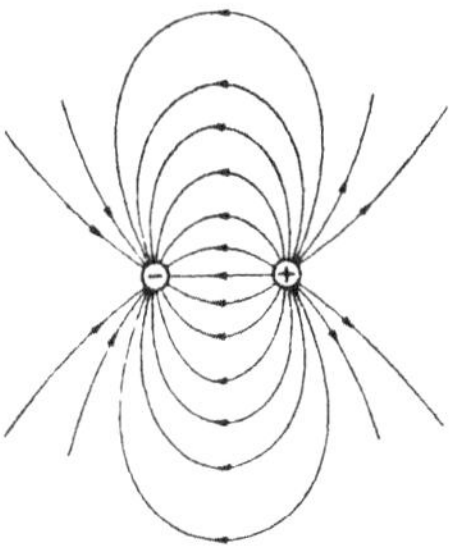

FIG. 8-17. Electric field of a dipole made up of two electric charges, $+q$ and $-q$.

from the particular structure of the molecule (a *permanent dipole*) as is the case, for example, with H_2O, NH_3, C_2H_5OH, etc., or it can be produced by the electric field of neighboring molecules (an *induced dipole*) as, for example, in H_2, N_2, O_2, CO, and CO_2. The electric field of a dipole decreases rapidly with the distance and effects only neighboring molecules. From the law that like electric charges repel while unlike charges attract, the oppositely charged ends of two dipoles are generally closer together than the ends having like charges. It can be shown that the force of attraction of the unlike charges out-weighs the force of repulsion of the like charges and gives rise to the van der Waals cohesive force.

It may be remarked at this point that atomic physics can even tell us the positions of the different kinds of charge in an atom. The positive charges are at the center of the atom, and are surrounded by a "cloud" of negative electrons.† This leads to the result that when two molecules approach each other very closely, it is their clouds of negative charge which collide and which give rise to forces of repulsion called Born repulsive forces. This points the way in which the van der Waals assumptions must be modified when we wish to deal with closely packed aggregations of molecules (liquids and solids).

The preceding description of molecular interactions is satisfactory only when the distance between molecules is large compared with the molecular radius, or when the molecules can rotate with an angular velocity which is great enough for their structural details to average out. If the rotation is *hindered*, as it is in the liquid state, or if the molecules are packed even more closely, as in a solid, the *space symmetry* of their electric fields must be considered. Preferred directions in space, and axes and planes of symmetry then occur. This enables us to understand the *crystalline structure* of matter in its manifold forms.

75-8 Deviations from the van der Waals State

It is not surprising, in view of the considerations above, that appreciable departures from the van der Waals equation occur in the neighborhood of the condensation point and in the liquid state. The covolume of water (see Table 8-4) is 30.5 cm^3/mole, while the actual molar volume of liquid water is 18 cm^3/mole. This discrepancy can be explained when we recall that in the calculation of the covolume it was assumed that the molecule was free to rotate, while this is not the case in the liquid state. In the liquid state the triangular water molecules are already partially ordered in their geometrical relationship to each other.

The existence of the Born repulsive forces makes itself evident in the liquid state by the great discrepancy between the values of internal pressure P_i

† This description is more accurate than the picture of the negative charges orbiting about the positive charges in analogy to the solar system.

as calculated from the van der Waals equation and from the second law of thermodynamics. From the kinetic viewpoint, we have in addition to the external pressure and the internal pressure, both of which tend to compress the gas, opposing pressures which maintain the other two in equilibrium. One of these is the so-called *thermal pressure* RT/V_M, resulting from the thermal motion. The other is the pressure P_r resulting from the repulsive forces between the electron clouds of the atoms, the so-called Born repulsive forces.

76. Sizes of Molecules. Collision Frequency. Mean Free Path

A knowledge of the constant b in the van der Waals equation is of especial interest since it enables us to estimate the size of a molecule. From the definition of the covolume as four times the volume of all the molecules in one mole, the radius r of a molecule can be estimated:

$$b = 4\frac{4\pi}{3}r^3N_0,$$

$$r = \sqrt[3]{\left(\frac{3b}{16\pi N_0}\right)} = 4.62 \times 10^{-9}\ \sqrt[3]{(b)}\ \text{cm}.$$

Two comments should be made at this point. On the one hand, we must recall that the molecular radius as defined above applies to a collision between two molecules. The quantity $2r$ is the distance of closest approach between the centers of gravity of two molecules in a collision. Since a molecule is not perfectly rigid, this distance depends on the impulsive force between the colliding molecules. Hence we must expect a small temperature dependence of b. This dependence is most marked for heavy molecules, since at constant temperature the impulsive force at a collision is proportional to the square root of the molecular mass.

On the other hand, it must be kept in mind that the quantity b has a physical meaning only for the gaseous state. If the molecules are not free to rotate, the shape of the molecule (rod, plane, pyramid, etc.) has a marked effect. Although in the gaseous state the quantity b represents the limiting volume approached at very high pressures, the volume in the liquid state may be, and in fact usually is, smaller than b. Volumes less than b are most often observed for molecules which exhibit association and particularly for those which are rod-shaped, plane, or have a flat pyramidal form.

When the radius of the sphere of exclusion of a molecule is known, it becomes possible to calculate the average distance traveled by a molecule between collisions with other molecules. This distance is called the *mean free path*, $\bar{\lambda}$. The mean free path is best computed from the fact that two like molecules will just make contact in their thermal motion when the distance between their centers of gravity is equal to $2r$. The number of collisions made by a given molecule would be unchanged if its radius were increased to $2r$ and if at the same time all the other molecules were to shrink to geometrical points at their centers of gravity. In a time t, the molecule under consideration will collide with all molecules whose centers of gravity lie in a cylinder of radius $2r$ and length ut, where u is the molecular velocity (Fig. 8-18). Since the number of molecules per unit volume is N_0/V_M, the

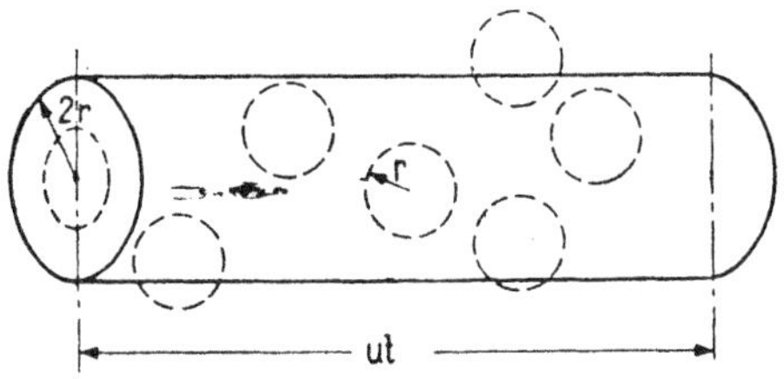

FIG. 8-18. Diagram for calculating the number of molecular collisions per unit time.

number of collisions per unit time, or the *collision frequency* $\bar{z}$, is

$$\bar{z} = \frac{\left(\frac{N_0}{V_M}\right) \cdot \pi \cdot (2r)^2 \cdot ut}{t} = \frac{4\pi r^2 u N_0}{V_M},$$

In this derivation only the motion of one of the colliding molecules has been considered. Since the other molecules are also in motion, we should actually use the *relative* molecular velocity rather than the velocity u. An exact calculation shows that u must be replaced by $\sqrt{(2)}u$. The average collision frequency $\bar{z}$ is therefore

$$\bar{z} = \frac{4\sqrt{(2)}\pi r^2 u N_0}{V_M} = \frac{u\sqrt{(2)}}{V_M} \sqrt[3]{\left(\frac{9b^2 \pi N_0}{4}\right)}.$$

The *mean free path* $\bar{\lambda}$ can now be found by dividing the distance moved per unit time, u, by the mean number of collisions per unit time, $\bar{z}$.

$$\bar{\lambda} = \frac{u}{\bar{z}} = \frac{V_M}{4\sqrt{(2)}\pi r^2 N_0} = \frac{1}{4\sqrt{(2)}\pi r^2 n}.$$

where $n = N_0/V_M$ is the number of molecules per unit volume. At a given temperature and pressure all quantities on the right side of this equation are known, assuming that r can be computed from b. Conversely, if $\bar{\lambda}$ is measured experimentally, the equation can be used to compute r.

For oxygen (O_2) at 760 mm Hg and 273°K, $u = 461$ m/sec and

$$\bar{z} = 4.74 \times 10^9 \text{ collisions/sec},$$

$$\bar{\lambda} = 9.73 \times 10^{-6} \text{ cm} \cong 10^{-5} \text{ cm}.$$

The dependence of these quantities on pressure and temperature is readily derived. From the equations

$$\bar{z} = 4\sqrt{(2)}\pi r^2 N_0 \frac{u}{V_M}, \quad u = \sqrt{\left(\frac{3RT}{M}\right)},$$

and

$$V_M = \frac{RT}{P},$$

it follows that

$$\bar{z} = \frac{4\sqrt{(6)}\pi N_0 r^2}{\sqrt{(MR)}} \frac{P}{\sqrt{T}} = c_1 \frac{P}{\sqrt{T}}$$

and

$$\bar{\lambda} = \frac{V_M}{4\sqrt{(2)}\pi r^2 N_0} = c_2 \frac{T}{P}.$$

Thus both $\bar{z}$ and $\bar{\lambda}$ vary with pressure and with temperature. On the other hand, the product $\bar{z}\bar{\lambda}$ is independent of pressure, since

$$\bar{z}\bar{\lambda} = u = \sqrt{\left(\frac{3R}{M}\right)}\sqrt{T} = c_3 \sqrt{T}.$$

Since diffusion and heat conduction in a gas depend only on the product $\bar{z}\bar{\lambda}$, both of these processes are independent of the pressure.

If the pressure in a vessel containing a gas is reduced to such a low value that the mean free path is of the same order of magnitude as the dimensions of the vessel, a new set of phenomena appear because intermolecular collisions are then too infrequent for the formation of representative mean values. The mean free path is inversely proportional to P, and is about 10^{-5} cm at atmospheric pressure. Hence, for vessels of ordinary size (dimensions ~10 cm) these deviations commence at pressure below 10^{-6} atm or about 10^{-3} mm Hg, where λ is of the order of 10 cm. This state is called a *high vacuum*.

77. Diffusion

Gases are miscible in all proportions. If two different gases A and B, at the same pressure and temperature, are separated by a wall and are then brought into contact by removing the wall, the gases will in the course of time become completely intermixed. A state of molecular chaos will eventually be reached in which the relative number of molecules per unit volume, n_A/n_B, is the same in every volume element. Each component has then expanded uniformly throughout the gas. This process is called *diffusion*.

The kinetic phenomenon which is the cause of this mixing process is the thermal motion of the molecules. Molecules of gas A make countless collisions with other molecules of their own type and also with those of gas B. The boundary surface between the gases, however, does not move with the average molecular velocity u, but with a much smaller velocity. It is evident that the mean free path $\bar{\lambda}$ is the determining factor. In this case, however, $\bar{\lambda}$ cannot be computed as easily as in a gas all of whose molecules are alike because both the molecular velocities and the molecular radii are different for the two kinds of molecules.

When the pressure is very low, the mixing process is extremely rapid and is essentially completed in a few seconds. A spontaneous unmixing, however, does not take place and hence the process is said to be *irreversible*.

Liquids which are miscible in one another also exhibit the phenomenon of diffusion, but the diffusion velocities are very much smaller than for gases and complete mixing may require hours or even days.

When a gas flows through a small opening into a vacuum, the amount of gas passing through the opening per unit time is proportional to the root mean square molecular velocity u, where

$$u = \sqrt{\left(\frac{3RT}{M}\right)} = \text{const}\sqrt{\left(\frac{T}{M}\right)}.$$

The velocity is proportional to $\sqrt{T}$ and inversely proportional to $\sqrt{M}$. Hence under the same conditions H_2 will flow four times as rapidly as O_2. This process can be used to compare the molecular weights of gases.

One important application of the diffusion process is the *diffusion pump*, shown in Fig. 8-19. Pumps of this sort are used for the production of high vacua. Gas diffuses from the system to be evacuated into a stream of oil vapor and is then carried by the vapor stream to the fore-vacuum opening where it is then removed by a mechanical vacuum pump called a forepump. The oil vapor is kept in circulation by a heater. The walls of the pump are cooled by a continuous flow of water through water coils or a jacket. The

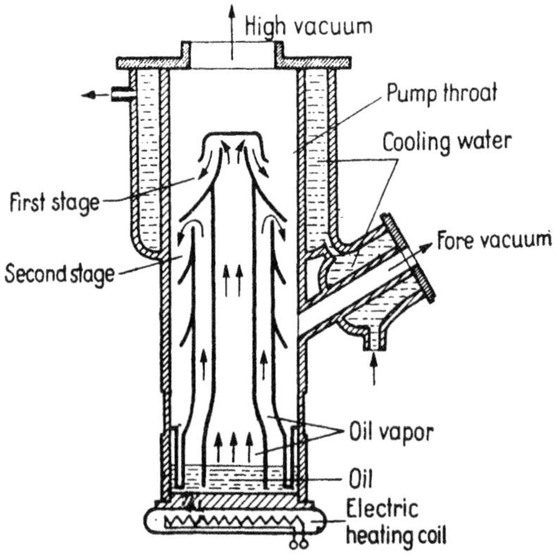

FIG. 8-19. Cross-section view of an oil diffusion pump.

vapor condenses on the walls and does not diffuse either into the high vacuum or the fore-vacuum. To obtain a good high vacuum it is necessary to use an oil with a low vapor pressure, such as Apiezon or silicone oil, whose vapor pressure at room temperature is of the order of 10^{-7} mm Hg.

The *ideal pumping speed* of a diffusion vacuum pump is defined as the number of liters of gas which diffuse per second out of the space to be evacuated, at the pressure existing in this space. The pumping speed depends chiefly on the area of the throat of the pump (see Fig. 8-19) and is given by the number of liters diffusing through the throat per second. The maximum speed of a diffusion pump can easily be calculated provided the mean free path in the pump is large compared with the dimensions of the pump, i.e. assuming a high vacuum is maintained in the pump. Let A represent the area of the throat and $\bar{v}$ the arithmetic mean molecular velocity. If all molecules traveled with the velocity $\bar{v}$, and all were headed directly toward the throat, the pumping speed would be given by $V = A\bar{v}$, since this volume of gas would stream through the throat per second. Actually, the molecules travel in all possible directions in space and have different velocities. An exact calculation shows that the pumping speed is only one-fourth of the value given above, that is,

$$V = \tfrac{1}{4} A\bar{v}.$$

Replacing $\bar{v}$ (see page 363) by

$$\bar{v} = \sqrt{\left(\frac{8RT}{\pi M}\right)},$$

we obtain

$$V = \frac{1}{4} A \sqrt{\left(\frac{8RT}{\pi M}\right)}.$$

The speed of the pump is independent of the pressure in the region to be evacuated. It depends only on the temperature and the molar mass. Hence the speed is small for heavy gases at low temperatures. The actual pumping speed of a diffusion pump is only one half to one tenth of the ideal speed, due to various effects such as the backstreaming of the pump vapor.

For air at room temperature ($T = 300$°K, $M = 29$ gm/mole) and for a throat area A of 10 cm^2, the ideal speed is 117 liters/sec.

77-1 The Diffusion Equation

Consider a diffusion process taking place along the axial direction x in a cylindrical tube of cross-sectional area A. At the beginning of the diffusion process the partial pressures and the corresponding molar volume concentrations q_v vary from point to point along the tube. (The molar volume concentration q_v is the number of gram-moles r of dissolved material per unit volume, $q_v = r/V$.)

(a) *The stationary state.* A stationary state is said to exist in the tube if the *concentration* q_v at each point remains constant in time. This means that at every point x of the tube the concentration gradient of the substance is constant in time. Such a state can be realized experimentally as follows. Two parallel channels (Fig. 8-20) are connected by a channel of cross-section A. A pure solvent (such as water) flows in one channel, while the other carries a solution of concentration q_v. At one end of the cross-channel the concentration of dissolved substance is always equal to that of the solution, q_v, while at the other end the concentration is zero. If the cross-section is constant, then $\Delta q_v/\Delta x$ is constant along the channel in the stationary state. The concentration decreases linearly from one end of the tube to the other (Fig. 8-21). The number of moles n of the substance which diffuse

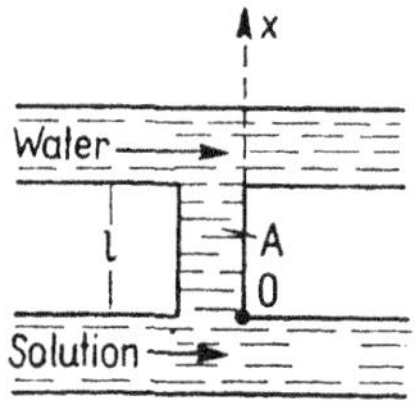

FIG. 8-20

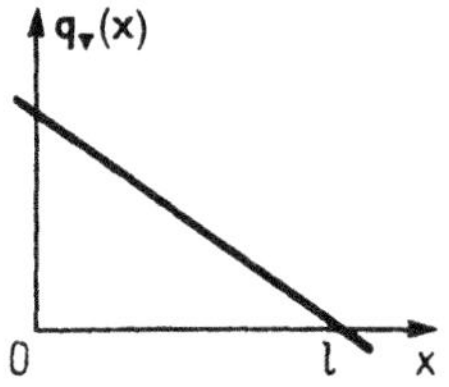

FIG. 8-21

FIG. 8-20. A stationary diffusion process.

FIG. 8-21. The linear decrease of concentration in the stationary state.

along the cross-channel in time t is proportional to the area A of the channel, to the time t, to the concentration gradient $\Delta q_v/\Delta x$, and depends also on the material through the quantity D,

$$n = -DA\frac{\Delta q_v}{\Delta x}t.$$

This equation applies only to diffusion in a channel of constant cross-section and to the stationary state. More complicated cases must be treated by the methods of calculus; see sub-section (b).

The empirical quantity D is called the *coefficient of diffusion.* It is numerically equal to the number of moles diffusing in unit time across unit cross-sectional area, when the concentration gradient is unity.

Dimensions: $[D] = \left[\frac{L^2}{T}\right]$.

Units: cgs system, 1 cm^2/sec.

Some values of D are given in Table 8-6.

EXAMPLE. What must be the concentration of a glycerin solution at 10°C, in the lower channel of Fig. 8-20, if 0.010 gm of glycerin are to diffuse in 10 hr into the pure water

TABLE 8-6

Substance	Diffusing in	Temp. (°C)	D
Gases			
H_2	Air	0	0.634 cm^2/sec
O_2	Air	0	0.178 cm^2/sec
CO_2	Air	0	0.139 cm^2/sec
Aqueous solutions (1 mole/liter)			
NaCl	Water	15	0.94 cm^2/day
Sugar	Water	12	0.254 cm^2/day
Glycerin	Water	10	0.339 cm^2/day

in the upper channel through a tube 10 cm long and of cross-section 5 cm^2? ($M_{glycerin}$ = 92 gm/mole).

$$\frac{\Delta q_v}{\Delta x} = \frac{n}{DAt} = 1.54 \times 10^{-4} \frac{\text{mole/cm}^3}{\text{cm}},$$

$$q_v = \frac{\Delta q_v}{\Delta x} l = 1.54 \times 10^{-3} \frac{\text{mole}}{\text{cm}^3} = 1.54 \frac{\text{mole}}{l}.$$

(b) *The nonstationary state.* The nonstationary state must be treated by the methods of calculus. The concentration at a given point varies with time. We shall consider only the special case illustrated in Fig. 8-22, in which

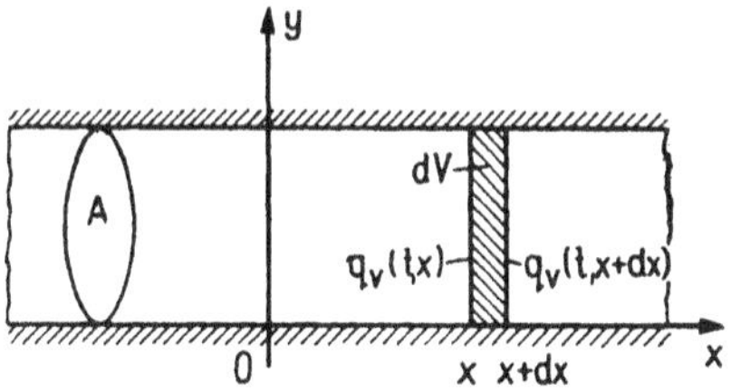

FIG. 8-22. Diagram for the derivation of the diffusion equation.

the concentration in the plane x at time t is $q_v(t, x)$, and in the neighboring plane at a distance dx from the first it is $q_v(t, x + dx)$. The concentration gradient at the coordinate x and the time t is

$$\left.\frac{q_v(t,x + dx) - q_v(t,x)}{dx}\right|_t = \left.\frac{\partial q_v(t,x)}{\partial x}\right|_t.$$

At the plane $x + dx$ the corresponding gradient is

$$\left.\frac{\partial q_v(t,x + dx)}{\partial x}\right|_t = \left.\frac{\partial q_v(t,x)}{\partial x}\right|_t + \left.\frac{\partial^2 q_v(t,x)}{\partial x^2}\right|_t dx.$$

We now investigate how the concentration changes in the volume element of area A and thickness dx between the planes considered above. The number of moles diffusing into this element across the plane x in a time interval dt is

$$dn_1 = -DA\frac{\partial q_v(t,x)}{\partial x}dt.$$

The number of moles diffusing out of the element across the plane $x + dx$ in the same time is

$$dn_2 = -DA\frac{\partial q_v(t,x+dx)}{\partial x}dt$$

$$= -DA\left[\frac{\partial q_v(t,x)}{\partial x} + \frac{\partial^2 q_v(t,x)}{\partial x^2}dx\right]dt.$$

The net change in the number of moles in the volume element is therefore

$$dn = dn_1 - dn_2 = DA\frac{\partial^2 q_v(t,x)}{\partial x^2}\bigg|_t dxdt.$$

The volume of the element is $dV = Adx$, so the change in the volume concentration in the element is

$$dq_v = \frac{dn}{Adx} = D\frac{\partial^2 q_v(t,x)}{\partial x^2}\bigg|_t dt.$$

However, we may fix our attention on a point x and determine q_v at times t and $t + dt$. The net change in q_v in a time dt is then

$$dq_v = \frac{q_v(t+dt,x) - q_v(t,x)}{dt}dt = \frac{\partial q_v(t,x)}{\partial t}\bigg|_x dt.$$

Therefore,

$$\frac{\partial q_v(t,x)}{\partial t}\bigg|_x = D\frac{\partial^2 q_v(t,x)}{\partial x^2}\bigg|_t.$$

This last equation is called the *diffusion equation*, and it is valid throughout the whole volume. It is a relation between the *time* rate of change of the concentration and its *space* rate of change. In contrast to simpler physical problems, this relation cannot be expressed by elementary mathematics. The phenomenon can only be described by this *partial differential equation.*

The diffusion equation is one example of the general type of differential equation encountered in all *equalization* processes (in this case, equalization of concentration differences). We encounter the same equation in *heat conduction* (equalization of temperature differences), in *viscous flow* (equalization of momentum differences), and in *electrical conduction*, (equalization of charge differences). In the problem of heat conduction, concentration is replaced by temperature and the coefficient of diffusion by the thermal conductivity.

78. Heat Transfer

The exchange of heat between two systems can take place in three different ways, by *radiation*, by *convection*, and by *conduction*.

78-1 Radiation

The transfer of heat by radiation belongs in the domain of optics and the laws of thermal radiation will be given in Vol. II.

78-2 Convection

Heat is transferred by convection whenever there is a mass motion of heated material from one place to another. The quantity of heat transported evidently depends on the mass displaced and on its heat capacity. Heat convection in liquids and gases is an important problem in hydrodynamics.

78-3 Conduction

Heat conduction consists of a flow of heat from one point to another in a body which is itself at rest. The flow is in the nature of a diffusion of thermal energy and the law of heat conduction is obtained from the *diffusion equation.*

For an elementary treatment it will suffice to consider only the special case in which the flow has reached a stationary state. A flow of heat is said to be stationary if the temperature at each point remains constant with time. In the stationary state a definite temperature gradient dT/dx exists at every point. The quantity of heat dQ which flows per unit time across an area dA is found by experiment to be proportional to the temperature gradient and to the area dA, and in addition to depend on the material,

$$\frac{dQ}{dt} = k\frac{dT}{dx}dA.$$

The quantity k is called the *thermal conductivity* of the material.

DEFINITION. The thermal conductivity k of a material is the quantity of heat flowing across a surface of unit area per unit time, and per unit temperature gradient.

Unit: $$1\,\frac{\text{cal/sec}}{\dfrac{{}^\circ\text{C}}{\text{cm}}\cdot\text{cm}^2} = 1\,\frac{\text{cal}}{{}^\circ\text{C}\cdot\text{cm}\cdot\text{sec}}$$

The problem of the *plane flow* of heat is particularly simple and is very important in practice. Let us imagine that a number of homogeneous plane sheets or layers of different materials are placed between two heat reservoirs kept at the constant temperatures T_A and T_B (Fig. 8-23). Let d_i be the thickness of the ith layer and k_i its thermal conductivity. We assume here for simplicity that the thermal conductivity is independent of the temperature. This is not generally a good approximation over a wide range of temperature. Initially, the flow is not steady since the temperature changes at every point as a result of the heat flux. This nonstationary state

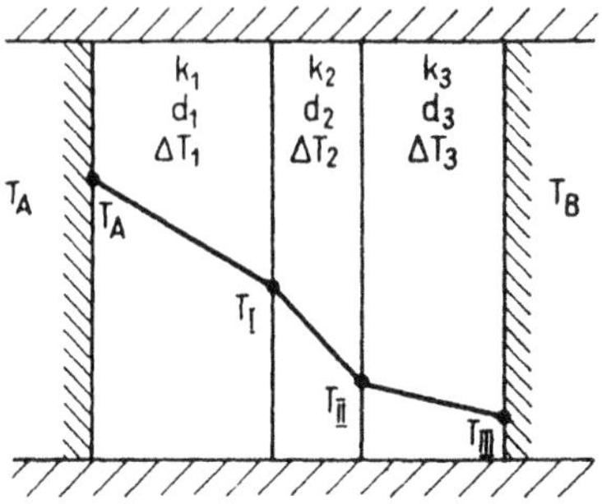

FIG. 8-23. Stationary planar heat conduction. The top and bottom bounding surfaces are thermally insulated.

continues until such time as the rate of flow of heat out of every volume element equals the rate of flow into it. In the plane problem, a volume element is bounded by two parallel planes. The heat flow through the entire system is in a direction normal to the layers of material. The requirement that

$$dQ_{x,\,\text{in}} = dQ_{x+dx,\,\text{out}}$$

shall hold in the stationary state leads to the result that at each surface of separation

$$k_{i,x}\left(\frac{dT}{dx}\right)_x = k_{i,x+dx}\left(\frac{dT}{dx}\right)_{x+dx} = \text{const.}$$

Since the conductivities k_1, k_2, etc. are assumed to be constants, this condition can only be satisfied if the temperature gradient within each layer is constant. That is the temperature decreases linearly with distance in each layer. From the condition that

$$k_i\left(\frac{dT}{dx}\right)_i = \text{const.}$$

we see that the temperature gradient is large in layers for which k is small, and vice versa. The fact that dT/dx is constant within each layer enables us to replace these gradients by the ratios of finite temperature differences and finite thicknesses:

$$\left(\frac{dT}{dx}\right)_{\text{layer 1}} = \frac{T_A - T_I}{d_1} = \frac{\Delta T_1}{d_1},$$

$$\left(\frac{dT}{dx}\right)_{\text{layer 2}} = \frac{T_I - T_{II}}{d_2} = \frac{\Delta T_2}{d_2},$$

etc., where T_I, T_{II}, etc. are the temperatures at the boundary surfaces between adjacent layers. We also have the following equations:

$$\sum \Delta T_i = T_A - T_B,$$

$$\frac{k_i \Delta T_i}{d_i} = \text{const.} \quad (i = 1, 2, \cdots).$$

This provides us with precisely the necessary number of equations to compute the unknown quantities ΔT_i, and the constant. The meaning of this constant,

$$\text{const.} = \frac{k_i \Delta T_i}{d_i} = \frac{1}{A}\frac{dQ}{dt},$$

is simply the quantity of heat, per unit time and per unit area, flowing through each layer.

Example. One wall of a room adjoins a refrigerated room at a temperature of −20°C. The wall is constructed of brick having a thermal conductivity $k_1 = 1.5 \times 10^{-3}$ cal/°C·cm·sec and a thickness $d_1 = 25$ cm. The wall is insulated with a layer of cork ($k_2 = 2.5 \times 10^{-4}$ cal/°C · cm · sec) having such a thickness d_2 that when the temperature of the room is 20°C no more heat is lost than would be lost through the uninsulated brick wall to the outside air when the outside temperature is 10°C. (We shall assume that the temperatures at the outer surfaces of the walls are the same as the temperatures of the air with which they are in contact. This does not correspond very closely to reality; actually there is a temperature drop in the boundary layer of air near the wall which may be of the same order of magnitude as that across the wall.)

The heat flowing through the brick wall per cm^2 and per sec is

$$\frac{1}{A}\frac{dQ}{dt} = \frac{k_1 \Delta T_1}{d_1} = 6.0 \times 10^{-4} \frac{\text{cal}}{\text{cm}^2 \cdot \text{sec}}.$$

The temperature drop across the insulated wall is $\Delta T_1 + \Delta T_2 = 40°\text{C}$ and the heat flow per unit area and per unit time is

$$\frac{1}{A}\frac{dQ}{dt} = \frac{k_1 \Delta T_1}{d_1} = \frac{k_2 \Delta T_2}{d_2} = 6 \times 10^{-4} \frac{\text{cal}}{\text{cm}^2 \cdot \text{sec}}.$$

From these equations we find:

$$\Delta T_1 = 10°\text{C},$$

$$\Delta T_2 = 40° - 10° = 30°\text{C},$$

$$d_2 = \frac{2.5 \times 10^{-4} \times 30}{6.0 \times 10^{-4}} = 12.5 \text{ cm},$$

$$\frac{1}{A}\frac{dQ}{dt} = 360 \frac{\text{cal}}{\text{m}^2 \cdot \text{min}},$$

$$\frac{\Delta T_1}{d_1} = 0.40 \frac{°\text{C}}{\text{cm}}, \quad \frac{\Delta T_2}{d_2} = 2.40 \frac{°\text{C}}{\text{cm}}.$$

For liquids at rest, the heat flow in the absence of convection is very small and may approach values of the same order as in solid thermal insulators.

Conduction in gases at rest is smaller by a factor of 10 or so. Some representative values of thermal conductivity at room temperature are given in the Table 8-7.

The large heat conductivity of metals is related to their good electrical conductivity. Both electric charge and heat are transported in metals largely by the conduction electrons.

TABLE 8-7

Material	$k \left(\frac{\text{cal}}{{}^\circ\text{C} \cdot \text{cm} \cdot \text{sec}} \right)$
Copper	0.918
Glass	2.5×10^{-3}
Compact snow	5.1×10^{-4}
Water (20°C)	1.43×10^{-3}
Air	5.7×10^{-5}

CHAPTER 9

CRYSTAL STRUCTURE OF SOLIDS

79. Fundamental Principles

The atoms or molecules of a solid substance may be arranged spatially with great regularity relative to each other, in which case this substance is said to have a *crystalline* structure. If the spatial arrangement is irregular (i.e. the atoms or molecules may still have regular relative spacings, but there is no long range order to their arrangement), then the substance has an *amorphous* structure. Glass and opal are both amorphous. In this chapter we shall consider only crystalline substances.

79-1 Force Functions

Since the van der Waals assumptions were shown to be insufficient to explain completely the properties of the liquid state, it was found necessary to extend kinetic theory to include the Born forces of repulsion as well as the van der Waals forces of attraction. At distances large compared with the molecular diameter, the van der Waals attractive forces predominate, and they decrease slowly with increasing distance. The Born repulsive forces, on the other hand, have an appreciable magnitude only in the immediate vicinity of the molecule and increase very rapidly with decreasing distance.

Similar forces also occur in solid bodies. However, for those solid bodies that can be considered to be composed of positive and negative ions, rather than molecules or neutral atoms, the forces of attraction are chiefly Coulomb forces, which decrease with the square of the distance.

In general, the resultant force exerted by one particle on another is given by the *Born force law*, which states that the resultant force is the sum of a repulsive force which decrease with a high power of the distance, and an attractive force which decreases with a lower power. That is, the resultant force can be expressed as

$$F = \frac{A}{r^n} - \frac{B}{r^m} \quad (n > m).$$

The force F is therefore equal to the difference between two curves, each of which is a hyperbola of high order. Figure 9-1 shows the form of the force curve for the case $m = 2$ and $n = 10$. If one ion is located at the origin and a second ion of the same kind is brought in along the r-axis, the full curve

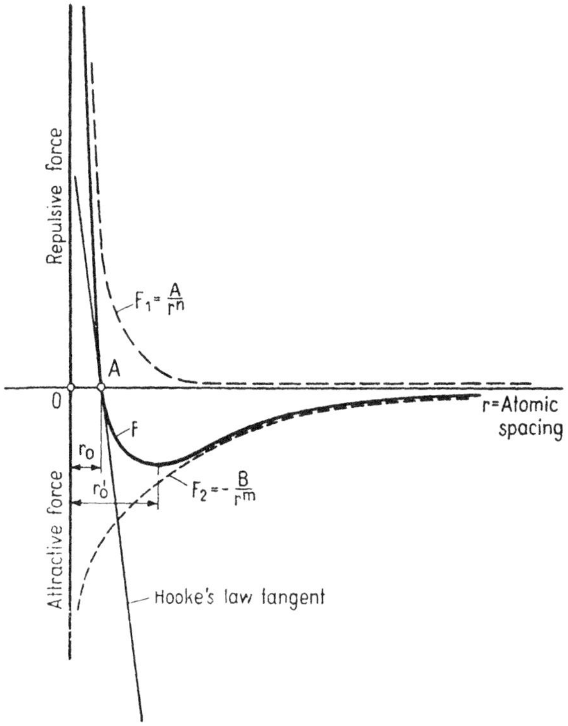

FIG. 9-1. Plot of the interatomic force in a crystal according to the Born force law.

represents the force exerted on it. At one particular distance r_0 the resultant force is zero. The second ion can remain in equilibrium at this point. Furthermore the point is one of *stable* equilibium. If the second ion approaches the first more closely, it is repelled; if it recedes, it is attracted. If a third ion is brought in form the right, the same conditions apply, to a first approximation. That is, this ion would be in eqilibrium at a coordinate $2r_0$. The neglect, to a first approximation, of the disturbing effect of more distant ions is easily justified. The effect of the more distant ions is only to decrease the magnitude of r_0 somewhat, since for all distances greater than r_0 the net force is one of attraction. The *equidistance* of the points of equilibrium is not affected.

79-2 Thermal Motion of Atoms

Only at a temperature of absolute zero are the atoms in a crystal at rest in their equilibrium positions.† Each atom has three degrees of freedom. At a temperature *T*, each degree of freedom has both kinetic and potential

† The quantum mechanical "zero-point energy" is disregarded in this discussion.

energy associated with it. If the crystal structure consists wholly of single ions, atoms, or of radicals or molecules which are so rigidly bound that their internal vibrations can be neglected, then by the equipartition theorem, each particle has its energy divided into six equal components, three corresponding to kinetic and three to potential energy. The increase in energy of N_0 particles, per degree rise in temperature (i.e. C_v), is then

$$6N_0 \frac{k}{2} = 3R \frac{\text{erg}}{\text{gm-ion} \cdot \text{C}^\circ} = 5.97 \frac{\text{cal}}{\text{gm-ion} \cdot \text{C}^\circ}.$$

This value does not apply to crystals containing molecules or radicals within which internal vibrations can take place.

For small displacements from its equilibrium position, the restoring force on a particle is proportional to the displacement. That is, a short segment of the curve in Fig. 9-1 can be replaced by its tangent, called the Hooke's law tangent. The particles therefore oscillate with *harmonic motion*, and the average displacement of each from its equilibrium position is zero. It follows that the coefficient of linear expansion is $\lambda = 0$. Crystalline bodies show this behavior in the neighborhood of absolute zero.

At higher temperatures, where the amplitude of oscillation is greater, the vibrations are *nonharmonic*. This is evident from the shape of the force curve in Fig. 9-1, since for a small decrease in separation the repulsive force increases much more rapidly than the attractive force. As a consequence, the coefficient of linear expansion λ is larger than zero because the oscillating particle spends more time at larger rather than at smaller separations. Large coefficients of expansion correspond to weak forces of attraction between the particles, with consequent large increases of amplitude for small elevations of temperature.

79-3 Hooke's Law

Hooke's law was stated in Chapter 5 simply as an experimental law. No reason could be given for the linear relation between stress and strain, nor could we explain why the same value of Young's modulus Y applied to both extension and compression. Both of these questions can now be answered directly from the Born law of force. For small displacements from the equilibrium position, the force curve can be replaced by its tangent. This means that the restoring force has the same magnitude for displacements to the right and to left of the equilibrium position. The resulting behavior is completely described by Hooke's law for the elastic expansion and compression of a substance. Suppose that a compressive force F is exerted on the particle at one end of a row of particles. This particle then moves toward the next particle until the force of repulsion equals the external force F, where it is again in equilibrium. From Newton's third law, a force equal to F now acts on the second particle and this process propagates along the entire row to the last particle, which transmits the force to any object with which it makes contact. Each of the original distances of separation, r_0, decreases by the same amount Δr_0. The total decrease in length is proportional to the number of particles, that is, to the original length. The change in length depends on the material, and is the same for both tension and compression. But this is precisely the content of Hooke's law.

When the change in length becomes very large, the force curve departs from linearity and in a different way for tension and compression (limit of proportionality). However, it must be kept in mind that these considerations apply strictly only for so-called single crystals. Most objects are not single crystals but have a *microcrystalline* structure, consisting of many small crystals with their crystalline axes oriented at random. Unless the crystals are isotropic, such bodies must be treated by special methods in order to average over the random orientations.

79-4 Lattices. Lattice Points

The equidistance of atoms or molecules in a crystal brings us to the concept of a lattice, which is basic to the study of crystal structure. A *lattice* is an idealized, regular infinite array of points called *lattice points.* The array may be in one, two, or three dimensions. (Only the three-dimensional lattice is used in crystallography.)

A one-dimensional lattice would consist of lattice points equidistantly spaced along a straight line (Fig. 9-2a). Any one of the points can be picked as an origin of coordinates. This lattice can be completely described by a vector $\boldsymbol{a}$ whose direction is parallel to the straight line, and whose length

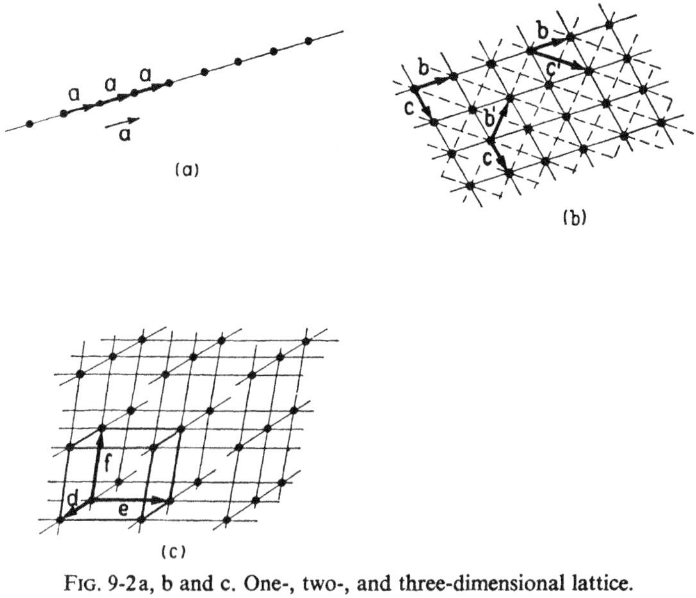

FIG. 9-2a, b and c. One-, two-, and three-dimensional lattice.

is the distance between any two neighboring lattice points. Given an origin, a complete lattice can be generated by repeated displacements of $\boldsymbol{a}$ by a distance $|\boldsymbol{a}|$.

A two-dimensional lattice is a planar array of lattice points (Fig. 9-2b). If we consider any straight line which can be drawn through a series of the points, the spacings between the points lying on the line will all be equal. Two non-parallel vectors $\boldsymbol{b}$ and $\boldsymbol{c}$ serve to give a complete description of a two-dimensional lattice, and repeated displacements of the vectors, starting from some origin, will generate the complete lattice. That is, the position of all lattice points is given by the following expression for the radius vector from some origin

$$\boldsymbol{r} = m\boldsymbol{b} + n\boldsymbol{c},$$

where m and n run over all integral values from $+\infty$ to $-\infty$. The choice of vectors is not unique, i.e. the lattice can also be described equally well by the vectors $\boldsymbol{b}$ and $\boldsymbol{c}'$ or $\boldsymbol{b}'$ and $\boldsymbol{c}$. A two-dimensional lattice can be thought of as a series of one-dimensional lattices (described by $\boldsymbol{b}$) displaced parallel to each other with equidistant separations (given by the vector $\boldsymbol{c}$).

If a two-dimensional lattice (given by $\boldsymbol{d}$ and $\boldsymbol{e}$) is repeatedly displaced parallel to itself, the displacement being given by a vector $\boldsymbol{f}$ (Fig. 9-2c), then a three-dimensional or *space lattice* is formed. Three non-parallel vectors $\boldsymbol{d}$, $\boldsymbol{e}$, and $\boldsymbol{f}$, one of which does not lie in the plane of the other two, are necessary to give a complete description of a three-dimensional lattice. The position of all lattice points is given by the following expression for the radius vector from some origin:

$$\boldsymbol{r} = m\boldsymbol{d} + n\boldsymbol{e} + p\boldsymbol{f},$$

where m, n, and p run over all integral values from $+\infty$ to $-\infty$. As in the two-dimensional case, the set of vectors is not unique. The parallelepiped whose edges are the vectors $\boldsymbol{d}$, $\boldsymbol{e}$, and $\boldsymbol{f}$ is called the *basis parallelepiped* (Fig. 9-3).

The essential characteristic of a lattice is that every lattice point has *identical surroundings*. The grouping of lattice points about any given point is identical with the grouping about any other lattice point in the lattice. The fact that all lattice points are equivalent means that only certain spatial arrangements of the points are allowed, and in fact there are only fourteen different types of space lattices (see Fig. 9-11). All other arrangements fail to give identical surroundings to each lattice point. The equivalence of the

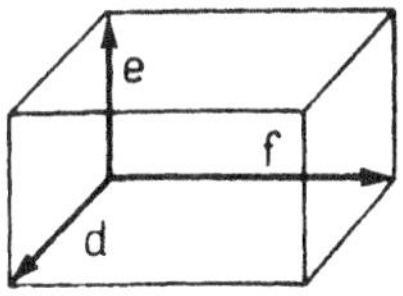

FIG. 9-3. A basis parallelepiped of a space lattice.

lattice points has the further consequence that there is a high degree of symmetry in most space lattices.†

It is clear that if we consider arrangements in two and three dimensions, the internal symmetry of the atoms or molecules making up the crystal will play a large part in determining the lattice symmetry of the crystal. If we assume, for example, that at small distances the molecule has the symmetry of a cube, then the law of force is the same along three mutually perpendicular directions. We therefore conclude that three equally spaced linear arrays of molecules will extend outward from any given molecule in three mutually perpendicular directions and that an infinite cubic structure will result. It is evident that the type of symmetry of the atoms or molecules (corresponding to their electric fields or force fields of other sorts) determines whether the crystal structure will be cubic, hexagonal, tetragonal, or of some lower degree of symmetry.

At this point, it is well worth stressing the difference between a lattice and a crystal structure. A lattice is an idealized array of mathematical points. The points are fixed in the array and are arranged with great regularity and symmetry. It is from the lattice that the crystal derives its symmetry classification. Lattice points are *not* the same as the atoms, ions, or molecules which make up an actual crystal structure, although they may coincide in some simple cases.

A crystal structure is made up of a regular array of atoms, ions or molecules. (In what follows, for brevity we shall often replace the phrase atoms, ions, or molecules by the word particles.) The particles are arranged in particular positions relative to the points of the lattice which is characteristic for that kind of crystal. The particles are not fixed in position as are the lattice points (see Section 79-2). Since all the particles are generally not identical, the crystal structure is often more complex than the associated lattice and has less symmetry.

79-5 Fundamental Concepts of Crystallography

Most elements and compounds are not ordinarily found in the form of well-developed single crystals, but instead have a micro-crystalline structure. They consist of aggregations of countless microscopic or submicroscopic crystals, whose axes are randomly oriented relative to each other. The Debye–Scherrer method of x-ray analysis of powdered crystals (1916) first proved that many substances are built in this way out of microcrystals, each of which has exactly the same crystallographic structure as a macroscopic crystals.

The science of crystallography is concerned chiefly with the description of both the external and internal symmetry of crystals and with the relation-

† This symmetry is best discussed using the methods of a branch of mathematics called group theory; however, this subject is beyond the scope of this book.

ships between the two. In order to describe the exterior of a crystal mathematically, the surfaces of the crystal are referred to a three-dimensional set of crystal axes. This coordinate system must be suitably chosen for the particular crystal under consideration. Six fundamentally different coordinate systems are necessary for a simple description of all crystals and their symmetry relations. For reasons of convenience, these six systems are usually increased to seven (the rhombohedral and hexagonal systems are considered as separate systems). All crystals which can be related to one of these sets of axes are grouped together in a so-called *crystal system*.

In general, a surface of a crystal cuts the three coordinate axes at three finite intercepts. The *relative* values of these intercepts uniquely determine the orientation of the surface in space. The absolute magnitudes are not required for our discussion. Ordinarily, the relative intercepts are specified with respect to the intermediate one, to which the value of unity is assigned, and are represented by lower-case letters, e.g. (Fig. 9-4)

$$a : b : c = 0.6713 : 1 : 1.1357.$$

The quantities a, b, and c are called the *parameters* of the surface. All mutually parallel surfaces have the same relative parameters or *parameter ratios*. These ratios, in general, are irrational and are constants for the crystal under consideration.

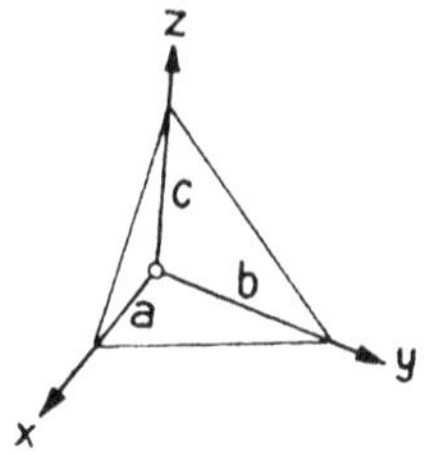

FIG. 9-4. The axis intercepts of a crystal plane.

From the standpoint of the lattice structure of crystals, these ideas are obvious. If the lattice points on the x-axis are all separated by the same distance a_0, and those on the y- and z-axes by the distances b_0 and c_0, these three fundamental units of length are in general related irrationally to one another and vary from substance to substance. Only in exceptional cases (as when two axes are equivalent) are the parameter ratios rational numbers. However, any plane which passes through lattice points on all axes necessarily has an x-intercept which is an integral multiple of a_0, and y- and z-intercepts which are integral multiples of b_0 and c_0.

This brings us to one of the fundamental laws of crystallography, the *law of rational coefficients*. This law states that if an arbitrary surface of a crystal is described by the parameter ratios $a : b : c$, then the parameter ratios of all other surfaces of the same crystal are given by

$$ma : nb : pc,$$

where m, n, and p are rational numbers (e.g., $m = 2/3$, $n = 2$, and $p = 3/4$) or infinity (written ∞, and used for a surface parallel to an axis). If a, b, and c are the lattice point separations a_0, b_0, and c_0 mentioned above, then m, n, and p are called the *coefficients* of the surface. Denoting the crystal surfaces by giving their coefficients is called the *Weiss system* of notation.

A more common system for specifying crystal surfaces is by the use of *Miller indices*. If we take the reciprocals of the Weiss coefficients of some surface and reduce them to their lowest common denominator, then the numerators are the Miller indices of that surface. The Miller indices of a surface with the general ratio $ma : nb : pc$ are written [hkl]. The rationality law also applies to the Miller indices.

EXAMPLE. Consider the surface mentioned above, $2/3a : 2b : 3/4c$. Taking the reciprocals of the Weiss coefficients, we have 3/2, 1/2, 4/3; and reduced to a least common denominator, these become 9/6, 3/6, 8/6. The Miller indices are [938], which is read, "nine, three, eight". A surface with parameter ratios $2a : 1b : \infty c$ has the Miller indices [120].

That the coefficients m, n, and p (and h, k, and l) are in general *small* integral or rational numbers results from the fact that "high index planes" are only sparsely populated with lattice points and hence have no great physical significance.

79-6 The Coordinate Systems and the Crystal Systems of Crystallography

We give here the seven coordinate systems used in crystallography. To each coordinate system, there corresponds a crystal system of the same name.

1. *The cubic system* (Fig. 9-5). This system includes all crystals which can be referred to three equivalent, mutually perpendicular, interchangeable axes. The fact that the axes are all equivalent means that the spacing of the lattice points is the same along all three axes:

$$a_0 = b_0 = c_0 .$$

The simplest surface has the parameter ratios $a : a : a$ and is called the octahedral surface. The hexahedral or cube surface has the parameter ratios

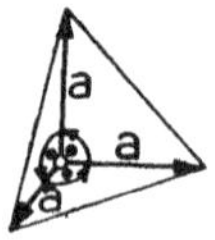

FIG. 9-5. Coordinate axes of the cubic system.

$a : \infty a : \infty a$. The dodecahedral surface is determined by the parameter ratios $a : a : \infty a$. The Miller indices of these surfaces are [111], [100], and [110] respectively.

2. *The tetragonal system.* This system includes all crystals which can be referred to three mutually perpendicular axes of which two are equivalent (Fig. 9-6). From the viewpoint of lattice geometry, this means that the lattice points are equally spaced along two axes, while the spacing along the third axis is either greater or less.

3. *The orthorhombic (or rhombic) system.* This system has three mutually perpendicular axes, no two of which are equivalent. The spacing of the lattice points is different along three axes (Fig. 9-7).

4. *The hexagonal system.* The hexagonal system includes crystals which have two equivalent axes at an angle of 120° with each other, and a third non-equivalent axis perpendicular to the plane of the first two (Fig. 9-8). The spacing of the lattice points is the same along the first two axes, but the internal symmetry of the lattice points is such that the axes make an angle of 120° with each other. The third axis is at right angles to both of the others, and the spacing of the lattice points is different.

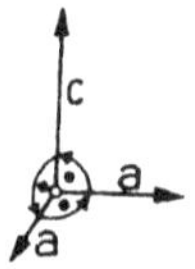

FIG. 9-6

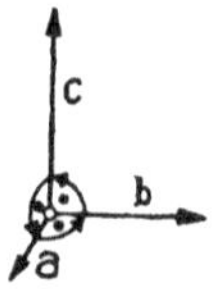

FIG. 9-7

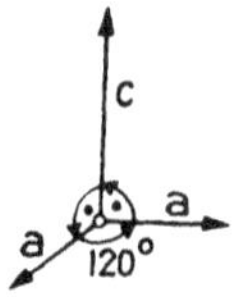

FIG. 9-8

FIG. 9-6. Coordinate axes of the tetragonal system.

FIG. 9-7. Coordinate axes of the orthorhombic system.

FIG. 9-8. Coordinate axes of the hexagonal system.

5. *The rhombohedral system.* The rhombohedral system has a lower degree of symmetry than the hexagonal and is often considered as a subsystem of the latter. For crystals belonging to this system, the three axes

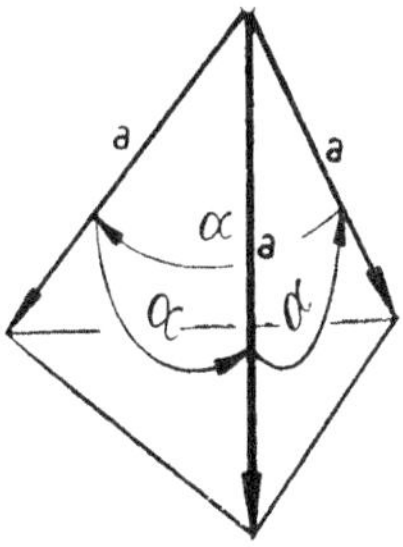

FIG. 9-9. Coordinate axes of the rhombohedral system.

are equivalent and make with each other equal angles α, different from 90° (Fig. 9-9).

6. *The monoclinic system.* The monoclinic system has three non-equivalent axes. One axis makes right angles with each of the other two. However, these latter two axes do not make a right angle with each other.

7. *The triclinic system.* This system has the lowest symmetry of all. All three axes are non-equivalent. None of the angles between axes is a right angle, and in fact all three angles are unequal.

For all questions which concern symmetry relationships in crystals a textbook of crystallography should be consulted.

79-7 The Binding Forces of Crystals

It is believed that all the forces binding together the atoms in a crystal are electrical in nature, though they are not always as simple as portrayed in Section 79-1. Four extreme types of bonds can be distinguished: The *ionic bond*, the *covalent bond*, the *metallic bond*, and the *molecular* or *van der Waals bond.* In many crystals, the bonds are predominantly of one type with a small admixture of some other type. There are also crystals in which there are two kinds of bond, each in rather pure form. For instance in a molecular crystal, the bond between the atoms in a molecule are often predominantly covalent, while that between the molecules is a van der Waals bond.

Ionic bonds are electrostatic in nature and act between electrically charged ions, such as Na^+ and Cl^- in a crystal of NaCl. As mentioned at the beginning of the chapter, in addition to the attractive force, there is a shorter range repulsive force which is also electrostatic and keeps the crystal from collapsing on itself.

Covalent (or *homopolar*) *bonds* are due to a sharing of electrons between atoms, and are present in pure form only when the atoms are alike, as in crystals of the elements. Thus, the carbon atoms in diamond are bound by covalent forces. Most of the force binding the atoms into the molecules in the molecular crystals mentioned above is covalent in nature.

A crystal having *metallic bonds* can be thought of as a regular array of positive ions immersed in a sea of electrons. The binding arises because the attractive forces between the ions and electrons are greater than the repulsive forces between ions and ions, and between electrons and electrons. These electrons are not fixed to any one ion, but are shared by all and are free to move throughout the crystal. They are known as conduction electrons because they are the medium for the conduction of heat and electricity in metals.

As is evident from the name, the *molecular bond* is in actuality due to van der Waals forces. That is, it is the result of attraction between dipoles, either permanent or induced. The strength of this bond is inversely proportional to the sixth power of the distance, which can be compared to an inverse square dependence for the ionic bond. For this reason, as well as others, the molecular bond is generally much weaker than any of the other three. This explains the relative softness and weakness of molecular crystals.

Crystals are often classified according to the dominant type of force which binds them together. Thus we can speak of *ionic crystals*, *covalent crystals*, *metallic crystals*, and *molecular crystals*. Another classification is by the state of aggregation of the atoms in the crystal into smaller subunits. A crystal with no aggregation, where each atom or ion stands alone and is bound to all other atoms by only one kind of bond, is called a *coordination crystal*. Crystals with only ionic, covalent, or metallic bonds are all examples of coordination crystals. If the atoms are bound into molecules, which themselves make up the crystal structure, the crystal is called a *molecular crystal*, as mentioned before. Finally, the atoms may be most strongly bound into chains or layers (e.g. mica, graphite), in which case we have *chain* or *layer crystals*.

79-8 Lattice Structure. Unit Cells

We now return to a consideration of the microscopic structure of a crystal.

DEFINITION. The smallest unit of a crystal lattice which retains all of the symmetry of the macroscopic crystal is called a *unit cell.*

The basis parallelepiped mentioned in Section 79-4 is a unit cell.

In a so-called *primitive* lattice, there are no lattice points in the interior of a basis parallelepiped, and it is then called a *primitive unit cell* (Fig. 9-3). The eight points at the corners of a primitive unit cell, however, do not belong to that cell alone since eight such cells come together at every lattice point. If follows that only $\frac{1}{8} \times 8 = 1$ lattice point is associated with each primitive unit cell. If more than one lattice point belongs to a unit cell, it is no longer a primitive unit cell, but instead is called a *multiple* or *nonprimitive* unit cell. In general, the number of lattice points associated with each unit cell can be found by displacing the framework of the cell by a short distance Δ in an arbitrary direction relative to the lattice. The number of lattice points of the original cell which lie in the interior of the displaced cell is the number of lattice points associated with the cell. It will be seen in Fig. 9-10 that only one lattice point (which has been encircled) belongs to this unit cell.

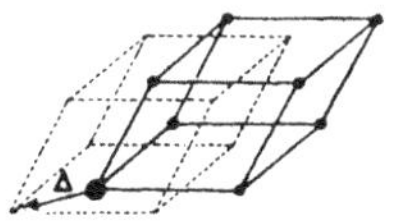

FIG. 9-10. Determination of the number of lattice points belonging to each unit cell.

The non-primitive unit cells are those belonging to the body-centered, face-centered, and base-centered lattice structures, of which there are seven (see Fig. 9-11 for examples). Each of these non-primitive lattices can be constructed from two or more primitive lattices of the same symmetry. The number of primitive lattices required is equal to the number of lattice points associated with one of the non-primitive unit cells.

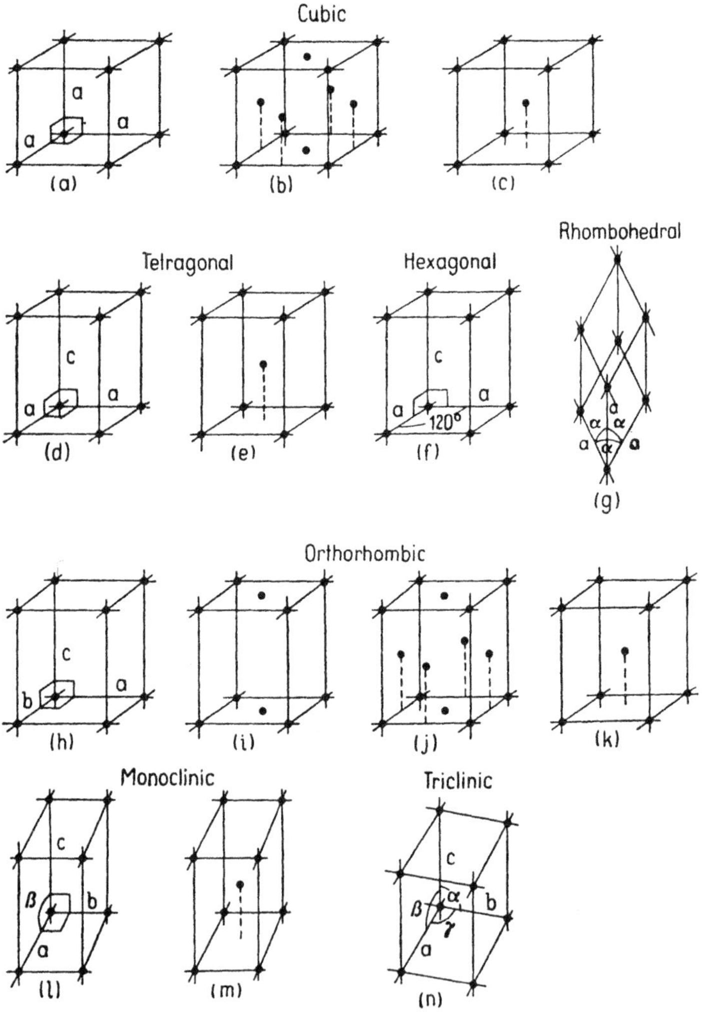

FIG. 9-11. The fourteen space lattices and their unit cells: (a) primitive cubic, (b) face-centered cubic, (c) body-centered cubic, (d) primitive tetragonal, (e) body-centered tetragonal, (f) hexagonal, (g) rhombohedral, (h) primitive orthorhombic, (i) base-centered orthorhombic, (j) face-centered orthorhombic, (k) body-centered orthorhombic, (l) primitive monoclinic, (m) body-centered monoclinic, (n) triclinic.

As an example, we will consider how a body-centered cubic lattice can be synthesized. We will need two primitive cubic lattices (Fig. 9-11 a). Both lattices must have the same edge length and orientation. Then if one of the lattices is fixed in such a position relative to the other that one of its lattice points lies at the midpoint of the body diagonal of a unit cell of the other lattice, the composite lattice will be a body-centered cubic lattice (Fig. 9-11 c). Each primitive unit cell of one lattice has a lattice point of the other in its exact center. The body-centered cubic unit cell thus contains two lattice points. All the other non-primitive lattices are constructed in a similar manner, with lattice points of one lattice lying at the *midpoints of the body or face diagonals* of the unit cell of another lattice.

79-9 Crystal Structure. Basic Cells

We now wish to relate the idealized lattice structures to the structure of actual crystals. This can be done in two ways. As a preliminary step, we define the *basic cell* of a crystal structure as the smallest cell which still has all the details of that structure. By means of infinite repetition of the basic cell, the whole crystal structure can be built up. In the basic cell there are particles rather than lattice points. Thus the basic cell is to the crystal structure what the unit cell is to the lattice structure. However, because the basic cell may contain many different kinds of particles, it is generally much more complicated than a unit cell.

The Bravais method is one of the ways of relating lattices to crystals. In this method we pick a point in each basic cell, called the Bravais point, all such points being in the same relative position in the cells (e.g. the center of gravity or one corner). These Bravais points make up a lattice which is congruent and parallel to the crystal structure and is called the *Bravais lattice*. We can describe the crystal structure by giving the coordinates of each particle in a basic cell relative to the Bravais point of the cell. This method is most useful for very complicated crystals having many kinds of particles. The configuration of the Bravais lattice determines the crystal system.

In the second method, we use one or more lattices to describe the positions of each kind of particle in the crystal structure, the atoms lying (only on the average, because of the thermal motion) at the lattice points. The several lattices are congruent and parallel to each other and to the crystal structure, and are considered to interpenetrate each other. We arbitrarily select one of the lattices as the primary one, and a corner point of one of its unit cells is taken as an origin. Then the relative positions of the other interpenetrating lattices are given by the coordinates of their corresponding corner points relative to this origin. These coordinates, $X_l = x_l a$, $Y_l = y_l b$, and $Z_l = z_l c$ are given in units of the lengths of the crystallographic axes a, b, and c (Fig. 9-12). The factors x_l, y_l, and z_l are all less than unity and are called the *basic coordinates* of the *i*th lattice relative to the first. They are written $\{x_i\ y_i\ z_i\}$. Of course, the basic coordinates of the primary lattice are $\{0,0,0\}$.

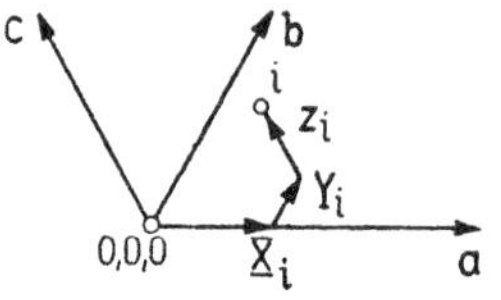

FIG. 9-12. Basic coordinates of the *i*th lattice.

The basic cell of a crystal structure made up of more than one *primitive* lattice will contain more than one particle. The number of particles can easily be determined using the displacement method of Fig. 9-10.

If the degree of symmetry is to remain unaltered by the superposition of several lattices, particles inside the basic cell must lie on the symmetry elements of the crystal system, or must be located symmetrically relative to them. Particles which are unsymmetrically located lead to *hemihedral* classes of crystals, i.e. crystals in which half of the crystal faces differ from the other half.

As a rule, it is easy to visualize a crystal structure whose basic cell is known because this cell repeats itself indefinitely in identical form. The basic cell is the fundamental element in the structure of a crystal and, for solid bodies, it plays a role similar to that of a molecule in the chemistry of gases. The mass associated with the basic cell is the same for all cells throughout the entire extent of the crystal.

Because of the fundamental significance of the concept of the basic cell, it is obviously important to undertake a description of the most important basic cell types that occur in nature. This is best done by first treating the simple basic cell types of the chemical elements and then proceeding to the many cell types of the chemical compounds. We will use the second method of describing crystal structures because its pictorial nature makes the structures easier to understand. This is true, at least, for the simple structures which we will discuss.

80. THE BASIC CELL TYPES OF THE ELEMENTS (A-TYPES)

Out of all the elements, most of which crystallize in several modifications, only one has a basic cell with but a single atom. This is mercury, which crystallizes in the rhombohedral system. For most elements, the lattice points are occupied by atoms; only a few have molecules. Most elemental crystals are held together by either metallic or covalent bonds, or by some hybrid which is a mixture of these two.

In describing any basic cell, the following details should be given if possible:

1. The number Z of particles in the basic cell.

2. The basic coordinates of all the lattices. This can be specified by giving the vector displacements from the origin of all the displaced unit cells or by a description in words.

3. The types of lattices, the kind of interlacing, and the lattice constants.

4. The environment or coordination; that is, the number of nearest neighbors of each particle, and their geometrical arrangement. For this purpose we introduce the *coordination number K*.

DEFINITION. The *coordination number* K of a particle is the number of nearest neighbors of the particle.

80-1 The Cubic System

In the cubic system all three axes are equivalent, that is, the lattice constant is the same on all three axes. The basic cell is therefore a cube of volume

$$V = a^3.$$

(a) *Primitive cubic structure* (Fig. 9-13).

For the basic cell of this structure, $Z = 1$, that is, only one atom belongs to the cell. All corners of the cube are occupied. The basic coordinates are $\{0\ 0\ 0\}$. The coordination number $K = 6$; i.e. every atom is surrounded octahedrally by 6 nearest neighbors. No known element crystallizes in a primitive cubic structure.

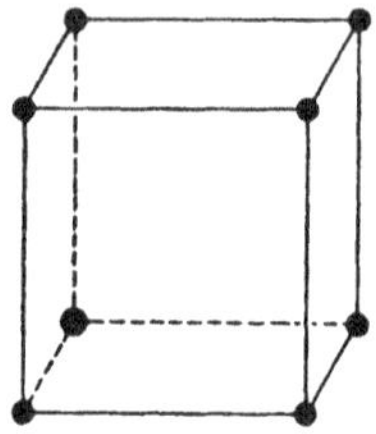

FIG. 9-13

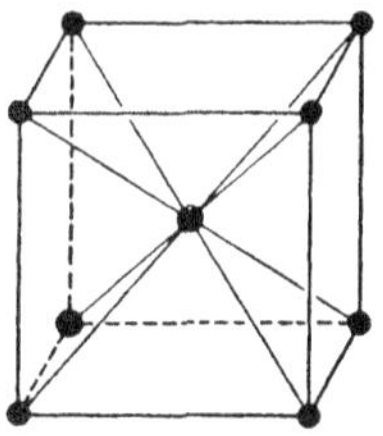

FIG. 9-14

FIG. 9-13. Primitive cubic basic cell.

FIG. 9-14. Body-centered cubic basic cell.

(b) *Body-centered cubic structure* (A2 type,† α-W type, Fig. 9-14).
$Z = 2$; two atoms in the basic cell.

Basic coordinates: $\left\{\begin{matrix} 0\ 0\ 0 \\ \frac{1}{2}\ \frac{1}{2}\ \frac{1}{2} \end{matrix}\right\}$

† The notation "A2 type" is part of a system of classification based on the chemical structure of a compound (see p. 420). This system is used in the crystallographic reference work, *Strukturberichte* and its successor, *Structure Reports*, Oosthoek, Utrecht.

The structure consists of two interpenetrating primitive cubic lattices with a relative translation of one-half a space diagonal, $K = 8$; each atom is surrounded *hexahedrally* by 8 others. Some crystals of this type and their lattice constants, a, are:

Element	Li	Na	K	Ba	α–Fe (20°C)	β–Fe (800°C)	δ–Fe (1425°C)	α–W (25°C)
a (Å)†	3.509	4.291	5.344	5.025	2.866	2.91	2.94	3.164

† At room temperature unless otherwise specified.

(c) *Face-centered cubic structure* (A1 type, Cu type; Fig. 9-15).

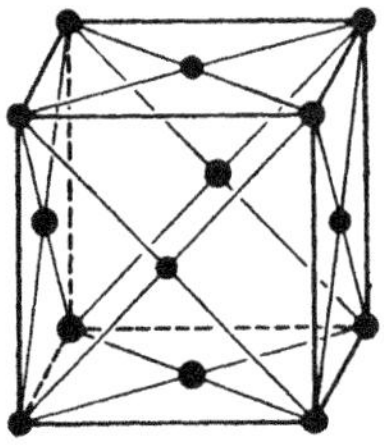

FIG. 9-15. Face-centered cubic basic cell.

$Z = 4$; i.e., four atoms in the basic cell.

$$\text{Basic coordinates:} \begin{Bmatrix} 0\ 0\ 0 \\ 0\ \tfrac{1}{2}\ \tfrac{1}{2} \\ \tfrac{1}{2}\ \tfrac{1}{2}\ 0 \\ \tfrac{1}{2}\ 0\ \tfrac{1}{2} \end{Bmatrix}.$$

The structure consists of four interpenetrating primitive cubic lattices. If the first gives the basic cube, then the corners of the other three are at the centers of the three faces of the basic cube which pass through the origin. Each is therefore translated along one-half a face diagonal. The centers of the other faces of the basic cube, and those of some of the faces of neighboring cells, are occupied by the remaining seven points of each of the displaced unit cells.

$K = 12$. Every atom has 12 nearest neighbors located in the form of a cubic octahedron.

This type of structure has a special significance since it corresponds to the three-layered, cubic closest packing of spheres, in contrast to the two-layered closest packing which leads to a hexagonal structure.

To understand this relationship imagine a face-centered cube to be oriented with one of its body diagonals vertical, as in Fig. 9-16. It will be seen that the atoms in the octahedral planes, [111], which are now horizontal, form a

FIG. 9-18. Two-layer or hexagonal closest packing of spheres in three stages of construction. This structure has hexagonal symmetry.

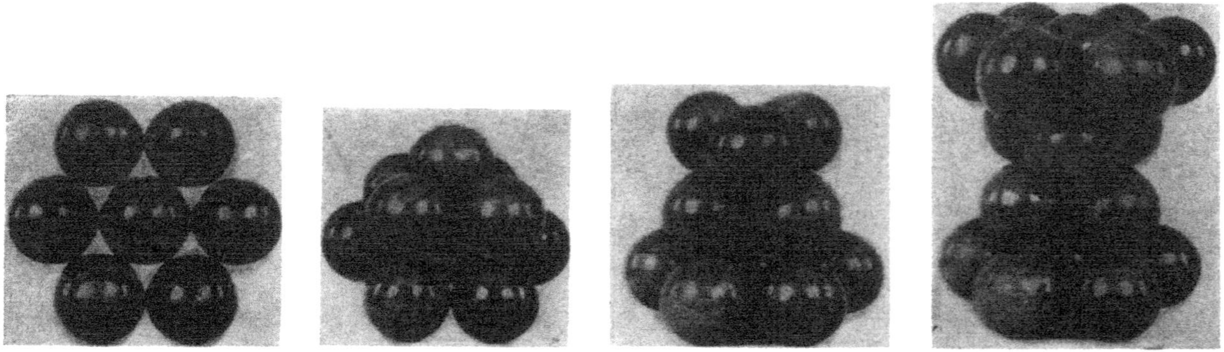

FIG. 9-19. Three-layer or cubic closest packing of spheres in four stages of construction. This structure has cubic symmetry, identical with that of the face-centered cubic type.

network of equilateral triangles. Figure 9-17 shows an extended network of such triangles. Now imagine that each atom is blown up like a balloon. The condition shown in a portion of the diagram will eventually result. Each sphere is surrounded, in this plane, by six similar spheres arranged in a ring. Around the central sphere are also six depressions alternately marked

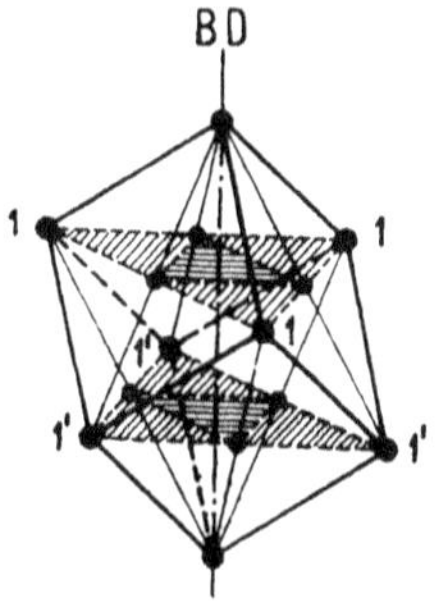

FIG. 9-16

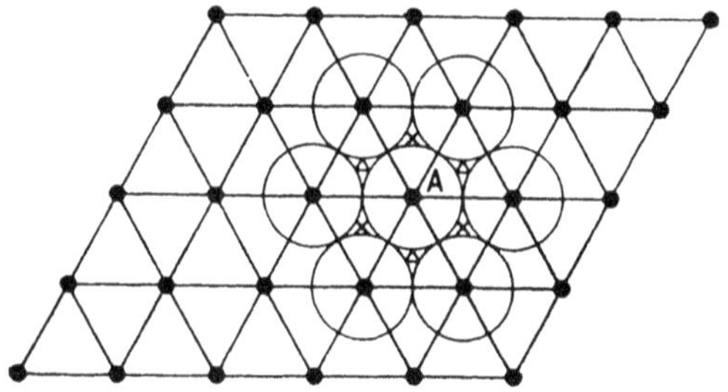

FIG. 9-17

FIG. 9-16. A face-centered cube with one of its body diagonals in a vertical direction. In the horizontal planes can be seen the typical triangular structure of closest spherical packing.

FIG. 9-17. Triangular structure, in a plane, of closest spherical packing.

Δ and x. To obtain the densest packing of spheres in space, the spheres in the next higher layer must rest in these depressions. It will be seen immediately that for lack of space only every other depression can be occupied. The next higher layer can therefore accommodate only three spheres which will touch sphere *A* and which we shall assume are placed in the depressions marked x. Their midpoints form an equilateral triangle with its apex pointing upward in the diagram. If the three depressions marked Δ had been filled, the apex of the triangle would point downward. The same considerations apply to the layer just below that shown in the diagram. Hence, in the closest three-dimensional packing of spheres, each sphere is surrounded by 12 nearest neighbors.

However, it is possible to continue the packing in two different ways. If the spheres in the third layer are placed vertically above those in the first, alternate layers are arranged in the same way (Fig. 9-18, Plate V). This arrangement is known as *two-layer* or *hexagonal closest packing* of spheres. If, on the other hand, we repeat in the third layer the neglected possible arrangement, there is no sphere vertically above *A* until the fourth layer is reached. We then have the *three-layer* or *cubic closest packing* of spheres. This type of structure is shown in four steps of construction in Fig. 9-19, Plate V. It can be seen that the face-centered cubic lattice, with a space diagonal of the cube vertical corresponds exactly to this type of packing since the triangles in Fig. 9-16 are arranged with their apices pointing alternately in opposite directions.

Both types of closest packing occur frequently in nature. Other types are theoretically possible, but seldom occur.

The following elements crystallize in a face-centered cubic lattice (room temperature unless otherwise given):

Element	Al	γ-Fe (1100°C)	β-Ni	Cu	Ag	Ir	Pt	Au	Pb
a (Å)	4.049	3.64	3.524	3.615	4.086	3.839	3.924	4.078	4.950

Note that the basic cells of the primitive cubic, body-centered cubic, and face-centered cubic crystal structures are identical to the unit cells of the space lattices of the same name (Fig. 9-12). This is true in the other crystal systems also.

(d) *The diamond-type structure* (A4 type, Fig. 9-20). $Z = 8$.

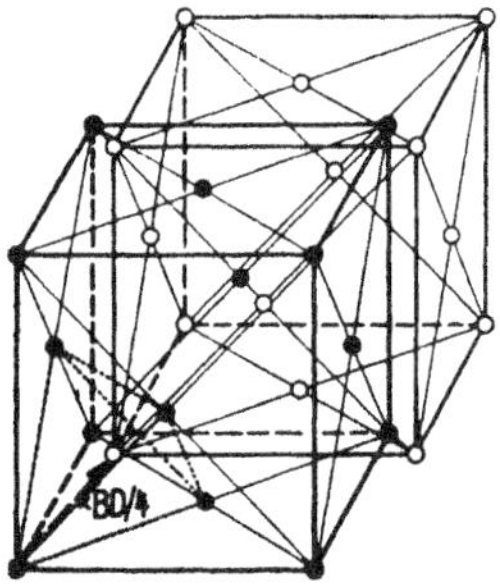

FIG. 9-20. Basic cell of the diamond type; two interlocking face-centered cubic lattices, with a displacement of one-fourth the body diagonal, BD/4.

$$\text{Basic coordinates:} \left\{ \begin{array}{c|c} 0\,0\,0 & \frac{1}{4}\,\frac{1}{4}\,\frac{1}{4} \\ 0\,\frac{1}{2}\,\frac{1}{2} & \frac{1}{4}\,\frac{3}{4}\,\frac{3}{4} \\ \frac{1}{2}\,\frac{1}{2}\,0 & \frac{3}{4}\,\frac{3}{4}\,\frac{1}{4} \\ \frac{1}{2}\,0\,\frac{1}{2} & \frac{3}{4}\,\frac{1}{4}\,\frac{3}{4} \end{array} \right\}$$

This structure consists of two interpenetrating face-centered cubic lattices, one of which is displaced relative to the other along $\frac{1}{4}$ of a body diagonal. Because of this displacement, the structure does not have the high degree of symmetry of the cubic system. There are three principal planes of symmetry. Each atom is surrounded by 4 others (all equidistant) arranged in a tetrahedron (one of which is shown by dotted lines). Since the face-centered cube consists of four interpenetrating elementary cubes, we have here an interlacing of 8 primitive cubic lattices. The coordination number $K = 4$.

Elements crystallizing in this way are (lattice constants measured at room temperature):

Element	α-C (Diamond)	Si	Ge	α-Sn (gray)
a (Å)	3.56696	5.431	5.658	6.491

80-2 The Hexagonal System

The basic cell of the hexagonal system is a parallelepiped (Fig. 9-21) whose base is a rhombus with sides of length a. The obtuse angle of the rhombus equals 120°, so that its shorter diagonal divides it into two equilateral triangles. The c-axis is not equivalent to the other two and the *axis ratio* $p = c/a$ is different for different substances. The volume of the basic cell is

$$V = \frac{a^2\sqrt{3}}{2}c = \frac{a^3\sqrt{3}}{2}p.$$

(a) *The primitive hexagonal structure* (Fig. 9-21).

$Z = 1$. Each corner of the cell is occupied, but only one atom belongs to the cell.

Basic coordinates: $\{0\ 0\ 0\}$.

No substance crystallizes in this type of lattice.

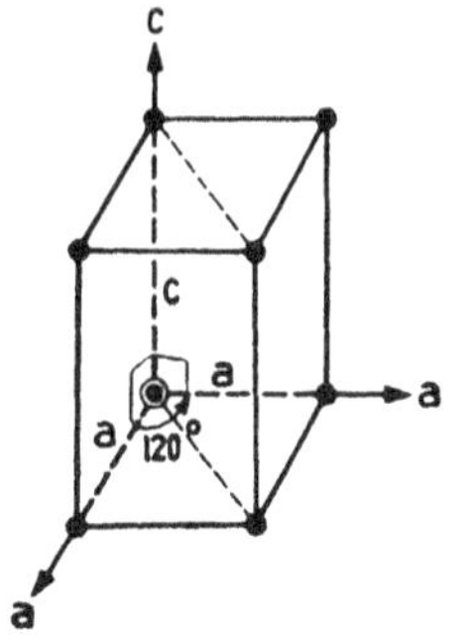

FIG. 9-21

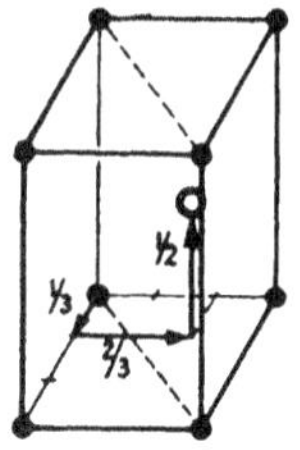

FIG. 9-22

FIG. 9-21. The basic cell of the primitive hexagonal structure.
FIG. 9-22. The basic cell for hexagonal closest packing.

(b) *Hexagonal close packed structure* (A3 type or Mg type; Fig. 9-22).

$Z = 2$. The structure is built up of two interpenetrating primitive hexagonal lattices.

Basic coordinates: $\left\{\begin{matrix} 0 & 0 & 0 \\ \frac{1}{3} & \frac{2}{3} & \frac{1}{2} \end{matrix}\right\}.$

The origin of the displaced lattice lies vertically above the center of gravity of one of the equilateral triangles into which the base can be divided.

Its height above the base is $c/2$. This type of structure is of special significance because it corresponds to the two-layer closest packing of spheres, as mentioned above. Because the second possible method of filling the depressions in a plane packing of spheres (see Fig. 9-17) is not utilized, the symmetry of the crystal is less than that of three-layer packing. The c-axis is perpendicular to the planes which contain the six-cornered structure, which leads automatically to hexagonal symmetry.

The axis ratio $p = c/a$ is easily calculated as follows. Above any three spheres of one layer lies a single sphere of the next layer, the centers of these spheres forming a regular tetrahedron $ABCD$ (Fig. 9-23a). Making use of

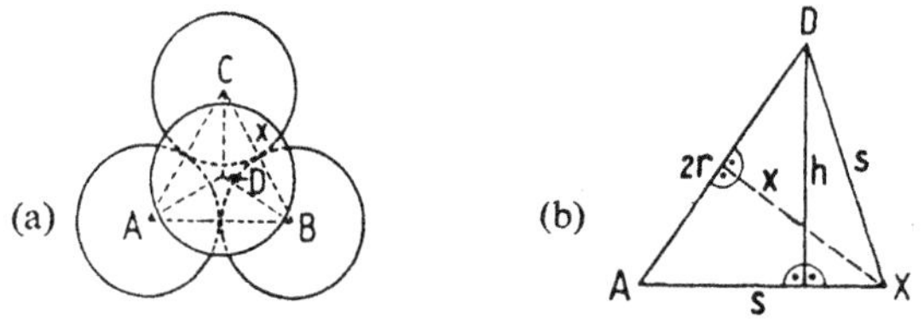

FIG. 9-23. Figure for the calculation of the axis ratio, $p = c/a$, of the two stage closest spherical packing ($s = r\sqrt{3}$).

the plane section ADX (Fig. 9-23b), and letting r be the radius of a sphere, the height h of the tetrahedron is seen to be

$$h = \frac{2r\sqrt{6}}{3}.$$

The length of the c-axis is $2h$. Since the lattice constant a in the base plane equals $2r$, it follows that

$$c = 2h = \frac{2a\sqrt{6}}{3}$$

and

$$\frac{c}{a} = \frac{2\sqrt{6}}{3} = 1.633.$$

Therefore if p is found to equal 1.633 for an A3 type crystal, we can conclude that the structure is packed with spherical atoms. On the other hand, if $p \neq 1.633$, the structure is packed with ellipsoids of rotation with a greater or less elongation or flattening in the c-direction.

Note: The basic cell for cubic closest packing of spheres has a height

$$c = 3h = 2r\sqrt{6} = a\sqrt{6} = w\sqrt{3},$$

where w is the length of an edge of the basic cell of the face-centered cubic structure (Fig. 9-15):

$$w = a\sqrt{2} = 2\sqrt{2}r.$$

This agrees with the fact that $a\sqrt{6}$ equals the length of the body diagona of a cube whose edges are $w = a\sqrt{2}$ (Fig. 9-16). We can see that indeed this is the cubic closest packing of spheres because the smallest distance between the centers of two spheres is equal to half the diagonal of a cube face, or

$$\frac{1}{2}\sqrt{(2)}w = \frac{w}{\sqrt{2}} = 2r.$$

The following substances crystallize in a lattice of the A3 or Mg type (lattice constants measured at room temperature):

Element	Mg	α-Ni	Zn
a (Å)	3.209	2.67	2.665
c (Å)	5.210	4.30	4.947
$p = c/a$	1.624	1.61	1.856

(c) *The α-graphite structure* (A9 type; Fig. 9-24).
$Z = 4$.

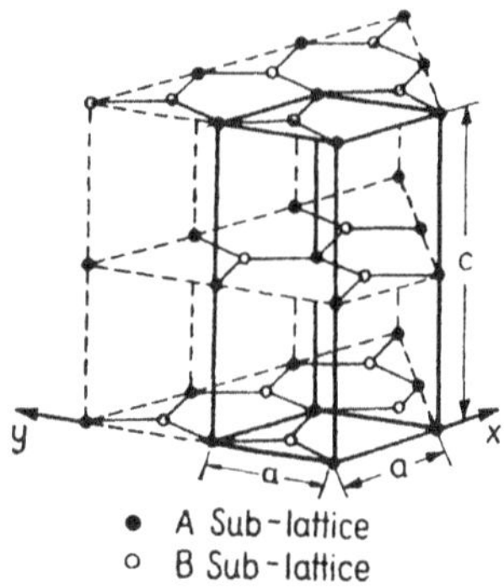

FIG. 9-24. Basic cell and layered structure of the α-graphite type structure.

This ordinary type of graphite crystallizes in a structure which has some similarities to hexagonal closest packing. The structure is formed from two similar interpenetrating lattices, each of which is itself made up of two interpenetrating primitive hexagonal lattices; or four interpenetrating primitive hexagonal lattices in all.

$$\text{Basic coordinates:} \left\{ \begin{array}{c|c} \text{(A)} & \text{(B)} \\ 0\ 0\ 0 & \frac{1}{3}\ \frac{2}{3}\ u \\ 0\ 0\ \frac{1}{2} & \frac{2}{3}\ \frac{1}{3}\ \frac{1}{2}+u \end{array} \right\}$$

The parameter u is small (<0.05), and hence this structure consists of parallel planes of linked hexagons of carbon atoms. The atoms in *alternate* planes of sub-lattice B lie directly over and under each other in analogy to the hexagonal close packed structure. However, *all* the planes of sub-lattice A are identical, the atoms lying directly over and under each other. The planes are more widely spaced (3.40 Å) along the c-axis than in hexagonal close packing. Within the planes the interatomic distances are all equal (1.42 Å) and are about equal to the $C-C$ spacings in the benzene ring. Since the parameter u is not zero, alternate atoms in each ring are displaced slightly up and down from each other, so that the planes are slightly wrinkled.

As mentioned in Section 79-7, the van der Waals force acts between the layers while covalent bonds hold the atoms together within a layer. The relative weakness of the former compared to the latter is the reason why the distance between layers is much larger than between atoms in a layer. This is also the cause of the pronounced cleavage and flaky nature of graphite. The lattice constants for graphite are:

$$a = 2.466\,\text{Å}$$

$$c = 6.721\,\text{Å}$$

The quantity $p = c/a = 2.725$ is much greater than in the case of hexagonal closest packing.

80-3 The Rhombohedral System

In this system there are three equivalent axes a, a, a, with equal angles $\alpha \neq 90°$ between each pair of axis. The basic cell is therefore of *rhombohedral* form (an inclined parallelepiped, each of whose faces is a rhombus having sides of length a, Fig. 9-25). At heights of one-third and two-thirds

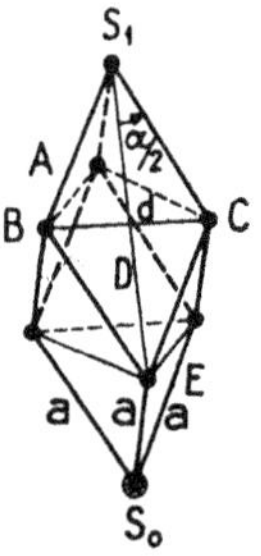

FIG. 9-25. Rhombohedral basic cell.

of the long body diagonal of the rhombohedron there are located three lattice points at the corners of equilateral triangles. The length d of one side of these triangles is equal to the shorter diagonal of a surface rhombus:

$$d = 2a \sin \frac{\alpha}{2}.$$

The length D of the longer diagonal of a surface rhombus is

$$D = 2a \cos \frac{\alpha}{2}$$

If the rhombohedron rests on one of its faces, and if β represents the angle between a side of length a and this face (Fig. 9-26a), the height h of the parallel face is

$$h = a \sin \beta$$

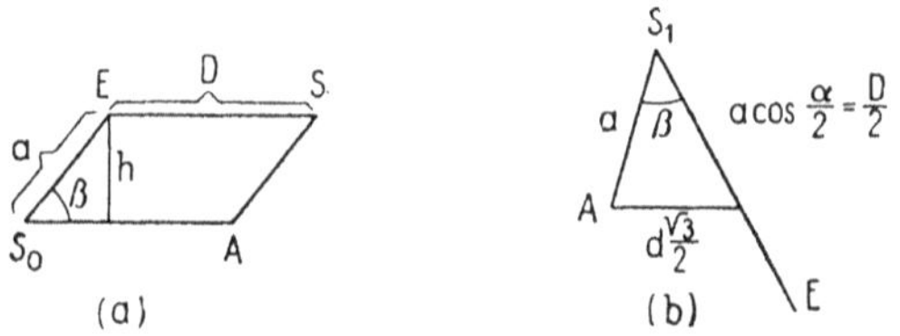

FIG. 9-26. Figure for the calculation of the volume of the rhombohedral basic cell.

and the volume of the unit cell is therefore

$$V = \frac{dD}{2} h = a^3 \sin \alpha \sin \beta .$$

The angle β can be found by considering a plane section through the tetrahedron $ABCS_1$ (Fig. 9-25), the section passing through the edge S_1A and the rhombus diagonal S_1E (Fig. 9-26b). From the cosine law,

$$\cos \beta = \frac{\cos \alpha}{\cos \frac{\alpha}{2}} .$$

(a) *The rhombohedral structure* (A 10 type, Fig. 9-24).

$Z = 1$. All corners of the rhombohedron are occupied, but only one atom is associated with the cell.

Basic coordinates: $\{ 0\ 0\ 0 \}$.

The rhombohedral structure is a layer structure. The distance between layers is one-third of the body diagonal of the rhombohedron. The layers consist of equilateral triangles with an atom at each corner. The centers of gravity of the triangles in the second layer lie vertically above the atoms in the first layer and the apices of the triangles point, say, forward. The centers of gravity of the triangles in the third layer are also vertically above the atoms in the first layer, but the apices of the triangles point in the opposite direction. The fourth layer is a repetition of the first. The structure is somewhat analogous to that of cubic closest packing.

Mercury crystallizes in this structure and has the following lattice constants at $-46°C$:

$$a = 3.005\ \text{Å},$$
$$\alpha = 70°\ 32'.$$

(b) *The β-graphite structure* (Fig. 9-27).

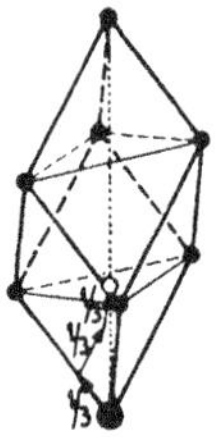

FIG. 9-27. The basic cell of the β-graphite type structure.

$Z = 2$. To a very good approximation, this type of structure can be considered as an interpenetration of two simple rhombohedral structures, displaced by one-third of the body diagonal of the rhombohedron.

$$\text{Basic coordinates:} \begin{Bmatrix} 0\ 0\ 0 \\ \frac{1}{3}\ \frac{1}{3}\ \frac{1}{3} \end{Bmatrix}.$$

The position of the atom enclosed in the basic cell is reached by displacements of $a/3$ along each of the three axis directions, as is indicated by the vectors in Fig. 9-27. This point lies on the body diagonal at a height $l/3$ where l is the length of the body diagonal of a basic cell. That is, the atom lies at the center of gravity of the lower triangle.

The structure of β-graphite and α-graphite are similar in that they both consist of layers of hexagonally bonded carbon atoms. Thus the individual layers of β-graphite are just like those shown in Fig. 9-24. However, where α-graphite shows a similarity to hexagonal closest packing in its two-layered structure, β-graphite is the analog of cubic closest packing and has a three-layered structure. The first two layers are oriented just like those of α-graphite, and the third is rotated about the c-axis through an angle of 60°. The distance between the layers and the $C-C$ spacings in each layer are the same as in α-graphite, 3.40 Å and 1.42 Å respectively. In terms of the basic cell dimensions the spacing between layers is $l/3$. The lattice constants for β-graphite are:

$$a = 3.642\ \text{Å},$$
$$\alpha = 39°\ 30'.$$

80-4 The Tetragonal System

The tetragonal system has two equivalent axes and a third non-equivalent axis, all three axes being mutually perpendicular. The basic cell is therefore a rectangular parallelepiped of volume

$$V = a^2c.$$

(a) *The primitive tetragonal basic cell* (Fig. 9-28).
$Z = 1$.
Basic coordinates: $\{0\ 0\ 0\}$.

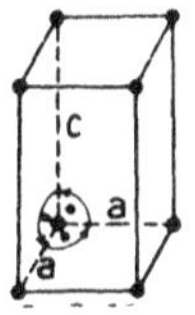

FIG. 9-28. The basic cell of the tetragonal system.

A crystal structure in which atoms occur only at the corners of the tetragonal basic cell never occurs in nature.

(b) *The indium type structure* (A6 type, also called face-centered tetragonal type, Fig. 9-29).

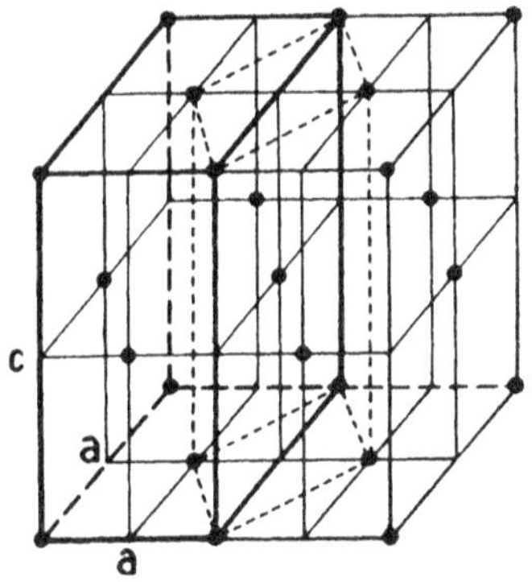

FIG. 9-29. The dotted lines show how the face-centered tetragonal type structure can also be represented as a body-centered structure with a lattice constant $\sqrt{2}/2$ times smaller.

$Z = 4$. Four interpenetrating primitive tetragonal basic cells with a translation of $\frac{1}{2}$ the face diagonals form the structure. All corners, and the mid-points of all faces, are occupied.

$$\text{Basic coordinates: } \left\{ \begin{matrix} 0\ 0\ 0 \\ 0\ \frac{1}{2}\ \frac{1}{2} \\ \frac{1}{2}\ \frac{1}{2}\ 0 \\ \frac{1}{2}\ 0\ \frac{1}{2} \end{matrix} \right\}.$$

The structure can also be represented as one with the a lattice constant $\sqrt{2}/2$ times smaller, and with the a axes rotated by 45°. It then becomes a body-centered tetragonal lattice with $Z = 2$ and basic coordinates

$$\left\{ \begin{matrix} 0\ 0\ 0 \\ \frac{1}{2}\ \frac{1}{2}\ \frac{1}{2} \end{matrix} \right\}.$$

In Fig. 9-29 this smaller cell is indicated by dotted lines. Elements crystallizing in this type are:

Element	In	γ-Mn
a (Å)	4.598	3.774
c (Å)	4.947	3.533
$p = c/a$	1.076	0.936

(c) *The white tin type structure* (A5 type, also called tetragonal diamond type, Fig. 9-30).

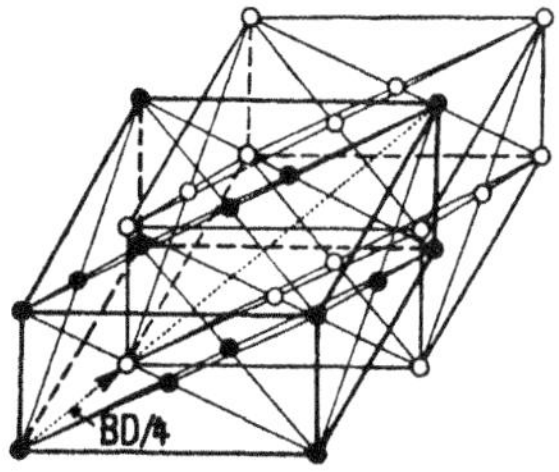

FIG. 9-30. The basic cell of the white tin type structure.

$Z = 8$. Two interpenetrating face-centered tetragonal structures with a displacement of $\frac{1}{4}$ of the body diagonal. The arrangement is very similar to the cubic diamond type and has a coordination number $K = 4$. Each point is surrounded tetrahedrally.

$$\text{Basic coordinates:} \left\{ \begin{array}{c|c} 0\,0\,0 & \frac{1}{4}\,\frac{1}{4}\,\frac{1}{4} \\ 0\,\frac{1}{2}\,\frac{1}{2} & \frac{1}{4}\,\frac{3}{4}\,\frac{3}{4} \\ \frac{1}{2}\,\frac{1}{2}\,0 & \frac{3}{4}\,\frac{3}{4}\,\frac{1}{4} \\ \frac{1}{2}\,0\,\frac{1}{2} & \frac{3}{4}\,\frac{1}{4}\,\frac{3}{4} \end{array} \right\}$$

White tin (β-tin) crystallizes in this lattice, and is stable in the temperature range from 13°C to 203°C, with lattice constants, $a = 8.228$ Å, $p = c/a = 0.3857$. Like indium, white tin can also be thought of as having a structure with the a-axes rotated 45° and decreased by the factor $\sqrt{(2)}/2$.

80-5 The Orthorhombic System

The orthorhombic system has three mutually perpendicular non-equivalent axes. The basic cell is a rectangular parallelepiped of volume

$$V = abc.$$

(a) *The primitive orthorhombic structure* (Fig. 9-31).
$Z = 1$.

Basic coordinates: $\{0\ 0\ 0\}$.

All corners of the cell are occupied, but only one point belongs to the cell. This type of structure is never observed in nature.

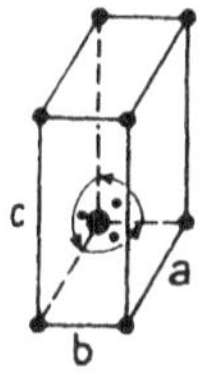

FIG. 9-31

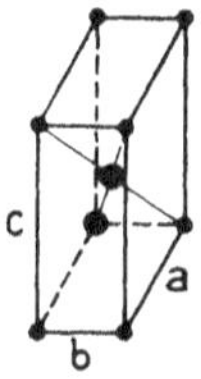

FIG. 9-32

FIG. 9-31. The orthorhombic basic cell.

FIG. 9-32. The basic cell of the body-centered orthorhombic structure.

(b) *The body-centered orthorhombic type* (Fig. 9-32).

$Z = 2$. Two interpenetrating primitive orthorhombic lattices with a displacement of $\frac{1}{2}$ a body diagonal.

$$\text{Basic coordinates:} \quad \left\{ \begin{matrix} 0\;0\;0 \\ \frac{1}{2}\;\frac{1}{2}\;\frac{1}{2} \end{matrix} \right\}.$$

This type does not occur alone.

(c) *The base-centered orthorhombic type* (Fig. 9–33).

$Z = 2$. Two interpenetrating primitive lattices with a displacement of $\frac{1}{2}$ the diagonal of the base surface.

$$\text{Basic coordinates:} \quad \left\{ \begin{matrix} 0\;0\;0 \\ \frac{1}{2}\;\frac{1}{2}\;0 \end{matrix} \right\}.$$

This type also does not occur alone.

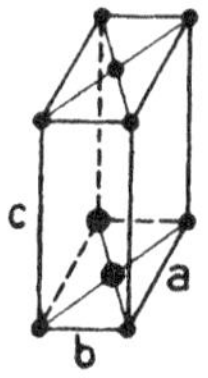

FIG. 9-33

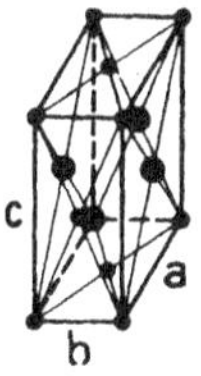

FIG. 9-34

FIG. 9-33. The basic cell of the base-centered orthorhombic structure.

FIG. 9-34. The basic cell of the face-centered orthorhombic structure.

(d) *The face-centered orthorhombic type* (A14 type, Fig. 9-34).

$Z = 4$. Four interpenetrating primitive lattices with displacements of $\frac{1}{2}$ face diagonal.

Basic coordinates: $\left\{\begin{matrix} 0\,0\,0 \\ 0\,\tfrac{1}{2}\,\tfrac{1}{2} \\ \tfrac{1}{2}\,\tfrac{1}{2}\,0 \\ \tfrac{1}{2}\,0\,\tfrac{1}{2} \end{matrix}\right\}$.

Iodine and bromine molecules crystallize in structures of this type. Three parameter are necessary to define the positions of the molecules:

Element	I_2 (26°C)	Br_2 (−150°C)
a (Å)	4.792	4.48
b (Å)	7.271	6.68
c (Å)	9.803	8.72

80-6 Relation of the Basic Cell to the Density

When we know the basic cell structure of a crystal, and the kind or kinds of atoms of which it is composed, the mass of the basic cell is determined. Since the basic cell is the same throughout the entire extent of the crystal, the density of the crystal can be found if the volume of the basic cell is known. Let m_i be the mass of the ith type of atom in the unit cell, Z_i the number of atoms of the ith type in the basic cell, and V_c the volume of the basic cell. The density is then

$$\varrho = \frac{\sum Z_i m_i}{V_c} = \frac{\sum Z_i A_i^*}{N_0 V_c},$$

since

$$m_i = \frac{A_i^*}{N_0}.$$

The volume of the basic cell can be found from the lattice constants along the three coordinate axes and the angles between the axes.

The relation between ϱ, Z, and V_c is very useful in the checking of lattice constants and occupation number. It also gives us a very good way to calculate Avogadro's number.

EXAMPLES. 1. Silver crystallizes in an Al type lattice. Compute the density ϱ, given that

$$a\,(18°\text{C}) = 4.086\ \text{Å},$$

$$A^* = 107.88\ \text{gm/mole},$$

$$Z = 4,$$

$$\varrho = \frac{ZA^*}{N_0 V_c} = 10.50\ \text{gm/cm}^3.$$

(The experimental value is $\varrho_{exp} = 10.50\ \text{gm/cm}^3$.)

2. Tantalum crystallizes in an A2 type lattice. Compute the lattice constant a, given that:

$$Z = 2,$$

$$\varrho = 16.6 \text{ gm/cm}^3,$$

$$A^* = 180.95 \text{ gm/mole},$$

$$a = \sqrt[3]{\left(\frac{ZA^*}{N_0 \varrho}\right)} = 3.307 \text{ Å}.$$

(The experimental value is $a_{exp} = 3.303$ Å.)

81. Some Basic Cell Types of Inorganic Compounds

81-1 General Remarks

Compounds crystallize in a wide variety of lattice structures. Space will permit us to give only a general survey of the problem, and to describe a few of the most common types.

According to the picture of crystal structure which we are using, each kind of atom (or ion, or radical) of a compound is arranged in a lattice of its own, and these partial lattices interpenetrate to form the complete lattice. For instance, in NaCl, the Na^+ ions and Cl^- ions each form a face-centered cubic lattice of their own, and these two lattices interpenetrate to form the complete crystal.

The reason for the existence of a great many different basic types of crystal structure lies in the structure of the atoms. Three of the determining factors of the type of structure are the *binding number n*, the *coordination number K*, and the *polarizability* of the particles.

The *binding number n* is the number of atoms of element B which can combine with an atom of element A to form a neutral molecule. This number is determined on purely chemical grounds by the ratio of the valences of A and B. If the formula of the compound is A_xB_y, the binding number of atom A, with respect to atom B, is

$$n_{A,B} = y/x\,.$$

The binding number of atom B, with respect to atom A, is

$$n_{B,A} = x/y\,.$$

The *coordination number K* of an atom is the number of its nearest neighbors. For most crystals, the number of atoms of element B which surround an atom of element A is the coordination number for atoms of element A, and vice versa. This number depends on the sizes of the atoms and, if these are considered as rigid particles, is determined by the ratio of the ionic or atomic radii. In some cases, it also depends on details of the binding in the crystal.

Finally, the *polarizability* of an atom or ion is a measure of the extent to which it is deformed by the electric fields of neighboring atoms.

If, in a given compound, an atom can bind to itself by its valence properties the same number of atoms as it can group around itself in the most efficient utilization of space, the resulting structure is said to be *saturated* and does not tend to any further extension. An example of such a case is CH_4. Such compounds usually exist as gases at ordinary temperatures, and if by lowering the temperature they are condensed to the solid state, they always form molecular crystals.

If, on the other hand, the binding number n is smaller than the maximum possible coordination number K_{max}, an ion surrounds itself with K_{max} oppositely charged ions even though it can neutralize the charges of only n of these ions. It therefore attracts additional like-charged ions in order to neutralize the electric charge. The continuation of this process results in the formation of an ionic lattice. The classical concepts of chemical binding no longer apply, in that one can no longer say that an ion is bound to any particular one of its oppositely charged neighbors.

The lattices formed by chemical compounds are classified on a basis of chemical structure. Some of the possible types are listed below.

B-type: Compounds of the form AB.
C-type: Compounds of the form AB_2.
D-type: Compounds of the form A_mB_n.
G-type: Compounds with a radical which contains four atoms.

81-2 Examples of B-type Structures

The configurations of these multiple cell structures are best described by a specification of the interpenetrating lattices and their relative displacements. This avoids the necessity of listing a large number of basic coordinates.

1. B1 type (NaCl type, Fig. 9-35).

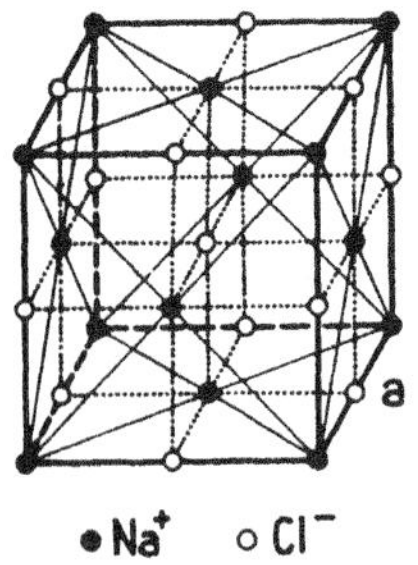

FIG. 9-35. NaCl type structure.

Ionic crystal. $Z = 4$ (i.e., 4 NaCl in the basic cell). Cubic symmetry.
Na^+: Face-centered cubic, lattice constant a.
Cl^-: Face-centered cubic, lattice constant a.
Displacement: $\frac{1}{2}$ a body diagonal or $\frac{1}{2}$ the length of one side.

Coordination: Each Na^+ is surrounded octahedrally by 6 Cl^-, and each Cl^- is surrounded octahedrally by 6 Na^+. $K = 6$ for both ions.

2. B2 type (CsCl type, Fig. 9-36).

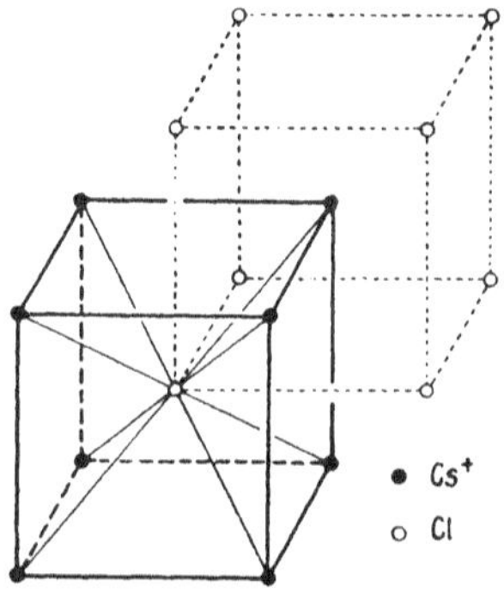

FIG. 9-36. CsCl type structure.

Ionic crystal. $Z = 1$. Cubic symmetry.
Cs^+: Primitive cubic, lattice constant a.
Cl^-: Primitive cubic, lattice constant a.
Displacement: $\frac{1}{2}$ body diagonal.

Coordination: Each Cs^+ is surrounded hexahedrally by 8 Cl^-, and each Cl^- is surrounded hexahedrally by 8 Cs^+. $K = 8$ for both ions.

3. B3 type (Zincblende type, Fig. 9-37).

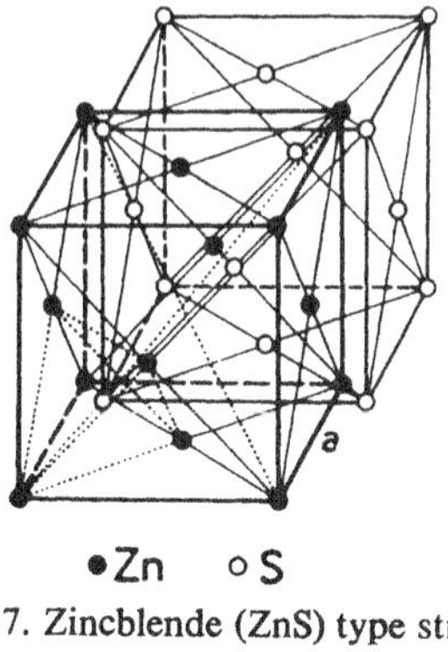

FIG. 9-37. Zincblende (ZnS) type structure.

Covalent crystal. $Z = 4$. Cubic symmetry.
Zn: Face-centered cubic, lattice constant a.
S: Face-centered cubic, lattice constant a.
Displacement: $\frac{1}{4}$ body diagonal.

This structure is the same as the diamond structure, except that the displaced lattice is composed of a different kind of ion (see Fig. 9-20).

Coordination: Each Zn is surrounded tetrahedrally by 4 S, and each S is surrounded tetrahedrally by 4 Zn. $K = 4$ for both atoms.

4. B4 type (Wurtzite type, Fig. 9-38).

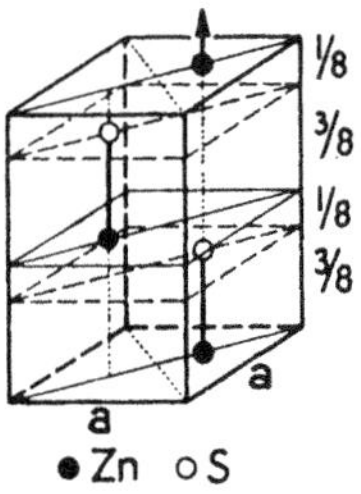

FIG. 9-38. Wurtzite (ZnS) type structure.

Covalent crystal. $Z = 2$. Hexagonal symmetry.

Zn: Hexagonal closest packing of spheres, lattice constants a, c.

S: Hexagonal closest packing of spheres, lattice constants a, c.

The two lattices are displaced relative to each other by a distance $u \cong \frac{3}{8}$ in the c-direction. Since the axis ratios are $c/a = 1.633$, the crystal consists of two sets of closely packed spheres. The basic coordinates of the four lattices in the hexagonal basic cell† are:

$$\text{Zn:}\left\{\begin{matrix} \frac{1}{2} & \frac{2}{3} & 0 \\ \frac{2}{3} & \frac{1}{3} & \frac{1}{2} \end{matrix}\right\}, \quad \text{S:}\left\{\begin{matrix} \frac{1}{3} & \frac{2}{3} & u \\ \frac{2}{3} & \frac{1}{3} & \frac{1}{2} + u \end{matrix}\right\}.$$

This sort of packing results in a layer structure in which, in the c-direction, there are alternating planes containing either S or Zn atoms arranged in

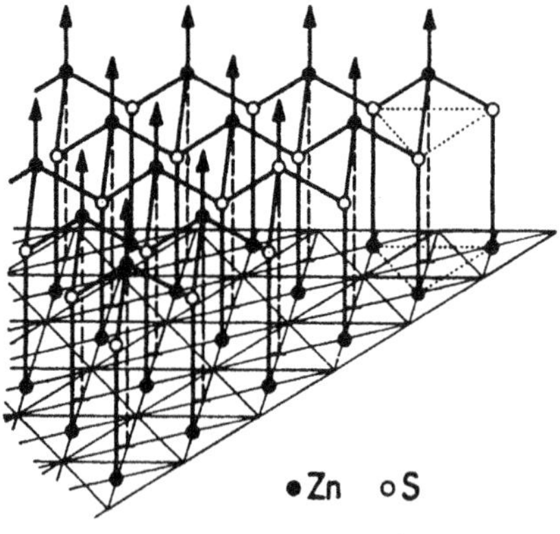

FIG. 9-39. Layer structure of Wurtzite type crystal.

† Figure 9-38 shows the basic cell with the two characteristic Zn–S rods located over the center of gravity of the half-rhombs in the base. The distances between layers are indicated at the right.

a typical triangular structure. The distances between consecutive planes are, to a good approximation, $\frac{3}{8}$, $\frac{1}{8}$, $\frac{3}{8}$, $\frac{1}{8}$, etc. Figure 9-39 shows a series of three such layers and their triangular networks.

Coordination: Each Zn is surrounded tetrahedrally by 4 S, and each S by 4 Zn. $K = 4$ for each atom.

5. B8 type (Nickel-arsenide type, Fig. 9-40).

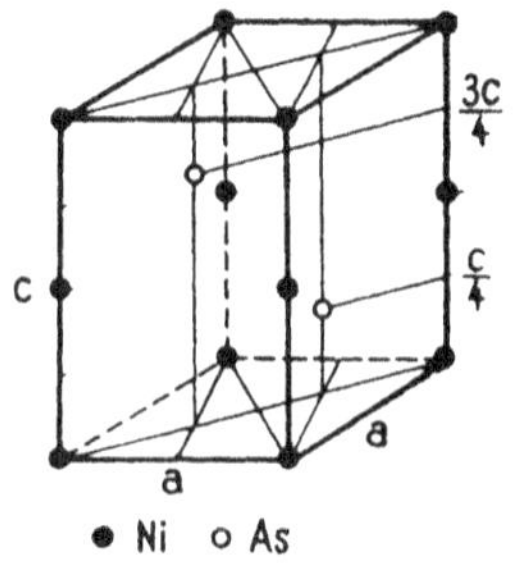

FIG. 9-40. Nickel-arsenide (NiAs) type structure (see Fig. 9-41).

Covalent crystal. $Z = 2$. Hexagonal symmetry.

Ni: Two hexagonal lattices, displaced relatively by a distance $c/2$ along the c-axis, contain the nickel atoms.

As: One hexagonal lattice has its origin at a height $u = \frac{1}{4}$ above the center of gravity of one triangle of the base rhombus. A second hexagonal lattice is located with its origin above the center of gravity of the other triangle, at a height $\frac{1}{2} + u = \frac{3}{4}$. The basic coordinates of the four lattices are:

$$\text{Ni}: \begin{Bmatrix} 0\,0\,0 \\ 0\,0\,\frac{1}{2} \end{Bmatrix}, \quad \text{As}: \begin{Bmatrix} \frac{1}{3}\,\frac{2}{3}\,u \\ \frac{2}{3}\,\frac{1}{3}\,\frac{1}{2}+u \end{Bmatrix}.$$

Coordination: Each Ni is surrounded octahedrally by 6 As; each As is surrounded by 6 Ni atoms arranged in a triangular prism. $K = 6$ for each atom.

The tables on the next page list some of the compounds which crystallize in the types described above.

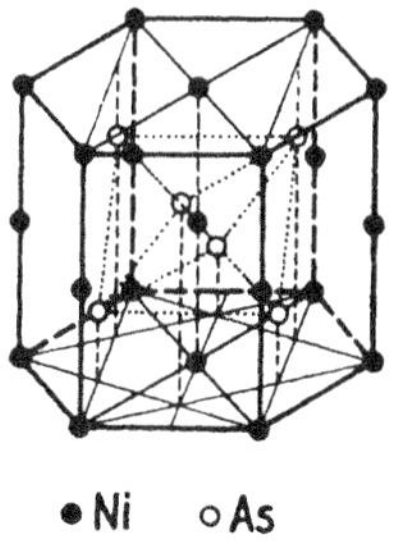

FIG. 9-41. Three basic cells of the NiAs type. The octahedron of As atoms surrounding one Ni atom is shown with dotted lines.

B1 type:

Substance	NaCl	NaBr	NaI	NH_4Cl	AgCl	MgO	CaO	FeO	PbS
a (Å)	5.640	5.973	6.473	6.53 250°C	5.556	4.203	4.797	4.294	5.936

B2 type:

Substance	CsCl	CsBr	CsI	NH_4Cl
a (Å)	4.121	4.296	4.567	3.876

B3 type:

Substance	HgS	SiC(β)	ZnS(β)
a (Å)	5.852 Metacinnabarite	4.439	5.423 Zincblende

B4 type:

Substance	ZnS(α)	NH_4F	ZnO	
a (Å)	3.819	4.39	3.249	$u \cong 3/8$
c (Å)	6.246	7.02	5.207	
c/a	1.636 Wurtzite	1.60	1.603	

B8 type:

Substance	NiAs	FeS	NiS	
a (Å)	3.602	3.438	3.428	$u = 1/4$
c (Å)	5.009	5.88	5.340	
c/a	1.391	1.71	1.558	

81-3 Examples of C-type Crystals

1. C1 type (Fluorite type, CaF_2, Fig. 9-42).

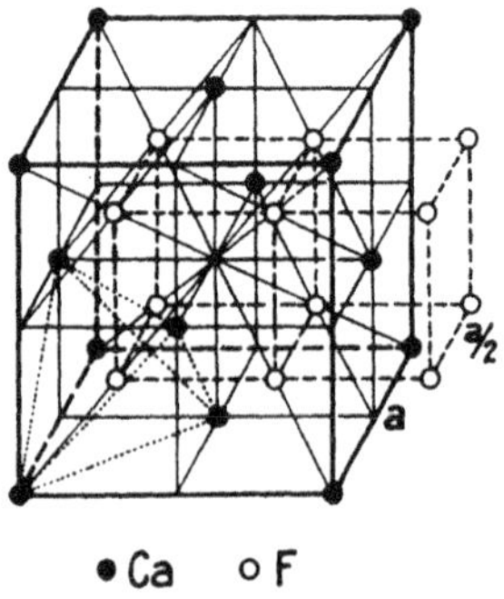

FIG. 9-42. Fluorite (CaF_2) type structure.

Ionic crystal. $Z = 4$. Cubic symmetry.

Ca^{++}: Face-centered cubic lattice, lattice constant a.

F^-: Primitive cubic lattice, lattice constant $a/2$.

Displacement: $\frac{1}{2}$ a body diagonal of the F^- cube, or $\frac{1}{4}$ a body diagonal of the Ca^{++} cube. That is, beginning at one-quarter of the body diagonal of a face-centered Ca^{++} cube, there is an F^- cube whose edges are half that of the Ca^{++} cube. Eight of the F^- cubes together form a cube of the same size as the Ca^{++} cube.

Coordination: Each Ca^{++} ion is surrounded hexahedrally by 8 F^- ions; each F^- ion is surrounded tetrahedrally by 4 Ca^{++} ions.

Coordination numbers: $K_{Ca} = 8$, $K_F = 4$. Figure 9-42 shows the Ca^{++} cube and two of the eight F^- cubes, and at the lower left, one of the tetrahedrons surrounding an F^- ion.

2. C2 type (Pyrite type, FeS_2, Fig. 9-43).

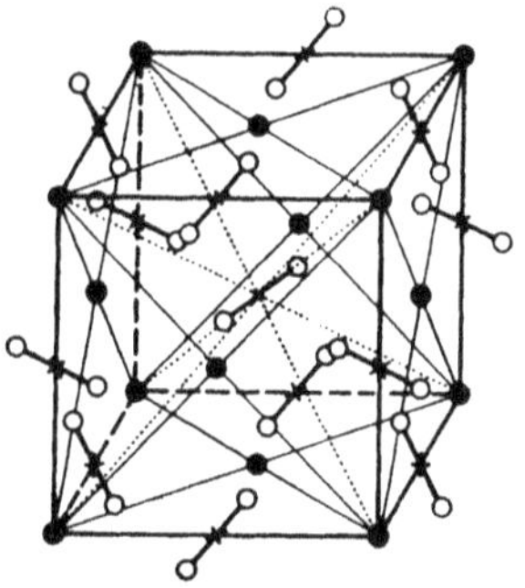

FIG. 9-43. Pyrite (FeS_2) type structure.

Covalent crystal. $Z = 4$. Cubic symmetry.

Fe: Face-centered cubic, lattice constant a.

S: The centers of gravity of pairs of S-atoms, which form short dumbbells, are also arranged in a face-centered cubic lattice, displaced relative to the Fe-lattice by one-half a body diagonal. The face-centered cubic lattice of dumbbell elements is in turn built up of four interpenetrating primitive cubic lattices. The dumbbells of a given primitive lattice are parallel to one another and are in the direction of one of the four body diagonals so that each of the four dumbbells associated with one single cell of pyrite lies in the direction of one of the four body diagonals. The parameter u given in the table below describes the relative displacement of the iron lattice and the four primitive sulfur lattices.

Coordination: Each iron atom is surrounded octahedrally by 6 S-atoms. Therefore $K_{Fe} = 6$. Each S-atom lies within a tetrahedron, at the apex of which is another S-atom but whose base consists of three Fe-atoms. The pyrite lattice consists of the interpenetration of an atomic lattice (Fe) and a molecular lattice (S_2).

3. C3 type (Cuprite type, Cu_2O, Fig. 9-44).

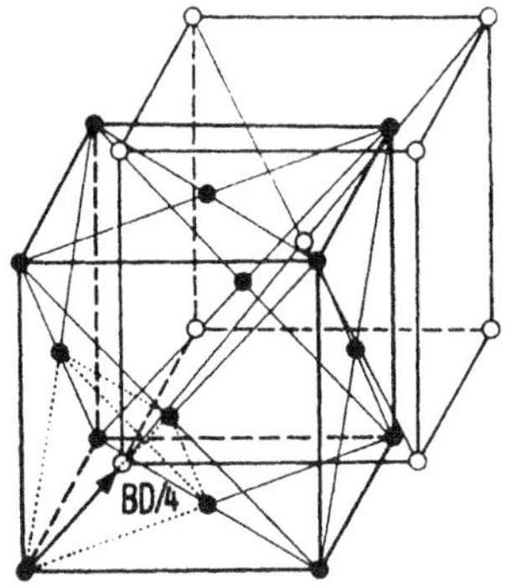

FIG. 9-44. Cuprite (CuO_2) type structure.

Covalent crystal. $Z = 2$. Cubic symmetry.

Cu: Face-centered cubic, lattice constant a.

O: Body-centered cubic, lattice constant a.

Displacement: $\frac{1}{4}$ body diagonal.

Coordination: Each O-atom is surrounded tetrahedrally by 4 Cu-atoms. $K_O = 4$. Each copper atom is flanked by two O-atoms, $K_{Cu} = 2$.

The following table lists some compounds which crystallize in C-type structure.

C 1 types:

Substance	CaF_2	BaF_2	UO_2	Na_2O	Na_2S
a (Å)	5.463	6.200	5.469	5.56	6.539

C 2 types:

Substance	FeS_2	MnS_2	NiS_2	CO_2	N_2O
a (Å)	5.418	6.101	5.676	5.575 (−190°C)	5.656 (−190°C)
u	0.386	0.4012	0.395	0.11	0.117

C 3 types:

Substance	Cu_2O	Ag_2O	Pb_2O
a (Å)	4.270	4.72	5.38

81-4 Radical Structure Types

A single example will serve to illustrate a simple radical type of structure. For the countless other types, the reader is referred to the literature and in particular to the excellent stereoscopic illustrations of von Laue and von Mises.†

G1 type (Calcite, $CaCO_3$, Fig. 9-45).

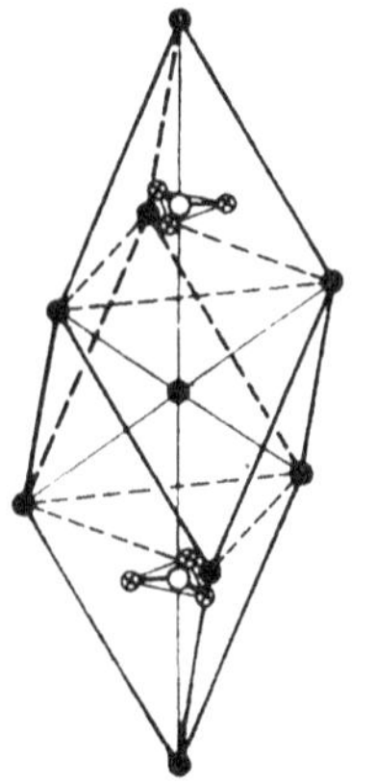

● Ca ○ C ⊗ O

FIG. 9-45

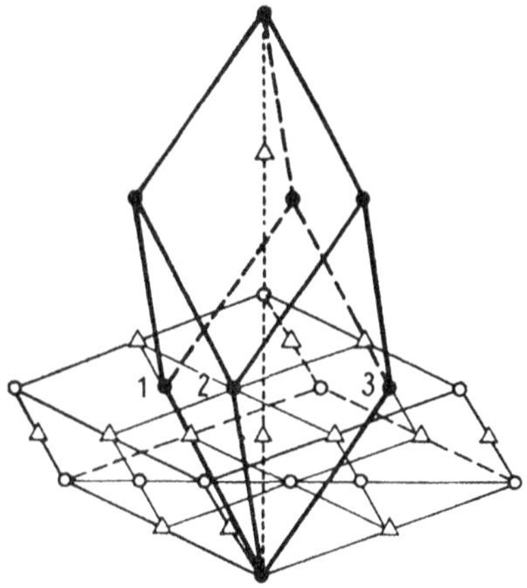

FIG. 9-46

FIG. 9-45. Calcite ($CaCO_3$) type structure.

FIG. 9-46. Connection between the basic cell of $CaCO_3$ (see Fig. 9-45) and the cleavage rhombohedron.

● Ca in the unit cell.

○ Ca, △ C } in the cleavage rhombohedron.

The points 1 and 3 lie in the centers of the back sides of the cleavage rhombohedron, while point 2 lies in the center of its upper face.

Ionic crystal. $Z = 2$. Rhombohedral symmetry.

Ca^{++}: The calcium ions form a body-centered rhombohedron.

CO_3^{--}: The radicals are positioned with the carbon atoms in a similar body-centered rhombohedron, displaced by one-quarter of the body diagonal. Three O-atoms surround each C-atom in the form of equilateral triangles whose plane is perpendicular to the body diagonal, so that $K_C = 3$. Note that the triangle of the upper radical is rotated 60° with respect to the lower one.

† M. von Laue and R. von Mises, *Stereoscopic Pictures of Crystal Structures*, Springer, Berlin, Vol. I, 1926, Vol. II, 1936.

Basic coordinates:

$$Ca^{++}: \left\{ \begin{matrix} 0\,0\,0 \\ \frac{1}{2}\,\frac{1}{2}\,\frac{1}{2} \end{matrix} \right\}, \qquad CO_3^{--}: \left\{ \begin{matrix} \frac{1}{4}\,\frac{1}{4}\,\frac{1}{4} \\ \frac{3}{4}\,\frac{3}{4}\,\frac{3}{4} \end{matrix} \right\}.$$

Another possible representation of this structure is shown in Fig. 9-46. Ca^{++} ions are placed in a face-centered rhombohedral structure (shown by light lines) formed by compressing a cube along one of its body diagonals. Interpenetrating this structure is a congruent one, populated with CO_3^{--} radicals and displaced by $\frac{1}{2}$ a body diagonal. The whole structure can be thought of as a deformed rocksalt type.† Some substances crystallizing in this way are:

Substance	$CaCO_3$	$MgCO_3$	$FeCO_3$	$NaNO_3$
a (Å)	6.373	5.62	5.83	6.32
α	46° 7′	48° 10′	47° 45′	47° 15′

81-5 Calculation of the Lattice Constant of Calcite from its Density

$$\varrho(CaCO_3) = 2.711 \text{ gm/cm}^3 \text{ at } 20°\text{C},$$

$$\alpha = 46°\,7', \quad M^* = 100.09 \text{ gm/mole}, \quad Z = 2.$$

$$\varrho = \frac{ZM^*}{N_0 a^3 \sin\alpha \sin\beta},$$

where

$$\cos\beta = \frac{\cos\alpha}{\cos\frac{\alpha}{2}}.$$

Then

$$\beta = 41°\,7',$$

$$a = \sqrt[3]{\left(\frac{ZM^*}{N_0 \varrho \sin\alpha \sin\beta}\right)} = 6.371 \text{ Å}.$$

$$(a_{exp} = 6.373 \text{ Å}.)$$

81-6 Concluding Remarks

In conclusion, it may be mentioned again that the geometrical lattices described above represent only idealized cases. As a result of thermal energy and zero-point energy, the atoms are continuously vibrating and hence are almost never exactly at the lattice points.

The structures described above are those of an *ideal crystal*, which does not actually exist. Because of the enormous number of single events which are required for the construction of a macroscopic crystal, many accidents take place during growth which result in departures from a perfectly regular

† This description neglects the rotation of the upper CO_3 radical.

structure. An actual crystal consists of a more or less regular macroscopic structure with many different kinds of microscopic and sub-microscopic *imperfections* scattered all through it. A simple example of such imperfections is a *vacancy*, in which one atom is absent from its proper site. The type and density of the imperfections in a crystal has a very strong effect on such crystal properties as plasticity and strength, but only a weak effect on elasticity and on the electrical conductivity of metals.

CHAPTER 10

THE LAWS OF THERMODYNAMICS

THE science of thermodynamics makes use of principles that are more general, and of wider scope, than those of kinetic theory. The laws of thermodynamics, like Newton's laws of motion, provide us with a powerful physical tool, but in using these laws we do not have the help of our physical intuition as we had in the consideration of kinetic theory. We conclude this book with a discussion of these laws.

In mechanics it was possible to summarize all factual material in the three laws of Newton. A similar crystallization took place in thermodynamics about the middle of the 19th century. The *equivalence of heat and energy* was recognized by Julius Robert Mayer in 1842, and by James Prescott Joule in 1843. Both of these men determined (although by very different means) the conversion factor by which energy could be incorporated into the theory of heat. In 1847, Hermann von Helmholtz formulated the completely general principle of conservation of energy (the *first law of thermodynamics*), which establishes a relation between all mechanical, thermal, and electrical phenomena. Even before this time, in 1824, Sadi Carnot, in his work entitled *Some Considerations regarding the Motive Power of Heat and its Application to Machines*, had discovered that the conversion of heat to mechanical work in a cyclic process is possible only when heat can flow from a higher to a lower temperature. The maximum amount of work that can be obtained in such a process depends only on the temperatures of the hot and cold heat reservoirs, and on the quantity of heat that is withdrawn from the warmer reservoir. Only a part of this heat can be transformed into work. The term "maximum" refers to the work obtained in the absence of all frictional losses, and of any heat transfer which does not result in the performance of work. These fundamental ideas were utilized by William Thomson (Lord Kelvin) in 1848 to generalize the concept of temperature and to replace the gas temperature scale by the so-called *absolute* temperature scale. In 1850–1852, Thomson and Clausius, with the help of Carnot's principle and the concept of absolute temperature, announced a principle that proved to be one of the foundations of thermodynamics (the *second law of thermodynamics*). Much later, in 1906, W. Nernst added to these two principles the capstone in his formulation of the *third law of thermodynamics* (the principle of the unattainability of absolute zero). We shall only discuss the first two laws in this book.

82. The First Law of Thermodynamics

The law of conservation of energy, in mechanics, is a consequence of Newton's laws of motion, provided no processes occur in which energy is converted to non-mechanical form. Such conversions, however, are of common occurrence. In every process in which frictional forces act, mechanical energy is converted to heat. On the other hand, it has been known for a long time that mechanical energy can be obtained from heat with the help of expanding gases or vapors. In such a process the expanding substance becomes cooled.

J. R. Mayer gave a theoretical, and J. P. Joule an experimental, proof that a unique relation existed between the amount of heat developed in a process and the mechanical work necessary to produce it. That the same relation also applies to the converse process of producing mechanical work from heat was first proved by Clausius in 1850.

Without going into the numerous methods of determining the mechanical equivalent of a calorie, we shall simply state that the frictional experiments of Joule and Rumford, as well as Mayer's method of determining the equivalent from the difference between the molar heat capacities of a gas, lead to the same value as the experiments of Hirn, in which the equivalent was determined from the work done in deforming plastic lead. The best experimental value of the mechanical equivalent of heat at present is

$$1 \text{ cal} = 4.185 \text{ joule}.$$

The following derivation of this relation is based on Mayer's method. For all quasi-ideal gases, it is found by experiment that

$$C_p - C_v = 1.987 \frac{\text{cal}}{\text{mole} \cdot {}^\circ\text{C}}.$$

The difference between C_p and C_v is equal in magnitude to the work done when a mole of gas expands at constant pressure. The magnitude of this work is

$$\Delta W = P \Delta V_M,$$

where ΔV_M is the increase in molar volume when the temperature of the gas is increased 1°C at a constant external pressure P. The work ΔW can be found from the ideal gas equation,

$$PV_M = RT.$$

When the temperature is increased by 1°C,

$$PV'_M = R(T + 1).$$

From these two equations, we have

$$P(V'_M - V_M) = P \Delta V_M = R = 8.317 \times 10^7 \frac{\text{ergs}}{\text{mole}} = 8.317 \frac{\text{joules}}{\text{mole}}.$$

This work is equivalent to the heat of 1.987 calories which is necessary to perform the work against the external pressure. That is,

$$1.987 \text{ cal} = 8.317 \text{ joule},$$

$$1 \text{ cal} = 4.185 \text{ joule}.$$

It has been shown by Joule, Hirn, and Mayer in many careful experiments, that the same relation holds in electrical, magnetic and physiological processes.

For the precise formulation of the first law, we require a new concept introduced by Lord Kelvin, namely that of the internal energy U of a system.

DEFINITION. The *internal energy* U of a system is the sum of the energies of all component parts of the system.

The internal energy U can consist of several different forms, such as mechanical, thermal, electrical, magnetic, or chemical energy. The internal energy U then corresponds to the sum of the mechanical and all other forms of energy.

The energy law is concerned only with the *difference* between the internal energy of a system in two different states, so that absolute values of the internal energy do not come into consideration. Thus, from the above definition, we see that the energy difference between two states of a system is

$$\Delta U = \Delta E_{\text{mech}} + \Delta E_{\text{therm}} + \Delta E_{\text{elec}} + \cdots$$

We have already discussed mechanical energy and we know how to compute its change. Changes in electrical energy occur in systems containing galvanic and electrolytic cells, for instance. The energy change is given by the product of the quantity of charge supplied to the system and the potential difference through which it is transferred. Magnetic energy will not be considered further at this point. Chemical energy is a special form of potential energy which is released or stored up as a result of changes in the relative positions of atoms, and occurs on a molecular level. If the chemical energy of a system decreases in a reaction the energy released (per mole) is distributed among N_0 individual molecules and makes itself evident by an increase in the temperature of the system.

Changes in the *form* of the internal energy of a system are of no importance in the calculation of the energy difference between two states of the system. It is only necessary to consider the interactions of the system with the surroundings; these consist solely of exchanges of *work* and *heat*. The entire subject of thermodynamics may be said to consist of nothing more than certain laws regarding the processes of work and heat exchange, and the applications of these laws. It turns out that every change of state of a system can be described from the energy standpoint by giving the magnitudes of two quantities added to or taken from the system. These are:

1. The work dW, where the term work has the meaning assigned to it in mechanics or electricity.

2. The quantity of heat dQ.

Work can be described as *ordered energy*, and heat as *disordered* or *random molecular energy.*

After these introductory remarks, the principle of conservation of energy can be stated as follows:

PRINCIPLE OF CONSERVATION OF ENERGY. *The increase in the internal energy of a system, in any process, is equal to the sum of the work done on the system and the heat flowing into the system:*

$$dU = dW + dQ.$$

As is discussed in textbooks on thermodynamics, of these three quantities only dU is a true differential in the mathematical sense; dQ and dW are not (see Section 82-2). However, for the purposes of our present discussion, it is unnecessary to distinguish between the differential characters of dU and of dQ and dW.

A system is said to be *isolated* if it can exchange neither heat nor work with its surroundings. The internal energy of an isolated system cannot change, but can only be transformed from one form to another,

$$dU_{\text{isolated system}} = 0.$$

The internal energy of an isolated system remains constant, no matter what processes take place within the system.

Two remarks are necessary in connection with the statements above:

1. We consider the system as "the bank" that is, work and heat are considered positive when they are received by the system.

2. In most thermodynamic processes no electrical work is done. The mechanical work most frequently takes the form of "volume work", that is, work done against a pressure when the volume of the system changes.

The work dW done in a *quasi-static* volume change dV against a pressure P is (see Fig. 10-1)

$$dW = -F\,ds = -PA\,ds = -P\,dV.$$

Unless otherwise noted, we shall only consider quasi-static volume changes.

The work is easily computed in an *isobaric* process (P =const) since no integration is necessary. When the pressure varies during a process, the work is given by

$$W = -\int_{V_0}^{V_1} P\,dV.$$

The integral can be evaluated with the help of the equation of state, when the process is specified.

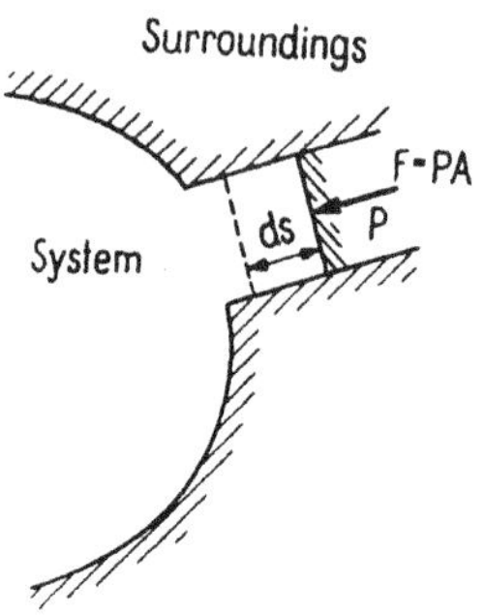

FIG. 10-1. Work done against an external pressure P.

EXAMPLE. Isothermal expansion of an ideal gas. From the universal gas equation,

$$P = n\frac{RT}{V},$$

where n is the number of moles. Then

$$W = -nRT\int_{V_0}^{V_1}\frac{dV}{V} = -nRT\ln\frac{V_1}{V_0}.$$

In an isothermal process,

$$\frac{V_1}{V_0} = \frac{P_0}{P_1},$$

so we can also write

$$W = -nRT\ln\frac{P_0}{P_1} = nRT\ln\frac{P_1}{P_0}.$$

If, in a chemical reaction, one mole of an ideal gas is produced at constant temperature and pressure, the volume change is just the molar volume since the volumes of the solid and liquid components of the reaction are negligible. The work done is

$$W_M = -PV_M.$$

For an ideal gas,

$$W_M = -RT.$$

82-1 Thermodynamic Processes

Out of the enormous number of possible thermodynamic processes, we single out for consideration a few which have a special practical or theoretical importance. We consider only mechanical forms of work.

(a) *Isothermal processes*, in which the temperature of the system remains constant.

$$dT = 0; \quad dW = -P\,dV$$

$$dQ = dU + P\,dV.$$

(b) *Isobaric processes*, in which the pressure is constant.

$$dP = 0; \quad dW = -P\,dV,$$
$$dQ = C_p^s dT,$$
$$dU = C_p^s dt - P\,dV.$$

(c) *Isochoric processes*, in which the volume is constant.

$$dV = 0; \quad dW = 0,$$
$$dQ = C_v^s\,dt,$$
$$dU = C_v^s\,dT.$$

(d) *Adiabatic processes*, in which there is no flow of heat between the system and its surroundings.

$$dQ = 0; \quad dW = -P\,dV,$$
$$dU = -P\,dV.$$

(e) *Cyclic processes.* A system is said to perform a cyclic process if, after an exchange of work and heat with its surroundings, it returns to its initial state. This requires that all thermodynamically significant properties of the system such as its pressure, temperature, concentration, etc., shall be the same in the final state as in the initial state. It is not necessary that every molecule of the system shall be in its initial position. In a cyclic process, the internal energy of a system is the same at the end as it was at the beginning of the process. Thus

$$W = -\oint P\,dV,$$
$$dU = 0,$$
$$Q = -W = \oint P\,dV.$$

82-2 State Variables and State Functions of a System

A system can change from a given state *A* to a second state *B* in different ways or, as we say, along different paths (Fig. 10-2). The work dW exchanged

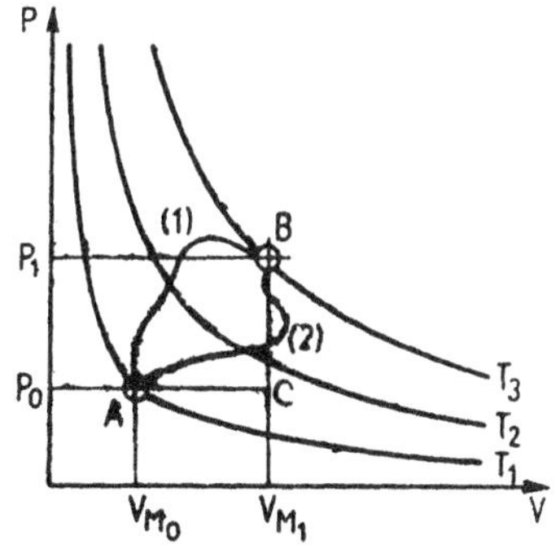

FIG. 10-2. Change of state of a system from state *A* to state *B*.

with the surroundings is, in general, different for each of these paths and the heat exchanged, dQ, is different also. For example, the work done per mole in the isobaric path AC is

$$dW_1 = -P_0(V_{M_1} - V_{M_0})$$

and the heat supplied per mole is

$$dQ_1 = C_p\,dT,$$

where dT is the change in temperature of the system between A and C.

In the path CB the work done is zero, since the volume remains constant, and the heat supplied per mole is

$$dQ_2 = C_v\,dT',$$

where dT' is the temperature change between C and B. The total change in internal energy per mole between A and B is

$$dU = C_p\,dT + C_v\,dT' - P_0(V_{M_1} - V_{M_0}).$$

From the first law, we can compute the change in internal energy for each of the infinite number of possible paths from A to B. However, the first law does much more than merely state that the change in internal energy on each path equals the sum of the heat dQ and the work dW. This law asserts that the sum of dQ and dW is the *same for all paths* between the states. To show this, let us consider a system S which can exchange heat and work with its surroundings. That is, heat flows into or out of the system from a series of heat reservoirs at various temperatures and mechanical work is done on or absorbed from the system by a work reservoir. (The work reservoir can be thought of as a spring, which can supply work or store up work.) Let us now extend the boundary of the system so as to include all reservoirs, as well as the original system S, in a combined system S_0. (The system S_0 then becomes an isolated system.) The principle of conservation of energy states that the total energy of the system S_0 can neither increase nor decrease. If the system S goes along path (1) in Fig. 10-3 from state A to state B, and in this process receives the heat dQ_1 and the work dW_1, then it must give up exactly the same *sum* of heat and work when it returns

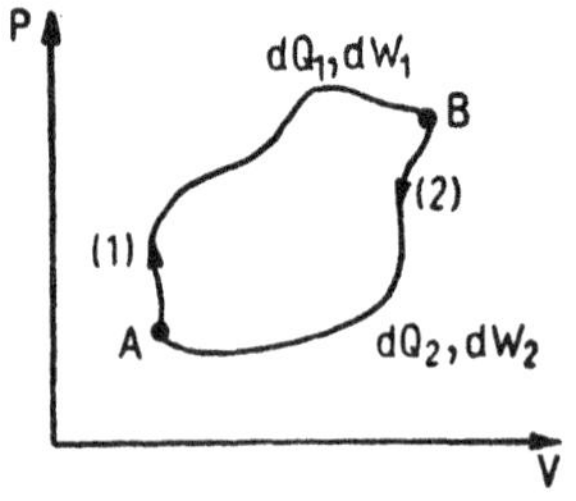

FIG. 10-3. A cyclic process.

to state A. Let us assume that a second path (2) exists for which the energy change is greater than along path (1), and let us take the system from A to B along path (1) and return it from B to A along path (2). It would then be possible to have changed the energy of the complete system while leaving it in the same state, which is contradictory to experience. The most general formulation of the principle of conservation of energy, or *the first law of thermodynamics*, is therefore:

The change in internal energy of a system between two arbitrary states is independent of the path by which the system is taken from one state to the other. The individual paths differ only in the way in which the total energy change dU is distributed between the work dW and the heat dQ. For every path,

$$dU = dW + dQ.$$

When expressed in this form, the principle of conservation of energy can readily be applied. Of particular significance are the facts that the work and heat exchanges depend on the choice of path and that these quantities are not determined by the initial and final states alone. On the other hand, the sum $dW + dQ$ depends only on the initial and final states. The change in internal energy dU is uniquely determined by the begin and end states and is equal to the difference between the internal energies in these states:

$$dU = U_2 - U_1.$$

The first law defines only energy differences. However, if we think of the internal energy as being measured from some arbitrary reference state, then a numerical value can be assigned to U which is uniquely determined by the variables such as P, V, and T that specify the state. Such quantities are called *state variables.* Quantities like U in contrast to Q and W are called *state functions* (see Section 91 for further details).

Many thermodynamic relations can be derived with the help of cyclic processes and state functions, without making use of the methods of calculus, as illustrated in the next section.

83. Application of the First Law to Cyclic Processes

The utility of a cyclic process is based on the following:

1. One can easily calculate the change in internal energy of a system, dU, between an initial and a final state.

2. Relationships between empirical constants and state variables can be derived if the energy difference dU is computed for two different paths and the two expressions for dU are set equal to each other.

To simplify the notation used in working with cyclic processes, we shall introduce the following abbreviations: (s), (l), or (g) represent the solid, liquid, and gaseous states. For example, $V(s)$ represents the volume of a system in the solid state. If no temperature is explicitly indicated, the substance is always considered to be at *normal temperature and pressure*, that is, at a temperature of 25°C (298°K) and 1 atmosphere. Modern thermodynamic literature makes use for the most part of this normal state (often abbreviated as *NTP* or *STP*).

83-1 Heat of Vaporization

We start with one mole of a substance B in the liquid state and at its boiling point, vaporize it at constant pressure, and condense it at constant pressure. Path (1) in Fig. 10-4 represents the vaporization process, and path (2) the condensation process.

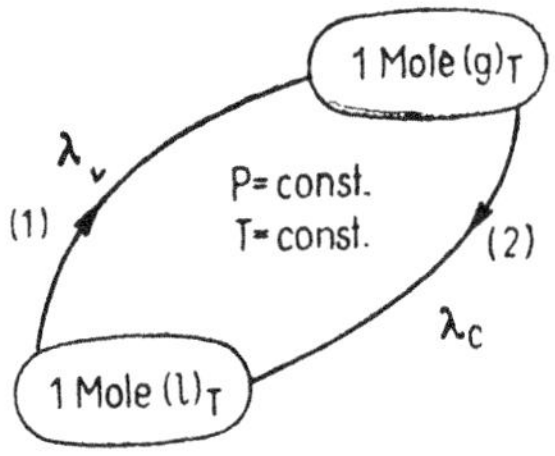

FIG. 10-4. A cyclic process at the boiling point of a liquid.

(a) The change in internal energy in the vaporization process is

$$dU = \lambda_v - P[V_M(g) - V_M(l)],$$

where λ_v is the molar heat of vaporization at the pressure P. The molar heat of vaporization is not equal to the increase in internal energy of the system, but exceeds it by an amount equal to the volume work done on the surroundings. A similar equation applies to the condensation process.

(b) In the cyclic process in Fig. 10-4, $dU = 0$ and the net volume work is zero since the quantities of work along the two paths are equal in magnitude and of opposite sign. It follows that the net heat supplied to the system (λ_v = heat of vaporization, λ_c = heat of condensation) in the cyclic process is also zero. Therefore

$$\lambda_v - \lambda_c = 0.$$

The heat of vaporization of a substance is equal to its heat of condensation.

83-2 Measurement of Heat of Solution with a Calorimeter; Temperature Dependence of Heat of Solution†

The heat of solution λ_t of one mole of a substance in n moles of solvent is defined as the quantity of heat that must be supplied to the system in

† Note that in all isobaric cyclic processes the net volume work is zero, since initial and final volumes are equal.

order that the process shall take place at *constant temperature t*. Heats of solution are often measured approximately by starting with substance and solvent at the same temperature t and mixing them in a calorimeter. If the temperature of the calorimeter decreases by Δt, the heat of solution can be calculated from the observed calorimetric data, assuming that the temperature dependence of the heat of solution is small enough to be neglected.

To calculate the heat of solution it is not necessary to know the details of the complicated process, represented by path (3) in Fig. 10-5, by which

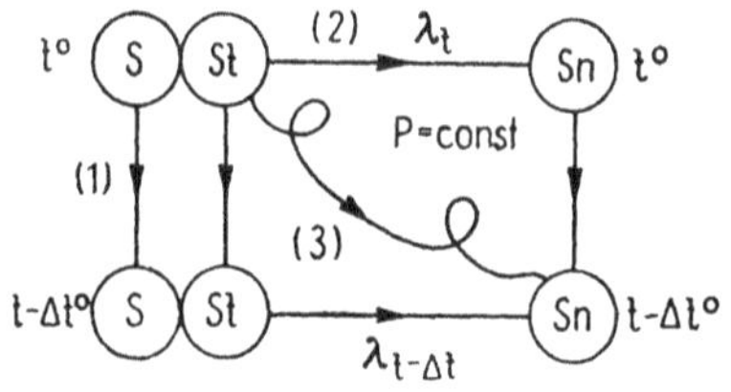

FIG. 10-5. Determination of the temperature dependence of the heat of solution λ_t.

the solution process actually takes place. The calorimeter equation can be applied to the two simple, although practically not realizable, paths shown in Fig. 10-5. This, however, leads to two different results:

Path (1). The initial components are cooled to $t - \Delta t$ and the substance is dissolved at this constant temperature. From the calorimetric equation,

$$\lambda_{t-\Delta t} = (M_1^* c_{p1} + nM_2^* c_{p2} + W_{\text{cal}})\,\Delta t,$$

where W_{cal} is the water equivalent of the calorimeter and the subscripts 1, 2 and 3 refer respectively to the substance, the pure solvent, and the solution.

Path (2). The substance is dissolved at the constant temperature t and the solution is cooled to $t - \Delta t$. Then

$$\lambda_t = (M_1^* + nM_2^*)\,c_{p3}\,\Delta t + W_{\text{cal}}\,\Delta t.$$

If we imagine the first path to be followed, we obtain an approximate value for the heat of solution at the lower temperature and, for its calculation need to know the specific heats of the components. In the second path, only the specific heat of the solution must be known and we obtain an approximate value for the heat of solution at the temperature t. As long as the temperature dependence of λ is small, these two values will be approximately equal to each other, and to the true value.

We can easily determine this temperature dependence from a consideration of these same two paths between the initial and final states. From the two preceding calorimetric equations we find the following identity:

$$-(M_1^* c_{p1} + nM_2^* c_{p2})\,\Delta t + \lambda_{t-\Delta t} = \lambda_t - (M_1^* + nM_2^*)\,c_{p3}\,\Delta t.$$

The change in the heat of solution with temperature is therefore

$$\frac{d\lambda}{dt} \cong \frac{\lambda_t - \lambda_{t-\Delta t}}{\Delta t} = (M_1^* + nM_2^*)\, c_{p3} - M_1^* c_{p1} - nM_2^* c_{p2}.$$

The rate of change of the heat of solution with temperature is thus uniquely determined by the specific heats and molar masses of the substance, solvent, and solution, and is independent of the heat of solution itself.

83-3 Temperature Dependence of the Heat of an Isobaric Chemical Reaction

The *heat of reaction q* at constant pressure is the quantity of heat that must by supplied during the reaction to produce at constant temperature one mole of the end product. According to our sign convention (system = "bank") q is positive in an endothermic reaction and negative in an exothermic one.

Suppose that a chemical reaction at a pressure $P = 1$ atm takes place according to the equation

$$n_1 A + n_2 B = n_3 E.$$

The symbols A and B represent the components and E the end product of the reaction. The reaction takes place in a calorimeter of water equivalent W_{cal}; the initial temperature is 25°C, the final temperature $(25 + \Delta t)$°C. To determine the temperature dependence of the heat of reaction we consider the two paths in Fig. 10-6 and determine the sum $dQ + dW$ for each.

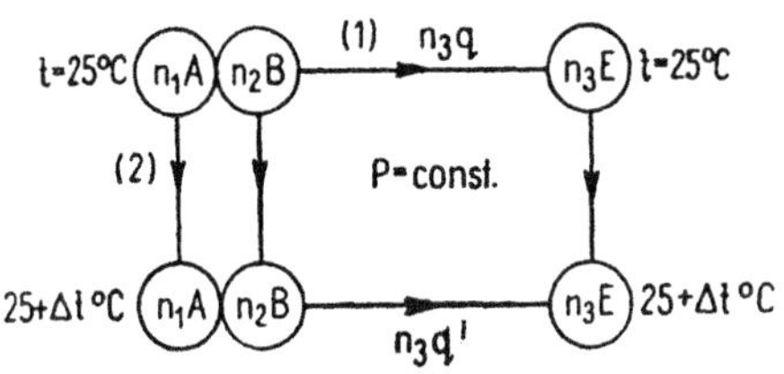

FIG. 10-6. Determination of the temperature dependence of the isobaric heat of reaction.

Initially both A and B are at 25°C. When the reaction is completed, the compound E is at $(25 + \Delta t)$°C. The actual reaction is a complex process during which parts of the system are at higher temperatures, but this does not need to be taken into account in setting up the energy balance. The volume works ΔW_1 and ΔW_2 are equal since the process is isobaric. Then, from the first law, the quantities of heat supplied along paths 1 and 2 are also equal. We therefore have

$$n_3 q + n_3 C_{pE} \Delta t = (n_1 C_{pA} + n_2 C_{pB})\, \Delta t + n_3 q',$$

where q and q' are the heats of reaction at the temperatures $t = 25°C$ and $t' = (25 + \Delta t)°C$ and C_{pE}, C_{pA}, and C_{pB} are the molar specific heats at constant pressure of the substance E, A, and B.

The rate of change of the heat of reaction with temperature is therefore

$$\frac{dq}{dt} \cong \frac{q' - q}{\Delta t} = \frac{1}{n_3}(n_3 C_{pE} - n_1 C_{pA} - n_2 C_{pB}).$$

EXAMPLE. Iron-sulfur reaction. The reaction

$$Fe_\alpha(s) + S_{\text{rhomb}}(s) \rightarrow FeS(s)$$

has a heat of reaction at 18°C of $q_{18°C} = -23{,}100$ cal/mole. What is the magnitude of dq/dt at 25°C and what is the heat of reaction at this temperature?

The molar specific heats at 25°C (298°K) are:

$$C_{pFeS} = 13.65 \text{ cal/mole} \cdot C°,$$
$$C_{pFe} = 6.03 \text{ cal/mole} \cdot C°,$$
$$C_{pS} = 5.54 \text{ cal/mole} \cdot C°,$$

$$\left(\frac{dq}{dt}\right)_{25°C} = (13.65 - 6.03 - 5.54) = +\ 2.08 \text{ cal/mole} \cdot C°.$$

Since the specific heats change little with temperature, dq/dt can be considered constant in the temperature interval from 18°C to 25°C and the heat of reaction at 25°C is

$$q_{25°C} = q_{18°C} + \frac{dq}{dt}\Delta t = -23{,}085 \text{ cal/mole}.$$

84. Molecular Interpretation of the First Law

The first law of thermodynamics makes assertions regarding the changes in the internal energy of a system which are valid without exception. On the other hand, the science of thermodynamics deliberately renounces any molecular interpretation of the first law. Thermodynamics has no curiosity; it makes no assumptions regarding either the form in which the energy acquired by a system is stored up, or the form in which the energy released by a system was formerly stored. Kinetic theory attempts to answer these questions.

Heat is disordered molecular energy. It is, however, a very concentrated form of energy. A cup of tea when heated through 10 C° (heat absorbed ≅ 800 cal) gains an amount of energy equal to that required to lift a 200 lb stone nearly 15 feet. Although this energy is present in the hot water, we are scarcely aware of it because it is randomly distributed among the water molecules. In practice, therefore, heat is considered as an unavailable form of energy because of the impossibility of controlling the motions of the individual atoms or molecules and assembling them in an ordered group.

Heat energy can be stored up by molecules in two forms:

(a) as kinetic energy,
(b) as potential energy.

(a) Molecular kinetic energy occurs in the following forms:

1. Kinetic energy of translation, $E'_{\text{kin}} = \frac{1}{2}mv^2$. According to the equipartition principle, each of the three translational degrees of freedom has an average kinetic energy $\frac{1}{2}kT$.

2. Kinetic energy of rotation, $E''_{\text{kin}} = \frac{1}{2}I\omega^2$. Depending on the form of a molecule, it may have 2 or at the most 3 rotational degrees of freedom, each with an energy $\frac{1}{2}kT$.

3. Kinetic energy of internal molecular vibration. The number of degrees of freedom can be very different in molecules of different structure and in some instances can take on high values. The number of degrees of freedom can change with temperature.

(b) Molecular potential energy appears in the following forms:

1. Real gases have a van der Waals potential energy. Because of the mutual attraction of two molecules, work must be done in an expansion against the internal pressure. As a result of these forces, the molecules have potential energy, which is a maximum at an infinitely large molar volume. When the molar volume of a real gas decreases from $V_{M,1}$ to $V_{M,2}$, the gas gives up random molecular energy, that is, energy that was uniformly distributed among its molecules. It appears in the form of an increased evolution of heat over and above that which would be developed in the compression of an ideal gas.

2. In solids and liquids the van der Waals forces of attraction, together with the Born forces of repulsion, give rise to a force field which can also store up potential energy. This energy can manifest itself in two ways:

(α) In the liquid state, the large forces of attraction can result in the formation of sizeable groups of molecules. We say that the liquid is *polymerized.* Water is one example, in which definite types of polymerization have been identified. The binding energy of the assembly of molecules represents a certain potential energy, which can be as great as 1000 cal/mole.

(β) The van der Waals and Born forces can also store potential energy when a body is compressed. This energy can be recovered as mechanical work when the pressure is relieved.

3. A third form of storage of molecular energy occurs when chemical reactions take place between different atoms or molecules. The forces which come into play here are electrical in nature and are much greater than the van der Waals forces. They make possible the storage of large amounts of energy in the form of chemical potential energy. The order of magnitude is from 10^4 to 10^6 cal/mole.

On the basis of these considerations we can now attempt to understand some thermodynamic processes from the molecular viewpoint.

84-1 Heat of Fusion of Ice

To change 1 mole of ice at 0°C to one mole of water at the same temperature requires 1436 calories of heat. It is well known that ice contracts on melting, the decrease in volume being 1.63 cm^3/mole. Some additional energy is therefore provided by the work done by the external pressure, although its magnitude is negligible. The work done by the van der Waals forces, however, cannot be neglected. The internal pressure, which is about 400 atm in ice, does an amount of work, $\Delta W \cong 400 \times 1.63$ atm · cm^3/mole = 15.7 cal/mole. Altogether, therefore, about 1452 cal/mole of energy are supplied to the ice in the melting process. What happens to all this energy?

In an ideal gas, the magnitude of C_v gives the number of degrees of freedom of the molecules (see Section 74-4). The same relation holds to a first approximation in liquids and solids except that the vibratory degrees of freedom must be counted twice. Ice has a molar heat capacity C_v of about 9 cal/mole-deg, while that of water at 0°C is about 18 cal/mole-deg. The number of degrees of freedom f therefore increases in the melting process by about $\Delta f = 9$, so that according to the equipartition principle a quantity of energy ΔU, given by

$$\Delta U = \Delta f \frac{RT}{2} \cong 2480 \text{ cal/mole},$$

must be added to the system.

However, we found above that only 1452 calories are supplied from outside and by the van der Waals forces. The remaining 1028 calories are obtained from the breaking down of the lattice structure of the ice, and the freeing of some potential energy. The energy is freed by the decrease in the angle between the sides of the water molecule from the tetrahedral angle of 109° 25′ in ice to one of 104° 34′ in the liquid state.

84-2 Kinetics of the Vaporization Process

Evaporation takes place from the free surface of every liquid and the temperature of the liquid is thereby decreased. Evaporation is possible because of the distribution of thermal velocities of the molecules of the liquid. The great majority of the molecules of the liquid have a mean kinetic energy which is not sufficient to enable them to penetrate the van der Waals boundary layer. However, there are always present in the liquid an appreciable number of molecules with kinetic energies greatly in excess of the average value.

When a liquid evaporates into an enclosed space, the number of molecules in the vapor phase increases until the density of the vapor reaches a value for which the number of molecules striking the liquid surface, per unit area and per unit time, is equal to the number penetrating the boundary layer from within the interior of the liquid. When this state has been reached a condition of equilibrium results and the vapor pressure remains constant at a value called the *saturation pressure.*

If the volume of the gas phase is decreased, the pressure of the vapor is momentarily increased. More molecules strike the surface of the liquid, per unit area and per unit time, than escape from it. Hence the rate of condensation exceeds the rate of evaporation until the pressure of the vapor decreases to its former value. The vapor pressure in the equilibrium state is determined wholly by the number of molecules per unit volume in the vapor phase and hence by the number of molecules striking the boundary surface per unit area and per unit time. Since the latter is equal to the number of molecules escaping from the liquid, the equilibrium pressure depends only on the temperature of the liquid. We can therefore formulate the *law of a saturated vapor*: A vapor in an enclosed space and in contact with its liquid has a vapor pressure which is independent of the volume occupied by the vapor. The vapor pressure depends only on the temperature of the liquid and increases rapidly with increasing temperature, reaching a value of 760 mm Hg at the *normal boiling point* of the liquid.

A vapor is said to be *saturated* at a temperature T if its pressure is equal to the saturation pressure at this temperature. If the pressure of a vapor is less than the saturation pressure, the vapor is said to be *unsaturated.* The temperature of an unsaturated vapor is greater than that at which its pressure is equal to the saturation pressure. Hence an unsaturated vapor is also called a *superheated* vapor. If a saturated vapor at the temperature T is compressed, or is cooled, some of it will condense. Condensation will take place immediately, however, only if liquid is present or if there are condensation nuclei such as dust, smoke, or ions in the vapor.

84-3 The Vapor Pressure of Water; Relative Humidity; Dewpoint

As a result of evaporation from the oceans, lakes, etc., water vapor is always present in the earth's atmosphere. The vapor pressure is in general smaller than the saturation pressure at the corresponding air temperature. The pressure of the water vapor in a given air mass depends on the previous history and on the temperature of the air mass, and a meteorological prediction regarding precipitation is not a simple problem.

In an enclosed space, on the other hand, the relations are greatly simplified. If a sufficient quantity of liquid is present, the vapor pressure will build up to the saturation pressure. If the vapor is unsaturated, a reduction in volume or a lowering of temperature will bring about a condition in which the vapor pressure equals the saturation pressure. Further compression or cooling will result in the precipitation of liquid. The *dewpoint* temperature t_d of an air mass is defined as that temperature at which the vapor pressure of the water vapor in the air mass is equal to the saturation pressure P_S. The amount of water vapour in the air determines the humidity.

The *absolute humidity* is defined as the partial pressure P_a of the water vapor (unit: 1 mm Hg). The *relative humidity* is defined as the quotient of the partial pressure of the water vapor and the saturation pressure at the existing temperature.

EXAMPLE. The vapor pressure in a room at a temperature of 22°C is 14.60 mm Hg (this is the absolute humidity). What is the relative humidity and what is the dewpoint? From a table of vapor pressures, we find:

1. At 22°C, $P_S = 19.63$ mm Hg.
2. The saturation pressure of water is 14.60 mm Hg at a temperature of 17.26°C.

The relative humidity is therefore

$$\text{Relative humidity} = \frac{14.60}{19.63} = 0.743 = 74.3\,\%.$$

The dewpoint is $t_d = 17.26$°C.

The amount of water vapor in the air can be computed from the ideal gas equation since at this low pressure (about 20 mm Hg) water vapor can be considered an ideal gas. In the example above, the molar volume is

$$V_M = \frac{RT}{P} = 1.26\ \text{m}^3/\text{mole}.$$

The mass of water per cubic meter (molar mass = 18 gm) is

$$m = 18 \times \frac{1}{V_M} = 14.3\ \text{gm}.$$

One method of measuring absolute humidity makes use of a dewpoint hygrometer. This consists of a metal container whose outer surface is highly polished. The container is partly filled with ether, through which a stream of air is slowly forced by a rubber bulb. The resulting evaporation lowers the temperature of the ether; its temperature can be read by means of a thermometer inserted in it. At the instant when the polished surface becomes clouded by condensed moisture the dewpoint has been reached and the temperature can be read from the thermometer to within 0.1°C. The absolute humidity can then be found from a table of vapor pressures, since it equals the saturation pressure at the dewpoint.

85. Isothermal and Adiabatic Processes in Ideal Gases

Both isothermal and adiabatic processes play an important part in the ideas to be worked out in the next section. We shall therefore consider them in some detail.

85-1 Isothermal Processes

An isothermal process is one in which the temperature remains constant: $\Delta T = 0$. Because an ideal gas cannot store any potential energy, its molar internal energy is independent of its density and hence of its volume. Except for the zero-point energy, the internal energy depends only on the temperature. Hence for an ideal gas in an isothermal process, we have

$$U = \text{const}, \quad \text{i.e.,} \quad dU = 0.$$

From the first law,

$$dU = 0 = dQ - P\,dV,$$

$$dQ = P\,dV.$$

This means that in an isothermal process all of the heat supplied to the gas is converted to work. This does not contradict the second law of thermodynamics, which states that in a *cyclic* process only a part of the heat supplied can be converted to work.

85-2 Adiabatic Processes

An adiabatic process is one in which there is no flow of heat into or out of a system. That is, in an adiabatic process $dQ = 0$ (see p. 435). Then, from the first law,

$$dU = dW = -P\,dV.$$

The increase in the internal energy of a system, in an adiabatic process, equals the work done on the system.

An ideal gas obeys the ideal gas equation in *any* process:

$$PV_M = RT. \tag{85-1}$$

In the special case of an adiabatic process, the adiabatic condition,

$$dU_M = dW = -P\,dV_M,$$

must also be satisfied. For an ideal gas,

$$dU_M = C_v\,dT$$

since there can be no storage of potential energy (Section 85-1).

The first law therefore becomes

$$dU_M = C_v\,dT = -P\,dV_M. \tag{85-2}$$

Eliminating P between Eqns. (85-1) and (85-2), we get

$$C_v\,dT = -\frac{RT}{V_M}\,dV_M.$$

But $R = C_p - C_v$, and representing the ratio C_p/C_v by γ, we obtain

$$\frac{dT}{T} = -\frac{C_p - C_v}{C_v}\frac{dV_M}{V_M} = -(\gamma - 1)\frac{dV_M}{V_M}. \tag{85-3}$$

This is the equation of an adiabatic process in differential form. It gives the change in temperature of an ideal gas in an adiabatic volume change.

Integration of this equation gives

$$\ln \frac{T}{T_0} = -(\gamma - 1) \ln \frac{V_M}{V_{M_0}} = \ln \left(\frac{V_{M_0}}{V_M} \right)^{\gamma-1},$$

or

$$\frac{T}{T_0} = \left(\frac{V_{M_0}}{V_M} \right)^{\gamma-1} = \left(\frac{V_0}{V} \right)^{\gamma-1}.$$

Hence in terms of the variables T and V, the equation of state for an ideal gas under adiabatic conditions is

$$TV^{\gamma-1} = T_0 V_0^{\gamma-1} = \text{const}. \qquad (85\text{-}4)$$

With the help of Eqn. (85-1), this equation for an adiabatic ideal gas system can be expressed in terms of P and V or P and T. From the relation $PV = nRT$, it follows that

$$T = \frac{PV}{nR}, \qquad V = \frac{nRT}{P}.$$

Introducing the expression for T or for V in Eqn. (85-4), we obtain for the variables P, V and P, T the equations

$$PV^\gamma = P_0 V_0^\gamma = \text{const}$$

and

$$\frac{T^\gamma}{P^{\gamma-1}} = \frac{T_0^\gamma}{P_0^{\gamma-1}} = \text{const}.$$

Adiabatic processes, like isothermal processes, can be represented graphically. If we select the form $PV^\gamma = \text{const}$ and construct a graph in the (P, V) plane, the adiabatics have a steeper slope than the isotherms since $\gamma > 1$. The adiabatics cut across the isotherms. Since the adiabatics are steeper, the temperature decreases in an adiabatic expansion. The pressure decreases, not only as a result of the increase in volume, but also because of the decrease in temperature.

If an ideal gas performs an infinitesimal adiabatic expansion, which will be the case if the system is thermally insulated, an amount of work $-P\,dV$ will be done on its surroundings. The energy is provided by the internal energy of the gas. The temperature of the gas decreases according to Eqn. (85-3), a relation which can be expressed in terms of the variables T and P. From the relation $PV^\gamma = \text{const}$, it follows that

$$\frac{dP}{P} = -\gamma \frac{dV}{V} \qquad (85\text{-}5)$$

and hence

$$\frac{dT}{T} = \left(1 - \frac{1}{\gamma}\right) \frac{dP}{P}. \qquad (85\text{-}6)$$

An adiabatic pressure increase results in a rise in temperature whose magnitude depends on the value of $\gamma = C_p/C_v$.

The adiabatic equations have many applications, two of which we shall describe.

1. *Warming of an air mass by compression.* When an extended atmospheric air mass is compressed, it cannot readily exchange heat with its surroundings since the outer surface of a large volume of air is too small for a rapid heat exchange. Hence if the pressure on a large mass of air is suddenly increased. the process can be considered as adiabatic. Equation (85-6) can then be used to find the increase in temperature. For diatomic gases like N_2 and O_2, $\gamma = 1.40$.

Suppose that an air mass is at a height of 2500 m where the pressure $P_0 = 550$ torr and the temperature is $T_0 = 273°K$. It descends in a short time interval to a height of 500 m, where the pressure $P_1 = 745$ torr. From Eqn. (85-6),

$$\Delta T = T_0 \frac{\gamma - 1}{\gamma} \frac{\Delta P}{P_0} = 27.7°\text{K}.$$

2. *Propagation of sound waves.* In the propagation of sound waves of sufficiently high frequency, there can be very little flow of heat in the interval between a compression and its neighboring rarefaction. The process is therefore adiabatic and the velocity of propagation of such a process is determined by the adiabatic properties of the substance in which the waves are travelling. This effect is very pronounced in gases, where the adiabatic compressibility,

$$\varkappa_{\text{ad}} = -\frac{1}{V}\left(\frac{dV}{dP}\right)_{\text{ad}},$$

is smaller by a factor γ than the isothermal compressibility $\varkappa_T$.

To show this we use the relation given in Eqn. (85-5),

$$\left(\frac{dP}{P}\right)_{\text{ad}} = -\gamma\left(\frac{dV}{V}\right)_{\text{ad}}.$$

It follows that

$$\left(\frac{dV}{dP}\right)_{\text{ad}} = -\frac{V}{P}\frac{1}{\gamma},$$

and hence

$$\varkappa_{\text{ad}} = \frac{1}{V}\frac{V}{\gamma P} = \frac{1}{\gamma P}.$$

The isothermal compressibility of an ideal gas (see p. 258) is

$$\varkappa_T = \frac{1}{P}.$$

We show in Vol. II that the velocity of propagation v of a sound wave is given by the expression

$$v = \sqrt{\frac{1}{\varkappa_{\text{ad}}\varrho}}.$$

It follows from the expression for $\varkappa_{ad}$ and the ideal gas equation that

$$v = \sqrt{\frac{1}{\varkappa_{ad}\varrho}} = \sqrt{\frac{\gamma P}{\varrho}} = \sqrt{\frac{\gamma RT}{M^*}}.$$

EXAMPLES.
Air at 18°C, $M^* = 28.8$ gm/mole, $\gamma = 1.40$: $v = 343$ m/sec.
H_2 at 18°C, $M^* = 2$ gm/mole, $\gamma = 1.40$: $v = 1300$ m/sec.

86. The Second Law of Thermodynamics

Physical processes cannot be completely understood solely on the basis of the first law of thermodynamics. For example, the most important thermal process, namely, the conversion of heat into work, remains unexplained if we consider only this one law. This conversion is subject to more stringent restrictions than the conversion of work into heat, although both processes are identical from the standpoint of the first law. The first law asserts only that if 1 calorie of heat is converted to work, 4.185 joules are obtained. The portion of a given amount of heat which *can* be converted to work, however, is not specified. It is here that the *second law of thermodynamics* enters. This law is related to the fact that certain natural processes are inherently *irreversible* or *one-way processes*. In every system that is left completely to itself, i.e. in every closed system, changes take place of such a nature that the system approaches a definite equilibrium state as its final state. Once a system has reached this end state it can never be returned to its original state without outside influence. This irreversible approach to a final state, which is called a state of *thermodynamic equilibrium*, is a law of nature. However, the state of equilibrium is only a macroscopic one. Microscopically the state is neither one of equilibrium nor of uniformity of temperature or distribution of matter. In our discussion, therefore, we shall restrict ourselves to those properties of a system which are not affected by fluctuations on a molecular scale. We then have as an experimental law: *An isolated system tends irreversibly toward a state of thermodynamic equilibrium.* This fact is the basis of the second law.

Examples of irreversible processes are: The swinging of a pendulum in air; the falling of a ball of putty; the dissolving of sugar in water; the rusting of iron in air; the attainment of temperature equilibrium; the self-demagnetization of a magnet; the streaming of gas out of a vessel.

86-1 Reversible and Irreversible Processes

On the basis of the first law alone, any of the irreversible processes above could proceed in the reverse direction. For example, a lump of putty, having fallen to the ground, might withdraw enough heat from its surroundings so that this energy would suffice to restore it to its original shape and its original elevation. This process, although it is compatible with the first law, does not occur in nature. For a complete understanding of thermodynamic processes, therefore, additional laws are required.

The initial state of a system can always be restored with the aid of *external* agencies. However, when this is done, we find that changes of state have taken place in these agencies. Thus we can lift the lump of putty to its original height and restore its original shape, but this requires that we expend energy. The temperature of the putty and the ground can be restored to their original values by a flow of heat from them into some other cooler body, but the state of this body is then altered. Every irreversible process produces an effect that can never be completely wiped out. Not only will an irreversible process not reverse itself, but its initial state cannot be restored by any means whatever without leaving behind some effect on the surroundings. Experience justifies the postulate: All *natural* processes are *irreversible.*

This postulate is equivalent to the statement (known as the Kelvin–Planck statement of the second law of thermodynamics): It is impossible to construct a *perpetual motion machine of the second kind.* This is a machine which, in the absence of temperature differences, would convert heat to work. Such a machine could withdraw heat from a single heat reservoir and convert it completely into work. With the help of a perpetual motion machine of the second kind it would be possible to restore the initial state of an irreversible process without any after effects. We shall demonstrate this for the example of the falling lump of putty. When it strikes the ground its original potential energy is completely converted to heat, which is conducted into the surroundings. A perpetual motion machine of the second kind would withdraw this heat from the surroundings (the single heat reservoir) and produce an amount of work which would just suffice to raise the lump to its original height. The initial state of the irreversible process would then be completely restored without any residual changes in other systems. The non-existence of a perpetual motion machine of the second kind is therefore equivalent to the postulate that all natural processes are irreversible.

Reversible processes do not occur in nature. (In making this statement, however, we exclude cosmic and atomic systems because of the difficulty of defining thermodynamic state variables for these systems.) On the other hand, reversible processes can be thought of as limiting cases and are often closely realizable. They must be carried out in such a way that the system remains in *complete equilibrium,* since otherwise the possibility of irreversible changes would exist. The following conditions must therefore be fulfilled in order to carry out a reversible process:

1. When work is done in a reversible process, all forces must be in equilibrium. Work is then done infinitely slowly. Friction forces are excluded, since all frictional processes are irreversible.

EXAMPLE. Let a spring be stretched by a force F which at every instant is only infinitesimally greater than the spring force F'. In the limiting case of a reversible process, $F' = F$.

2. Reversible heat flow between two bodies can only take place when temperature equilibrium exists ($\Delta T = 0$). If heat were to flow from a higher temperature body to one at lower temperature, then the performance of the work required to restore this heat to the higher temperature body would make the final state of the system different than the initial state, thus making the process irreversible.

EXAMPLE. Consider a system composed of three heat reservoirs R_1, R_2, and R_3 with temperatures $T_1 > T_2$ and $T_2 > T_3$. Let heat flow from R_1 to R_3. We now attempt to restore the initial state by returning this heat from R_3 to R_1. To do this, we do work by running the heat engine, supplying it with energy from R_2. When R_1 has been restored to its original state, the heat content of R_2 is less than at the start. Hence the original heat flow is irreversible.

Reversible processes offer the opportunity to restore the initial state of a system with only infinitesimal changes to the state of the surroundings.

EXAMPLE. The spring considered above can be relaxed subject to the condition that the force F' is infinitesimally greater than the force F. Thus the initial state can be completely restored by an infinitesimally small decrease in the force F. In this sense the process can be described as reversible. A reversible process is a succession of equilibrium states.

86-2 Distinction between Reversible and Irreversible Processes

No work is done in a *completely* irreversible process. For example, when a lump of putty falls to the earth its entire potential energy becomes converted to heat.

When a process is carried out reversibly, on the other hand, the maximum possible amount of work is obtained. Thus when the putty falls reversibly it can lift an equal weight to its original height. Its entire decrease of energy appears as work done on the body lifted and no heat is developed.

Reversible processes are very important in the development of the theory of thermodynamics. They set an *upper limit* to the amount of work that can be obtained in a process between two given states. A reversible process is the optimum with respect to the work obtained, and determines the *maximum work* which any process between two states can provide.

In spite of the fact that the maximum work is obtained in a reversible process, such processes are not utilized in practice. Reversible processes proceed infinitely slowly and their practical utility is therefore zero. Hence in all real processes some irreversibility occurs.

87. THE REVERSIBLE CARNOT CYCLE USING AN IDEAL GAS

A *Carnot cycle* is a cyclic process, all stages of which are reversible and in which heat is converted to work with the aid of *two heat reservoirs* at constant temperatures. The cycle was first introduced into thermodynamic reasoning by Sadi Carnot in 1824. For simplicity we shall consider that the cycle is performed with 1 mole of an ideal gas.

To carry out the process we imagine one mole of an ideal gas to be enclosed in a cylinder provided with a movable piston (Fig. 10-7). The process consists of four steps (Fig. 10-8):

1. An isothermal expansion at the temperature T_1 of the warmer reservoir.
2. An adiabatic expansion to the temperature T_2 of the cooler reservoir.
3. An isothermal compression at the temperature T_2 of the cooler reservoir.
4. An adiabatic compression to the temperature T_1 of the warmer reservoir.

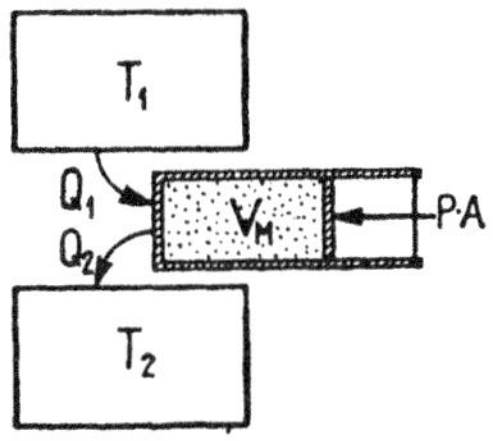

FIG. 10-7

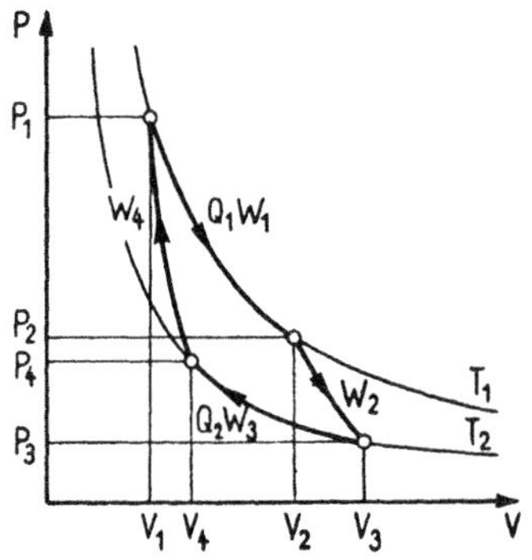

FIG. 10-8

FIG. 10-7. Piston, cylinder, and two heat reservoirs for a Carnot cyclic process.

FIG. 10-8. $P-V$ diagram of a Carnot cyclic process.

By a proper choice of the end point of the third step of the process, the initial state of the gas will be completely restored at the end of the final adiabatic compression.

The Carnot cycle contains two isothermal processes. During each of these there is a flow of heat between the gas and the heat reservoirs. During the adiabatic processes, on the other hand, no flow of heat takes place. These processes serve merely to adjust the temperature of the gas to equal that of the heat reservoirs, so that a flow of heat between the gas and the reservoirs can take place with no difference of temperature. Work is done on or by the gas in all four stages of the cycle.

With the help of the first law and the ideal gas equation we can make the following statements for the individual processes.

1. *Isothermal expansion.* The gas is in contact with the warmer reservoir. In spite of the expansion, the temperature and the internal energy of the gas remain constant; $T_1 = \text{const}$, $U_1 = \text{const}$.

During the expansion a quantity of heat Q_1 is withdrawn from the reservoir, flows into the gas, and is completely converted to the amount of work†

$$W_1 = \int dW_1 = -\int P\,dV = -Q_1\,.$$

The pressure of the gas does not remain constant in an isothermal expansion however:

$$P = \frac{RT_1}{V},$$

and therefore

$$dW_1 = -RT_1\frac{dV}{V}.$$

For a finite volume change, integration of this equation gives

$$W_1 = -RT_1 \ln\frac{V_2}{V_1} = -Q_1\,.$$

2. *Adiabatic expansion.* In order that the gas may exchange heat with the cooler reservoir, without a difference of temperature between them, its temperature must be lowered to T_2 by an adiabatic expansion. We therefore insulate the cylinder against any heat flow and expand the gas within it.

From Eqn. (85-4), p. 447, we have

$$T_1 V_2^{\gamma-1} = T_2 V_3^{\gamma-1}.$$

The volume V_3 is therefore determined.

In this adiabatic process, $Q = 0$ and the work done by the gas is provided by its internal energy. Since ideal gases can store no potential energy internally, the work is supplied from the thermal energy of the gas and is

$$W_2 = -\int P\,dV = \Delta U_1' = \tfrac{3}{2}R\,\Delta T = C_v(T_2 - T_1),$$

where $\Delta U_1'$ is the decrease in internal energy.

3. *Isothermal compression.* The gas is now brought into contact with the cooler reservoir. During the compression process the temperature remains constant at the value T_2, and the entire work of compression appears as heat $-Q_2$ flowing out of the gas into the reservoir. The internal energy of the gas remains constant:

$$T_2 = \text{const.},\qquad U_2 = \text{const.},$$

$$W_3 = -RT_2 \ln\frac{V_4}{V_3} = RT_2 \ln\frac{V_3}{V_4} = -Q_2\,.$$

† All quantities refer to 1 mole of an ideal gas.

The final volume V_4 of this compression process is so chosen that the initial temperature T_1 will be reached when the volume has been decreased adiabatically to the initial volume V_1.

4. *Adiabatic compression.* This adiabatic process is given by the equation

$$T_2 V_4^{\gamma-1} = T_1 V_1^{\gamma-1}.$$

This equation determines V_4. Work is done on the gas in the process and is completely converted to an increase in internal energy. Since the temperature increases from T_2 to T_1, we have

$$Q = 0,$$
$$W_4 = \Delta U_2' = C_v(T_1 - T_2).$$

The system has now been restored to its initial state, so that the net change in its internal energy is zero. The changes in internal energy $\Delta U_1'$ and $\Delta U_2'$, which took place during the two adiabatic processes, are equal in magnitude and of opposite sign and their sum is zero. We can therefore set up the following "balance sheet" of energy.

TABLE 10-1

Process	Q	W	ΔU
Isothermal expansion	$Q_1 \geqq 0$	$W_1 = -RT_1 \ln \frac{V_2}{V_1}$	0
Adiabatic expansion	0	$W_2 \leqq 0$	$\Delta U_1' \leqq 0$
Isothermal compression	$Q_2 \leqq 0$	$W_3 = +RT_2 \ln \frac{V_3}{V_4}$	0
Adiabatic compression	0	$W_4 = -W_2$	$\Delta U_2' = -\Delta U_1'$
Cyclic process	$\sum Q_i = Q_1 - Q_2$	$\sum W_i = W_1 + W_3 = W$	$\Delta U = 0$

The first law states that for the complete cycle,

$$\sum Q_i + \sum W_i = Q_1 - Q_2 + W_1 + W_3 = 0,$$

that is,

$$Q_1 - RT_1 \ln \frac{V_2}{V_1} - Q_2 + RT_2 \ln \frac{V_3}{V_4} = 0,$$

or

$$Q_1 - Q_2 = RT_1 \ln \frac{V_2}{V_1} - RT_2 \ln \frac{V_3}{V_4}.$$

The difference between the heat Q_1 extracted from the warmer reservoir and the heat Q_2 delivered to the cooler reservoir equals the net work done on the surroundings.

From the adiabatic equations

$$V_1^{\gamma-1}T_1 = V_4^{\gamma-1}T_2,$$

$$V_2^{\gamma-1}T_1 = V_3^{\gamma-1}T_2,$$

it follows that

$$\frac{V_2}{V_1} = \frac{V_3}{V_4}.$$

Therefore

$$Q_1 - Q_2 = R(T_1 - T_2)\ln\frac{V_2}{V_1} = -\sum W_i = -W.$$

Here the work W is negative because we are considering it relative to the Carnot engine. With respect to the surroundings, the work has the same magnitude and is positive. In the remainder of this section we will consider the work with respect to the surroundings, which is more natural than with respect to the engine.

A quantity of work W, equivalent to the difference $Q_1 - Q_2$, is produced in a reversible Carnot process, since the heat Q_2 given to the reservoir at the lower temperature is smaller than the heat Q_1 withdrawn from the reservoir at the higher temperature. From the practical point of view, the heat Q_2 is "lost" as far as its availability for the production of work is concerned.

Any device for producing mechanical work at the expense of heat is called a *heat engine*. The *thermal efficiency* of a heat engine is defined as the ratio of the mechanical work W done *by* the engine to the heat Q_1 supplied *to* it. We therefore have for the thermal efficiency of a Carnot engine

$$\eta_C = \frac{W}{Q_1} = \frac{Q_1 - Q_2}{Q_1} = \frac{T_1 - T_2}{T_1}. \tag{87-1}$$

The efficiency of a Carnot engine operating between two reservoirs at different temperatures is given by the ratio of the temperature difference between the reservoirs to the temperature of the warmer reservoir. This result is remarkable for its generality. It contains no empirical constants and is completely independent of the details of construction or operation of the engine. Only in the values of the temperatures, which are determined by a gas thermometer (absolute gas thermometer temperatures), is there any dependence on the properties of matter.

It follows from Eqn. (87-1) that

$$\frac{Q_2}{Q_1} = \frac{T_2}{T_1}. \tag{87-2}$$

That is, the ratio of the heat absorbed to the heat rejected is equal to the ratio of the temperatures of the corresponding heat reservoirs. This relation provides us with a new means of defining temperature (see Section 87-1).

By means of a Carnot process, heat can be transferred from a higher to a lower temperature in a completely *reversible* manner. In the process a part of the heat supplied is converted to mechanical work. By employing a reversed Carnot process (a *refrigerator* or *heat pump*), this work can be utilized to remove from the reservoir at lower temperature the heat that it absorbed and to return the heat to the reservoir at higher temperature.

In the irreversible process of heat conduction, heat also flows from a higher to a lower temperature, but no work is done and the difference $T_1 - T_2$ sinks to $T_1' - T_2$, where $T_1' < T_1$. The work obtained in an irreversible process is always smaller than if the process is reversible.

Every step of a Carnot cycle is carried out so slowly that at every point the gas is in pressure and temperature equilibrium with its surroundings. The entire cycle can therefore be completely reversed, in which case all processes take place in the opposite direction. In the isothermal expansion at the lower temperature T_2 a quantity of heat Q_2 is withdrawn from the heat reservoir and in the subsequent isothermal compression the reservoir at the higher temperature receives heat Q_1. The adiabatic stages of the process again serve to change the temperature of the system. The net result is the transfer of heat from a lower to a higher temperature by means of the expenditure of external work. The machine is then called a *heat pump* or a *refrigerator*, depending on the particular reservoir in which one is interested. The corresponding *coefficients of performance* η^* are defined as follows:

$$\eta^*_{\text{heat pump}} = \frac{Q_1}{W} = \frac{Q_1}{Q_1 - Q_2} = \frac{T_1}{T_1 - T_2} = \frac{1}{\eta_C},$$

$$\eta^*_{\text{refrigerator}} = \frac{Q_2}{W} = \frac{Q_2}{Q_1 - Q_2} = \frac{T_2}{T_1 - T_2}.$$

The preceding results are not completely general in two respects:

1. They do not refer to an arbitrary cyclic process but to a Carnot cycle, consisting of two isothermal processes and two adiabatic processes.

2. The working substance is an ideal gas. The results are therefore a consequence of the first law and the equation of state of an ideal gas. However, we will now show that they can be applied quite generally to other cyclic processes and working substances:

1′. Any *arbitrary* cyclic process can be decomposed into infinitesimally small Carnot cycles. However, an infinite number of heat reservoirs at different temperatures are required.

Figure 10-9 illustrates the subdivision of an arbitrary cyclic process *ABCA* (smooth curve) into infinitesimal Carnot cycles (outlined by the

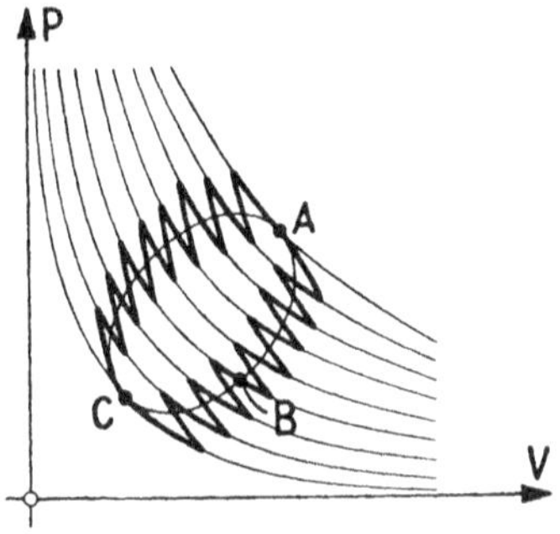

FIG. 10-9. Decomposition of an arbitrary cyclic process into many infinitesimal Carnot cyclic processes.

zigzag curve). We have superposed on the diagram of the process *ABCA* a network of isotherms, and have connected neighbouring isotherms by two short adiabatics to form a Carnot cycle. If the subdivision is sufficiently fine, the same portion of every isotherm is traversed twice in opposite directions. The quantities of work and heat absorbed or expended in crossing the loop along the isotherms cancel and only the process *ABCA* remains. Q.E.D.

2'. The efficiency of a Carnot engine is independent of the working substance.

Suppose there existed some material which, when used as the working substance in a Carnot engine, had the effect of making the efficiency η_C' greater than the efficiency η_C of a Carnot engine using an ideal gas. Let us consider what would happen if we ran the ideal gas engine as a heat pump which derived its work from the other engine operating between the same two heat reservoirs (Fig. 10-10). Then since

$$\eta_C' = \frac{W}{Q_1'},$$

$$\eta_C = \frac{W}{Q_1},$$

$$\eta_C < \eta_C',$$

we find that

$$Q_1 > Q_1'.$$

That is, we have constructed a self-acting device whose only effect is to transfer heat from a cooler heat reservoir to a hotter one. But, as we have noted before (Section 86-1), this is impossible. Therefore $\eta_C' \leqq \eta_C$. Now by reversing the mode of operation of both engines so that the non-ideal gas engine operates as the heat pump, assuming $\eta_C' < \eta_C$, and following the same line of reasoning we can show that $\eta_C' \geqq \eta_C$. The only possibility is that

$$\eta'_C \equiv \eta_C$$

which proves the statement 2′.

As an important corollary to statement 2′, we have the following.

2″. In performing a reversible, cyclic process, no engine has an efficiency greater than the Carnot engine.

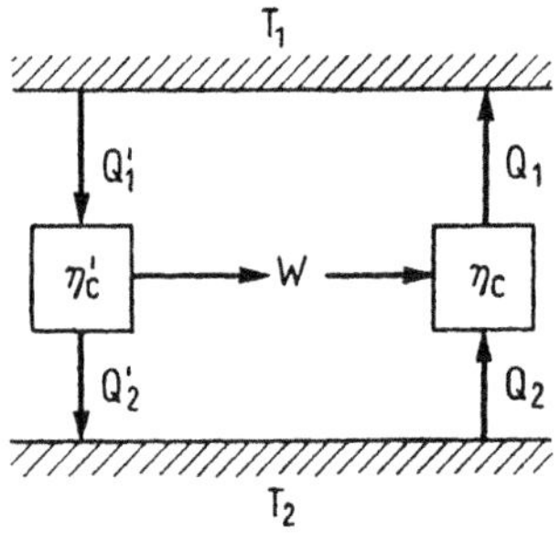

FIG. 10-10. Combination of Carnot engines used to prove that the efficiency of a Carnot engine is independent of the working substance.

The proof follows the same line of argument as used in establishing statement 2′. Suppose there existed some material M which, when used as the working substance in a reversible cyclic process resulted in an efficiency η_M, greater than the efficiency η_C of a Carnot engine. We could then construct a perpetual motion machine of the second kind (PM II) as follows (Fig. 10-11). The Carnot engine could be used as a heat pump and the machine

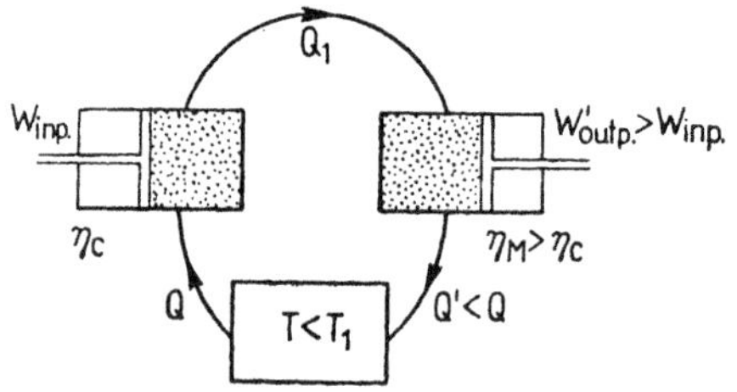

FIG. 10-11. A perpetual motion machine of the second kind (PM II) which could be built with a heat engine which is more efficient than a Carnot engine.

of efficiency η_M as a heat engine. The high temperature reservoir could then be dispensed with, since the heat pump would provide as much heat at the temperature T_1 as is absorbed by the heat engine. As a result, we would obtain a net amount of work $W'_{out} - W_{in}$ at the expense of a quantity of heat $Q - Q'$ withdrawn from a single heat reservoir. The combination would then be equivalent to a PM II and with its help the result of irreversible processes could be eliminated. But this is impossible according to the fundamental postulate regarding PM II.

Having shown that in a reversible process a Carnot engine has the maximum efficiency, an upper limit is set to the efficiency of an arbitrary heat engine operating between two reservoirs at given temperatures. Since a reversible cycle has the maximum possible efficiency, it follows that

$$\eta_{\text{actual engine}} < \eta_{\text{Carnot engine}}.$$

The original formulation of the second law of thermodynamics, though rigorous, was qualitative. We can now give an alternative quantitative formulation:

The experimental fact that the result of an irreversible process can never be completely eliminated is equivalent to the assertion that a Carnot engine has the highest possible efficiency for the conversion of heat into mechanical work:

$$\eta_{\max} = \eta_C = \frac{T_1 - T_2}{T_1}$$

The efficiency η is zero even for a Carnot engine, if no temperature difference $T_1 - T_2$ is available. This is equivalent to the non-existence of a PM II.

EXAMPLES. 1. The second law shows the superiority of an internal combustion engine to a steam engine. A steam engine operating at a pressure of about 20 atm and with a small amount of superheating takes in steam at about 250°C, and exhausts it at about 80°C. The maximum theoretical thermal efficiency is

$$\eta_{\max} = \frac{170}{523} = 32.4 \text{ per cent.}$$

In an internal combustion engine, on the other hand, the combustion temperature can reach 2000°C while the exhaust temperature is about 400°C. The maximum theoretical efficiency is then

$$\eta_{\max} = \frac{1600}{2273} = 70 \text{ per cent.}$$

In practice, efficiencies of no more than 40 per cent are generally obtained.

2. If an ideal heat engine takes in a quantity of heat Q_1 from a high temperature reservoir, it always rejects a portion of this to the low temperature reservoir. The heat rejected, according to Eqn. (2) on page 455, is

$$Q_2 = Q_1\left(\frac{T_2}{T_1}\right).$$

If we adjust the temperature so that $T_2 = T_1 - 1°$, we see that the amount of work obtainable for a temperature difference of 1°C is smaller the higher the temperature of the high temperature source.

The amount of work obtained from 1 cal with a Carnot engine, for a source at a temperature of 500°K and with a temperature difference of 1°C, is

$$W = \frac{1}{500} \text{ cal} = 8.37 \times 10^{-3} \text{ joule}.$$

If, on the other hand, the temperature of the reservoir is $T_1 = 273°K$, the maximum work obtainable is

$$W = \frac{1}{273} \text{ cal} = 1.53 \times 10^{-2} \text{ joule}.$$

We can also see that if $T_2 = 0$,

$$Q_2 = Q_1 \left(\frac{T_2}{T_1}\right) = 0,$$

so that in this case the heat Q_1 would be completely converted to work; i.e. as

$$T_2 \to 0, \quad \eta \to 1.$$

87-1 The Thermodynamic Temperature Scale

It was shown in the preceding discussion of the Carnot cycle that

$$\frac{Q_1}{Q_2} = \frac{T_1}{T_2}.$$

This relationship is completely independent of the working substance in the Carnot cycle. The temperature T is the absolute temperature indicated by a gas thermometer using an ideal gas.

Lord Kelvin made use of this property of a Carnot cycle to set up a physically satisfying definition of the concept of temperature. Kelvin proposed that the *thermodynamic temperature* θ should be *defined* by the Carnot cycle, that is, that the ratio of two temperatures θ_1 and θ_2 should be defined as the ratio of the quantities of heat Q_1 and Q_2 taken in and rejected by a Carnot engine (using any working substance) operating between reservoirs at these temperatures. Then by definition,

$$\frac{\theta_1}{\theta_2} = \frac{Q_1}{Q_2}.$$

If, in addition, the temperature of the triple point of water is defined as 273.16°K, the thermodynamic scale thus defined approaches very closely the absolute gas temperature scale we have used so far. However, the *concept* of temperature is now completely independent of the properties of any thermometric substance and accordingly is defined in a way that is much more satisfactory physically. For the experimental measurement of Kelvin temperature, it suffices to measure the hydrogen (or helium) temperature with a constant-volume gas thermometer and to convert its reading to those that would be obtained with an ideal gas thermometer with the help of the van der Waals constants for hydrogen (or helium). This avoids the complicated procedure of actually measuring temperature by means of a Carnot cycle. The differences between the constant-volume hydrogen scale and the thermodynamic scale, in the temperature range between $-100°C$ and $1000°C$, are extremely small as the following table shows.

TABLE 10-2

θ °C	$\theta - T_{H_2}$ °C	θ °C	$\theta - T_{H_2}$ °C
−240	+0.18	200	+0.003
−200	+0.06	300	+0.007
−100	+0.008	400	+0.013
0	0	500	+0.02
20	−0.000 46	800	+0.04
40	−0.000 55	1000	+0.05
60	−0.000 52		
100	0		

Relation between the constant-volume hydrogen scale and the thermodynamic scale, for $P_0 = 1000$ mm Hg. If $P_0 = 760$ mm Hg, each value of $\theta - T_{H_2}$ is to be multiplied by 0.76 (table from Landolt-Börnstein).

88. Measure of the Irreversibility of a Process. Entropy

Although all actual processes are irreversible, they may differ from one another in the degree or extent of irreversibility. A falling pin and an avalanche are both irreversible processes, but the amount of irreversibility in the two is different. The characteristics which determine the extent of irreversibility are those changes which cannot be eradicated by any means whatever.

To obtain a quantitative measure of the irreversibility of a process, we proceed in the same way as in the measurement of other physical quantities. Two requirements are necessary:

1. Definition of the *unit* of irreversibility.
2. Establishment of the *method of measurement*.

1. For the definition of a unit of irreversibility we choose out of the many possible irreversible processes a certain so-called *standard process*. Our *standard system* consists of a heat reservoir at a temperature T and a coil spring. The temperature T refers to the thermodynamic temperature as defined by Lord Kelvin. Its value does not depend on the properties of any material. The *standard irreversible process* consists of a contraction of the spring and the dissipation of the spring's stored energy by means of some friction device (such as a piston with a hole in it) which can be attached to the spring (Fig. 10-12). All of the energy released by the spring is given to

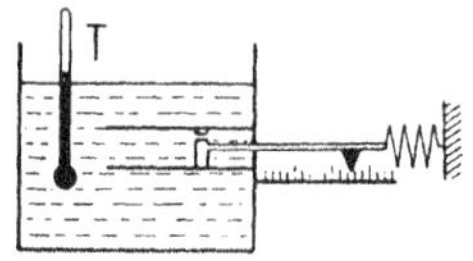

FIG. 10-12. A standard system for measuring irreversibility. When the spring contracts, an amount of heat ΔQ is given to the heat reservoir.

the heat reservoir at the temperature T. The extent of the irreversible process (the quantity of energy given up by the spring) can be found either by measuring the change in length of the spring or by measuring the quantity of heat ΔQ delivered to the reservoir. The definition of irreversibility, however, is not yet complete. If the spring gives its energy to a reservoir at a temperature T_1, as in Fig. 10-13, the process is less irreversible than if the

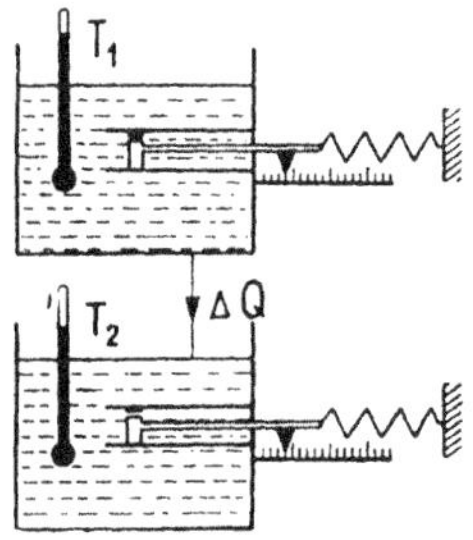

FIG. 10-13. Two standard systems at different temperatures.

same contraction had taken place at a reservoir at a lower temperature T_2, because the heat ΔQ can still flow from the higher to the lower temperature. This last process introduces an additional amount of irreversibility. The conversion of the energy of the spring into heat, at the higher temperature T_1, is thus less irreversible than the conversion of the same quantity of energy at the lower temperature T_2.

A *unique* measure of the irreversibility of the standard process is obtained, however, if we divide the quantity of heat dissipated by the temperature of the reservoir:

$$\frac{\Delta Q}{T} = \frac{\text{heat given to heat reservoir}}{\text{temperature of the reservoir}}.$$

The choice of this particular expression will be justified later.

We call the quantity $\Delta Q/T$ the *increase of entropy* ΔS_{St} of the standard system,

$$\Delta S_{St} = \frac{\Delta Q}{T}.$$

Unit of entropy: 1 cal/degree.

Only *differences* of entropy (of the standard system) are defined by the equation above. The absolute value of the entropy of the system remains undetermined. It can, however, be defined from the third law of thermodynamics.†

Entropy changes are now defined for the standard system. Since ΔS_{St} is a unique measure, it is reasonable to assume that there also exists a unique measure of the irreversibility for an arbitrary system (see Section 91). We now make this assumption and denote the quantity as the change in entropy for the arbitrary system, ΔS. The units of ΔS are those of ΔS_{St}, 1 cal/degree. We will discuss later the following two basic properties of S. First, S is a state function and does not depend on the path of the process. Second, S is an additive quantity: $S = \sum S_i$, where i runs over all components of the arbitrary system.

2. *Method of measurement.* We now wish to determine the extent of irreversibility for an arbitrary system undergoing an irreversible process. We shall do this by ultimately measuring the change of entropy of a standard system, which we have already stated is a unique measure of irreversibility. We proceed as follows.

Let a system undergo an arbitrary irreversible process from state A to state B, as indicated schematically in Fig. 10-14. With the aid of one or

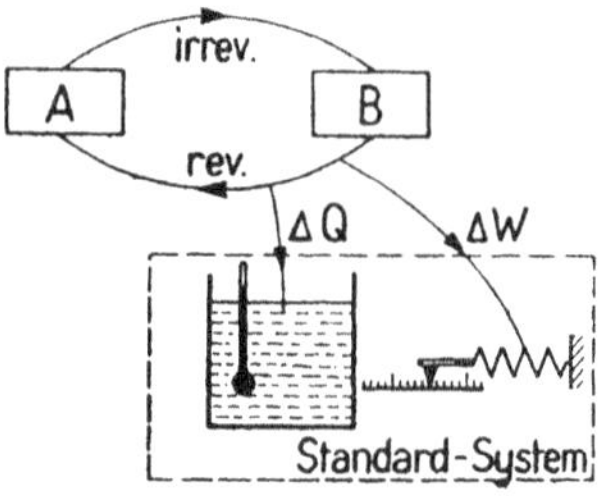

FIG. 10-14. Measuring the irreversibility of the process $A \rightarrow B$.

more standard systems, the original system can always be restored to its original state. In this restoration, heat and work are to be provided by (or given to) the standard systems only, so that the combination of original and standard systems form an isolated system. (Note that the friction device of Fig. 10-12 is no longer attached to the spring.) Out of all possible paths by which the original system may be restored to its initial state, we choose a reversible path, which we assert can always be found. On this path, *all* changes in the original system and in the standard systems, as well as the exchanges of work and heat between them, take place reversibly during the restoration. In this case no new irreversible changes occur in the isolated

† G. N. Lewis and M. Randall, *Thermodynamics*, 2nd edition, Chap. 12, McGraw-Hill, 1961.

system. Hence, the change of entropy of the isolated system, which is a measure of the irreversibility of the restoration process, is zero.

Now in the arbitrary irreversible process $A \to B$, let the entropy change of the original system be $\Delta S^{AB}_{\text{Sys}}$. Since the entropy change in the isolated system is zero during the restoration, then the change in entropy in the standard system ΔS_{St} is the negative of the entropy change of the original system $\Delta S^{BA}_{\text{Sys}}$:

$$\Delta S_{\text{St}} = -\Delta S^{BA}_{\text{Sys}},$$

which follows from the additive property of S. But S is state function and since the original system has been restored to its initial state, its entropy returns to the initial value and

$$\Delta S^{BA}_{\text{Sys}} = -\Delta S^{AB}_{\text{Sys}}.$$

Combining these equations, we obtain

$$\Delta S_{\text{St}} = \Delta S^{AB}_{\text{Sys}},$$

and hence the entire change of entropy of the original system in the irreversible process $A \to B$ is transferred to the standard system.

We are thus led to the following definition:

DEFINITION. The increase in entropy ΔS_{system} of a given system in an *arbitrary* process is defined as the increase in entropy of the standard system (or systems) when the given system is restored reversibly to its original state at the expense of reversible changes in the standard systems.

$\Delta S_{\text{St}} = \Delta S_{\text{Sys}}$ in a *reversible restoration* of the given system to its original state.

The change in entropy of a standard system ΔS_{St}, when the original state of the given system is restored, is equal to the heat ΔQ_{rev} given to the heat reservoir, divided by the temperature T of this reservoir:

$$\Delta S_{\text{St}} = \frac{\Delta Q_{\text{rev}}}{T}.$$

The subscript "rev" serves to remind us that the heat is absorbed reversibly by the reservoir. We therefore have as a measure of the irreversibility of the arbitrary process

$$\Delta S_{\text{Sys}} = \frac{\Delta Q_{\text{rev}}}{T}.$$

88-1 Change in Entropy of a Closed System when a Part of the System undergoes a Reversible Process

We consider a large system, in some part of which a reversible process takes place. This partial system is in contact with its surroundings and can exchange heat and work with them. That part of the surroundings which interacts directly with the partial system during the course of the process can be taken together with the partial system and the combination can be thought of as a closed or isolated system. We will now seek to determine the entropy change of the closed system, if a reversible process takes place in some part of it.

1. *Reversible Process*. Suppose a reversible process takes place in the partial system A. The process is reversible at all instants of time, so that for every interval of time the partial system and its surroundings experience changes of entropy having equal magnitudes, but opposite signs. We now replace the surroundings by a standard system, so that partial system A and the standard system then make up the closed system. Between the two only reversible processes occur, and hence the closed system experiences no net change in entropy:

$$\Delta S_A = -\Delta S_{\mathrm{St}}.$$

The change in entropy of the standard system is

$$\Delta S_{\mathrm{St}} = \frac{\Delta Q_{\mathrm{St}}}{T_{\mathrm{St}}}.$$

Since for a reversible process

$$\Delta Q_{\mathrm{St}} = -\Delta Q_A,$$

and

$$T_{\mathrm{St}} = T_A,$$

then

$$\Delta S_A = \frac{\Delta Q_A}{T_A}.$$

The total change in entropy of the partial system A for a reversible process is the sum of the quotients given by the quantities of heat ΔQ_{rev} supplied to the system divided by the temperatures at which they were supplied:

$$\Delta S_A = \sum \frac{\Delta Q_{\mathrm{rev}}}{T}.$$

Reversible adiabatic processes produce no entropy change in either partial system since $\Delta Q = 0$. If the temperature is variable during the process $1 \to 2$, then the above sum must be evaluated as an integral:

$$\Delta S_A = \int_1^2 \frac{dQ_{\mathrm{rev}}}{T}.$$

The entropy of a system is peculiar in that it is a state function which can be evaluated only for reversible paths. Every reversibly supplied quantity of heat is accompanied by an entropy increase of amount $\Delta Q_{\mathrm{rev}}/T$. A reversible loss of the same amount of heat at the same temperature T entails a decrease in entropy of the same magnitude. During reversible processes, the entropy of a closed system can neither increase nor decrease. Entropy therefore obeys a *conservation law*, but only for reversible processes.

2. *Reversible Cyclic Process.* If the partial system A undergoes a reversible cyclic process (Fig. 10-15), there is no net change of entropy for the partial

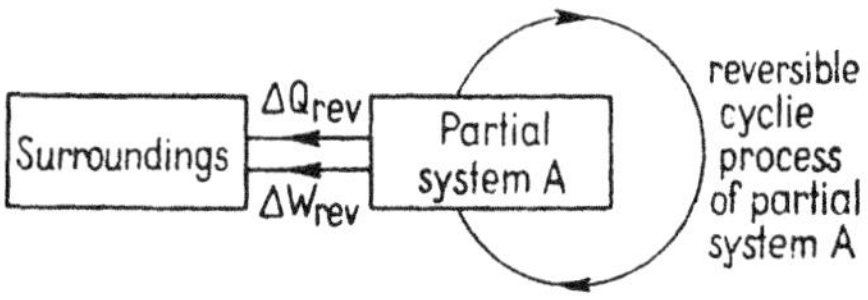

FIG. 10-15. Reversible cyclic process of a partial system. The partial system A can exchange heat and work with its surroundings.

system nor for its surroundings, since the entropy is a state function for a reversible process:

$$\sum_{\substack{\text{rev}\\\text{cyclic}\\\text{process}}} \Delta S_A = \sum_{\substack{\text{rev}\\\text{cyclic}\\\text{process}}} \Delta S_{\text{surr}} = \sum_{\substack{\text{rev}\\\text{cyclic}\\\text{process}}} \left(\frac{\Delta Q_{\text{rev}}}{T}\right)_{\text{St}} = 0\,.$$

The above equations can give relationships between the quantities characterizing the properties of the materials of the interacting systems, just as was possible through use of the law of conservation of energy.

The Carnot process is a reversible cyclic process. The change in entropy of the surroundings is, according to the discussion on p. 464,

$$\sum \Delta S_{\text{surr}} = \sum \left(\frac{\Delta Q_{\text{rev}}}{T}\right)_{\text{St}} = -\frac{Q_1}{T_1} + \frac{Q_2}{T_2}.$$

For the Carnot process, we know this sum is zero [see p. 455, Eqn. (87-2)] in agreement with the above discussion. This is a justification for our previous choice of the quantity which serves as the measure of irreversibility

89. Irreversible Processes

The change in the entropy of a system in a reversible process is

$$\Delta S_{\text{rev}} = \frac{\Delta Q_{\text{rev}}}{T}\,.$$

The change in entropy depends only on the heat flowing reversibly into or out of the system and on the temperature of the system.

If a system undergoes an *irreversible* process, whose end states are the same as for a reversible process, the change in entropy is the same as in the reversible process. This can easily be understood since entropy is a function of the state of the system. When the system is restored reversibly to the initial state of the irreversible process, the same entropy change will be measured by the standard system as if the initial state of the reversible process were being restored. However, the reversible process is characterized by the fact that no irreversibilities occur at any stage. Whenever a given system undergoes a reversible process, other systems are always involved whose changes in entropy are equal in magnitude and opposite in sign to those of the given system, so that the total entropy of all systems taking part in the process does not change. For irreversible processes this is not so. When a process proceeds irreversibly, the total entropy of all systems taking part in the process always increases. Let us illustrate this by an example.

Consider a system consisting of one mole of an ideal gas at a temperature T_1 and a volume V_1. Let the process be an isothermal expansion of the gas to a volume V_2, so that the end state is determined by V_2 and the temperature T_1. This change of state can take place either reversibly or irreversibly.

(a) *Reversible process.* We bring the gas in contact with a heat reservoir at the temperature T_1, and perform a reversible isothermal expansion

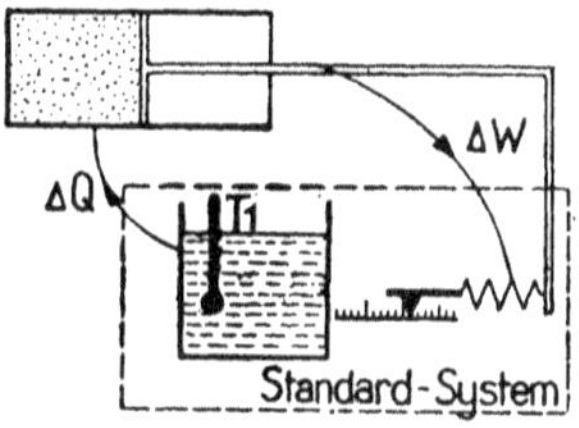

FIG. 10-16. Reversible expansion of an ideal gas.

(Fig. 10-16). A quantity of heat ΔQ flows into the gas from the reservoir and the gas does work ΔW, where (see page 453)

$$\Delta W = -RT_1 \ln \frac{V_2}{V_1}.$$

From the first law, $\Delta W = -\Delta Q$. Since the process is reversible, the increase in entropy of the gas can be found from the quantity of heat and the temperature,

$$\Delta S_{V_1 \to V_2} = \frac{\Delta Q}{T_1} = R \ln \frac{V_2}{V_1}.$$

The heat reservoir undergoes an equal and opposite change in entropy, since the heat ΔQ is transferred reversibly at the temperature T_1,

$$\Delta S_{\text{res}} = \frac{-\Delta Q}{T_1} = -R \ln \frac{V_2}{V_1}.$$

There is no change in the total entropy of the combined system, as must be the case for a closed system in which only reversible processes take place.

(b) *Irreversible process.* The irreversible process is carried out as in Fig. 10-17, in which the gas expands through a suddenly opened stopcock into an evacuated enclosure of such a volume V_1' that $V_1 + V_1' = V_2$. This process is called a free expansion. The walls of the whole container are insulated so that no heat transfer takes place during the process. No work is done because the gas expands into a vacuum. Thus the internal energy of the gas remains constant during the process and since it is an ideal gas, the temperature is also unchanged. Finally, since there is no heat transfer or performance of work, there is no change of state of entropy of the surroundings during the process.

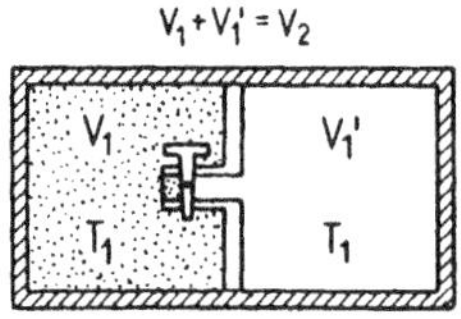

FIG. 10-17. Irreversible expansion of an ideal gas.

To compute the entropy change in this irreversible process we must return the gas reversibly to its original state with the help of a standard system. To do this we imagine the gas to be contained in a cylinder provided with a piston; we then compress it isothermally and reversibly. The necessary work ΔW is provided by the spring of the standard system; the heat developed in the gas, ΔQ, flows into the heat reservoir of the standard system at the temperature T_1 causing the entropy of the standard system to increase. Then

$$\Delta Q_{St} = +\Delta W_{V_2 \to V_1} = RT_1 \ln \frac{V_2}{V_1}.$$

After its compression to the volume V_1, the original state of the gas is completely restored. The entropy change ΔS of the gas, in its original irreversible expansion, has been completely passed on to the standard system and is given by

$$\Delta S_{V_1 \to V_2} = \Delta S_{St} = \frac{\Delta Q_{St}}{T_1} = R \ln \frac{V_2}{V_1}.$$

The change in entropy of the system is therefore the same in both processes. The difference in the nature of the two processes does not appear as a difference in the changes of entropy of the ideal gas system. But if all systems taking part in each process are grouped together and in each case considered as a single closed system, the total entropy remains constant in the reversible process but increases in the irreversible process. Since every natural process is irreversible, we have the following alternative formulation of the second law:

In a closed system, all natural processes proceed in such a way that the total entropy of the system increases. This is known as the *principle of the increase of entropy.*

All natural processes are therefore "one-way" processes. They can take place only in such a way that the entropy in the final state is greater than that in the initial state.

EXAMPLES. 1. The falling of a ball of putty is a possible process, since an increase of entropy results. The heat produced flows irreversibly to a body at a lower temperature. The reversed process — cooling of the ground and return of the putty to its original elevation — is impossible since it would result in a decrease of entropy.

2. When compressed air streams out of a tank, its entropy increases. It is for this reason that compressed air has an economic value, and not only because the compressed air represents a storage of energy. The reversed process cannot take place of itself, since it would result in a decrease of entropy.

We therefore see that natural phenomena are governed not by the *energy law* alone (first law of thermodynamics) but also by the *entropy law* (second law of thermodynamics). Natural processes which are consistent with the first law can take place spontaneously only if they involve an increase of entropy.

A state of thermodynamic equilibrium is characterized by a maximum value of the entropy. Any process in such a system would have to cause a decrease of entropy and therefore could not occur spontaneously.

89-1 Limits of Validity of the Second Law

In conclusion, we add a few remarks concerning the limitations of the second law. According to this law natural processes take place in one direction only, without exception. However, the validity of the law is closely related to the condition of complete molecular disorder. If this condition is not fulfilled, then it is possible for natural processes to proceed in any direction.

The principle of complete molecular disorder is not satisfied for elements of volume so small that they contain only a few molecules. If, for example, the number of molecules in such a region is of the order of 1000, it becomes possible to detect *fluctuations* in the density, the pressure, the number of molecules, and the mean kinetic energy (the temperature). These quantities have a definite meaning only when very large numbers of molecules (of the order of Avogadro's number) are involved because then we can calculate very precise average values for them. When the number of molecules is small, the concepts of density, etc., are no longer precise. The study of fluctuation phenomena requires the methods of *statistical mechanics* and lies outside the scope of this book. The phenomenon of Brownian motion, for example, lies in the field.

90. A Special Cyclic Process for a van der Waals Gas

It was mentioned in Section 75-3 that when a van der Waals gas is compressed at a temperature below its critical temperature it eventually reaches an unstable state in which it does not follow the van der Waals isotherm. Instead, in the presence of condensation nuclei, condensation sets in, with the formation of drops of liquid, and the pressure remains constant at the saturation vapor pressure P_S until all of the gas is condensed. The statement was also made that the horizontal line P_S = const was located at such a height in the P-V plane that the areas above and below this line, bounded by the van der Waals isotherm, were equal. The proof of this statement, first given by Maxwell, follows with the help of the second law.

We imagine the gas to be carried through a cyclic process (Fig. 10-18), first along the van der Waals isotherm from A to B to C to D, and then back

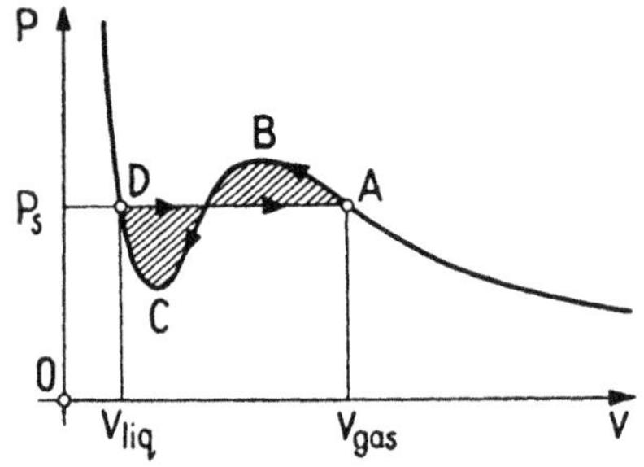

FIG. 10-18. Determination of the saturation vapor pressure of a van der Waals isotherm in the condensation region.

from D to A along the condensation line DA. The work of compression W_1 along the first path is

$$W_1 = \int_{A \to B \to C \to D} P\,dV.$$

The expansion work along the second path is

$$W_2 = -P_s(V_{\text{gas}} - V_{\text{liquid}}).$$

The temperature remains constant during the entire process. But the Kelvin–Planck formulation of the second law states that in an isothermal process, in which heat is exchanged with a single heat reservoir only, the net work is zero. Therefore

$$\int_{A \to B \to C \to D} P\,dV - P_s(V_{\text{gas}} - V_{\text{liquid}}) = 0.$$

Geometrically, the integral corresponds to the area between the van der Waals curve, the V-axis, and the ordinates at the points A and D. The expression $P_s(V_{gas} - V_{liquid})$ represents the area of a rectangle between the horizontal line AD, the ordinates at A and D, and the V-axis. Since the equation above states that these areas are equal, it follows that the line AD must lie at such a height that the shaded areas in Fig. 10-18 are equal.

We shall not consider the application of the second law to other cyclic processes, since the mathematical methods to be developed in the next chapter provide a much more convenient method of applying the first and second laws in such a way that the requirements of both laws are simultaneously fulfilled.

CHAPTER 11

APPLICATIONS OF THE FIRST AND SECOND LAWS

JUST as one may use Newton's laws and kinematic quantities to obtain new useful concepts such as angular momentum or energy, so the first and second laws of thermodynamics, with the aid of thermodynamic quantities, may be used to introduce new quantities. The derivation of relations between the phenomenological properties of materials and this extended set of thermodynamic quantities forms one of the most important applications of thermodynamics. These relations are obtained most conveniently from a consideration of infinitesimal cyclic processes. Certain mathematical techniques, such as partial differentiation and simple integration, are indispensible for this purpose. The necessary techniques in partial differentiation are explained as they are needed.

91. THERMODYNAMIC PROPERTIES; STATE VARIABLES AND STATE FUNCTIONS

A thermodynamic *property* of a system is defined as follows:

DEFINITION. A thermodynamic property of a system is a quantity which depends only on the state of the system.

The value of a property is independent of the path or process by which a particular state of the system was reached. If the system performs a cyclic process, the value of a property is the same at the end of the cycle as at the beginning.

According to the above definition, the quantities P, V, T, U, and S are all thermodynamic properties. Thus we see that the class of quantities which are properties include both the quantities which we called *state variables* in the last chapter and those we called *state functions*. The reason for this division of properties into these particular groups is easy to understand. Pressure, volume, and temperature are quantities which are easily measured with simple instruments, but there is no such thing as an internal energy meter or an entropy gauge. However we do have mathematical relationships between all these properties, and it is then quite natural to think of any two of P, V, and T as independent *variables* and U and S as *functions* of these variables. On the other hand, these designations are only historical and are quite arbitrary. Since all these quantities are thermodynamic properties, it is just as correct, both thermodynamically and mathematically, to call U and T the independent or *state variables* and P, V, and S the *state functions*; and any other subdivision of the properties would do as well.

For the historical and practical reasons given above, we will often consider P, V, and T to be the state variables of a system.

We have used the concept of the state of a system without so far defining it. We shall say that a system is in a definite state when there exists a *unique* functional relationship between any one thermodynamic property and two (or more) of the other properties: $\Phi_1 = \Phi_1(\Phi_2, \Phi_3, \cdots)$. The familiar equation of state for a gas can thus be expressed as $P = P(V, T)$. However, equally valid expressions are $U = U(S, V)$ or $S = S(U, V)$. Since P and T are defined only when the system is at equilibrium, and since these quantities may be any one of Φ_1, Φ_2, or Φ_3, it follows that *the state of a system is uniquely defined only at equilibrium.* Unless otherwise stated, we shall generally refer to systems in thermal equilibrium. More of the mathematical properties of the equation of state are discussed in Section 91-2.

91-1 State Variables

The macroscopic state of a homogeneous substance can be specified in most instances by the values of two *state variables.* For such a case, the substance and its surroundings are referred to as a simple system. For mixtures and solutions, the concentrations or amounts of the components enter as additional state variables. In a consideration of phenomena which are related to a change of surface area, this area must also be included.

A state variable which remains constant during a process is called a *state parameter.* Thus in an isothermal process, for example, the temperature is a state parameter.

We are already familiar with the quantities P, V, and T. In considering concentrations, a number of different quantities are used which are defined in Appendix I.

To characterize a reaction equation, the *reaction variables* ν_i are introduced:

$$\nu_1 A + \nu_2 B \rightarrow \nu_3 C + \nu_4 D.$$

The ν_i are also called *stoichiometric coefficients* and are proportional to the number of moles of the various substances. It is convenient to write the equation of a reaction proceeding in one direction only in such a form that the products of the reaction have positive reaction variables, while the substances that disappear have negative variables:

$$-\nu_1 A - \nu_2 B + \nu_3 C + \nu_4 D = 0.$$

The sum $\mathfrak{J} = \sum \nu_i$ is called the *reaction index*. If it is zero, the number of moles does not change as a result of the reaction. *A* positive value of $\mathfrak{J}$ indicates that the number of moles increases by $\mathfrak{J}$.

EXAMPLES.

$$H_2 + Cl_2 \rightarrow 2HCl \quad (\mathfrak{J} = 0).$$

$$2H_2 + O_2 \rightarrow 2H_2O \quad (\mathfrak{J} = -1).$$

91-2 State Functions and their Mathematical Properties

All thermodynamic quantities relating to a system in equilibrium can be expressed as functions of the state variables *P*, *V*, and *T*. In mixtures and solutions, the concentrations must also be included. We find by experiment that it is necessary to give the values of only a few measurable quantities to specify completely the state of a system in thermodynamic equilibrium. For 1 mole of a gas, for example, a knowledge of only two quantities such as *P* and *V* is sufficient. All other properties of the system then either have no thermodynamic significance or are dependent quantities which can be computed with the help of the equation of state (for an ideal gas, $PV_M = RT$).

As discussed earlier, it is a matter of choice which quantities are to be considered independent and which dependent. Certain processes can be described in a particularly simple way in terms of properly chosen variables. It may well be an advantage in some instances to use the state functions as independent variables, instead of the directly measurable quantities *P*, *V*, and *T*.

In order to be quite general, we will speak of thermodynamic properties rather than state functions. For any property Φ,

$$\sum_{\text{cyclic process}} \Delta\Phi = 0.$$

For simplicity, we shall consider only those properties Φ which are functions of two independent state variables *x* and *y*:

$$\Phi = \Phi(x, y),$$

where *x* and *y* might, for example, be the pressure and volume of a system. The following results can easily be extended to functions of three or more variables. A property which is a function of only two state variables can be represented graphically by a three-dimensional surface. The values of the independent variables *x* and *y* (that is, variables which can be given arbitrary values) determine a point in the *x*-*y* plane (Fig. 11-1). The value of the property $\Phi(x, y)$ is now laid off in the *z*-direction, vertically above this point, thus determining a point in space. When this is done for all possible values of *x* and *y*, the points form a surface as shown, which represents geometrically the value of the property Φ as a function of *x* and *y*. The requirement that the property shall be uniquely determined by *x* and *y* means that only one point of the surface can lie above a given point (x, y). Hence there can be no vertical breaks in the surface, and no overhanging portions. For, if the surface could be of this nature, as shown in Fig. 11-2, then two values of Φ would correspond to the same point (x, y). It would then be possible to start from the point (x, y) with a value of the property $\Phi = z_1$, and by a cyclic process arrive at a different value $\Phi = z_2$. This is impossible, however if Φ is a property.

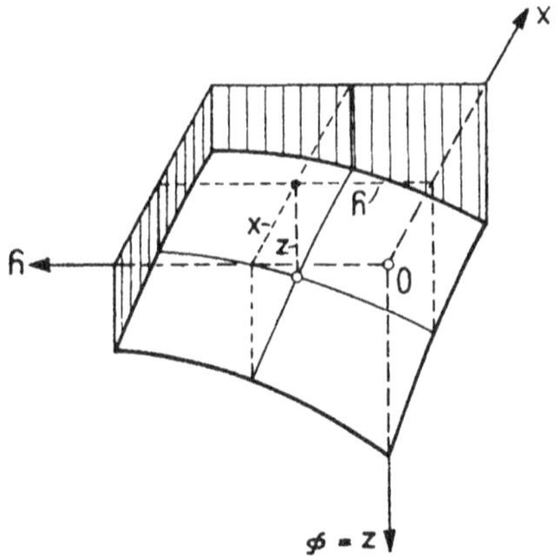

FIG. 11-1. Three-dimensional graphical representation of the property of a system $\Phi(x, y)$ as a function of x and y.

Quantities which are *not* uniquely determined by the state variables of a system are the *work* done on the system and the *heat* flowing into the system. *Work and heat are not properties of a system.* On the other hand, the internal

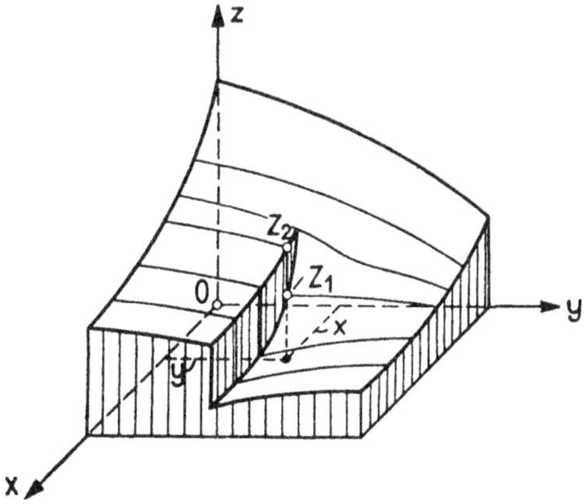

FIG. 11-2. Example of a surface which is two-valued for certain values of the independent variables, x and y.

energy of a system is a property since (by the first law) it is uniquely determined at equilibrium by the values of the state variables. Its value does not change in a cyclic process (a perpetual motion machine of the first kind (see Section 29-1) is impossible). The second law shows that the entropy of a system is also a property of the system.

If the variables x and y are changed to new values $x + dx$, $y + dy$, the property Φ changes also. The methods of partial differentiation give for the total change $d\Phi$ of the property Φ,

$$d\Phi = \left(\frac{\partial \Phi}{\partial x}\right)_y dx + \left(\frac{\partial \Phi}{\partial y}\right)_x dy. \tag{91-1}$$

The symbols $\left(\frac{\partial \Phi}{\partial x}\right)_y$ and $\left(\frac{\partial \Phi}{\partial y}\right)_x$

are called the *partial derivatives* of Φ with respect to x and y respectively. The term "partial" refers to the fact that the change in Φ results from the change in one variable only, the other being kept constant. The partial derivative thus is an ordinary derivative with respect to the free variable. The total change in Φ is given by the sum of the partial changes, as shown by Eqn. (91-1). These partial derivatives are defined only in an infinitesimal neighborhood of equilibrium. Their existence and uniqueness follows from the existence and uniqueness of Φ at equilibrium.

EXAMPLE. For an ideal gas,

$$V_M = \frac{RT}{P}.$$

The molar volume of the gas is uniquely determined by the state variables P and T, and hence is a state function or a property. The total change in V_M in an infinitesimal process is given by

$$dV_M = \left(\frac{\partial V_M}{\partial P}\right)_T dP + \left(\frac{\partial V_M}{\partial T}\right)_P dT.$$

The term $\left(\frac{\partial V_M}{\partial P}\right)_T dP$ corresponds to the change of molar volume when the pressure changes by dP at constant temperature, and $\left(\frac{\partial V_M}{\partial T}\right)_P dT$ corresponds to the change in volume when the temperature changes by dT at constant pressure. The partial derivatives can be computed from the equation of state:

$$\left(\frac{\partial V_M}{\partial P}\right)_T = \frac{\partial}{\partial P}\left(RT\frac{1}{P}\right)_T = RT\frac{d}{dP}\left(\frac{1}{P}\right) = -RT\frac{1}{P^2},$$

$$\left(\frac{\partial V_M}{\partial T}\right)_P = \frac{R}{P}\frac{d}{dT}(T) = \frac{R}{P}.$$

Hence

$$dV_M = -\frac{RT}{P^2}dP + \frac{R}{P}dT. \tag{91-2}$$

As a special case of Eqn. (91-1), suppose that the changes in x and y are so related that the property Φ remains constant. Then

$$d\Phi = 0 = \left(\frac{\partial\Phi}{\partial x}\right)_y dx + \left(\frac{\partial\Phi}{\partial y}\right)_x dy.$$

Let us divide this equation through by dy and, since Φ = const, write the ratio $\frac{dx}{dy}$ as $\left(\frac{\partial x}{\partial y}\right)_\Phi$. Then

$$\left(\frac{\partial x}{\partial y}\right)_\Phi = -\frac{(\partial\Phi/\partial y)_x}{(\partial\Phi/\partial x)_y}, \tag{91-3}$$

a relation we shall use shortly.

In the application of the first and second laws, the *second* derivatives of the properties of a system play an important role. Two of these are represented by the mathematical symbols

$$\frac{\partial}{\partial x}\left(\frac{\partial\Phi}{\partial y}\right) = \frac{\partial^2\Phi}{\partial x\,\partial y}$$

and

$$\frac{\partial}{\partial y}\left(\frac{\partial\Phi}{\partial x}\right) = \frac{\partial^2\Phi}{\partial y\,\partial x}.$$

They represent a second partial differentiation of the partial derivatives $\partial\Phi/\partial y$ and $\partial\Phi/\partial x$, with respect to x and y respectively, and hence are called the *mixed second partial derivatives.* It is shown in texts on mathematics that these mixed partial derivatives are equal to each other. This means that the same result is obtained if we differentiate first with respect to one variable and then with respect to the other, or vice versa. Starting with the expression for the total change in a property,

$$d\Phi = \left(\frac{\partial\Phi}{\partial x}\right) dx + \left(\frac{\partial\Phi}{\partial y}\right) dy,$$

we see that the partial derivative of the coefficient of dx, with respect to y, equals the partial derivative of the coefficient of dy with respect to x. Conversely, it can be shown that if the mixed second partial derivatives of Φ are equal, then Φ is a property.

EXAMPLE. The total change in V_M for an ideal gas, from Eqn. (91-2), is

$$dV_M = -\frac{RT}{P^2}\,dP + \frac{R}{P}\,dT.$$

But

$$\frac{\partial}{\partial T}\left(-\frac{RT}{P^2}\right) = -\frac{R}{P^2}$$

and

$$\frac{\partial}{\partial P}\left(\frac{R}{P}\right) = -\frac{R}{P^2},$$

in agreement with the statement above.

We now show that the equality of $\partial^2\Phi/\partial x\,\partial y$ and $\partial^2\Phi/\partial y\,\partial x$ corresponds physically to the performance of an infinitesimal cyclic process. Consider the cyclic process shown in Fig. 11-3. The change in Φ along the path 012 must equal, the change along the path 01′2.

Path 012:

$$\Phi_2 - \Phi_0 = (\Phi_1 - \Phi_0) + (\Phi_2 - \Phi_1)$$
$$= \frac{\partial\Phi(x, y)}{\partial x}dx + \frac{\partial\Phi(x + dx, y)}{\partial y}dy.$$

Path 01′2:

$$\Phi_2 - \Phi_0 = (\Phi_1' - \Phi_0) + (\Phi_2 - \Phi_1')$$
$$= \frac{\partial\Phi(x, y)}{\partial y}dy + \frac{\partial\Phi(x, y + dy)}{\partial x}dx.$$

Since the changes must be equal, it follows that

$$\frac{\partial\Phi(x, y)}{\partial x}dx + \frac{\partial\Phi(x + dx, y)}{\partial y}dy = \frac{\partial\Phi(x, y)}{\partial y}dy + \frac{\partial\Phi(x,y + dy)}{\partial x}dx,$$

or, after rearranging terms,

$$\frac{\partial}{\partial x}[\Phi(x, y + dy) - \Phi(x, y)]\,dx = \frac{\partial}{\partial y}[\Phi(x + dx, y) - \Phi(x, y)]\,dy.$$

But

$$\Phi(x, y + dy) - \Phi(x, y) = \left(\frac{\partial\Phi}{\partial y}\right)_x dy,$$

$$\Phi(x + dx, y) - \Phi(x, y) = \left(\frac{\partial\Phi}{\partial x}\right)_y dx.$$

Therefore

$$\frac{\partial}{\partial x}\left(\frac{\partial\Phi}{\partial y}\right)_x dx\,dy = \frac{\partial}{\partial y}\left(\frac{\partial\Phi}{\partial x}\right)_y dx\,dy.$$

The equality of $\frac{\partial}{\partial x}\left(\frac{\partial\Phi}{\partial y}\right)_x$ and $\frac{\partial}{\partial y}\left(\frac{\partial\Phi}{\partial x}\right)_y$ therefore corresponds to carrying out an infinitesimal cyclic process. Knowing this fact, we can now carry a system through a cyclic process in an elegant way by simply equating the mixed second partial derivatives of a property. Many useful thermodynamic relations can be derived in this way. It is important, however, to know what the properties are which correspond to the process in question. The fundamental functions, in terms of which the first and second laws are expressed, are the internal energy U and the entropy S. From these, and the state variables P, V, and T, five additional properties can be formed which are characteristic of certain specific processes. These are:†

† In many texts, the Helmholtz function is denoted by A and the Gibbs function by F. The above notation is the most commonly used in applications to solid state physics and statistical mechanics.

1. The enthalpy H.
2. The Helmholtz function F.
3. The Gibbs function G.
4. The Planck function Y.
5. The Massieu function J.

Each of these properties has its appropriate field of application. They are suitable for application to specific processes and they implicitly include the statements of both the first and second laws.

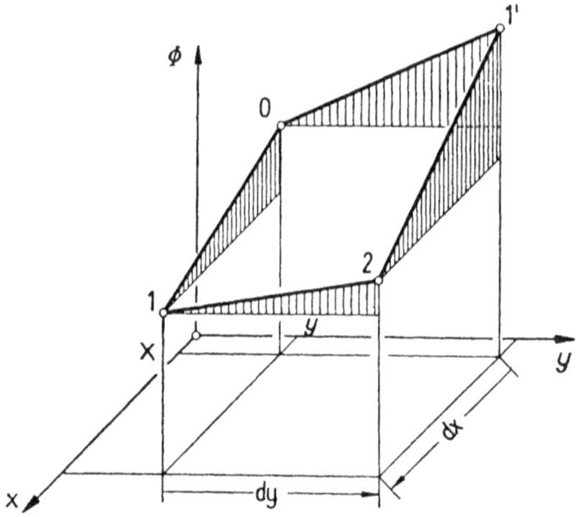

FIG. 11-3. Diagram of an infinitesimal cyclic process for the proof of the equality of the mixed second partial derivatives, $\partial^2\Phi/\partial x\,\partial y$ and $\partial^2\Phi/\partial y\,\partial x$.

92. Phenomenological Properties of Matter

In addition to the specific heats and thermal conductivity, some other important thermodynamic properties of matter whose experimental values can be found from tables are the following:

1. The isobaric coefficient of volume expansion,

$$\alpha_p = \frac{1}{V}\left(\frac{\partial V}{\partial T}\right)_P. \tag{92-1}$$

2. The isothermal compressibility,

$$\varkappa_T = -\frac{1}{V}\left(\frac{\partial V}{\partial P}\right)_T. \tag{92-2}$$

3. The isochoric pressure coefficient,

$$\beta_v = \frac{1}{P}\left(\frac{\partial P}{\partial T}\right)_V. \tag{92-3}$$

This third property is not as frequently tabulated, but can easily be computed from the two above. The relation between these quantities can be derived as follows. Every substance, solid, liquid, or gas, has an equation of state which can be written as

$$V = V(P, T).$$

From Eqn. (91-3),

$$\left(\frac{\partial P}{\partial T}\right)_V = -\frac{(\partial V/\partial T)_P}{(\partial V/\partial P)_T},$$

and when the expressions for α_p, $\varkappa_T$, and β_v are inserted in this equation we obtain

$$\frac{\beta_v \varkappa_T P}{\alpha_p} = 1. \qquad (92\text{-}4)$$

EXAMPLE. The tabulated values of α_p and $\varkappa_T$, for water at 1 atm and 20°C, are

$$\alpha_p = 2.07 \times 10^{-4}\,°\mathrm{C}^{-1}.$$

$$\varkappa_T = 49.1 \times 10^{-6}\,\mathrm{atm}^{-1}.$$

It follows that

$$\beta_v = \frac{\alpha_p}{\varkappa_T P} = 4.21\,°\mathrm{C}^{-1}.$$

That is, when water at a temperature of 20°C is heated by 0.1°C at constant volume, the pressure increases by 0.42 atm.

93. Thermodynamic Functions and Maxwell's Relations

If we apply the first and second laws to an *infinitesimal* and *reversible* change of state, then

$$dU = \delta Q + \delta W \qquad \text{(first law)}$$

$$dS_{\text{rev}} = \frac{\delta Q_{\text{rev}}}{T} \qquad \text{(second law).}$$

In writing these equations, we have used the symbols d and δ to distinguish between quantities like U and S on the one hand, whose changes depend only on the initial and final states, and those like Q and W whose changes depend on the path. Thus dU means "a small *change* in U", but δQ means "a small *quantity* of heat". This distinction will be made from now on.

By combining the above equations, and assuming a reversible process in which only volume work is done (and this is the only kind of work we shall consider), we obtain a concise statement comprising both the first and second laws. This statement, and others that we shall derive with its aid, relates infinitesimal changes between state properties and is valid for reversible infinitesimal processes between two equilibrium states. From these statements, we can derive a number of useful relations using the method of infinitesimal cyclic processes and its consequence that

$$\frac{\partial^2 \Phi}{\partial x \, \partial y} = \frac{\partial^2 \Phi}{\partial y \, \partial x}.$$

At first glance, it will appear that such relations cannot be applied to an irreversible process in an isolated system, even if it connects equilibrium states. The utility of the method, however, lies in the fact that at least in principle we can enlarge the system to include a sufficient number of heat and work reservoirs to effect the change in the isolated system by means of a *reversible* process. The relations we shall derive are valid for this reversible process. But all Φ's are state properties and are independent of the path of the process. Thus, we may apply any results obtained for the imagined reversible process to the actual irreversible process between the same states.

In this way the first and second laws and related statements, which themselves do not involve any properties of matter, can be used to derive relations between such properties, and to obtain useful results applicable to all processes.

93-1 The Internal Energy Function expressed in General Form

The *total* change of the internal energy of a system, in any infinitesimal arbitrary process, is given by the first law,

$$dU = \delta Q + \delta W$$

and for a quasi-static process (i.e. pressure equilibrium), this becomes

$$dU = \delta Q - P\,dV.$$

If the process is reversible, then from the second law

$$\delta Q = T\,dS,$$

where T is the temperature at which the heat δQ is transferred. Both laws can now be combined in the single equation,

$$dU = T\,dS - P\,dV.$$

This extremely important equation will be used in many future derivations. It can be considered to express the internal energy as a function of the variables S and V, a relation which is valid only in a region which is infinitesimally close to an equilibrium state. Since S and V are state properties, their differentials are exact, and hence we can apply the techniques of partial differentiation to them.

Since the total change in U is given in general by

$$dU = \left(\frac{\partial U}{\partial S}\right)_V dS + \left(\frac{\partial U}{\partial V}\right)_S dV$$

we have

$$\left(\frac{\partial U}{\partial S}\right)_V = T, \qquad \left(\frac{\partial U}{\partial V}\right)_S = -P.$$

We now have an explicit relationship between the statement that P and T are defined only for equilibrium states, and the fact that $dU = TdS - PdV$ is valid only in the neighborhood of an equilibrium state. Then since

$$\frac{\partial}{\partial V}\left(\frac{\partial U}{\partial S}\right)_V = \frac{\partial}{\partial S}\left(\frac{\partial U}{\partial V}\right)_S,$$

it follows that

$$\left(\frac{\partial T}{\partial V}\right)_S = -\left(\frac{\partial P}{\partial S}\right)_V. \tag{93-1}$$

Equation (93-1) is known as the *first Maxwell relation.* It states: The rate of change of temperature with volume at constant entropy (reversible adiabatic process) equals the negative rate of change of pressure with entropy at constant volume (reversible isochoric process).

The change in entropy in a reversible isochoric process is given by

$$(dS)_V = \left(\frac{dU}{T}\right)_V = \frac{C_v\,dT}{T}.$$

Therefore

$$\left(\frac{\partial T}{\partial S}\right)_V = \frac{T}{C_v}.$$

Now, from the theory of partial differentiation, we have

$$\left(\frac{\partial P}{\partial S}\right)_V = \left(\frac{\partial P}{\partial T}\right)_V\left(\frac{\partial T}{\partial S}\right)_V$$

and it therefore follows that

$$\left(\frac{\partial P}{\partial S}\right)_V = \left(\frac{\partial P}{\partial T}\right)_V\frac{T}{C_v}. \tag{93-2}$$

This expression can also be obtained if the denominator in the expression $(\partial P/\partial S)_V$ is replaced by $C_v dT/T$. Thus

$$\left(\frac{\partial P}{\partial S}\right)_V = \frac{(dP)_V}{(dS)_V} = \frac{(dP)_V}{C_v(dT)_V/T} = \left(\frac{\partial P}{\partial T}\right)_V\frac{T}{C_v}.$$

From Eqns. (93-1) and (93-2) we have for $(\partial T/\partial V)_S$

$$\left(\frac{\partial T}{\partial V}\right)_S = -\frac{T}{C_v}\left(\frac{\partial P}{\partial T}\right)_V. \tag{93-3}$$

Making use of Eqns. (92-3) and (92-4), we can write

$$\left(\frac{\partial T}{\partial V}\right)_S = -\frac{T\alpha_p}{C_v\varkappa_T},$$

a relation which applies to any homogeneous phase in a reversible adiabatic process.

EXAMPLE. For an ideal gas,

$$PV_M = RT,$$

$$\left(\frac{\partial P}{\partial T}\right)_{V_M} = \frac{R}{V_M}.$$

From this equation and Eqn. (93-3),

$$\left(\frac{\partial T}{\partial V}\right)_S = -\frac{T}{C_v}\frac{R}{V_M} = -\frac{P}{C_v},$$

and

$$-P\,dV = C_v\,dT,$$

a relation we have already derived for an adiabatic process of an ideal gas.

93-2 The Entropy Function in General Form

The second law states that

$$dS_{\text{rev}} = \frac{\delta Q_{\text{rev}}}{T}.$$

From the first law,

$$\delta Q_{\text{rev}} = dU + P\,dV,$$

and considering U now as a function of V and T,

$$dU = \left(\frac{\partial U}{\partial V}\right)_T dV + \left(\frac{\partial U}{\partial T}\right)_V dT. \qquad (93\text{-}4)$$

From these three equations we obtain the general differential expression for the entropy function in terms of the variables T and V:

$$dS = \frac{1}{T}\left(\frac{\partial U}{\partial T}\right)_V dT + \left[\frac{P}{T} + \frac{1}{T}\left(\frac{\partial U}{\partial V}\right)_T\right] dV.$$

This equation relates reversible *isochoric* and *isothermal* changes of state.

From the equality of the mixed second derivatives $\partial^2 S/\partial T\partial V$ and $\partial^2 S/\partial V\partial T$ we may derive a further relation:

$$\frac{\partial}{\partial V}\left[\frac{1}{T}\left(\frac{\partial U}{\partial T}\right)_V\right] = \frac{\partial}{\partial T}\left[\frac{P}{T} + \frac{1}{T}\left(\frac{\partial U}{\partial V}\right)_T\right],$$

or

$$\frac{1}{T}\frac{\partial^2 U}{\partial V\partial T} = \frac{1}{T}\left(\frac{\partial P}{\partial T}\right)_V - \frac{P}{T^2} + \frac{1}{T}\frac{\partial^2 U}{\partial T\partial V} - \frac{1}{T^2}\left(\frac{\partial U}{\partial V}\right)_T.$$

It follows from this equation that

$$\left(\frac{\partial U}{\partial V}\right)_T = T\left(\frac{\partial P}{\partial T}\right)_V - P. \qquad (93\text{-}5)$$

Equation (93-5) leads to important results which were used in earlier sections and which we can now derive.

1. *Relation between C_p and C_v in a homogeneous phase.* The molar heat capacities at constant pressure and at constant volume are defined by the equations

$$C_p = \frac{\delta Q_P}{dT_P},$$

$$C_v = \frac{\delta Q_V}{dT_V} = \left(\frac{\partial U}{\partial T}\right)_V.$$

From the first law and Eqn. (93-4),

$$\delta Q = dU + P\,dV = \left(\frac{\partial U}{\partial T}\right)_V dT + \left[\left(\frac{\partial U}{\partial V}\right)_T + P\right] dV.$$

We therefore have for C_p,

$$C_p = \left(\frac{\partial U}{\partial T}\right)_V + \left[\left(\frac{\partial U}{\partial V}\right)_T + P\right]\left(\frac{\partial V}{\partial T}\right)_P$$

$$= C_v + \left[\left(\frac{\partial U}{\partial V}\right)_T + P\right]\left(\frac{\partial V}{\partial T}\right)_P. \quad (93\text{-}6)$$

Combining this with Eqn. (93-5), we get

$$C_p - C_v = T\left(\frac{\partial P}{\partial T}\right)_V\left(\frac{\partial V}{\partial T}\right)_P. \quad (93\text{-}7)$$

But from Eqns. (92-1), (92-3), and (92-4),

$$\beta_v P = \left(\frac{\partial P}{\partial T}\right)_V,$$

$$\alpha_p V = \left(\frac{\partial V}{\partial T}\right)_P,$$

$$\beta_v \varkappa_T P = \alpha_p,$$

and hence

$$\left(\frac{\partial P}{\partial T}\right)_V = \frac{\alpha_p}{\varkappa_T}.$$

Eqn. (93-7) can therefore be written

$$C_p - C_v = \frac{\alpha_p^2 T V}{\varkappa_T}, \quad (93\text{-}8)$$

a relation which applies to any homogeneous phase whatever. We note that T, V, and $\varkappa_T$ are always positive. α_p can be positive, negative, or zero, but $\alpha_p^2 \geqq 0$. Hence $C_p \geqq C_v$.

EXAMPLE. For an ideal gas,

$$\alpha_p = \frac{1}{V}\left(\frac{\partial V}{\partial T}\right)_P = \frac{R}{PV},$$

$$\varkappa_T = -\frac{1}{V}\left(\frac{\partial V}{\partial P}\right)_T = +\frac{1}{V}\frac{RT}{P^2} = \frac{1}{P},$$

and

$$C_p - C_v = \frac{R^2}{P^2V^2} \cdot TV \cdot P = R.$$

2. *The internal pressure of a homogeneous phase.*

In Eqn. (93-6), the term $(\partial U/\partial V)_T$ represents the internal pressure P_i, which must be added to the external pressure P (Section 75-3). The sum is called the *total pressure* P_t of the system,

$$P_t = P + P_i = P + \left(\frac{\partial U}{\partial V}\right)_T .$$

From Eqns. (93-6) and (93-7),

$$P_t = \frac{C_p - C_v}{\left(\frac{\partial V}{\partial T}\right)_P} = \frac{T\alpha_p}{\varkappa_T}. \tag{93-9}$$

EXAMPLES. (1) For an ideal gas,

$$P_t = \frac{T(1/T)}{1/P} = P.$$

The total pressure equals the external pressure and the internal pressure is zero.

(2) For a van der Waals gas,

$$\alpha_p = \frac{RV_M^2(V_M - b)}{RTV_M^3 - 2a(V_M - b)^2},$$

$$\varkappa_T = \frac{V_M^2(V_M - b)^2}{RTV_M^3 - 2a(V_M - b)^2},$$

$$P_t = \frac{T\alpha_p}{\varkappa_T} = \frac{RT}{V_M - b} = P + \frac{a}{V_M^2},$$

in agreement with the results of Section 75-4.

We have already made use of Eqns. (93-8) and (93-9) in computing the difference $C_p - C_v$ for a homogeneous phase and in estimating the internal pressure in liquids and solids. The total pressure P_t in a solid is usually

TABLE 11-1

Substance	T (°K)	α_p (°t)$^{-1}$	$\varkappa_T$ (atm^{-1})	P_t (atm)
H_2O	293	2.07×10^{-4}	4.91×10^{-5}	1,235
Ice	273	1.125×10^{-4}	8.1×10^{-5}	375
Hg	293	1.82×10^{-4}	3.95×10^{-6}	13,500
C_2H_5OH	293	11.2×10^{-4}	1.12×10^{-4}	2,930
$CHCl_3$	293	12.73×10^{-4}	1.02×10^{-4}	3,657
C_6H_6	293	12.37×10^{-4}	9.1×10^{-5}	3,980
α-Fe	293	0.335×10^{-4}	5.9×10^{-7}	16,636
Pb	293	0.84×10^{-4}	2.25×10^{-6}	10,939
Mg	293	0.762×10^{-4}	2.85×10^{-6}	7,834

computed from Eqn. (93-9), using tabulated values of α_p and $\varkappa_T$. The total pressure can have very different values in different materials, as is shown in Table 11-1.

93-3 Enthalpy

The *enthalpy* H or *heat content* is defined as the sum of the internal energy U and the product PV:

$$H = U + PV.$$

Both the internal energy U and the product PV are properties of a system, so the enthalpy H is a property also. In an infinitesimal process

$$dH = dU + P\,dV + V\,dP,$$

and, for a reversible process using $dU = T\,dS - P\,dV$, we obtain

$$dH = T\,dS + V\,dP.$$

The enthalpy function therefore relates reversible *adiabatic* processes (S = const) and reversible *isobaric* processes (P = const). From the relation

$$\frac{\partial^2 H}{\partial S\,\partial P} = \frac{\partial^2 H}{\partial P\,\partial S},$$

it follows that

$$\left(\frac{\partial T}{\partial P}\right)_S = \left(\frac{\partial V}{\partial S}\right)_P. \tag{93-10}$$

This is known as the *second Maxwell relation.*

Since

$$dS = \frac{\delta Q_{\text{rev}}}{T}$$

and

$$(\delta Q)_P = C_p^s\,dT,$$

we have

$$\left(\frac{\partial V}{\partial S}\right)_P = T\frac{(dV)_P}{(\delta Q)_P} = \frac{T}{C_p^s}\left(\frac{\partial V}{\partial T}\right)_P.$$

Making use of Eqn. (92-1), the second Maxwell relation can be written in the form

$$\left(\frac{\partial T}{\partial P}\right)_S = \frac{T}{C_p^s}\left(\frac{\partial V}{\partial T}\right)_P = \frac{T\alpha_p V}{C_p^s}. \tag{93-11}$$

This equation gives the change in temperature in an adiabatic pressure change for any homogeneous phase.

EXAMPLE. For an ideal gas,

$$\alpha_p = \frac{1}{T}.$$

Hence from Eqn. (93-11), for an ideal gas,

$$dT\Big|_S = \frac{V}{C_p}\,dP\Big|_S = \frac{RT}{PC_p}\,dP\Big|_S.$$

$$\frac{dT}{T}\Big|_S = \frac{R}{C_p}\frac{dP}{P}\Big|_S = (1-\gamma)\frac{dP}{P}\Big|_S,$$

which will be recognized as the differential form of the equation of state for an ideal gas under adiabatic conditions.

1. *Isobaric processes.* An important property of the enthalpy function is revealed if we consider an isobaric process. Then $dP \equiv 0$ and

$$(dH)_P = (\delta Q)_P.$$

The increase in the enthalpy of a system in an isobaric process equals the heat flowing into the system.

A very important class of isobaric processes are chemical reactions. In general, a chemical reaction involves either the absorption or evolution of heat. The enthalpy function is particularly suited for studying isobaric chemical reactions. The study of the heat effects accompanying chemical changes is called *thermochemistry*, and in what follows we discuss some of the basic principles of this field.

We start by considering the enthalpy of a single chemical element. For one mole of an element we have from above

$$(dH)_P = (\delta Q)_P = C_p\,dT,$$

and

$$\left(\frac{\partial H}{\partial T}\right)_P = C_p.$$

Hence, the molar enthalpy of the element as a function of temperature is

$$H(T) = H_0 + \int C_p(T')\,dT'.$$

The quantity H_0 is an integration constant which can be determined if the value of H is known at any one temperature T_0. It can be seen that H_0 is simply a mathematical quantity and is without physical significance. Since for most substances C_p is not known as a function of temperature all the way down to absolute zero, the enthalpy of all chemical elements is arbitrarily set equal to zero at normal temperature and pressure (25°C, 1 atm), provided the element can exist in a stable modification in this state. In the thermodynamics of reactive systems we are interested only in changes in the function H, never in its absolute value, and these changes can be measured experimentally since the heat absorbed at constant pressure is equal to the increase in enthalpy ΔH.

In Chapter 7 we saw that the heat capacity at constant pressure of gases as a function of temperature could be represented as a power series in the temperature T (see Table 7-4). This is true in general for solids and liquids also. Thus, for a mole of element D, we have

$$C_{p,D} = A_0 + A_1T + A_2T^2 + \cdots.$$

Then the enthalpy of D as a function of the temperature T is

$$H_D(T) = H_{0D} + A_0T + A_1\frac{T^2}{2} + A_2\frac{T^3}{3} + \cdots,$$

where

$$H_{0D} = -298A_0 - A_1\frac{(298)^2}{2} - A_2\frac{(298)^3}{3} - \cdots.$$

The next step is to consider the formation of simple compounds. These are formed by a chemical reaction which is either endothermic or exothermic, as stated above. The amount of heat evolved or absorbed at constant pressure is called the *heat of formation.* In more quantitative terms, the heat of formation is the quantity of heat which must be supplied when 1 mole of a substance is formed from its elements. Hence the heat of formation is negative if the reaction is exothermic, positive if it is endothermic.

Consider the reaction

$$\nu_1A + \nu_2B \rightarrow \nu_3C,$$

and let the molar enthalpies or heat contents of elements A, B, and C be H_A, H_B, and H_C respectively. In the reaction, ν_1 moles of A and ν_2 moles of B vanish, and ν_3 moles of C are formed. Since enthalpy is a thermodynamic property, the change in enthalpy in going from the elements to the compound is

$$\Delta H = -\nu_1H_A - \nu_2H_B + \nu_3H_C$$

and hence ΔH *is the heat of formation of* C.

We also have

$$\left(\frac{\partial H_A}{\partial T}\right)_P = C_{p,A},$$

$$\left(\frac{\partial H_B}{\partial T}\right)_P = C_{p,B},$$

$$\left(\frac{\partial H_C}{\partial T}\right)_P = C_{p,C}.$$

Thus, for the reaction we are considering

$$\left[\frac{\partial(\Delta H)}{\partial T}\right]_P = -\nu_1C_{p,A} - \nu_2C_{p,B} + \nu_3C_{p,C} = \Delta C_p.$$

The quantity ΔH is thus given by

$$\Delta H(T) = \Delta H_0 + \int \Delta C_p\,dT,$$

where

$$\Delta H_0 = -\nu_1 H_{0A} - \nu_2 H_{0B} + \nu_3 H_{0C}$$

is the constant of integration.

EXAMPLE.

$$\tfrac{1}{2}H_2(g) + \tfrac{1}{2}Cl_2(g) \rightarrow 1HCl(g),$$

$$\nu_1 = -\tfrac{1}{2}, \quad \nu_2 = -\tfrac{1}{2}, \quad \nu_3 = 1.$$

$$C_{p,H_2} = 6.52 + 7.8 \times 10^{-4}\,T + \cdots,$$

$$C_{p,Cl_2} = 8.85 + 1.6 \times 10^{-4}\,T + \cdots,$$

$$C_{p,HCl} = 6.50 + 10 \times 10^{-4}\,T + \cdots,$$

$$\Delta C_p = -1.18 + 5.3 \times 10^{-4}\,T + \cdots.$$

$$\Delta H = \Delta H_0 - 1.18\,T + 2.65 \times 10^{-4}\,T^2 \cdots.$$

This equation expresses the heat of reaction ΔH as a function of temperature. The magnitude of the integration constant ΔH_0 can be found from the tabulated value of the heat of reaction at 298°C:

$$\Delta H_{298°C} = -22{,}060 \text{ cal/mole}.$$

It follows on inserting this value in the series expansion for ΔH that

$$\Delta H_0 = -21{,}732 \text{ cal/mole}$$

and hence that

$$\Delta H = -21{,}732 - 1.18\,T + 2.65 \times 10^{-4}\,T^2 + \cdots.$$

The temperature dependence of the heat of reaction is

$$\frac{\partial(\Delta H)}{\partial T} = -1.18 + 5.3 \times 10^{-4}\,T + \cdots.$$

We can generalize to any reaction, for instance,

$$\nu_1 A + \nu_2 B + \nu_3 C \rightarrow \nu_4 D + \nu_5 E$$

where now the substances A, etc. may themselves be compounds. Just as before

$$\Delta H = -\nu_1 H_A - \nu_2 H_B - \nu_3 H_C + \nu_4 H_D + \nu_5 H_E,$$

but now we give ΔH the more general name, *heat of reaction*. The same sign convention applies as before, and also

$$\left[\frac{\partial(\Delta H)}{\partial T}\right]_P = \Delta C_p,$$

$$\Delta H(T) = \Delta H_0 + \int \Delta C_p\, dT.$$

With the help of tabulated values of ΔH it is possible to compute many quantities such as heats of transformation, heats of solution, and heats of dilution.

EXAMPLE. The *Chemical Engineers Handbook* lists the following values for $CaCl_2$:

$$CaCl_2(s): \quad \Delta H_{18^\circ C} = -190{,}600 \text{ cal/mole}$$

$$CaCl_2(aq\ \infty): \Delta H_{18^\circ C} = -208{,}690 \text{ cal/mole}.$$

The heat of solution for an infinitely dilute solution is therefore

$$\lambda(aq\ \infty) = -18{,}090 \text{ cal/mole}.$$

The process is exothermic since λ is negative.

93-4 The Helmholtz Function

The Helmholtz function F (sometimes called free energy) is defined by the equation

$$F = U - TS.$$

Therefore

$$dF = dU - T\,dS - S\,dT.$$

But for a reversible process,

$$dU = T\,dS - P\,dV$$

and hence

$$dF = -S\,dT - P\,dV.$$

The change in F during an isothermal reversible process equals the work done on the system.

The Helmholtz function is expressed in terms of the independent variables T and V and relates reversible *isothermal* and *isochoric* processes. Since isochoric processes are not of common occurrence, this function does not play an important role in the practical applications of thermodynamics in industry. It is, however, often used in research since its change can be measured for a reaction taking place in a bomb calorimeter. It also plays an important role in both solid state physics and statistical mechanics. Once a physical model is constructed, F can often be determined, at least approximately, from statistical considerations. From the expression for F, we can then find P and S via

$$P = -\left(\frac{\partial F}{\partial V}\right)_T,$$

$$S = -\left(\frac{\partial F}{\partial T}\right)_V.$$

From the condition

$$\frac{\partial^2 F}{\partial T\,\partial V} = \frac{\partial^2 F}{\partial V\,\partial T},$$

it follows that

$$\left(\frac{\partial S}{\partial V}\right)_T = \left(\frac{\partial P}{\partial T}\right)_V,$$

which is the *third Maxwell relation.*

Making use of Eqns. (92-3) and (92-4), and the fact that $dS = \delta Q_{rev}/T$, this can be written as

$$\left.\frac{\partial S}{\partial V}\right|_T = \frac{(\delta Q)_T}{(\partial V)_T} = T\left(\frac{\partial P}{\partial T}\right)_V = T\beta p = \frac{T\alpha_p}{\varkappa_T}. \qquad (93\text{-}12)$$

EXAMPLE. For a van der Waals gas,

$$P = \frac{RT}{V_M - b} - \frac{a}{V_M^2},$$

$$\left(\frac{\partial P}{\partial T}\right)_V = \frac{R}{V_M - b}.$$

The heat flowing into a van der Waals gas in a reversible isothermal process is therefore

$$(\delta Q)_T = \frac{RT}{V_M - b}\, dV_M = \left(P + \frac{a}{V_M^2}\right) dV_M.$$

The quantity $\delta Q_1 = P\, dV_M$ represents the heat which must be supplied because of the external work done by the gas. The heat $\delta Q_2 = \frac{a}{V_M^2}\, dV_M$ equals the increase in internal energy.

93-5 The Gibbs Function

The Gibbs function G is defined by the equation

$$G = H - TS = U + PV - TS.$$

In an infinitesimal process,

$$dG = dU + P\, dV + V\, dP - T\, dS - S\, dT.$$

But, for a reversible process,

$$dU = T\, dS - P\, dV,$$

and hence

$$dG = -S\, dT + V\, dP. \qquad (93\text{-}13)$$

The Gibbs function is expressed in terms of the variables T and P, and provides a relation between reversible *isobaric* and *isothermal* processes.

From the equation

$$\frac{\partial^2 G}{\partial T\, \partial P} = \frac{\partial^2 G}{\partial P\, \partial T},$$

it follows that

$$-\left(\frac{\partial S}{\partial P}\right)_T = \left(\frac{\partial V}{\partial T}\right)_P, \qquad (93\text{-}14)$$

which is the *fourth Maxwell relation*. Replacing $(\partial S/\partial P)_T$ by $\left.\frac{1}{T} \cdot \frac{\delta Q_{rev}}{dP}\right|_T$ and using Eqn. (92-1), we get:

$$\left(\frac{\delta Q_{rev}}{dP}\right)_T = -T\left(\frac{\partial V}{\partial T}\right)_P = -TV\alpha_p.$$

EXAMPLE. For an ideal gas, this equation reduces to a result already obtained.

$$\delta Q_{\text{rev},T} = -TV_M \frac{1}{T} dP_T = -V_M\,dP = -RT\frac{dP}{P}.$$

Then the heat absorbed in a reversible isothermal expansion is

$$Q_T = -RT \ln \frac{P_2}{P_1}.$$

93-6 The Planck Function

The Planck function Y is defined by the equation

$$Y = -\frac{G}{T} = -\frac{U + PV - TS}{T} = -\frac{H}{T} + S.$$

Therefore

$$dY = -d\left(\frac{G}{T}\right) = -\frac{dH}{T} + \frac{H}{T^2}dT + dS.$$

But, for a reversible process,

$$dH = T\,dS + V\,dP,$$

so

$$dY = \frac{H}{T^2}dT - \frac{V}{T}dP.$$

The Planck function is expressed in terms of the variables T and P and connects reversible *isothermal* and *isobaric* processes.

From the relation

$$\frac{\partial^2 Y}{\partial T\,\partial P} = \frac{\partial^2 Y}{\partial P\,\partial T}$$

it follows that

$$\frac{\partial}{\partial P}\left(\frac{H}{T^2}\right)_T = -\frac{\partial}{\partial T}\left(\frac{V}{T}\right)_P. \tag{93-15}$$

The utility of the Planck function lies in the fact that it can be used to find the change in the Gibbs function G. The partial derivative of Y with respect to T at constant pressure P is

$$\left(\frac{\partial Y}{\partial T}\right)_P = -\frac{\partial}{\partial T}\left(\frac{G}{T}\right)_P = \frac{H}{T^2}.$$

Hence

$$G = -T\left[\int \frac{H}{T^2}dT + I_0\right],$$

where the integration is to be performed at constant pressure and I_0 is an integration constant.

Before applying this relation to a chemical reaction, a few remarks are necessary. Consider the reaction

$$\nu_1 A + \nu_2 B \rightarrow \nu_3 C.$$

The quantity ΔG is defined as the change in the Gibbs function of the system as a result of the reaction. That is,

$$\Delta G = -\nu_1 G_A - \nu_2 G_B + \nu_3 G_C,$$

where G_i is the molar Gibbs function for the ith component. As before, ΔH is defined as the change in enthalpy, or heat of reaction,

$$\Delta H = -\nu_1 H_A - \nu_2 H_B + \nu_3 H_C.$$

Hence if the reaction above proceeds at constant pressure, the change in the Gibbs function is

$$\Delta G = -T\left[\int \frac{\Delta H}{T^2}\, dT + \Delta I_0\right].$$

The integral can be evaluated if the enthalpies of the components and products are known as functions of temperature over a range of temperatures.

EXAMPLE. Let us consider the reaction discussed earlier, namely,

$$\tfrac{1}{2}H_2(g) + \tfrac{1}{2}Cl_2(g) \rightarrow HCl(g).$$

We showed on p. 489 that

$$\Delta H = -21{,}732 - 1.18T + 2.65 \times 10^{-4}T^2.$$

Hence

$$\frac{\Delta H}{T^2} = -\frac{21{,}732}{T^2} - \frac{1.18}{T} + 2.65 \times 10^{-4},$$

and

$$\int \frac{\Delta H}{T^2}\, dT = \frac{21{,}732}{T} - 1.18 \ln T + 2.65 \times 10^{-4}T.$$

The change in the Gibbs function in the reaction above is therefore

$$\Delta G = -21{,}732 + 1.18T \ln T - 2.65 \times 10^{-4}T^2 - \Delta I_0 T.$$

Hence, if the value of ΔG is known at any one temperature, it can be found at any other temperature as long as the series expansion of ΔH is valid. The values of ΔG at normal temperature and pressure are tabulated.† For the reaction above, we find from the tables that at a temperature of 298°K,

$$\Delta G = -22{,}770 \text{ cal/mole}.$$

The integration constant is therefore

$$\Delta I_0 = 10.13 \text{ cal/°C} \cdot \text{mole}$$

and the change in the Gibbs function in the HCl reaction is

$$\Delta G = -21.732 + 1.18T \ln T - 2.65 \times 10^{-4}T^2 - 10.13T.$$

A knowledge of this function is important because it can be used to compute the *mass action constant* K_p (see Section 95-2) for any temperature in which the series expansion is valid.

† J. H. Perry, *Chemical Engineers Handbook*, New York, 1941, pp. 515–524. Here ΔG is called ΔF.

94. The Joule–Kelvin Effect

Suppose that a gas flows from one container into another through a porous ceramic wall or a cotton plug, and the pressure difference $P_1 - P_2$ between the containers is kept constant. This experiment was first performed by Joule and Kelvin and it is known as the Joule–Kelvin or porous plug experiment. The process is called a throttling process.

The flow through the plug is not affected if we imagine two pistons to be inserted in the flow channel as in Fig. 11-4, the pistons being moved

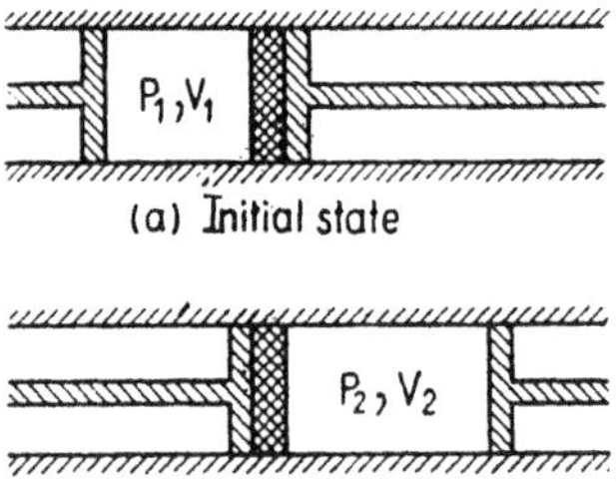

Fig. 11-4. The Joule–Kelvin effect. The walls of the cylinders are thermally insulated.

with a velocity equal to that of the gas at the corresponding point in the channel so as to maintain the pressures P_1 and P_2. We shall assume the system to be thermally insulated so that the process is adiabatic. It is irreversible and therefore is not isentropic. We wish to discover if the temperature of the gas changes when its pressure decreases from P_1 to P_2. To compute the change in temperature, let us imagine that 1 mole of gas flows through the plug. The net work done by the system is then

$$W = P_1V_1 - P_2V_2.$$

Since the process is adiabatic, $\delta Q = 0$ and

$$U_2 - U_1 = W = P_1V_1 - P_2V_2,$$

or

$$U_1 + P_1V_1 = U_2 + P_2V_2.$$

That is, the enthalpy $H = U + PV$ is the same for both the initial and the final states. It is not defined during the process, since the gas is not in equilibrium.

We now wish to apply some of the relations previously derived to compute dT for this process. Since it is irreversible, neither an equation of state $H = H(T, P)$ nor the Maxwell (or associated) relations exist during the occurrence of the process. We may imagine the system to be enlarged by the addition of heat and work reservoirs so that we can go from (P_1, V_1) to (P_2, V_2) reversibly. Since 1 and 2 are equilibrium states, all the results we obtain for the equivalent "*gedanken*" reversible process will also apply to the irreversible process.

In terms of the independent variables T and P,

$$dH = \left(\frac{\partial H}{\partial T}\right)_P dT + \left(\frac{\partial H}{\partial P}\right)_T dP,$$

and since $dH = 0$ for the reversible path,

$$dT = -\frac{(\partial H/\partial P)_T}{(\partial H/\partial T)_P} dP.$$

This formula gives the change in temperature of the gas in a throttling process.

We showed on page 487 that

$$\left(\frac{\partial H}{\partial T}\right)_P = C_p.$$

From Eqn. (93-15),

$$\left(\frac{\partial H}{\partial P}\right)_T = -T^2 \frac{\partial}{\partial T}\left(\frac{V}{T}\right)_P = -T\left(\frac{\partial V}{\partial T}\right)_P + V.$$

The change in temperature resulting from a drop in pressure dP is therefore also given by

$$dT = \frac{T(\partial V/\partial T)_P - V}{C_p} dP. \tag{94-1}$$

To evaluate this expression, we must know $(\partial V/\partial T)_P$ for the gas. This derivative can be obtained from the equation of state.

(a) *Ideal gas.* From the equation

$$PV = RT,$$

we have

$$\left(\frac{\partial V}{\partial T}\right)_P = \frac{R}{P},$$

and hence

$$T\left(\frac{\partial V}{\partial T}\right)_P = \frac{RT}{P} = V.$$

It follows that $dT = 0$, and that an ideal gas would experience no change in temperature when flowing through a porous plug. Physically, the reason for this is that the internal energy U is independent of volume.

(b) *Van der Waals gas.* The van der Waals equation is

$$\left(P + \frac{a}{V_2}\right)(V - b) = RT.$$

When this equation is differentiated with respect to T at constant pressure, we get

$$\left(P + \frac{a}{V^2}\right)\left(\frac{\partial V}{\partial T}\right)_P + (V - b)\left[-\frac{2a}{V^3}\left(\frac{\partial V}{\partial T}\right)_P\right] = R,$$

and since

$$P + \frac{a}{V^2} = \frac{RT}{V - b},$$

this reduces to

$$\left(\frac{\partial V}{\partial T}\right)_P = \frac{R(V - b)\,V^3}{RTV^3 - 2a(V - b)^2}.$$

When this is inserted in Eqn. (94-1) the following somewhat complicated expression for dT results

$$dT = \frac{\dfrac{2a}{RT}\left(\dfrac{V - b}{V}\right)^2 - b}{C_p\left[1 - \left(\dfrac{V - b}{V}\right)^2 \dfrac{2a}{RTV}\right]}\,dP.$$

Simplification is possible if $V \gg b$ and if $V \gg \dfrac{a}{RT}$. The second condition is equivalent to assuming that the internal pressure is negligible compared with the external pressure. Then

$$dT \approx \frac{\dfrac{2a}{RT} - b}{C_p}\,dP.$$

If $dP < 0$ $(P_1 > P_2)$, the temperature decreases provided $2a/RT > b$, that is, provided $T < 2a/Rb$. If $T > 2a/Rb$, the temperature increases. The particular temperature $T = 2a/Rb$, at which there is neither cooling nor heating, is called the *inversion temperature.* For a van der Waals gas, it equals twice the Boyle temperature.

EXAMPLE. Find the change in temperature when oxygen (assumed to be a van der Waals gas) at a temperature of 273°K flows through a porous plug across which there is a pressure drop of 1 atm. We have

$$T = 273°\text{K}$$

$$a = 1.36 \text{ liter}^2 \cdot \text{atm/mole}^2$$

$$b = 0.0319 \text{ liter/mole}$$

$$RT = 22.4 \text{ liter} \cdot \text{atm/mole}$$

$$C_p = 0.288 \text{ liter} \cdot \text{atm/mole} \cdot °\text{K}$$

$$dP = -1 \text{ atm}.$$

We find from these data that

$$dT = -0.31°\text{K}.$$

The Joule–Kelvin effect finds an important technical application in the liquefaction of gases. In order that the gas shall be cooled, it is necessary that the initial temperature of the gas shall be less than the inversion temperature, which is calculated to be 224.2°K for hydrogen and 35.1°K for helium. In both of these cases, the gas would become warmer if it was initially at room temperature. A reduction of temperature will result only if these gases are initially cooled below their inversion temperatures.

95. Equilibrium Conditions of a Chemical Reaction

When a natural process takes place spontaneously in a closed system, then according to the second law the entropy of the system increases. A state of *thermodynamic equilibrium* is reached when the entropy of the system has its maximum value and hence the *equilibrium condition* for a closed system is

$$S = \text{maximum}.$$

That is, $dS = 0$ for any process at equilibrium.

However, this condition, because of its extreme generality, is not sufficient by itself for a discussion of the equilibrium of systems in which isothermal, isobaric, or isochoric processes take place.

A very important group of processes in practice are those which take place at constant temperature and pressure, which includes a great many chemical reactions. We will now proceed to derive an equilibrium condition which is very useful for all processes taking place under such conditions.

It has been mentioned several times earlier that if a process is reversible, then the system in which the process takes place is at equilibrium. But according to Eqn. (93-13), for a reversible process,

$$dG = -S\,dT + V\,dP.$$

Therefore, for a system at constant temperature ($dT = 0$) and constant pressure ($dP = 0$), the condition for equilibrium is

$$dG = 0.$$

We now must determine whether G has a maximum or a minimum value at equilibrium.

Let us consider a system of constant mass that is in contact with a heat reservoir at temperature T and a work reservoir at pressure P. We assume that the system undergoes an irreversible process in which there is an exchange of heat δQ and work $\delta W = -P\Delta V$ with the reservoirs. We use Δ's in this derivation because many of the thermodynamic quantities are not exact differentials during an irreversible process. However, these quantities *are* properties for the initial and final equilibrium states of the process, and hence Δ may refer to either infinitesimal or macroscopic changes. P is the pressure of the work reservoir, and of the system at the beginning and end of the process. The pressure of the system is not defined during an irreversible process. δQ and δW are both considered with respect to the system. The process may involve a change in phase, a chemical reaction, or both.

Because of the irreversible process, the total entropy of the system plus reservoir increases,

$$\Delta S_{\text{total}} > 0.$$

But

$$\Delta S_{\text{total}} = \Delta S + \Delta S_R$$

where ΔS is the entropy change of the system and ΔS_R is the entropy change of the heat reservoir,

$$\Delta S_R = \frac{-\delta Q}{T}.$$

Therefore

$$-\frac{\delta Q}{T} + \Delta S > 0,$$

or

$$\delta Q - T\Delta S < 0.$$

Applying the first law of thermodynamics (and noting that all work is exchanged with the reservoir), we have

$$\delta Q = \Delta U + P\Delta V$$

where ΔU is the change in internal energy of the system and $P\Delta V$ the work done on the system during the irreversible process. Thus we have

$$\Delta U + P\Delta V - T\Delta S < 0,$$

which holds for any irreversible and hence real process. We note that because of the irreversibility of the process, the above inequality has replaced the usual equation of state.

If, in addition, the temperature and pressure are held constant, we have

$$\Delta U + P\Delta V + V\Delta P - T\Delta S - S\Delta T < 0$$

or

$$\Delta(U + PV - TS) = \Delta G < 0.$$

Thus we have found that the Gibbs function decreases during any spontaneous isothermal, isobaric process. *At equilibrium,* $\Delta G = 0$, *and G has a minimum value.*

Just like entropy and enthalpy, the Gibbs function is additive. Hence, for the reaction

$$\nu_1 A + \nu_2 B \rightarrow \nu_3 C$$

(which is a macroscopic, rather than an infinitesimal, process), the change in the Gibbs function of the system is

$$\Delta G = -\nu_1 G_A - \nu_2 G_B + \nu_3 G_C.$$

Under isobaric and isothermal conditions, the reaction is in equilibrium when G is a minimum and

$$\Delta G = 0.$$

95-1 The Clausius–Clapeyron Equation

A pure substance, at a given pressure, melts at a definite temperature, the *melting point* T_m. We wish to investigate the dependence of the melting point on the pressure. Since the melting process takes place at constant temperature and pressure, the solid and liquid can be in equilibrium only if $\Delta G = 0$. Let us suppose that 1 mole of ice is converted to one mole of water at constant pressure. The change in the molar Gibbs function for the system during the melting process is then

$$\Delta G = -G_{\text{ice}} + G_{\text{water}}$$

and for equilibrium at temperature T and pressure P

$$G_{\text{ice}} = G_{\text{water}}.$$

Let us now consider the same phase change taking place at slightly different temperature and pressure, $T + dT$ and $P + dP$. For equilibrium under these new conditions we have

$$(G + dG)_{\text{ice}} = (G + dG)_{\text{water}}$$

and hence

$$dG_{\text{ice}} = dG_{\text{water}}.$$

The change in the Gibbs function of one component with pressure and temperature is given by Eqn. (93-13):

$$dG_i = -S_i\,dT + V_i\,dP.$$

Hence for the melting process

$$S_{\text{ice}}\,dT - V_{\text{ice}}\,dP = S_{\text{water}}\,dT - V_{\text{water}}\,dP,$$

which gives after a slight rearrangement

$$(V_{\text{water}} - V_{\text{ice}})\,dP = (S_{\text{water}} - S_{\text{ice}})\,dT.$$

The change in melting point with pressure is therefore

$$dT = \frac{V_{\text{water}} - V_{\text{ice}}}{S_{\text{water}} - S_{\text{ice}}}\,dP.$$

The entropy difference $S_{\text{water}} - S_{\text{ice}}$, in a reversible change of phase, is

$$\Delta S = S_{\text{water}} - S_{\text{ice}} = \frac{\Delta Q}{T} = \frac{\lambda_m}{T_m},$$

where λ_m is the heat of fusion and T_m the melting point temperature. Generalizing, we then have for the melting of any arbitrary substance,

$$dT = \frac{[V(l) - V(s)]\, T_m}{\lambda_m}\, dP.$$

If $V(l) > V(s)$, an increase of pressure raises the melting point; if $V(l) < V(s)$, the melting point is lowered.

Exactly the same considerations apply to the equilibrium between liquid and vapor, and between solid and vapor. We can therefore write in general for any reversible phase change

$$\frac{dP}{dT} = \frac{S_f - S_i}{V_f - V_i} = \frac{\lambda}{T(V_2 - V_1)},$$

where λ is the heat of *transformation* (fusion, vaporization, or sublimation), T is the equilibrium temperature and $V_2 - V_1$ is the difference between the final and the initial volumes. This is known as the *Clausius–Clapeyron equation.*

EXAMPLE. 1. For water at 0°C

$$V(l) - V(s) = -1.63\ \text{cm}^3/\text{mole},$$

$$S(l) - S(s) = \frac{\lambda}{T} = 5.26\ \text{cal/mole} \cdot {}^\circ\text{K}.$$

It follows (1 cal = 41.3 atm · cm^3) that

$$\frac{dT}{dP} = -7.5 \times 10^{-3}\ {}^\circ\text{K/atm}.$$

EXAMPLE. 2. For water at 100°C,

$$V(g) - V(l) = 30{,}060\ \text{cm}^3/\text{mole},$$

$$\lambda_v = 18 \times 539.6 = 9717\ \text{cal/mole}.$$

It follows that

$$\frac{dP}{dT} = \frac{\lambda_v}{T(V_2 - V_1)} = 3.57 \times 10^{-2}\,\frac{\text{atm}}{{}^\circ\text{K}} = 27.2\,\frac{\text{mm Hg}}{{}^\circ\text{K}}.$$

95-2 Reactions of Ideal Gases at Constant Pressure. The Mass-action Law

Let us consider a chemical reaction between ideal gases given by the equation

$$\nu_1 A + \nu_2 B \rightarrow \nu_3 C.$$

When the reaction takes place at a pressure P the change in the Gibbs function of the system is

$$\Delta G = -\nu_1 G_A - \nu_2 G_B + \nu_3 G_C = \sum \nu_i G_i,$$

where G_i is the molar Gibbs function of the ith component.

For an isothermal process in an ideal gas,

$$\left(\frac{\partial G_i}{\partial P}\right)_T = V_{M_i} = \frac{RT}{P_i},$$

or

$$(dG_i)_T = RT\left(\frac{dP_i}{P_i}\right)_T,$$

where P_i is the ith partial pressure. Upon integration, this gives an expression for the molar Gibbs function of the ith component

$$G_i - G_i^0 = RT \ln \frac{P_i}{P_i^0},$$

where G_i^0 is the value of the molar Gibbs function at some standard pressure P_i^0. It is customary to express P in atmospheres and let the standard pressure be one atmosphere. Then

$$G_i = G_i^0 + RT \ln P_i.$$

For the above reaction

$$\begin{aligned} \Delta G &= -\nu_1 G_A^0 - \nu_2 G_B^0 + \nu_3 G_C^0 - \nu_1 RT \ln P_1 \\ &\quad - \nu_2 RT \ln P_2 + \nu_3 RT \ln P_3 \\ &= \Delta G^0 + RT \ln \frac{P_3^{\nu_3}}{P_1^{\nu_1} P_2^{\nu_2}}. \end{aligned}$$

Since $\Delta G = 0$ at equilibrium, we have as the equilibrium condition for a reaction among ideal gases at constant temperature and total pressure,

$$\Delta G^0 = -RT \ln \frac{P_3^{\nu_3}}{P_1^{\nu_1} P_2^{\nu_2}}. \tag{95-1}$$

ΔG^0 is called the *standard Gibbs function change* and it is a function of temperature only. Thus we have

$$\ln \frac{P_3^{\nu_3}}{P_1^{\nu_1} P_2^{\nu_2}} = -\frac{\Delta G^0(T)}{RT}$$

or

$$\frac{P_3^{\nu_3}}{P_1^{\nu_1} P_2^{\nu_2}} = e^{-\frac{\Delta G^0(T)}{RT}} = K_p(T).$$

At constant temperature $K_p(T)$ is a constant. Under this condition this equation expresses the famous *mass-action law* of Guldberg and Waage:

$$\frac{P_3'^{\nu_3}}{P_1'^{\nu_1} P_2'^{\nu_2}} = \frac{P_3^{\nu_3}}{P_1^{\nu_1} P_2^{\nu_2}} = \text{"constant"} = K_p(T). \tag{95-2}$$

The quantity $K_p(T)$ is called the *equilibrium constant* or *mass action constant* at constant pressure. It is a function of temperature and is independent of pressure. If the equilibrium constant of a given reaction is K_p, that of the inverse reaction is $K_p^* = 1/K_p$. The law may be applied to reactions among real gases as long as the density is low.

EXAMPLE.

$$2H_2O(g) \rightarrow 2H_2(g) + O_2(g),$$

$$K_p = \frac{P_{H_2}^2 P_{O_2}}{P_{H_2O}^2}.$$

$$2H_2(g) + O_2(g) \rightarrow 2H_2O(g),$$

$$K_p^* = \frac{P_{H_2O}^2}{P_{H_2}^2 P_{O_2}} = \frac{1}{K_p}.$$

The term "*mass action law*" arises from the fact that a change in the mass, and hence in the partial pressure, of one component of a system of gases in equilibrium results in a change in the partial pressures of the other components. The equation applies only to reactions which are in a true state of equilibrium.

If some of the components of the reaction are present in the solid or liquid phases, their vapor pressures are constant at constant temperature and have no influence on the equilibrium.

The mass action law plays a dominant role in the theory of gaseous reactions. It is often uesful to introduce the mole fractions or the molar concentrations in place of the partial pressures. If this is done, the value of the equilibrium constant changes.

95-3 The Mass Action Law in Terms of Mole Fractions

For ideal gases, the mole fraction x of any component is equal to the ratio of the partial pressure of that component to the total pressure. The mole fraction of an ideal gas is also identical with the partial volume concentration q_v (see Appendix I),

$$x_k = \frac{P_k}{P_t} = \frac{V_k}{V_t} = q_{vk}.$$

If we substitute mole fractions for the partial pressures, the mass action law for the reaction

$$\nu_1 A + \nu_2 B \rightarrow \nu_3 C$$

takes the form

$$\frac{P_3^{\nu_3}}{P_1^{\nu_1} P_2^{\nu_2}} = K_p(T) = \frac{P_t^{\nu_3} x_3^{\nu_3}}{P_t^{\nu_1+\nu_2} x_1^{\nu_1} x_2^{\nu_2}},$$

that is,

$$\frac{x_3^{\nu_3}}{x_1^{\nu_1} x_2^{\nu_2}} = \frac{P_3^{\nu_3}}{P_1^{\nu_1} P_2^{\nu_2} P_t^{\nu_3-\nu_1-\nu_2}} = \frac{K_p(T)}{P_t^{\vartheta}} = \text{const},$$

where ϑ is the reaction index (see p. 474):

$$\vartheta = \nu_3 - (\nu_1 + \nu_2).$$

All information regarding the equilibrium of a mixture of reaction gases can be determined from the mass action law, provided the equilibrium constant $K_p(T)$ and the temperature are known. A knowledge of the temperature dependence of this constant is of great importance for the understanding of reactions of gases. We shall return to this question later, after first illustrating the application of the law by an example.

EXAMPLE. The reaction

$$CO + H_2O \leftarrow CO_2 + H_2$$

has a mass action constant $K_p(T) = 1.596$ at a temperature $T = 1020°K$. Suppose that a mixture of these gases at 720 mm Hg and 20°C has the following composition: $x_1 = 80$ per cent CO, $x_2 = 16$ per cent H_2O, $x_3 = 4$ per cent CO_2, $x_4 = 0$ per cent H_2. The mixture is heated at constant pressure to a temperature of 1020°K. We wish to calculate the composition of the system after equilibrium has been reached and determine the effect of pressure on the reaction.

The reaction is independent of pressure since the reaction index $\mathfrak{V} = 0$. To calculate the new equilibrium assume that the mole fraction x_1 of CO decreases by an amount Δx when the temperature is increased to 1020°K. Then from the reaction equation x_2 also decreases by Δx, and x_3 and x_4 each increase by Δx. Hence for the new equilibrium

$$\frac{(x_3 + \Delta x)\Delta x}{(x_1 - \Delta x)(x_2 - \Delta x)} = \frac{1}{1.596},$$

$$0.596x^2 + 1.024x - 0.128 = 0.$$

This quadratic equation has the two solutions

$$\Delta x = \begin{cases} -1.835 \\ +0.117 \end{cases}.$$

The second solution is impossible since the decrease in the mole fraction of a component cannot exceed its original value. The new composition is therefore:

$$x_1' = 68.3 \text{ per cent CO}$$
$$x_2' = 4.3 \text{ per cent } H_2O$$
$$x_3' = 15.7 \text{ per cent } CO_2$$
$$x_4' = 11.7 \text{ per cent } H_2$$
$$\sum x_i' = 100.0 \text{ per cent}$$

95-4 Calculation of the Temperature Dependence of the Equilibrium Constant

One remaining point, still to be explained, is the dependence of the equilibrium constant on the temperature. For the reaction

$$\nu_1 A + \nu_2 B \rightarrow \nu_3 C,$$

we have from Eqns. (95-1) and (95-2):

$$-\Delta G^0 = RT \ln \frac{P_3^{\nu_3}}{P_1^{\nu_1} P_2^{\nu_2}} = RT \ln K_p(T) = 2.303\, RT \log K_p(T).$$

If the standard Gibbs function change for a reaction is known as a function of temperature, the equilibrium constant is determined for every temperature and the equilibrium state can be computed at all temperatures and pressures.

We showed on p. 493 that ΔG could be found as a function of temperature by integration of the series expansion of ΔH. The series expressing ΔG has the form

$$\Delta G = -T \int_{T_0}^{T} \frac{\Delta H}{T^2} dT = \Delta H_0 - \alpha T \ln T - \beta T^2 - \cdots - \Delta I_0 T.$$

The constant ΔI_0 can be computed from the magnitude of ΔG at some known temperature. The quantities α and β are the constants in the first terms of the series development of ΔH. The Gibbs function of substances at normal temperature can be found from tables.

EXAMPLE. For the reaction $\frac{1}{2} H_2 + \frac{1}{2} Cl_2 \rightarrow HCl$, the above coefficients are (p. 493):

$$\alpha = -1.18 \text{ cal/mole °K}$$

$$\beta = 2.65 \times 10^{-4} \text{ cal/mole °K}^2$$

$$\Delta H_0 = -21{,}732 \text{ cal/mole}$$

$$\Delta I_0 = 10.13 \text{ cal/mole °K}.$$

Converting to common logarithms and using $R = 1.987$ cal/mole °K, we then have

$$\log K_p(T) = \frac{-\Delta G}{2.303 RT} \cong \frac{-1}{4.576} \left[\frac{\Delta H_0}{T} - \alpha \ln T - \beta T - \Delta I_0 \right]$$

Putting in the above constants, we obtain for the HCl-forming reaction

$$\log K_p(T) \cong \frac{4749}{T} - 0.258 \ln T + 5.79 \times 10^{-5} T + 2.21.$$

Table 11-2 gives some values of the equilibrium constant for the formation of HCl at various temperatures calculated from the above equation. These values are in good agreement with the direct results of experiments.

TABLE 11-2

T (°K)	$\log K_p(T)$	$K_p(T)$	$\log K_p(T)$ (exp)
298°	16.66	4.58×10^{16}	16.69
500°	10.12	1.32×10^{10}	
1000°	5.23	1.70×10^{5}	
1500°	3.57	3.70×10^{3}	

APPENDIX I

DEFINITIONS OF CONCENTRATION

VERY often the colloquial use of concentration is either incomplete or fails to indicate the proper units of measure. Both are sources of countless mistakes. We define some of the various kinds of concentrations below.

1. The Mass Concentration p_m

The mass concentration p_m (often denoted as percentage weight) is the ratio of the mass m_c of a constituent C to the total mass m_t of the solution, mixture or compound.

$$p_m = \frac{m_c}{m_t} \text{ (pure number, temperature independent).}$$

EXAMPLES.

(a) NH_4Cl $\quad$ N + 4H + Cl

$$\begin{array}{rccc} M & A_1 & 4A_2 & A_3 \\ p_m : 1 & x & y & z \end{array}$$

$$x = \frac{A_1}{M} = \frac{14.008}{53.497} = 26.18 \text{ per cent}$$

$$y = \frac{4A_2}{M} = \frac{4.032}{53.497} = 7.54 \text{ per cent}$$

$$z = \frac{A_3}{M} = \frac{35.457}{53.497} = 66.28 \text{ per cent}$$

(b) 5.884 gm of NaCl were dissolved in 812.0 gm H_2O.

$$p_m = \frac{5.884}{817.88} = 7.194 \cdot 10^{-3} \cong 0.72 \text{ per cent}$$

The sum of all $p_{m,i}$ of a solution, etc., is always equal to unity.

2. The Mass-Volume Concentration q_m.

The mass-volume concentration q_m is the ratio of the mass m_c of a constituent C to the total volume of the solution or mixture V_t:

$$q_m = \frac{m_c}{V_t} \text{ (temperature dependent).}$$

Unit: 1 gm/cm^3.

The relation between q_m and p_m is:

$$q_m = \frac{m_c}{V_t} = \frac{m_c \varrho_t}{m_t} = p_m \varrho_t$$

where ϱ_t denotes the *density* of the solution or compound. The densities of the most important solutions have been tabulated.

It is impossible to make a statement about the volume of a solution if only the volumes of the solute and the solvent are given. In the course of dissolving, processes occur in which the solvent and the solute change to entirely new states (dissociation, hydration).

The mass-volume concentration is closely related to the normality of a solution (see below).

EXAMPLES.

(a) 14.50 gm of NH_4Cl were dissolved at 18°C to make 400 cm³ of solution. Then

$$q_m = \frac{14.50}{400} = 0.03625 \text{ gm/cm}^3 \text{ at } 18°\text{C}.$$

(b) 100 cm³ of a p_m = 10 per cent NaCl solution at 20°C is to be made into a solution with a value of $q_m = 4 \times 10^{-2}$ gm/cm³. The problem can be solved only if ϱ_t of the original solution is known. It is

$$\varrho^{20°\text{C}}_{10\%}\ \text{NaCl}_{\text{soln}} = 1.0707 \text{ gm/cm}^3.$$

Then

$$m_t = V_t \varrho_t = 107.07 \text{ gm}$$

$$m_{\text{NaCl}} = p_m m_t = 10.707 \text{ gm}$$

$$V_{t'} = \frac{m_{\text{NaCl}}}{q_m} = 267.7 \text{ cm}^3.$$

That is, one should add enough water at 20°C to the original solution so that the volume of the final solution is 267.7 cm³.

3. *The Partial Volume Concentration* q_v

The partial volume concentration q_v is given by the ratio of the volume V_c of the constituent C at a given pressure to the volume of the total mixture V_t:

$$q_v = \frac{V_c}{V_t}$$

(pure number, temperature independent for ideal gases). The sum of all $q_{v,i}$ is always equal to unity for ideal gases. This concentration expression has rigorous meaning only in the case of ideal gas mixtures. It is not useful for liquid mixtures since the total volume is not the sum of the partial volumes.

EXAMPLE. An analysis of air has shown that the composition is as follows:

$$\begin{array}{ll} N_2 & q_v = 78 \text{ per cent} \\ O_2 & q_v = 20 \text{ per cent} \\ CO_2 & q_v = 2 \text{ per cent} \\ \hline & \sum q_v = 100 \text{ per cent} \end{array}$$

In 1 liter of air at a given pressure there are 780 cm^3 N_2, 200 cm^3 O_2 and 20 cm^3 CO_2 present. The masses can be calculated from the ideal gas equation if the pressure and temperature are known.

4. The Molar Volume Concentration c_V

This quantity is used extensively in thermodynamics. It is closely related to the molar volume of gases, to the partial pressures of gas mixtures, to the concentration q_v, and to the normality of solutions. Because of these numerous connections it is the most frequently used concentration expression in thermodynamics.

The molar volume concentration c_V is the ratio of the number n_c of moles of a constituent C to the total volume V_t of the solution, compound, or mixture:

$$c_V = \frac{n_c}{V_t} \text{ (temperature dependent).}$$

Unit: 1 mole/cm^3.

(a) For an ideal gas

$$c_V = \frac{n_c}{V_t} = \frac{1}{V_M}.$$

The molar volume concentration is identically equal to the reciprocal of the molar volume V_M.

(b) For ideal gas mixtures,

$$c_{V,i} = \frac{n_i}{V_t} = \frac{1}{V_{M,i}}.$$

$c_{V,i}$ is the reciprocal of the molar volume of the ith component. The partial pressure of this component is

$$P_i = \frac{n_i RT}{V_t} = RTc_{V,i}.$$

EXAMPLE. In a reacting mixture of N_2 and H_2 at 500°K the following partial pressures were measured:

$$\begin{array}{ll} P_{NH_3} & = 0.839 \text{ atm.} \\ P_{N_2} & = 1.681 \text{ atm.} \\ P_{H_2} & = 5.043 \text{ atm.} \\ \hline P_t & = 7.563 \text{ atm.} \end{array}$$

The values of the concentrations $c_{V,i}$ and $q_{v,i}$ are then:

$$q_{v_{NH_3}} = \frac{P_{NH_3}}{P_t} = 0.1109, \quad c_{V,NH_3} = \frac{P_{NH_3}}{RT} = 2.046 \times 10^{-5} \text{ mole/cm}^3$$

$$q_{v_{N_2}} = \frac{P_{N_2}}{P_t} = 0.2223, \quad c_{V,N_2} = \frac{P_{N_2}}{RT} = 4.099 \times 10^{-5} \text{ mole/cm}^3$$

$$q_{v_{H_2}} = \frac{P_{H_2}}{P_t} = 0.6668, \quad c_{V,H_2} = \frac{P_{H_2}}{RT} = 1.230 \times 10^{-4} \text{ mole/cm}^3.$$

(c) For solutions

$$c_V = \frac{n_c}{V_t}.$$

The relation between q_m and c_V is

$$q_m = \frac{m_c}{V_t} = \frac{M_c^* n_c}{V_t} = M_c^* c_V$$

EXAMPLE. 15.00 gm of KOH were dissolved to form 300 cm³ of solution at 18°C. c_V and q_m are to be determined.

$$n_c = \frac{15.00}{56.104} = 0.2675 \text{ mole KOH}$$

$$c_V = \frac{0.2675}{300} = 8.912 \cdot 10^{-4} \text{ mole/cm}^3 \text{ at } 18^\circ\text{C}$$

$$q_m = M_c^* c_V = 5.00 \times 10^{-2} \text{ gm/cm}^3 \text{ at } 18^\circ\text{C}.$$

5. *The Molar Mass Concentration c_m*

The molar mass concentration c_m is the ratio of the number of moles n_c of a constituent C to the total mass of the solution m_t:

$$c_m = \frac{n_c}{m_t} \text{ (temperature independent).}$$

Unit: 1 mole/gm

c_m is closely related to p_m:

$$p_m = \frac{n_c M_c^*}{m_t} = c_m M_c^*.$$

EXAMPLE. 2.186 gm of HCl were dissolved in 210.0 gm of water. Then:

$$n_c = \frac{2.186}{36.456} = 5.995 \times 10^{-2} \text{ mole HCl}$$

$$c_m = \frac{5.995 \times 10^{-2}}{212.2} = 2.825 \times 10^{-4} \text{ mole/gm}.$$

6. *The Normality N*

The normality N of a solution is given by the ratio of the number of gram-equivalents Z_c of the solute to the total volume of the solution:

$$N = \frac{Z_c}{V_t} \text{ (temperature dependent).}$$

Unit: 1 gram-equivalent/liter

The normality has a close connection with q_m:

$$N = \frac{Z_c}{V_t} = \frac{q_m}{E^*} 10^3,$$

where E^* denotes the gram-equivalent weight. The factor 10^3 occurs because N is given per liter while q_m is given per cm^3.

EXAMPLE. 22.70 gm of $CuSO_4$ is dissolved to form 600 cm^3 of solution at 18°C. We wish to calculate the normality.

$$Z_c = \frac{22.70}{79.81} = 0.2844 \text{ gm-eq } CuSO_4$$

$$N = \frac{0.2844}{0.600} = 0.474 \text{ gm-eq/liter at 18°C } (0.474N).$$

7. *The Mass Molarity μ_m*

The mass molarity μ_m is the ratio of the number of moles n_c of a constituent C to the mass of the solvent m_{sol}:

$$\mu_m = \frac{n_c}{m_{sol}} \text{ (temperature independent).}$$

Unit: 1 mole/gm

EXAMPLE. 0.2186 kg of $C_{10}H_8$ (naphthalene) is dissolved in 16.10 kg C_6H_6. Then:

$$n_c = \frac{218.6}{128.16} = 1.706 \text{ mole}$$

$$\mu_m = \frac{1.706}{16.10} = 0.1059 \text{ mole/kg.}$$

8. *The Volume Molarity μ_v*

The volume molarity μ_v is the ratio of the number of moles n_c of a constituent C to the volume of the solvent V_{sol} (usually in liters):

$$\mu_v = \frac{n_c}{V_{sol}} \text{ (temperature dependent).}$$

Unit: 1 mole/liter

EXAMPLE. 42.22 gm of cane sugar $C_{12}H_{22}O_{11}$ ($M^* = 342.30$ gm/mole) were dissolved in 430 gm of water at 20°C.

$$\varrho_{H_2O}^{20°C} = 0.9982 \text{ gm/cm}^3$$

$$n_c = \frac{42.22}{342.30} = 0.1233 \text{ moles sugar}$$

$$\mu_v = \frac{0.1233}{0.4292} = 0.2873 \text{ mole/liter at } 20°\text{C}.$$

These concentrations play a role in the calculation of the lowering of the freezing point and the raising of the boiling point.

9. *The Atomic Percentage A per cent*

The atom percentage of a constituent C in a compound is the ratio of the number of gram-atoms n_c of the constituent C to the total number of gram-atoms of the compound:

$$A \text{ per cent} = \frac{n_c}{\sum n_i} \text{ (pure number, temperature independent).}$$

These quantities are used in chemistry for determining empirical formulas by elementary analysis.

EXAMPLE. Consider $Cu(NO_3)_2$.

Cu: $n = 1$	A per $\text{cent}_{Cu} = \frac{1}{9} = 11.11$ per cent
N: $n = 2$	A per $\text{cent}_{N} = \frac{2}{9} = 22.22$ per cent
O: $n = 6$	A per $\text{cent}_{O} = \frac{6}{9} = 66.67$ per cent
$\sum n_i = 9$	

10. *The Mole Fraction x_c*

The mole fraction x_c of the component C of a solution or mixture is the ratio of the number of moles n_c of this component to the total number of moles of all the components:

$$x_c = \frac{n_c}{\sum n_i} \text{ (pure number, temperature independent).}$$

The sum of all mole fractions of a solution or mixture is always unity:

$$\sum x_i = 1.$$

The mole fraction is one of the most frequently used concentrations in thermodynamics. For ideal gas mixtures the mole fraction is equal to the partial volume concentration. Since all ideal gases at a given pressure and temperature have the same molar volume, then

$$V_c = n_c V_M$$

$$V_{\text{total}} = V_M \sum n_i$$

$$q_v = \frac{n_c}{\sum n_i} = x_c.$$

EXAMPLE. We wish to find the mole fractions of NaOH and H_2O for an NaOH solution with $p_m = 5.51$ per cent.

$$n_1 = \frac{55.10}{40.005} = 1.377 \text{ mole NaOH/1000 gm solution}$$

$$n_2 = \frac{944.9}{18.016} = 52.45 \text{ mole H}_2\text{O/1000 gm solution}$$

$$\sum n_i = 53.83 \text{ mole}$$

$$x_{\text{NaOH}} = 0.0256$$

$$x_{\text{H}_2\text{O}} = 0.9744.$$

APPENDIX II

CALCULATION OF MOMENTS OF INERTIA

IN THE calculations below, we always assume that the density is uniform throughout the body.

1. The Thin Cylindrical Rod

Length l, cross section A, density ϱ (Fig. A2-1).

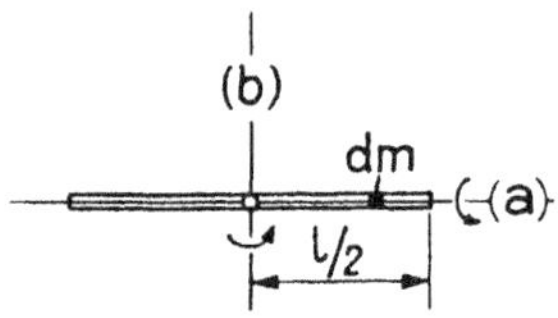

FIG. A 2-1. Moment of inertia of a thin cylindrical rod.

(a) For an axis of rotation a which lies along the axis of the cylinder the moment of inertia is $I_a \cong 0$.

(b) For an axis b through the center of gravity perpendicular to the axis of the rod it follows from $dm = A\varrho\, dr$ that the moment of inertia is

$$I_b = 2\int_0^{l/2} A\varrho r^2\, dr = 2A\varrho \frac{r^3}{3}\bigg|_0^{l/2} = \frac{A\varrho l^3}{12}.$$

Since the total mass of the rod is $m = Al\varrho$,

$$I_b = m\frac{l^2}{12}.$$

For the radius of gyration k one obtains

$$k = \frac{l\sqrt{3}}{6}.$$

2. The Thin Circular Ring

Cross section A, radius r, density ϱ.

(a) For an axis through the center of gravity perpendicular to the plane of the ring, it follows immediately that $I = mr^2$ since all mass points are the same distance from the axis of rotation. The radius of gyration is $k = r$.

(b) For an axis d along a diameter (Fig. A2-2), we have:

$$dm = \varrho A r\, d\varphi$$

$$dI = \varrho A r\, d\varphi \cdot r^2 \sin^2 \varphi$$

$$I_d = 2Ar^3\varrho \int_0^\pi \sin^2 \varphi\, d\varphi = \pi A \varrho r^3 .$$

$m = 2\pi r A \varrho$ gives

$$I_d = \frac{mr^2}{2} \quad \text{and} \quad k = \frac{r\sqrt{2}}{2} .$$

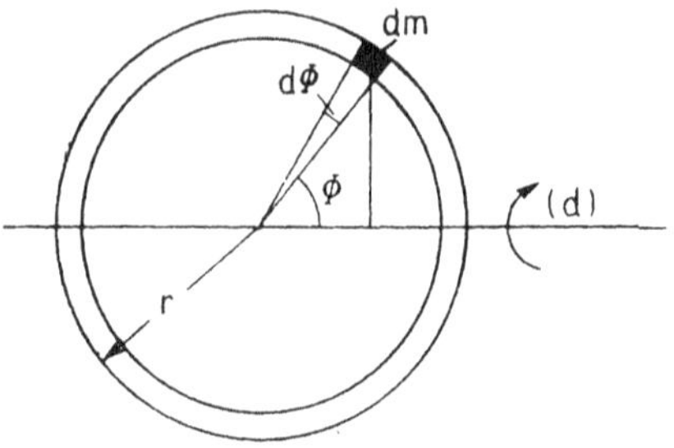

FIG. A 2-2. Moment of inertia of a thin circular ring about a diametral axis.

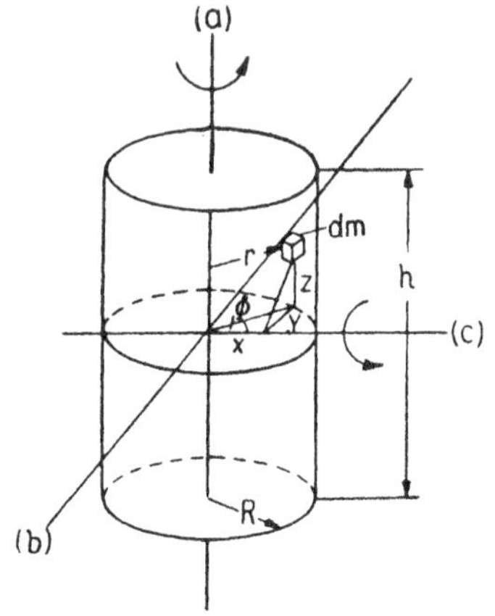

FIG. A 2-3. Moments of inertia of a right circular cylinder about its axis and about an axis through the center of gravity normal to the axis of the cylinder.

3. The Right Circular Cylinder

(a) For rotation about the cylinder axis a (Fig. A2-3):

$$dI_a = 2\pi r\,dr\,h\varrho r^2$$

$$I_a = 2\pi h\varrho \int_0^R r^3\,dr = \frac{\pi h\varrho R^4}{2}$$

$$I_a = \frac{mR^2}{2}, \quad k = \frac{R\sqrt{2}}{2}.$$

(b) For an axis of rotation (b or c) through the center of gravity and perpendicular to the cylinder axis (Fig. A2-3):

$$dI_c = \varrho r\,d\varphi\,dr\,dz(y^2 + z^2), \text{ where } y = r\sin\varphi$$

$$I_c = 2\varrho \int_{\varphi=0}^{2\pi} \int_{r=0}^{R} \int_{z=0}^{h/2} r(y^2 + z^2)\,dz\,dr\,d\varphi$$

$$= 2\varrho \int_0^{2\pi} \int_0^{R} \int_0^{h/2} (r^3 \sin^2\varphi + rz^2)\,dz\,dr\,d\varphi$$

$$= 2\varrho \int_0^{2\pi} \int_0^{R} \left(r^3 \sin^2\varphi \frac{h}{2} + \frac{rh^3}{24}\right) dr\,d\varphi$$

$$= 2\varrho \int_0^{2\pi} \left(\frac{R^4}{4}\frac{h}{2}\sin^2\varphi + \frac{R^2}{2}\frac{h^3}{24}\right) d\varphi$$

$$= R^2\pi h\varrho\left(\frac{R^2}{4} + \frac{h^2}{12}\right).$$

$$m = R^2\pi h\varrho \quad \text{gives} \quad I_c = m\left(\frac{R^2}{4} + \frac{h^2}{12}\right).$$

The moment of inertia of a hollow cylinder can be calculated by using the combination principle.

4. The Sphere

If the axis of rotation passes through the center of the sphere (Fig. A2-4),

$$dI = 2\cdot 2\pi r\,dr\,z\varrho r^2 \quad \text{and} \quad z = \sqrt{(R^2 - r^2)}.$$

Using the substitution $R^2 - r^2 = u^2$ and $r\,dr = -u\,du$, we have:

$$I = 4\pi\varrho \int_0^R r^3\sqrt{(R^2 - r^2)}\,dr = 4\pi\varrho \int_0^R (R^2 - u^2)\,u^2\,du$$

$$= 4\pi\varrho\left(\frac{R^5}{3} - \frac{R^5}{5}\right) = \frac{8}{15}\pi\varrho R^5.$$

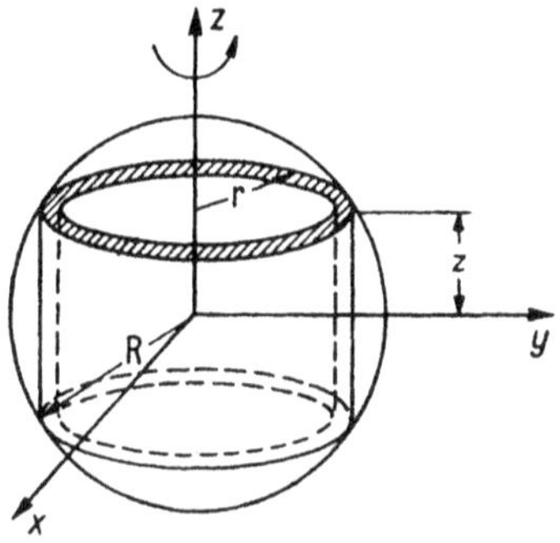

FIG. A 2-4. Moment of inertia of a sphere.

Using $m = \frac{4}{3}\pi R^3 \varrho$, we find

$$I = \frac{2mR^2}{5}, \quad k = \frac{R\sqrt{10}}{5}.$$

The calculation of the moment of inertia of a hollow sphere can be carried out by using the combination principle.

5. *The Rectangular Parallelepiped* (Fig. A2-5)

Let the axis of rotation be the x-axis which passes through the center of gravity. Then,

$$dI_x = \varrho \, dx \, dy \, dz (y^2 + z^2)$$

$$I_x = \varrho \int_{-a/2}^{+a/2} \int_{-b/2}^{+b/2} \int_{-c/2}^{+c/2} (y^2 + z^2) \, dx \, dy \, dz$$

$$= \varrho a \int_{-b/2}^{+b/2} \int_{-c/2}^{+c/2} (y^2 + z^2) \, dy \, dz = \varrho a \int_{-c/2}^{+c/2} \left(\frac{b^3}{12} + z^2 b \right) dz$$

$$= \frac{\varrho}{12} (ab^3 c + abc^3) = \frac{\varrho abc}{12} (b^2 + c^2).$$

$$I_x = \frac{m}{12} (b^2 + c^2)$$

$$I_y = \frac{m}{12} (c^2 + a^2)$$

$$I_z = \frac{m}{12} (a^2 + b^2).$$

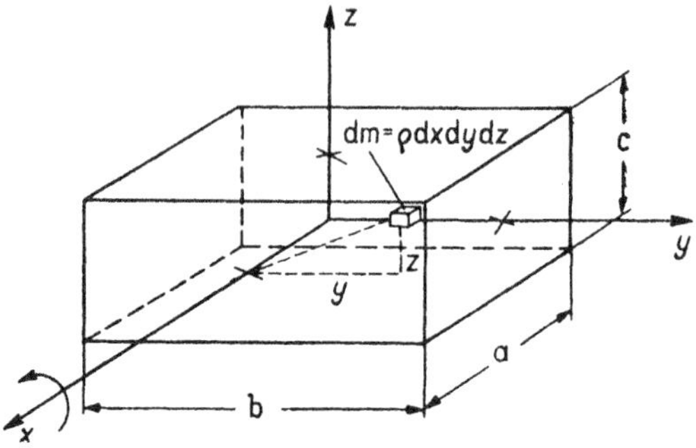

FIG. A 2-5. Moment of inertia of a rectangular parallelepiped about an axis parallel to one side, and passing through the center of gravity.

6. *The Right Circular Cone* (Fig. A 2-6)

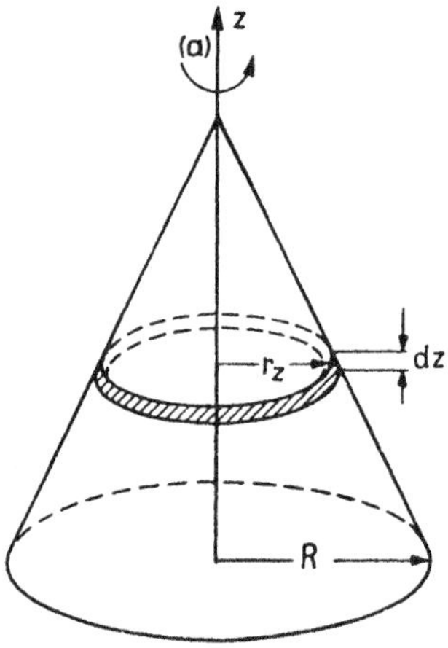

FIG. A 2-6. Moment of inertia of a right circular cone about the cone axis.

Let the axis of rotation be the axis of the cone *a*. The moment of inertia of a disk of radius *r* and thickness *dz* is:

$$dI_a = \int_0^{r_z} 2\pi r\, dr\, \varrho r^2\, dz = \frac{\pi \varrho r_z^4}{2} dz.$$

The moment of inertia of the cone is:

$$I_a = \int_0^h \frac{\pi \varrho r_z^4}{2} dz.$$

Now $\frac{r_z}{R} = \frac{h-z}{h}$ and $r_z = \frac{h-z}{h} R$. Thus

$$I_a = \frac{\pi \varrho R^4}{2h^4} \int_0^h (h-z)^4 \, dz.$$

With the substitution $h - z = u$, we find

$$I_a = \frac{\pi \varrho R^4}{2h^4} \int_0^h u^4 \, du = \frac{\pi \varrho R^4 h}{10} = R^2 \frac{\pi h \varrho}{3} \frac{3R^2}{10} = \frac{3}{10} m R^2.$$

$$I_a = \frac{3}{10} m R^2, \quad k = \frac{R\sqrt{30}}{10}.$$

7. *The Elliptical Plate* (Fig. A2-7)

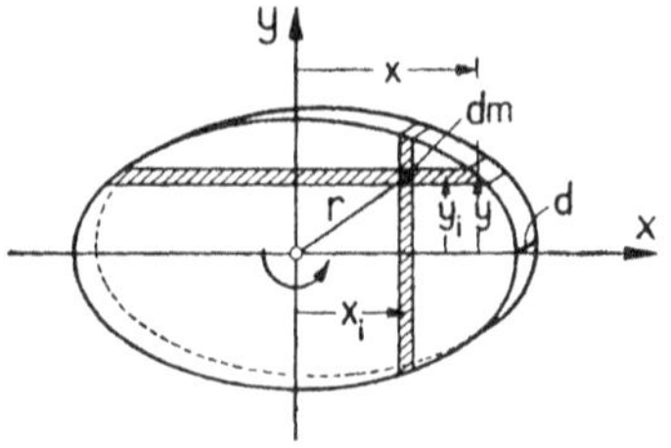

FIG. A 2-7. Moment of inertia of an elliptical plate about a normal axis.

The axis of rotation is perpendicular to the plate through the center of gravity.

$$I = \int r_i^2 \, dm = \int x_i^2 \, dm + \int y_i^2 \, dm$$
$$= I_y + I_x$$

$$I_x = 4\varrho d \int_0^b x y^2 \, dy.$$

From the parametric equations of an ellipse, $x = a \cos \varphi$ and $y = b \sin \varphi$ we find for I_x:

$$I_x = 4\varrho a\, d b^3 \int_0^{\pi/2} \sin^2 \varphi \cos^2 \varphi \, d\varphi = \frac{\varrho a\, d b^3 \pi}{4}$$

$$I_x = \frac{m}{4} b^2.$$

Similarly, it follows that

$$I_y = \frac{m}{4} a^2$$

and hence

$$I = I_x + I_y = \frac{m}{4}(a^2 + b^2).$$

8. The Ellipsoid (Fig. A2-8)

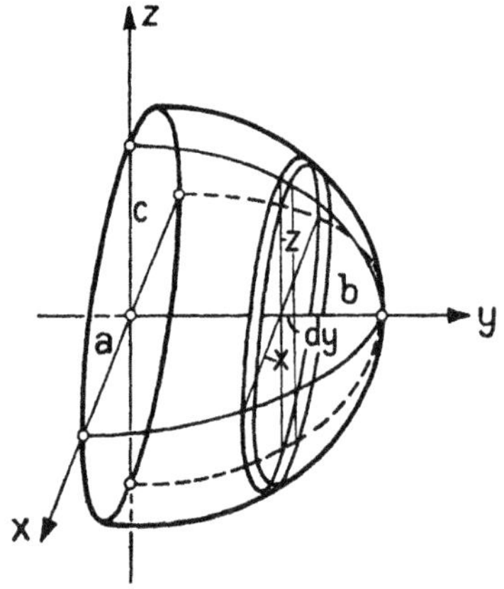

FIG. A 2-8. Moment of inertia of an ellipsoid about one of the symmetry axes.

The equation of the surface of the ellipsoid is

$$\frac{x^2}{a^2} + \frac{y^2}{b^2} + \frac{z^2}{c^2} = 1,$$

where a, b and c are the lengths of the semi-axes.

Consider an elliptical plate perpendicular to the y-axis. It has a mass

$$dm_y = \pi xz\varrho\, dy$$

and a moment of inertia for rotation around the y-axis equal to

$$dI_y = \frac{dm_y}{4}(x^2 + z^2),$$

where x is calculated from the *x-y principle section*:

$$\frac{x^2}{a^2} + \frac{y^2}{b^2} = 1; \quad x^2 = a^2\left(1 - \frac{y^2}{b^2}\right),$$

and z is calculated from the y-z principle section:

$$\frac{z^2}{c^2} + \frac{y^2}{b^2} = 1; \quad z^2 = c^2\left(1 - \frac{y^2}{b^2}\right).$$

Therefore it follows that

$$I_y = \frac{1}{4}\int dm_y(x^2 + z^2)$$

$$= \frac{\pi\varrho}{4}\int_{-b}^{+b} ac\left(1 - \frac{y^2}{b^2}\right)^2 a^2\,dy + \frac{\pi\varrho}{4}\int_{-b}^{+b} ac\left(1 - \frac{y^2}{b^2}\right)^2 c^2\,dy$$

$$= \frac{\pi\varrho}{4}(a^3c + ac^3)\int_{-b}^{+b}\left(1 - \frac{2y^2}{b^2} + \frac{y^4}{b^4}\right)dy$$

$$= \frac{\pi\varrho}{2}(a^3c + ac^3)\left(b - \frac{2}{3}\frac{b^3}{b^2} + \frac{b^5}{5b^4}\right) = \frac{\pi abc\varrho}{2}(a^2 + c^2)\frac{8}{15}.$$

$$m = \frac{4\pi}{3}abc\varrho.$$

For the moment of inertia we obtain the formulas

$$I_y = \frac{m}{5}(c^2 + a^2),$$

$$I_z = \frac{m}{5}(a^2 + b^2),$$

$$I_x = \frac{m}{5}(b^2 + c^2).$$

APPENDIX III

ABSOLUTE DENSITY OF WATER

DENSITY in grams per cm^3, calculated from the relative values of Thiesen, Scheel, and Diesselhorst (1900) and the absolute value at 3.98°C from "International Bureau of Weights and Measures" (1910).

Temperature	0	1	2	3	4	5	6	7	8	9
0	0.999841	847	854	860	866	872	878	884	889	895
1	900	905	909	914	918	923	927	930	934	938
2	941	944	947	950	953	955	958	960	962	964
3	965	967	968	969	970	971	972	972	973	973
4	973	973	973	972	972	972	970	969	968	966
5	965	963	961	959	957	955	952	950	947	944
6	941	938	935	931	927	924	920	916	911	907
7	902	898	893	888	883	877	872	866	861	855
8	849	843	837	830	824	817	810	803	796	789
9	781	774	766	758	751	742	734	726	717	709
10	700	691	682	673	664	654	645	635	625	615
11	605	595	585	574	564	553	542	531	520	509
12	498	486	475	463	451	439	427	415	402	390
13	377	364	352	339	326	312	299	285	272	258
14	244	230	216	202	188	173	159	144	129	114
15	099	084	069	054	038	023	007	*991	*975	*959
16	0.998943	926	910	893	877	860	843	826	809	792
17	774	757	739	722	704	686	668	650	632	613
18	595	576	558	539	520	501	482	462	444	424
19	405	385	365	345	325	305	285	265	244	224
20	203	183	162	141	120	099	078	056	035	013
21	0.997992	970	948	926	904	882	860	837	815	792
22	770	747	724	701	678	655	632	608	585	561
23	538	514	490	466	442	418	394	369	345	320
24	296	271	246	221	196	171	146	120	095	069
25	044	018	*992	*967	*941	*914	*888	*862	*836	*809
26	0.996783	756	729	703	676	649	621	594	567	540
27	512	485	457	429	401	373	345	317	289	261
28	232	204	175	147	118	089	060	031	002	*973
29	0.995944	914	885	855	826	796	766	736	706	676
30	646	616	586	555	525	494	464	433	402	371

APPENDIX IV

APPROXIMATE VALUES OF SOME IMPORTANT PHYSICAL CONSTANTS

Acceleration due to gravity (standard)	$g_s = 980.665 \text{ cm} \cdot \text{sec}^{-2}$
Atmosphere (standard)	$1.013249 \times 10^6 \text{ dyne} \cdot \text{cm}^{-2}$
Avogadro's number	$N_0 = 6.023 \times 10^{23} \text{ mole}^{-1}$
Boltzmann constant	$k = R/N_0 = 1.3804 \times 10^{-16} \text{ erg } (°\text{C})^{-1}$
Density of mercury (0°C)	$13.59509 \text{ gm} \cdot \text{cm}^{-3}$
Gas constant	$R = 8.31696 \times 10^7 \text{ erg } (°\text{C})^{-1} \text{ mole}^{-1}$ $= 1.9872 \text{ cal}_{15} (°\text{C})^{-1} \text{ mole}^{-1}$ $= 0.082082 \text{ liter} \cdot \text{atm } (°\text{C})^{-1} \text{ mole}^{-1}$
Gravitational constant	$G = 6.664 \times 10^{-8} \text{ dyne} \cdot \text{cm}^2 \cdot \text{gm}^{-2}$
Ice point	$T_0 = 273.15°\text{K}$
Joule's equivalent (mechanical equivalent of heat)	$J_{15} = 4.1852 \text{ joule} \cdot \text{cal}_{15}^{-1}$

PROBLEMS FOR THE STUDENT

CHAPTER II

1. A man of weight P stands on a horizontal plane and supports by his weight with the aid of a fixed pulley a load of weight Q (Fig. A.1). Find the force F with which the man presses on the plane.

2. A lamp of weight $P = 10$ kg is supported by a cable above the middle of a street of width $l = 10$ m. The permissible cable tension $p = 50$ kg. If the point at which the lamp is fixed to the cable must be at a height $h = 5$ m, at what height must the ends of the cable be held?

FIG. A.1.

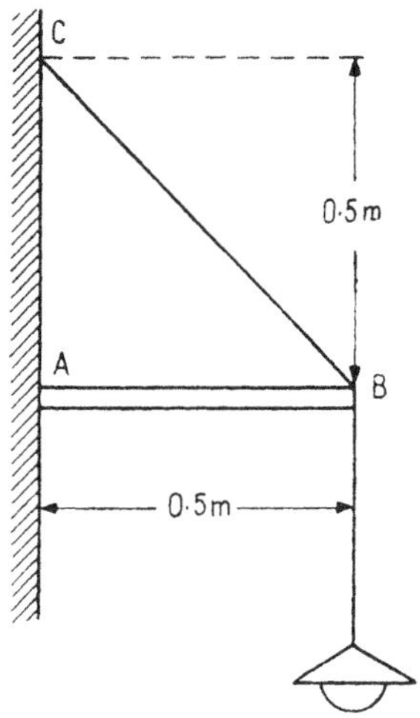

FIG. A.2.

3. The suspension of a lamp is as illustrated in Fig. A.2. The weight of the lamp is 5 kg. Find the forces acting on the girder AB and the wire CB (the dimensions are shown on the figure).

4. A rope is attached to a hook A and passes round a pulley C (Fig. A.3). A load of 20 kg is fastened to the rope at the point D, which cannot move along the rope. What load Q must be attached to the end of the rope in order for the rope tension over the piece AD to be twice that in the remaining part, and the angle ADC to be 90°? Find the force F pulling out the pul-ley C.

5. A wedge is driven into a beam. What must be the coefficient of friction for the wedge not to slip out from the beam? The angle of the wedge at the vertex is 30°.

6. With what force f must a man pull on a rope in order to support the platform on which he stands (Fig. A.4), if the weight of the man is $P_1 = 60$ kg and the weight of the platform $P_2 = 30$ kg? With what force F does the man press on the platform? What is the maximum weight of platform $(P_2)_{max}$ that the man can support?

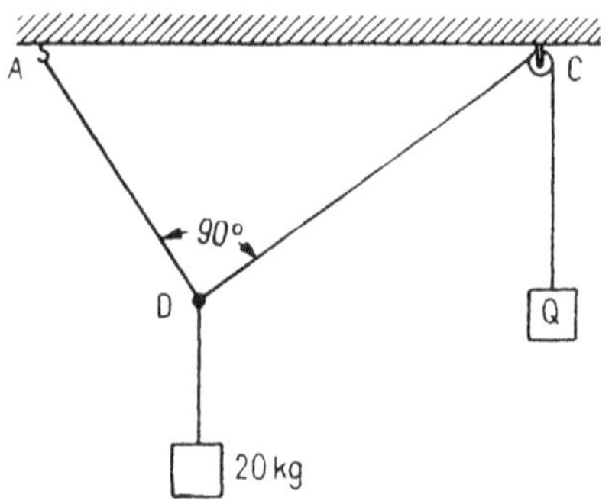

FIG. A.3.

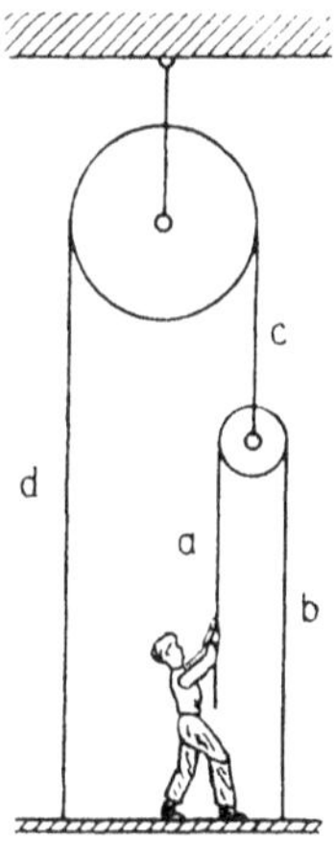

FIG. A.4.

7. The construction and dimensions of a hoist are illustrated in Fig. A.5. Find the pull F in the cable AB and the force T stretching the rod BC when the hoist lifts a 1 ton load. The points B, C and D are assumed to be hinged freely.

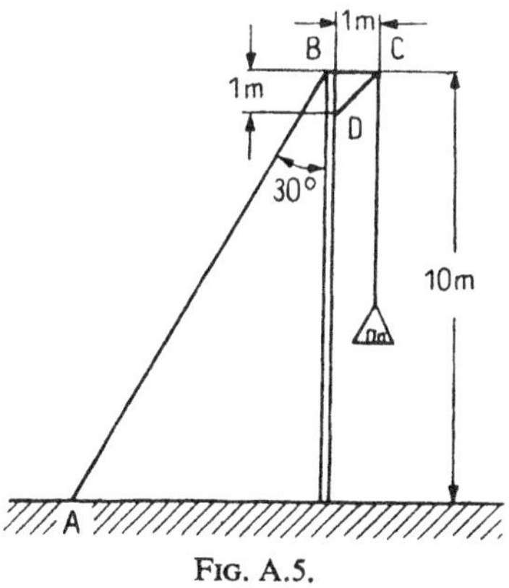

FIG. A.5.

8. What minimum coefficient of friction k is required between the wall of a cube and a horizontal plane in order for it to tilt about its rib as the result of a horizontal force F being applied to the upper face? What must be the value of the applied force?

9. A uniform lamina is bounded by a semicircle of radius R and an isosceles triangle with base and height $2R$ (Fig. A.6). Find the position x_c of the centre of mass C of the lamina.

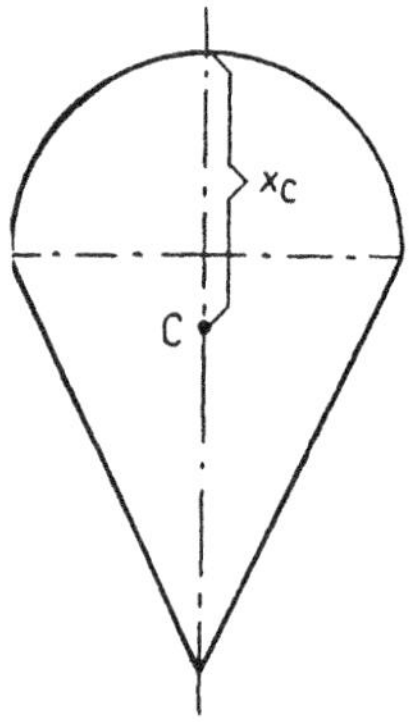

FIG. A.6.

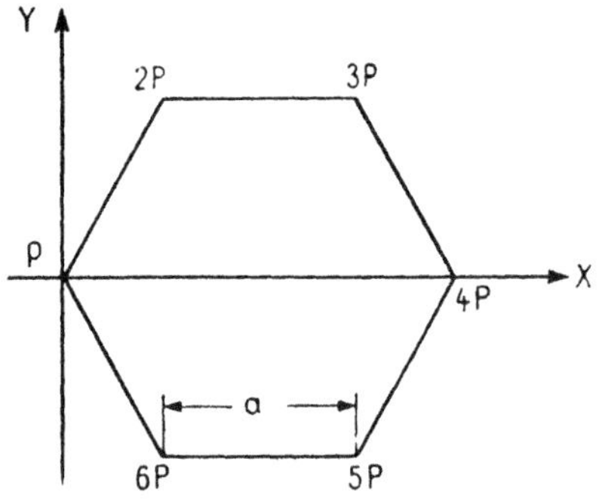

FIG. A.7.

10. A semicircular plate is lifted at four points, two at the ends of the diameter and the other two on the circumference. What must be the distances of the last two points from the diameter in order for a quarter of the weight of the plate to be supported by each point?

11. Loads weighing P, $2P$, $3P$, $4P$, $5P$ and $6P$ are suspended from the six vertices of a horizontal regular hexagonal of side a (Fig. A.7). Find the magnitude and point of application $M(x, y)$ of the resultant. The coordinate axes are as illustrated in the figure.

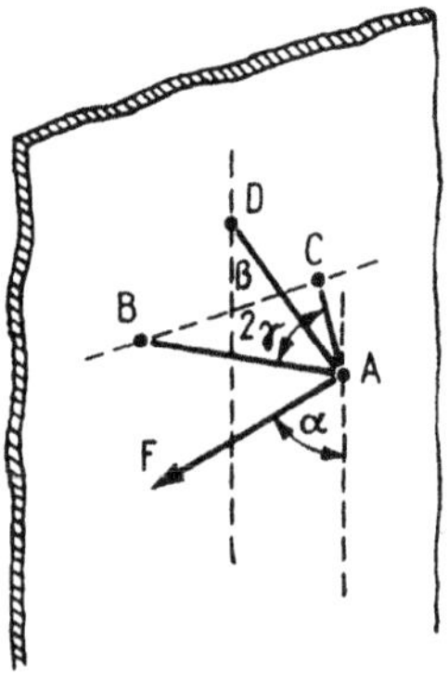

FIG. A.8.

12. A bracket, illustrated in perspective in Fig. A.8, consists of three rods AB, AC and AD. The ends B, C, D of the rods are fixed to the wall by hinges, whilst the other ends are welded together at the point A. The rods AB, AC lie in a horizontal plane and form an angle 2γ. The vertical plane through the rod AD bisects the angle ABC. The rod AD forms an

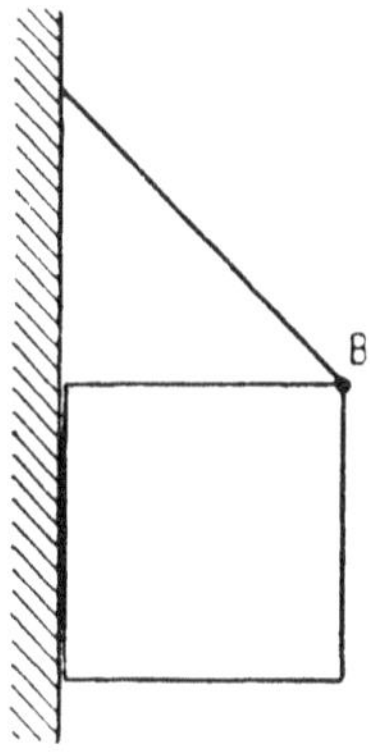

FIG. A.9.

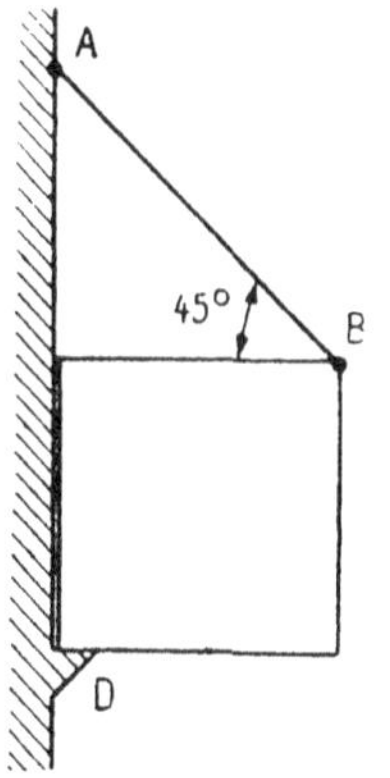

FIG. A.10.

angle β with the wall. A force F acts at the point A, in a plane parallel to the wall and forming an angle α with the vertical. (1) Find the tension forces in the rods. (2) Find the conditions in which there is no tension in the rod AC.

13. A cube of weight 1 ton is supported by the rib D on a projection in a vertical wall and by a cable AB from the rib B to the wall (Fig. A.10). The cable forms 45° with the wall. Find the force F exerted by the cube on the projection D.

14. Two identical beams rest on supports at their ends as shown in Fig. A.11. There is no friction between the beams and the supports. A cylinder A is squeezed between the beams and held by friction forces, whilst the beams are joined from below by a rope attached to spikes driven into the beams. Find the tension T in the rope and the pressure F of the cylinder on the beams, given that the distance h between the axis of cylinder A and the rope is 20 cm. The length of each beam is $l = 1{\cdot}5$ m, and the weight of each $P = 220$ kg, whilst the weight of the cylinder is $p = 20$ kg.

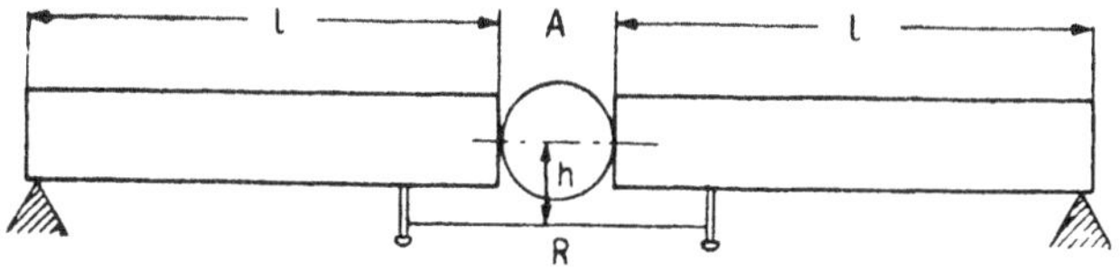

FIG. A.11.

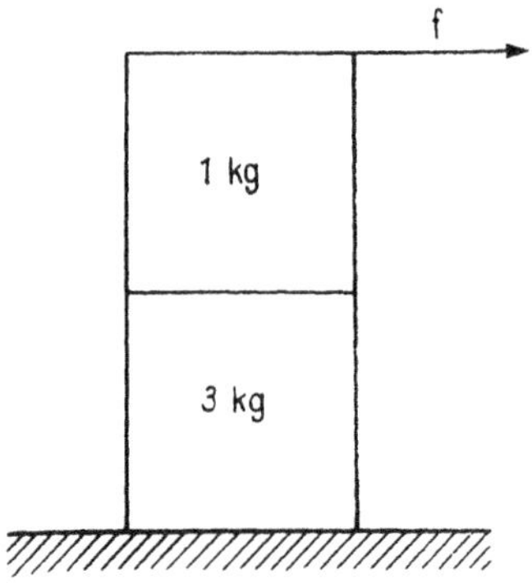

FIG. A.12.

15. Two cubes of edge length 10 cm are welded over their faces to form a prism; the weight of one cubes is 1 kg, of the other 3 kg. The prism stands on a rough horizontal plane (Fig. A.12). What horizontal force f must be applied to the upper face of the prism perpendicularly to its side in order to tilt it about the bottom rib? Does this force f depend on whether the lighter or heavier cube is at the top?

CHAPTER III

16. Two ferry landing-stages are situated opposite one another one the banks of a river, the speed of flow of which is 0·5 m sec^{-1}. What course should a ferry-boat take in order to cross the river along the straight line from one landing-stage to the other? In these conditions, what is the speed v_1 with which the ferry-boat will cross the river, given that the speed it develops relative to the water is 0·8 m sec^{-1}?

17. A tube is mounted on a flat cart which is in uniform motion on a horizontal plane. How must the tube be orientated on the cart in order for rain drops, falling vertically, to miss hitting the walls when falling inside the tube? The motion of the rain drops is assumed uniform.

18. Two aeroplanes are flying simultaneously from a given point in mutually perpendicular directions. The speed of one is $v_1 = 300$ km hr^{-1}, of the other $v_2 = 400$ km hr^{-1}. How does the distance between the aeroplanes increase with time? What will this distance S be at the instant when the first aeroplane has flown a distance $S_1 = 900$ km?

19. Two ships are travelling parallel to each other in opposite directions with speeds v_1 and v_2. One ship fires on the other. At what angle φ should the gun be aimed at the target ship in order to make a hit, if the shot is fired at the instant when both vessels are on the straight line perpendicular to their course? The shell velocity v_0 is assumed constant.

20. A fisherman is sailing up-stream; when passing under a bridge, he drops a boat-hook in the water. After half an hour he discovers this, turns back, and overtakes the boat-hook 5 km below the bridge. What is the speed of the current if the fisherman rows at the same speed up and down the river?

21. A body moving with constant acceleration travels consecutively over two equal 10 m segments of its path S. Find the acceleration a of the body and its speed v_0 at the start of the first segment, if the first segment is traversed in time $t_1 = 1·06$ sec and the second in $t_2 = 2·2$ sec.

22. Draw the graphs of the speeds and paths as functions of time, if the graphs of the acceleration of certain bodies are as shown in Fig. A.13 (the initial speed of the bodies is always zero).†

† The accelerations are schematised as functions of time on these graphs, i. e., it is assumed that the acceleration changes with a jump at certain instants. This is done to simplify the problems. An acceleration can in fact change very rapidly, thoug never with a jump, i.e., it must be a continuous function of time. The assumption of jump-type accelerations is equivalent to the graphs of the velocities having breaks. Similar remarks apply to Problem 14.

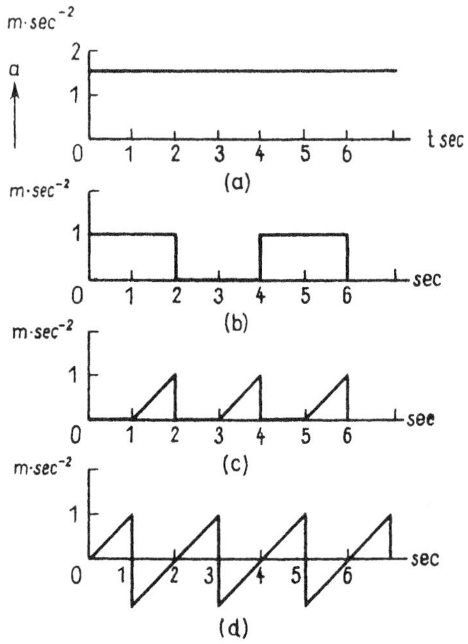

FIG. A.13.

23. What is the permissible limiting landing speed v of a parachutist, if a man can safely jump from a height $h = 2$ m?

24. Two bodies are thrown simultaneously from a tower with the same initial speed v_0: one vertically upwards, the other vertically downwards. How will the distance S between the bodies vary with time? (Air resistance is neglected.)

25. Jets of water flow at angles 60, 45 and 30° respectively from three pipes located at ground level. Find the ratio of the maximum heights h reached by the jets from each pipe, and the ratio of the ranges L over the ground. (Air resistance to the flow of the jets is neglected.)

26. An aeroplane flies horizontally along a straight course with speed v. The pilot has to drop a bomb on a target lying ahead of the plane. At what angle to the vertical must he see the target at the instant of dropping the bomb? At this instant, what is the distance from the target to the point immediately below the aeroplane? (The air resistance to the bomb is neglected.)

27. Loads start to slide without friction simultaneously from the top end of the vertical diameter of a circle along grooves which occupy different chords of the circle. Show that all the loads arrive at the circle at the same time.

28. A boat crosses a river with a constant velocity v relative to the water, perpendicular to the current. The width of the river is d and the speed of the current is zero at the banks and increases linearly as the centre of the river is approached, at which point its value is u. Find the trajectory of the boat, and the distance x_0 that it goes down along the current, from the point where it leaves one bank to the point where it reaches the other.

29. A particle moves uniformly along the plane trajectory illustrated in Fig. A.14. At what point is the acceleration of the particle a maximum?

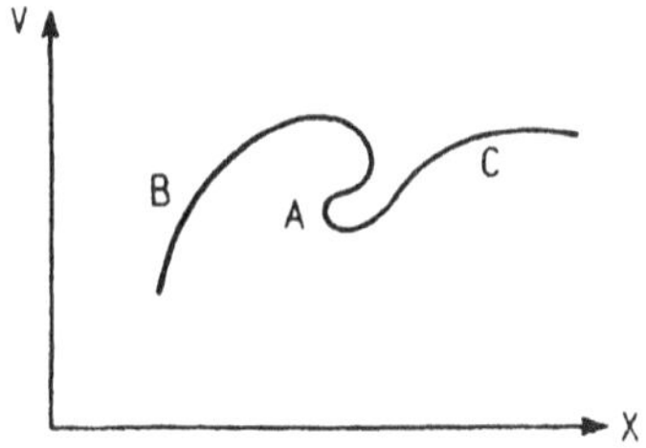

FIG. A.14.

30. The moon revolves round the earth with a period $T = 27$ days. The mean radius of the moon's orbit is $R = 4 \times 10^5$ km. Find the linear velocity v of the moon round the earth and its normal acceleration a.

31. Find the mean angular velocity of the third Soviet sputnik, if the period of its rotation in orbit round the earth was 105 min on 6 July 1958.

32. Find the linear velocity v of a point on the earth's surface on the geographical latitude φ, produced by the diurnal rotation of the earth about its axis. The radius of the terrestrial sphere is $R \approx 6400$ km.

33. A car is travelling at 60 km hr^{-1}. How many revolutions per second are made by its wheels if they roll over the road without slipping and the outer diameter of the tyres is 60 cm.

34. Given the conditions of the previous problem, find the normal acceleration of the outer tread of the car tyres.

35. A rope that unwinds without slipping from a rotating shaft carries a bucket that drops into a well with an acceleration 1 m sec^{-2}. What is the angular acceleration of the shaft? How does the angle of rotation of the shaft depend on time? The shaft radius is 25 cm.

36. A car travelling at 40 km hr^{-1} reaches a bend in the road with a radius of curvature of 200 m. The driver brakes the machine on the bend and gives it a deceleration of 0·3 m sec^{-2}.

37. A wheel of radius R rolls uniformly without slip along a horizontal path with velocity v. Find the coordinates x and y of any given point A on the circumference of the wheel, expressed as functions of time t or of the angle of rotation of the wheel φ, on the assumption that $\varphi = 0$, $x = 0$,

$y = 0$ at $t = 0$ (Fig. A.15). Use the expressions obtained for x and y to draw the graph of the trajectory of a point on the circumference.

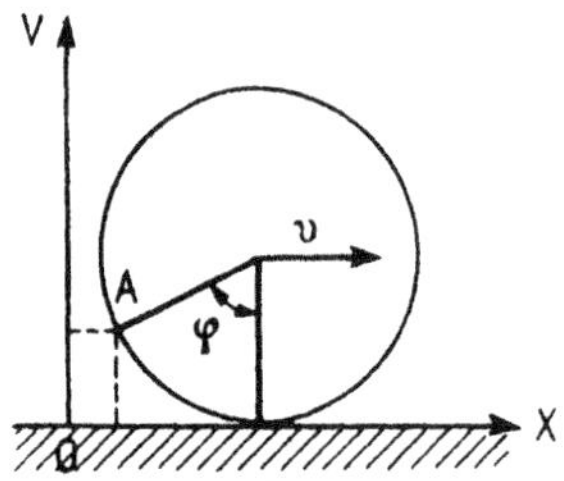

FIG. A.15.

38. A car with wheels of radius R moves with velocity v along a horizontal road, where $v^2 > Rg$ and g is the acceleration due to gravity. What is the maximum height h to which dirt picked up by the wheels can be thrown? Find the point on the tyre from which dirt will be thrown up highest for a given speed of the car. Neglect air resistance to the movement of the dirt through the air.

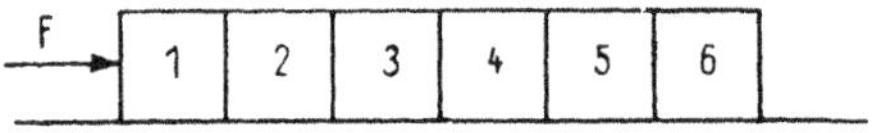

FIG. A.16.

39. Six equal cubes of mass $m = 1$ kg lie on a smooth horizontal table. A constant force $F = 1$ kg acts on the first cube in the direction indicated by the arrow (Fig. A.16). Find the resultant force f acting on each cube. Indicate by arrows on a sketch the forces acting on adjacent faces of each pair of cubes. With what force f does the fourth cube act on the fifth?

40. A body of mass M is placed on a smooth horizontal plane (Fig. A.17). Another mass m is suspended from a string passing over a pulley and attached to the mass M. Find the accelerations of the masses M and m and the tension in the string. Neglect the friction of the mass M on the plane and the friction in the pulley, also the masses of the pulley and string.

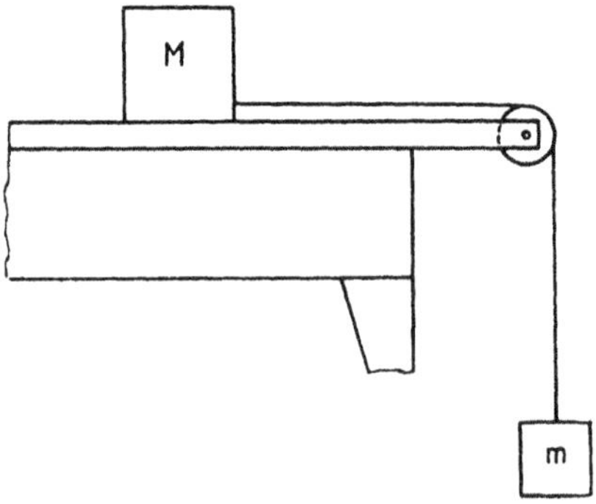

FIG. A.17.

41. Two identical bodies are joined by a string and lie on a smooth horizontal table, so that the string forms a straight line (Fig. A.18). The string can withstand a tension not exceeding 2 kg. What horizontal force F must be applied to one of the bodies in order to break the string?

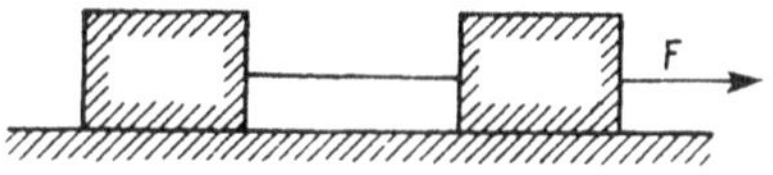

FIG. A.18.

42. Two plates of masses m_1, m_2 are joined by a spring (Fig. A.19). What force must be applied to the upper plate in order for it to raise the lower plate when it bounces up after the action of the force has ceased? Neglect the mass of the spring.

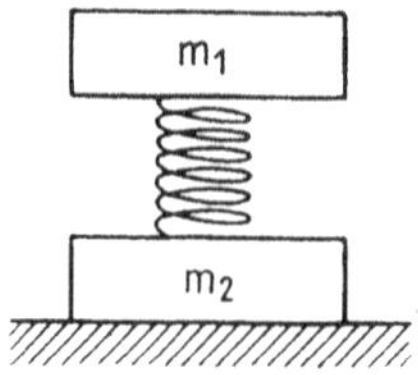

FIG. A.19.

43. What must the minimum coefficient of friction k be between the tyres of the drive wheels of a car and the road, if the car, with weight $2T$ and load $4T$, has an acceleration $a = 0{\cdot}2$ m sec^{-2}?

Consider the problem for two cases: (1) all the wheels are driven; (2) only the rear wheels are driven; assume that the centre of gravity of the car lies at the mid-point between its wheel axes, and the centre of gravity of the load above the rear axle.

44. A board of mass $M = 1$ kg lies on a table, and a weight of 2 kg on the board. What force F must be applied to the board in order for it to be slipped out from under the load? The coefficient of friction between the load and board is 0·25, and between the board and table 0·5.

45. A pendulum of mass m hangs from a support fixed to a trolley (Fig. A.20). Find the direction of the string of the pendulum, i.e. its angle α with the vertical, and its tension T in the following cases: (1) the trolley moves uniformly on a horizontal plane; (2) the trolley moves horizontally with acceleration a; (3) the trolley rolls firely down an inclined plane at an angle φ to the horizontal; (4) the trolley rolls up the plane with acceleration b directed along the plane; (5) the trolley rolls down the plane with the same acceleration b.

46. A load A of mass m is fixed on springs in a shell fired vertically upwards (Fig. A.21). What will be the force exerted by the springs on the load

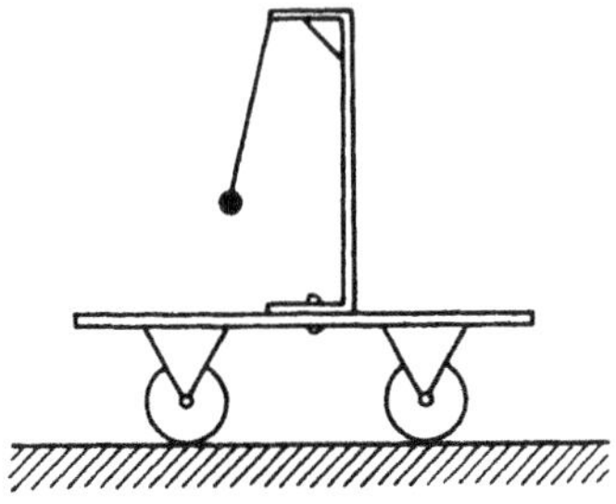

FIG. A.20.

when the shell is rising and falling? Consider the problem, first neglecting, then taking account of air resistance to the motion of the shell.

47. A string passes round a pulley at the top edge of a smooth, inclined plane (Fig. A.22). A load of mass m_1 is attached to one end of the string and lies on the plane. A load of mass m_2 is suspended from the other end. What is the acceleration a of the loads and what is the string tension T? The inclined plane is at an angle α to the horizontal.

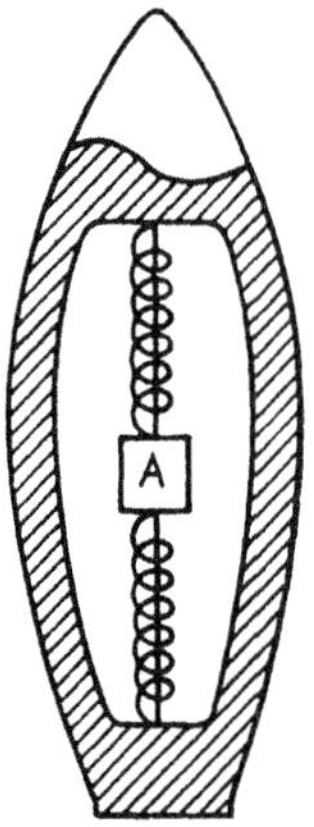

Fig. A.21.

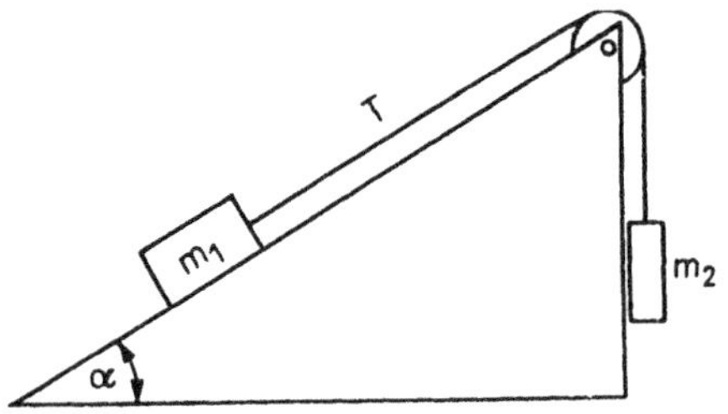

FIG. A.22.

48. Find the accelerations a_1, a_2 of the masses M_1, M_2 and the string tension T in the system illustrated in Fig. A.23(b). Neglect the weight of the pulleys and string.

49. Find the acceleration of the mass m_1 in the system illustrated in Fig. 23(c). Find the string tensions T_1 and T_2. Neglect friction and the weights of the pulleys and string.

50. A string of length l passes round a pulley whose axis is horizontal (see Fig. A.23(a) for the arrangement of the pulley). Two monkeys hang from the ends of the rope at the same distance $l/2$ from the pulley. The monkeys start simultaneously to climb upwards; one climbs with a speed v relative to the rope, the other with a speed $2v$. How long does it take each monkey to reach the pulley? Neglect the weights of the rope and pulley; the monkeys both have the same weight.

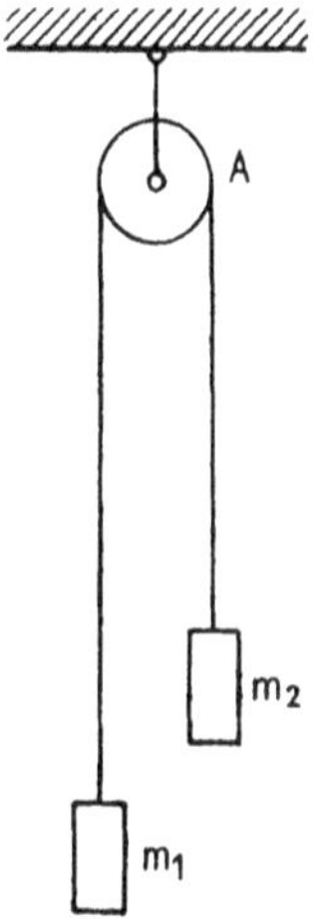

FIG. A.23(a).

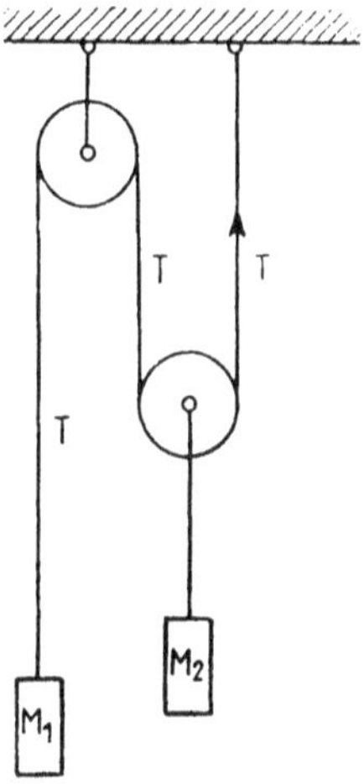

FIG. A.23(b).

51. A steel ball of radius 0·05 mm falls into a wide vessel filled with glycerine. Find the speed v of the equilibrium motion of the ball. The coefficient of internal friction in glycerine is $\eta = 14\ \text{g cm}^{-1}\ \text{sec}^{-1}$, the density of glycerine is $d_1 = 1{\cdot}26\ \text{g cm}^{-3}$, the density of steel $d_2 = 7{\cdot}8\ \text{g cm}^{-3}$.

Hint: To solve this problem, use has to be made of a well-known formula of hydrodynamics, expressing the resistive force exerted on the ball moving in a viscous fluid:

$$f = 6\pi rv\eta .$$

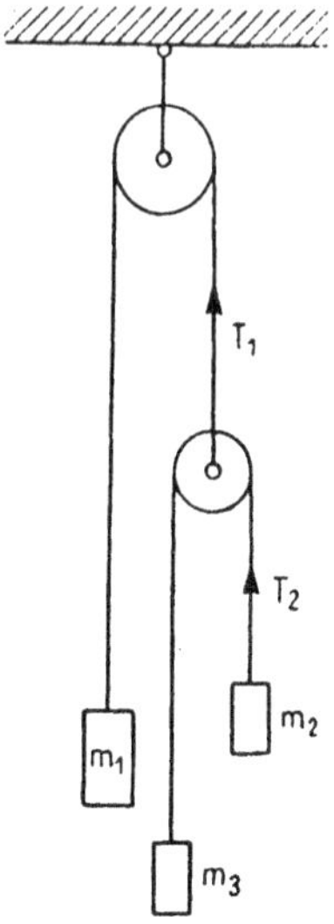

FIG. A.23 (c).

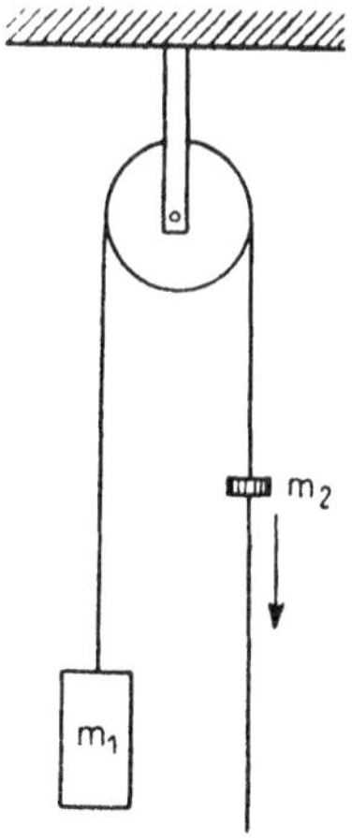

FIG. A.24.

52. A string passes over a light pulley that rotates without friction. A weight of mass m_1 is attached to one end of the string. A ring of mass m_2 slides along the other end of the string with constant acceleration a_2 relative to the string (Fig. A.24). Find the acceleration a_1 of the mass m_1 and the friction force R of the ring about the string. The weight of the string can be neglected.

53. What will be the period of small oscillation of an ideal pendulum of length l, if it oscillates in a truck moving in a horizontal direction with acceleration a?

54. What will be the period of small oscillation of a pendulum in a lift, falling with constant acceleration a? What is the period when $a = g$? How does the pendulum behave if $a > g$?

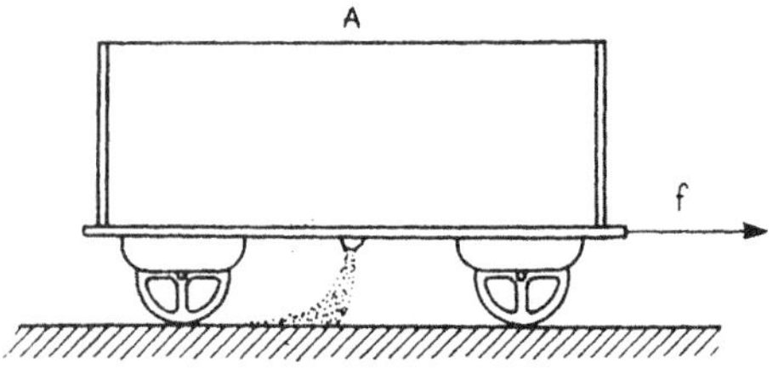

FIG. A.25.

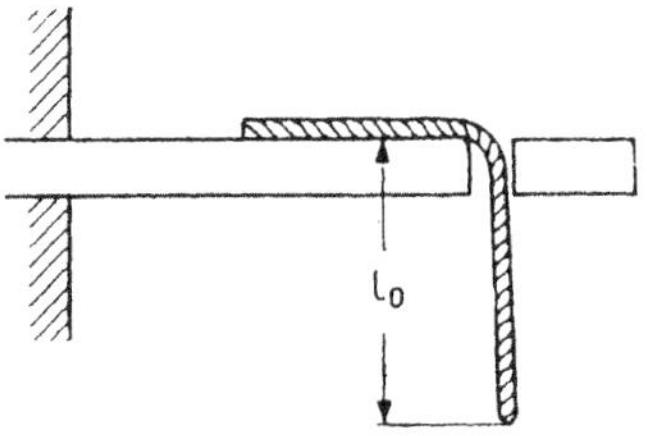

FIG. A.26.

55. Find expressions for the accleration and speed of a trolley A under the action of a constant horizontal force f (Fig. A.25) if it contains sand that pours out through the floor. The weight of sand pouring out in 1 sec is Δm; at time $t = 0$, the speed of the trolley was $v = 0$, and the weight of sand plus trolley was M.

56. A rope lies on a board and one end passes through a hole drilled in the board (Fig. A.26). Find the speed v with which the end slides away from the board, given that the total length of the rope is l and the length hanging initially from the hole is l_0. Find the length that has passes through the hole as a function of time. Neglect the friction between the rope and board.

51. A wedge of mass M lies on a horizontal plane (Fig. A.27). A body of mass m is placed on the face of the wedge. All the contacting surfaces of

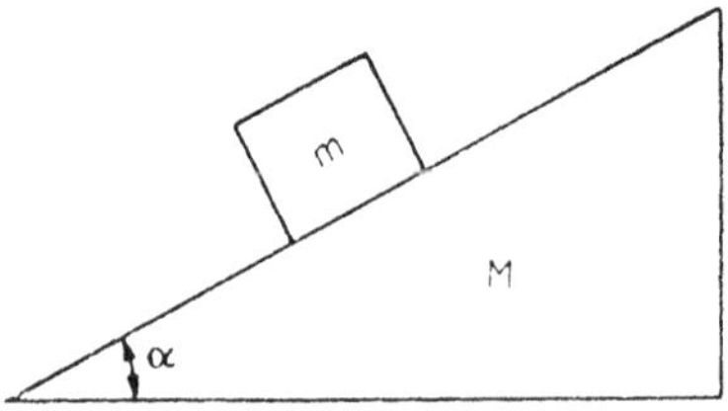

FIG. A.27.

the bodies are smooth. Find the horizontal accelerations of both bodies and the forces N and R with which the body presses on the wedge and the wedge presses on the plane.

58. Liquid is poured into a bent tube (Fig. A.28), such that the two halves of the bend form angles α and β with the horizontal; the length of the liquid column is l. If the liquid is taken from the equilibrium position, its levels in the tube start to vibrate. Find the period of the vibration. Neglect capillary forces and the viscosity of the liquid.

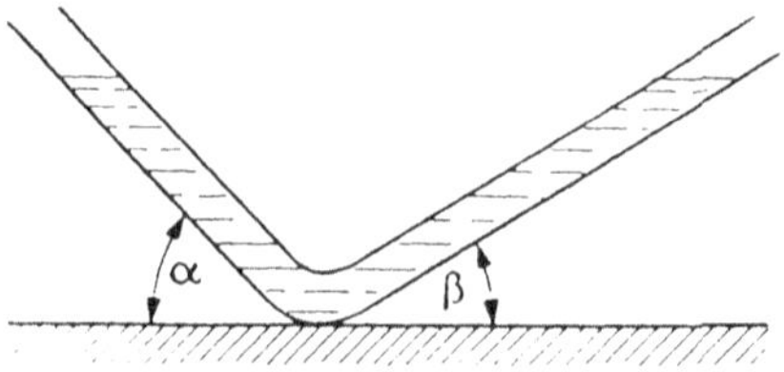

FIG. A.28.

59. Find the period of free small vibration of a load of mass m, attached at the mid-point of a thin string of length L (Fig. A.29(a)). The mass of the string can be neglected; the string tension is constant and equal to P.

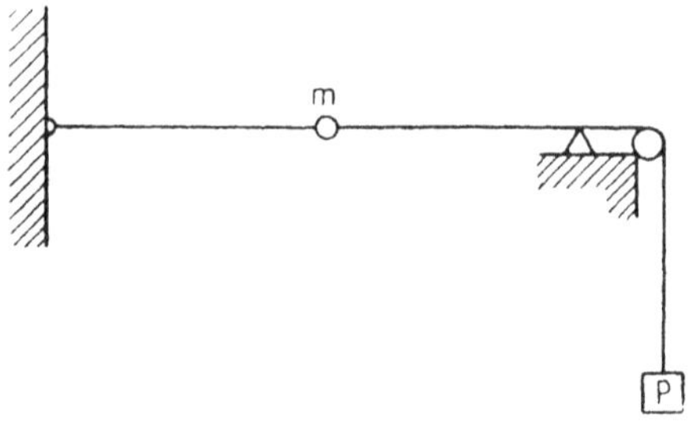

FIG. A.29(a).

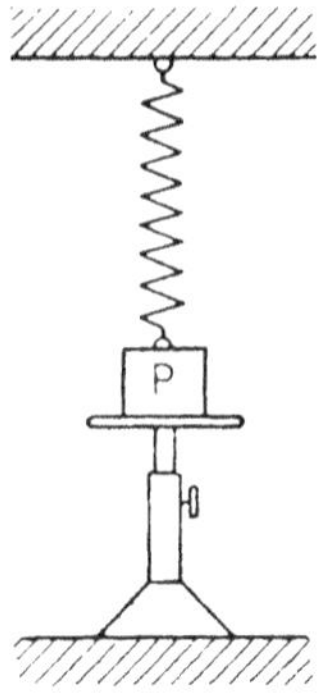

FIG. A.29(c).

60. One end of a spring is clamped, and the other end is attached to a load of weight P, which lies on a support in such a way that the spring is not extended (Fig. A.29(b)). The support is removed without jolting. Find the motion of the load and the maximum tension in the spring. The coefficient of stiffness of the spring is k.

61. A load of mass m falls from a height h on to the scale of a balance, hanging from a spring (Fig. A.30(a)); the load does not bounce relative to the scale but stays on it. The scale starts to vibrate. The coefficient of stiffness of the spring is k. Find the amplitude A of the vibration (the mass of the scale plus spring can be neglected compared with the mass of the load).

62. A body is suspended from a spring and has a characteristic period of $\frac{1}{2}$ sec (Fig. A.30(b)). A sinusoidal force of amplitude $F = 100$ dyne acts vertically on the body, in addition to a certain friction force. Find the amplitude F_{fr} of the friction force and the coefficient of friction (the friction force is proportional to the speed of the motion), if the amplitude of vibration at resonance A_{res} is 5 cm.

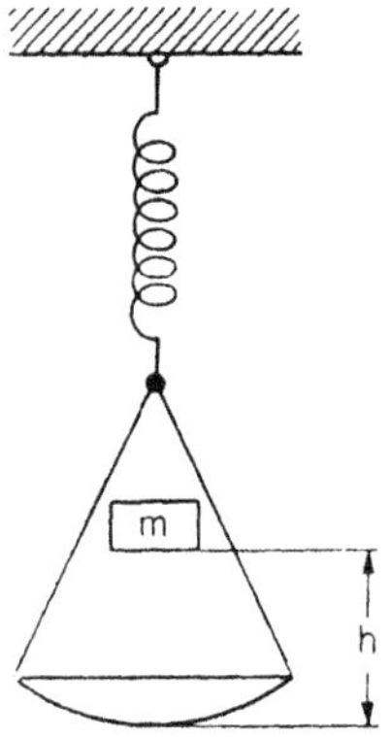

FIG. A.30(a).

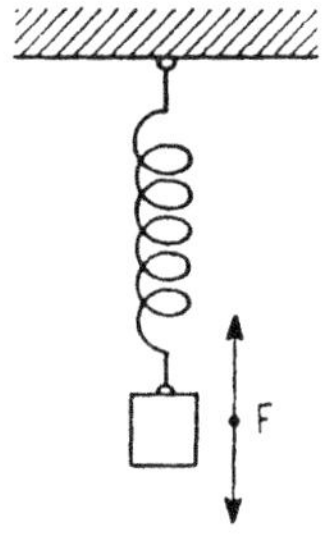

FIG. A.30(b).

63. A load weighing 10 kg is raised through 10 m by the action of a constant force of 20 kg. How much work is done? What is the potential energy of the raised load?

64. The coefficient of friction between a body and a plane inclined at 45° to the horizontal is 0·2. To what height will the body slide up the plane if it is given a velocity of 10 m sec^{-1} directed upwards along the plane? What will be the speed of the body when it returns to its initial starting point?

65. A window blind weighing 1 kg and 2 m long rolls round a thin shaft at the top of the window. How much work is done in rolling it up? Neglect friction.

66. Find the mean useful power in a shot from a smooth-bored gun, given that the bullet of mass m leaves the bore with speed v_0, and the length of the bore is l (the pressure of the cartridge gases is assumed constant throughout the time the bullet is in the bore).

67. A plumb-line is held vertically in a truck travelling along a horizontal path with constant acceleration a, then is suddenly released. Find: (1) an expression for the potential energy U of the plumb-line when its angle of deviation from the vertical is α; (2) an expression for the work A done by the force deviating the plumb-line through the angle α; (3) the maximum value of the angle of deviation α_{max} in the conditions of the experiment. (4) Show that α_{max} is twice the angle to the vertical of the equilibrium direction of the plumb-line when the truck is an accelerated motion (see also Problem 82). (5) Describe the motion of the plumb-line after being freed from the vertical position.

68. A plumb-line is held vertical in a railway truck until the latter is travelling with constant speed. When the train brakes the plumb-line starts to swing, its maximum deviation from the vertical being 3°. What distance S does the train travel until it is at rest if its deceleration is assumed constant throughout and its speed at the initial instant of braking was 47 km hr^{-1}.

69. A meteorite falls from a very great distance to the earth's surface. At what speed would the meteorite strike the earth if the atmosphere did not brake its motion?

70. What power W is expended by a horse in pulling a sledge up-hill with constant speed v? The weight of the sledge is p kg, the friction between the sledge and the hill is constant, the coefficient of friction is k. The angle of the slope of the hill is α.

71. Find the potential energy U of a compressed spring as a function of its deformation, on the assumption that the force of deformation is proportional to the third power of the size of the deformation, with coefficient of proportionality β.

72. A fly-wheel of radius R m does n revolutions per minute and transmits a power of P horse-power via a belt drive. Find the tension T (in kg) in the belt if it runs without slipping.

73. Find expressions for the potential, kinetic and total energies of a particle of mass m, performing harmonic oscillation in accordance with the law $A \cos \omega t$.

74. With what velocity v will a man start to move if he fires a shot horizontally whilst standing on very smooth ice? The weight of the man with rifle and bullet is 70 kg, whilst the weight of the bullet is 10 g and its initial velocity 700 m sec^{-1}.

75. Find the force exerted by a rifle on the shooter's shoulder on the assumption that a constant force acts towards the rifle and the shoulder is displaced $S = 1{\cdot}5$ cm, whilst the bullet leaves the barrel instantaneously. The weight of the rifle is 5 kg, the weight of the bullet 10 g, and its exit speed is $v = 500$ m sec^{-1}.

76. A shell breaks into two equal parts at the top point of its trajectory at a height $h = 19{\cdot}6$ m. One second after breaking, one part falls to earth at a point below that at which the breaking occurred. What is the distance S_2 from the firing point at which the second part of the shell falls, if the first part falls at a distance $S_1 = 1000$ m from the fring point? Neglect air resistance when solving this problem.

77. Three boats of the same weight P travel in line ahead (one after the other) with the same speed v. Loads of weight P_1 are thrown simultaneously from the middle boat to the front and rear boats with the same speed u relative to the boat. What are the speeds of the boats after the loads have been thrown across?

78. A sphere of mass m_1 moving with speed v_1 is struck by a second sphere of mass m_2, which overtakes the first in the same direction with speed v_2. Assuming that the collision is completely inelastic, find the speeds of the spheres after the collisions and their kinetic energy.

79. A ball falls vertically on to a wedge forming an angle of 45° with the horizontal. What will be the trajectory of the ball after striking the wedge? The surface of the wedge is smooth, the collision is complete elastic.

80. A body of mass m_1 collides inelastically with a body of mass m_2. Find the fraction q of the kinetic energy lost if the body m_2 was at rest prior to the collision.

81. A ballistic pendulum is used for finding the speed of a shell. Its principle of operation consits in the shell whose speed is to be measured striking the bob of the pendulum (Fig. A.31). Given the conditions of impact and the masses of the shell and pendulum, the speed v of the shell on collision can be determined from the angle of deviation α of the pendulum.

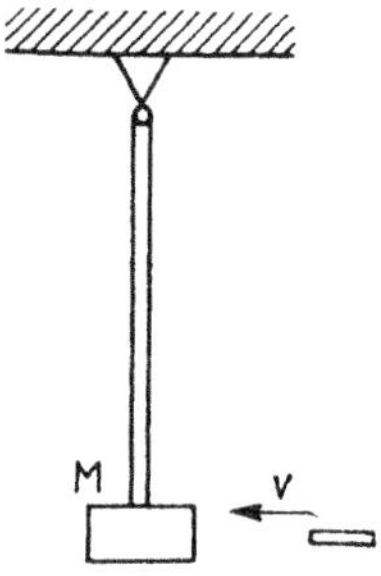

FIG. A.31.

Show how this is done for the following different cases: (1) the shell remains stuck in the bob after impact; (2) the shell rebounds after impact with speed v'; (3) the shell falls down, having lost its speed. The mass of the bob is M kg and the mass of the shell m kg; the pendulum can be regarded as ideal, its length being l.

82. Two spherical bobs, masses m_1, m_2, hang freely from strings of different lengths l_1, l_2, so that the bobs are in contact. The first bob is pulled through an angle α from its initial position in the plane of the strings and then released. A central collision of the spheres occurs. What are the angles α_1, α_2 of deviation from the vertical of the bobs after impact (the angles are assumed small and the impact elastic)?

83. The last truck gets uncoupled from a train travelling at constant speed and traverses a distance l before coming to rest. What will be the distance from the train to the wagon at the instant when the latter comes to rest, if the locomotive traction is constant, and the friction of each part of the train is independent of its speed and proportional to its weight?

84. A boat of mass M has a man of mass m standing on it and is at rest in a calm sea. The man starts to move along the boat with speed $\boldsymbol{u}$ relative to the boat. What will be the speed $\boldsymbol{w}$ of the man relative to the water? At what speed $\boldsymbol{v}$ will the boat now move relative to the water? Neglect the resistance of the water to the movement of the boat*.

85. Two men of the same mass m are standing at opposite ends of the boat described in Problem 185 and throw a ball of mass Δm to one another. The speed of the ball relative to the water is $\boldsymbol{u}$. Find (1) the speed of the boat $\boldsymbol{v}$ during the time the ball is in the air; (2) the displacement $\boldsymbol{S}_1$ of the boat and $\boldsymbol{S}_2$ of the ball relative to the water after each throwing of the ball across the boat, if the path length of the ball is $\boldsymbol{l}$.

86. A particular form of the equation for the motion of a body of variable mass can be derived from a consideration of a simple case of the motion of a rocket. Let the rocket obtain its acceleration from the issue of a continuous gas jet, laving the rocket at constant speed $\boldsymbol{u}$ relative to the rocket. The mass of gas issuing in unit time is μ, the mass of the rocket at a given instant is M. Find the equation of motion of the rocket.

87. Find the force P with which a truck of mass m, moving with velocity v, presses on a bridge in each of the following cases: (1) a horizontal bridge; (2) a convex bridge (Fig. A.32); (3) a concave bridge. (In cases (2) and (3) find P for the highest and lowest point of the bridge).

88. A ball of mass m slides from a height h down the smooth inner surface of a bowl in the form of a paraboloid of revolution with vertical z axis. The equation of the paraboloid is $z = k(x^2 + y^2)$. Find the acceleration a of the ball and force F it exerts on the bottom of the bowl at its lowest point.

89. With what initial velocity v_0 must a bullet travel from a gun in a horizontal direction in order for it to go round the earth without dropping on to it? What will be the acceleration of the bullet in this case? (The radius of the earth $R \approx 6{\cdot}4 \times 10^3$ km).

90. A trolley of mass m slides without friction down a concave track of the shape illustrated in Fig. A.33. (1) What is the minimum height H from which the trolley must start sliding in order for it not to leave the track throughout its length? (2) What forces act on the trolley at the highest point B of the loop? (3) What is the motion of the trolley if it slides from a height less than H? (When solving the problem, assume that the trolley wheels are of small size and small mass, and neglect their rotatory motion.)

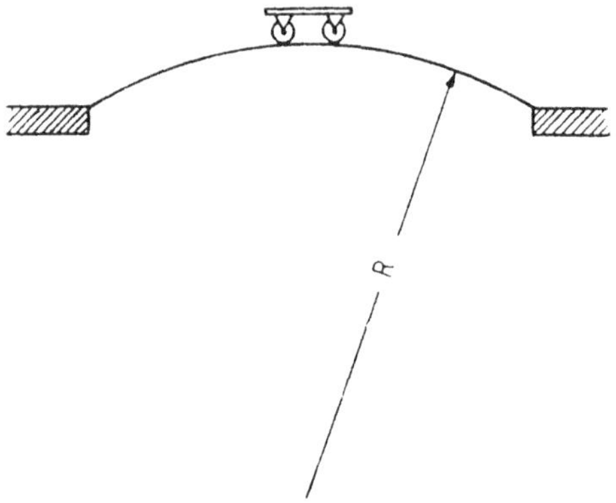

FIG. A.32.

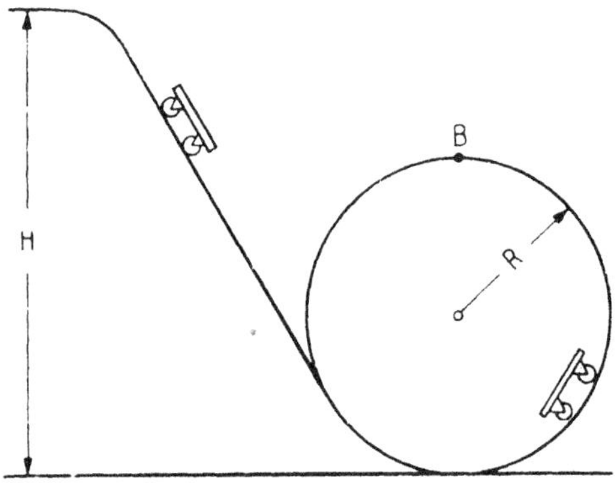

FIG. A.33.

91. What minimum coefficient of sliding friction k is required between the tyres of a car and the asphalt in order for the car to go round a bend of radius $E = 200$ m at a speed $v = 100$ km hr^{-1}?

92. A ball of radius R hangs on a thread of length l and touches a vertical cylinder of radius r, mounted along the axis of a centrifugal device (Fig. A.34). At what angular velocity of the device will the ball cease to apply pressure to the cylinder wall?

93. A body is weighed on a spring balance in the wagon of a train travelling round a bend in the track which is banked in the usual way. The balance shows an n per cent increase in weight of the body as compared with the reading obtained in the train when travelling at constant speed in a straight line. The balance can revolve freely about its support point and remains hanging at right angles to the floor of the wagon when rounding the bend. Find the radius of curvature R of the track, if the train is travelling at speed v.

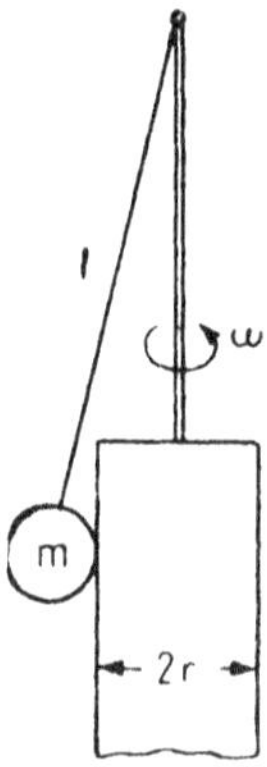

FIG. A.34.

94. An aeroplane loops a loop of radius $R = 100$ m, travelling at $v = 280$ km hr^{-1}. What will be the pressure of the pilot's body, weighing 80 kg, on the pilot's seat at the highest and lowest points of the loop?

95. A pendulum is hanging in an aeroplane that loops the loop. Find the direction of the pendulum at the various points of the loop for different aircraft speeds v and a loop radius R.

96. In the device described in Fig. A.35, a mass $K = 100$ g is connected by a spring with the axis. The mass K is free to slide along CD. The whole system rotates about AB. What is the coefficient of elasticity of the spring l, given that it is extended 50 per cent of its original length when the angular velocity of rotation $\omega = 120$ r.p.m.?

97. A small ball hangs like a bob from a string of length L, which is attached at the other end to the point O (Fig. A.36). The bob is lifted to

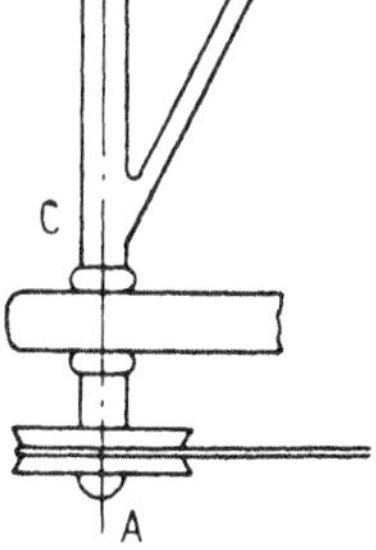

FIG. A.35.

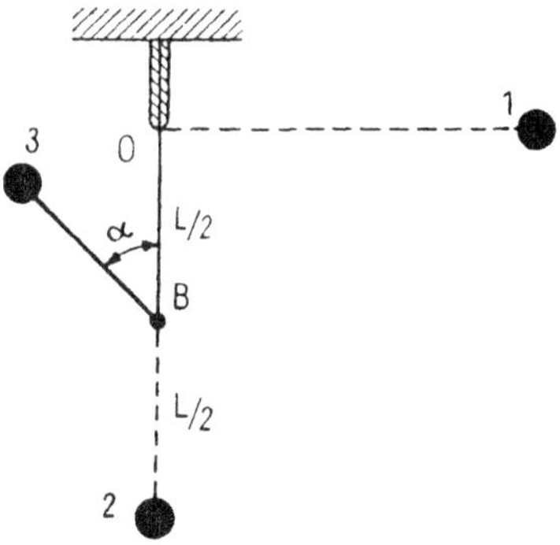

FIG. A.36.

the right to the horizontal position 1 then released. At the point B, located at a distance $\frac{1}{2}L$ from the point O, there is a nail on which the string catches as it returns to the vertical position 2. Find: (1) the angle α to the vertical formed by the bob string at the instant when its tension becomes zero; (2) the velocity v of the ball at this point of its trajectory; (3) the geometrical curve of the bob trajectory after the string tension has become zero.

98. The daily rotation of the earth leads to a deviation of artillery shells and rifle bullets from the original direction of firing, specified in a horizontal plane with respect to the earth's orientation. Calculate the transverse displacement x of a bullet, fired in the meridian plane in a horizontal direction, after it has been travelling for one second. The shot is fired at the latitude of Moscow (55°45′), the initial speed of the bullet is 1000 m sec^{-1}. Find the direction in which the bullet deviates if the rifle barrel was directed southwards at the instant of firing. Neglect the air resistance to the bullet. Solve the problem in a coordinate system fixed in the earth.

99. The daily rotation of the earth causes a deviation of falling bodies to the east. Find the distance x to which a body, falling freely from a height $h = 100$ m at the equator, deviates on the earth's surface from the radius of the earth that passes through the initial position of the body. Solve the problem in a coordinate system fixed in the earth.

100. Solve the previous problem by using a coordinate system not fixed in the rotating earth, but in the solar system instead. It is recommended that two methods be used: (1) direct application of the laws of dynamics to the falling body and (2) application to the body of the law of conservation of angular momentum. What change is there in the result if the experiment is carried out at a point at the geographical latitude φ?

101. Find the acceleration g due to gravity at a height 20 km above the earth, taking the acceleration due to gravity at the earth's surface as $g = 981$ cm sec^{-2}, and the radius of the earth $R \approx 6400$ km.

102. Find the acceleration due to gravity on the moon g_M if its radius is 1738 km, and the mean density is 0·6 times the density of the earth.

103. Calculate the acceleration a of a freely falling body on to the surface of the sun, if $R \approx 150 \times 10^6$ km is the radius of the earth's orbit, $r \approx 7 \times 10^5$ km is the radius of the sun and T is the time taken for the earth to revolve round the sun.

104. Jupiter takes 12 times as long as the earth to revolve round the sun. What is the distance in kilometres from Jupiter to the sun, if the distance of the earth from the sun is 150×10^6 km. Regard the orbits of the planets as circular.

105. Find the acceleration g_s due to gravity on the surface of the sun, given that the length of the terrestrial year is T, the distance from the earth to the sun is R ($\approx$ 8·3 light minutes) and the angle subtended by the diameter of the sun is α ($\approx 32'$).

106. Find the potential energy of a body (point) of mass m at different distances R from the centre of the earth. Regard the potential energy at an infinitely great distance as zero.

107. Find the radius R of the orbit of a "stationary earth satellite". A satellite is said to be stationary if it moves in a circular orbit round the earth in such a way that one revolution takes 24 hours. A stationary satellite moving in the plane of the equator in the direction of rotation of the earth will remain fixed relative to it. Express R in terms of the earth's radius R_0, the angular velocity ω of the earth's rotation and the acceleration g due to gravity on its surface.

108. Determine the second cosmic velocity, i.e. the velocity that must be communicated to a body for it to travel to an infinite distance from the earth. What must be the direction of the velocity relative to the vertical?

109. The maximum distance of the comet Gallea from the sun is $h = 35{\cdot}4$, its least distance $l = 0{\cdot}59$ (taking the distance of the earth from the sun as unit). The linear velocity of the comet is $v_1 = 0{\cdot}91$ km sec^{-1} at its point of maximum distance from the sun at the aphelion. What is the linear velocity v_2 of the comet when it is closest to the sun at the perihelion?

110. A simple analysis of mechanical motions enables us to discover whether the rings of the planet Saturn are continuous formations or accumulations of small satellites. All we need to know for this is whether the inner or outer edge of the ring moves the faster. What are the laws on which this analysis can be based?

111. Neglecting the resistance of the atmosphere, find the minimum work that must be done in order to raise a mass of 1 kg from the surface of the earth to that of the moon. The radius of the earth is 6400 km and of the moon 1740 km; the acceleration due to gravity on the moon, caused by its own attraction, is 0·6 g, where $g = 9{\cdot}8$ m sec^{-2} is the acceleration due to gravity on the earth's surface. Neglect the influence of the sun and other planets.

CHAPTER IV

112. Find the accelerations of the loads and the tension in the string in the device illustrated in Fig. A.37, taking into account the moment of inertia J of the rotating pulley, on the assumption that there is no slip between string and pulley. Find the stress in the support A if the mass of the pulley is M.

113. A uniform cylinder of mass M and radius R (see Fig. A.38) rotates without friction about a horizontal axis under the action of the weight of a load P, fixed to a fine string wound on to the cylinder. Find the angle φ through which the cylinder rotates as a function of time, if $\varphi = 0$ and $\dot{\varphi} = 0$ at $t = 0$.

114. Two fine strings, loaded by masses m_1 and m_2 (Fig. A.39) are wound in opposite directions round a stepped cylindrical pulley. Find the angular acceleration of the pulley and the tensions T_1 and T_2 in the strings, taking account of the moment of inertia J of the pulley.

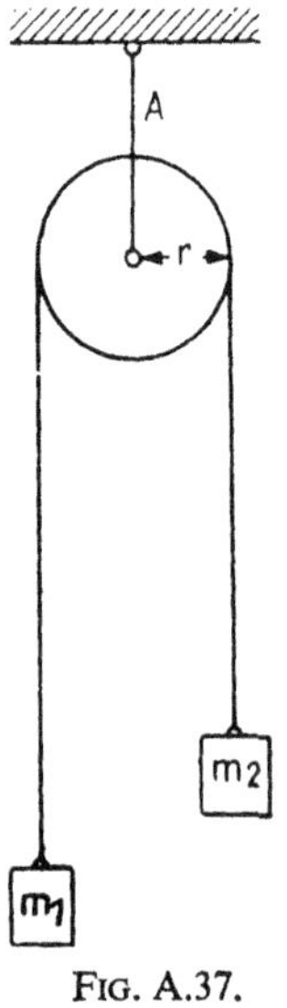

FIG. A.37.

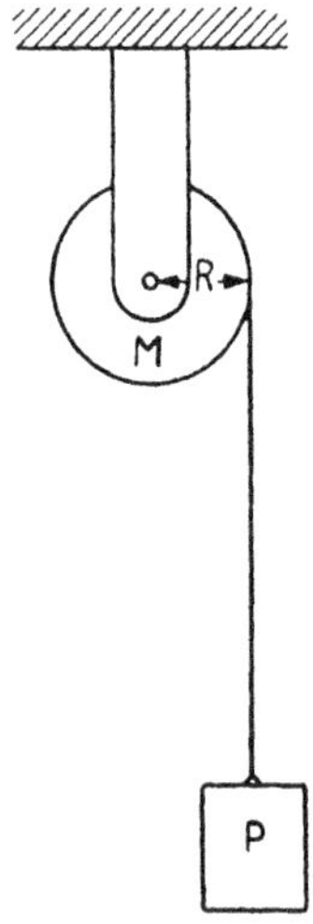

FIG. A.38.

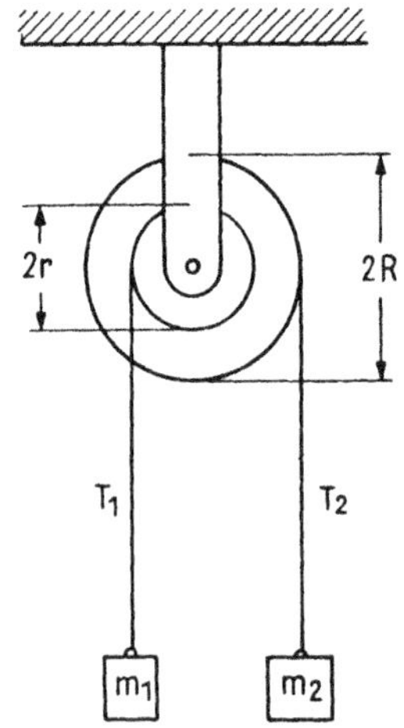

FIG. A.39.

115. A model of a capstan is clamped to one scale of a balance (Fig. A.40). A string carrying a load of mass m is wound round the capstan, the moment of inertia of which is J. The scales are balanced whilst the capstan is braked and the string is not unwinding. What change must be made in the weights in the second scale of the balance in order to re-establish the equilibrium when the capstan rotates due to the load being allowed to fall downwards?

116. The arrangement of a demonstration device (Maxwell disc) is illustrated in Fig. A.41. A solid disc of radius R and weight P is rigidly fixed to a shaft of radius r. The shaft and disc are made from the same material, the weight of the parts of the axis projecting from the disc being p.

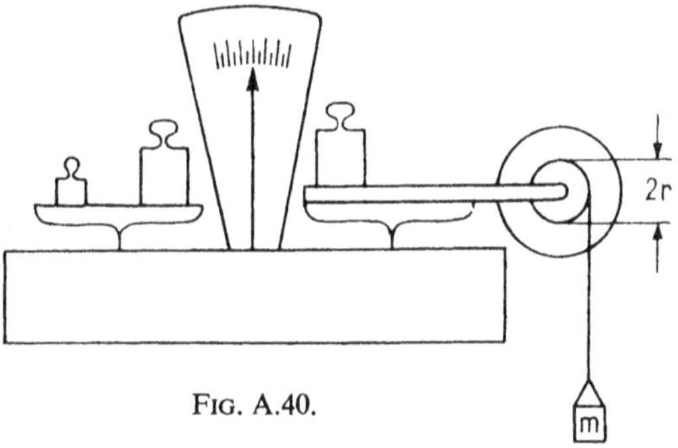

FIG. A.40.

The device is suspended from a support with the aid of strings of equal length attached to the shaft. The strings are wound up symmetrically on the shaft so as to raise the disc, then the latter is allowed to fall freely. Find the acceleration with which the disc drops.

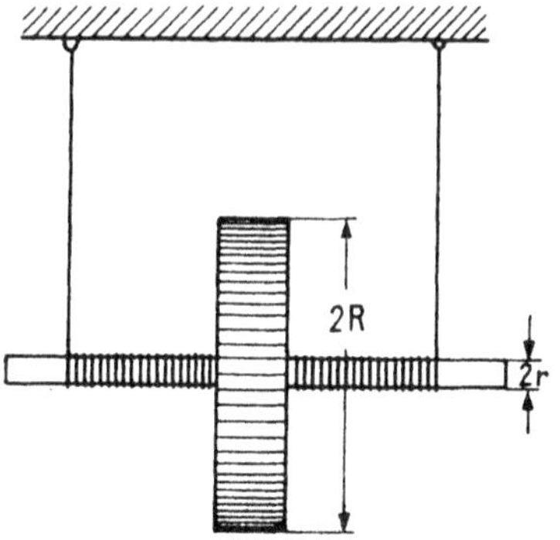

FIG. A.41.

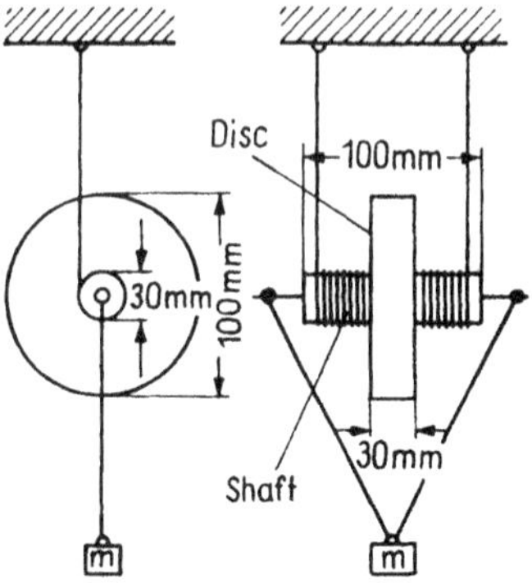

FIG. A.42.

117. What is the acceleration a with which a coil of mass M and moment of inertia J relative to the axis of symmetry falls, if it is suspended in the same way as the disc and shaft in Problem 116 (Fig. A.43). Two further strings are wound on the coil, and from these a load of mass m is suspended. Find the tension in the strings.

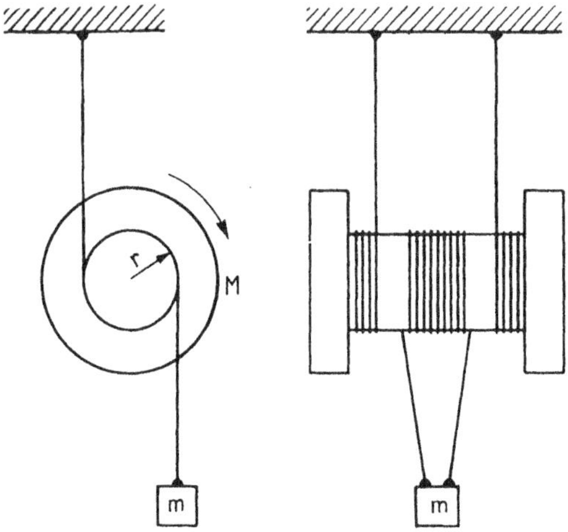

FIG. A.43.

118. Find the acceleration a of the centre of a uniform sphere rolling without slip along an inclined plane, forming an angle α with the horizontal. What is the friction force between the sphere and plane?

119. Find the kinetic energy T of a hoop of mass M that rolls without slip; the thickness of the hoop can be regarded as very small compared with its radius.

120. A solid cylinder and solid sphere of the same radii start to roll down an inclined plane simultaneously from the same level. (1) Which body will have the greater speed at a given level? (2) How much greater? (3) What will be the ratio of their speeds at a given instant?

121. A string is attached to a truck standing on a horizontal plane and passes round a pulley fastened to the edge of a table; a load of mass $m_3 = 500$ g is attached to the end of the string. Find the acceleration a of the truck, given that the mass of the body of the truck is $m_1 = 1{\cdot}4$ kg, the mass of each wheel is $m_2 = 400$ g and the wheels are solid discs. The wheels roll without slip over the surface of the table, and there is no rolling friction.

122. A roller consists of a solid cylinder of radius R and mass M_1 (Fig. A.45) and a frame of mass M_2. A mass M_3 is attached to the frame by a string. The entire system rests on an inclined plane forming an angle α with the horizontal. Find the acceleration a of the system, given that the coefficient of friction between the body M_3 and the plane is k. (The roller rolls without slip). How should the system be arranged—with the mass M_3 in front and the roller behind (as shown in the figure), or vice versa, in order for the string to be stretched during the motion?

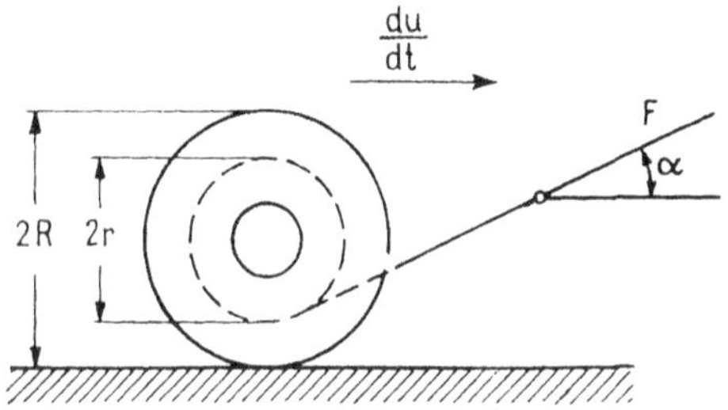

FIG. A.44.

123. Find the angular momentum N of the earth relative to its polar axis. Assume that the earth is a sphere of radius $R = 6000$ km, of density $d = 5{\cdot}5$ g cm^{-3}.

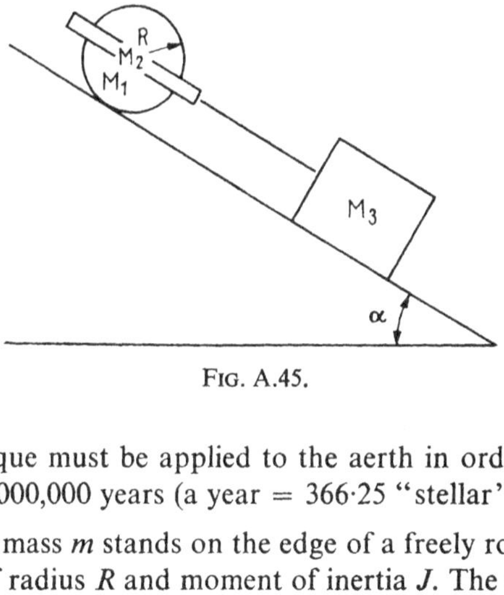

FIG. A.45.

124. What torque must be applied to the aerth in order for its rotation to stop after 100,000,000 years (a year = 366·25 "stellar" days).

125. A man of mass m stands on the edge of a freely rotating fairly large horizontal disc of radius R and moment of inertia J. The disc does n r.p.m. How does the speed of rotation of the disc vary if the man goes from the edge to the centre? How does the energy of the system vary in this case? Neglect the dimensions of the man compared with the radius of the disc.

126. A man of mass m stands on a uniform horizontal disc of mass M and radius R which is at rest. The disc can be rotated without friction about a vertical axis through its centre. The man starts to move at a certain instant. What is the angular velocity of the disc ω, when the man goes along a circle of radius r concentric with the disc, with velocity v relative to the disc?

127. An experimenter stands on a special stool and holds in his hands the vertical axis of a freely rotating wheel, having the moment of inertia J_1 relative to this axis AA (Fig. A.46). The wheel rotates in a horizontal plane with angular velocity ω_1. The stool is so made that it can rotate freely about a vertical axis. The moment of inertia of the stool and experimenter about the vertical axis is J_2. The axis of the wheel and axis of the stool lie along a straight line. What is the change ΔE_k in the kinetic energy E_k of the entire system of stool and wheel, if the experimenter revolves the wheel axis through (1) 180°, (2) 90°?

128. With what part of a sabre should a stick be slashed in order not to feel the impact on the hand? Regard the sabre as a uniform lamina of length l.

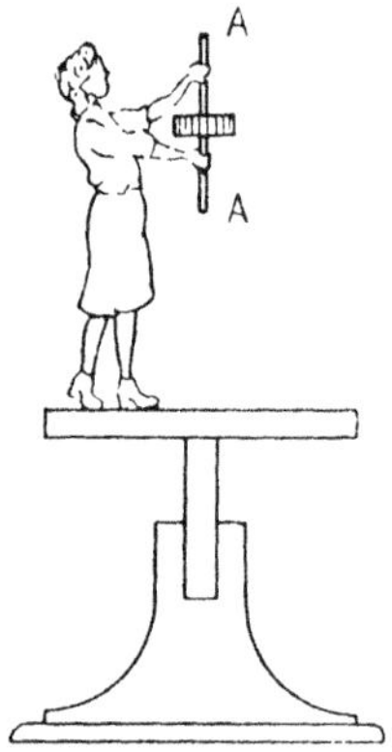

FIG. A.46.

129. A thin square uniform lamina of mass m_0 can rotate freely about a vertical axis (Fig. A.47). A sphere of mass m, moving with speed v, strikes the plate normally at the point A, which is $2a/3$ from the axis. How do the plate and sphere move after impact, which occurs according to the law of elastic collision.

130. A solid cylinder of radius R, mass m, on which a string is wound, lies on two parallel horizontal beams. A force F, equal to half the weight

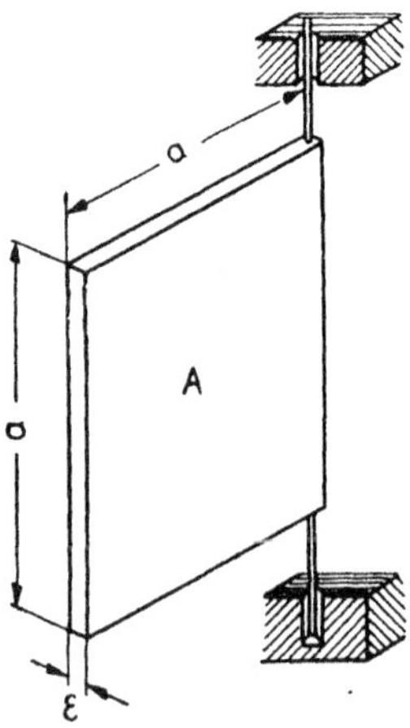

FIG. A.47.

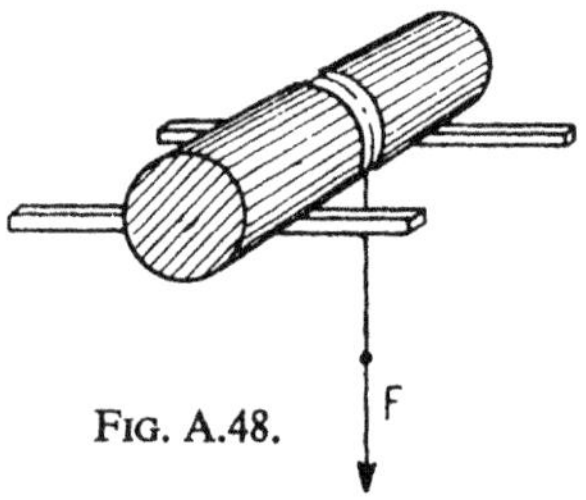

FIG. A.48.

of the cylinder, is applied to the end of the string that hangs down (Fig. A.48). Find the horizontal acceleration of the cylinder and the minimum value of the coefficient of friction between the cylinder and beams such that rolling occurs without slip. The axis of the cylinder is perpendicular to the beams, its centre of gravity and the force F lie in the vertical plane through the mid-point between the beams.

131. Two rollers, connected by a rod S, roll without slipping down a plane inclined at 30° to the horizontal (Fig. A.49). The rollers have the same mass $m = 5$ kg and the same radii $R = 5$ cm, the moment of inertia of the first is $J_1 = 80 \text{ kg cm}^2$, and of the second $J_2 = 40 \text{ kg cm}^2$. The masses of the roller frames and rod can be neglected. Calculate the angular acceleration with which the rollers roll without slipping down the plane. Find the force transmitted by the rod, if the roller with the greater moment of inertia moves in front, and vice versa.

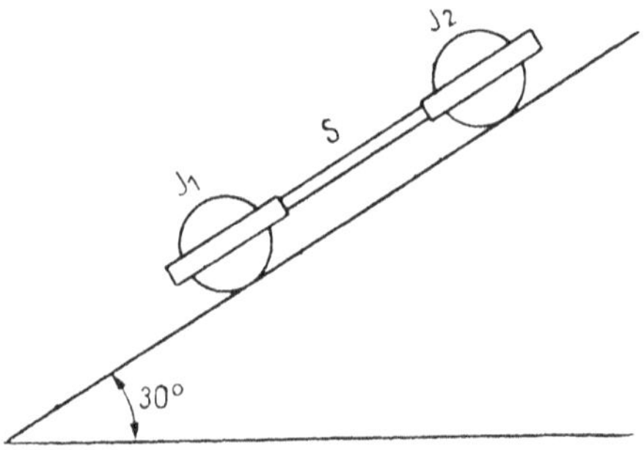

FIG. A.49.

132. A car with a track width $2b$ and centre of mass at a height h above the ground travels round a horizontal bend of radius R. (1) Show that, when the speed of the car is $v > \sqrt{(bRg/h)}$, it turns over unless the wheels slide in a direction perpendicular to the car's motion. (2) Assuming that the speed of the car is sufficient for it to be capable of turning over, find the minimum coefficient of friction between the wheels and the road surface at which this can occur.

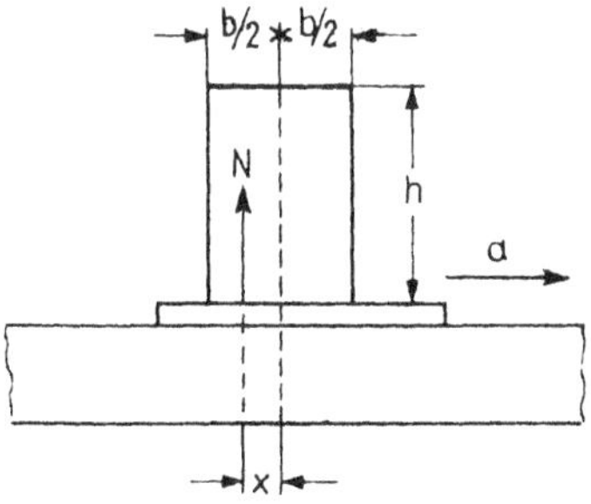

FIG. A.50.

133. A cylinder stands on a horizontal rotating disc. Find at what angular velocity ω the cylinder falls off the disc, if the distance between the axes of the disc and cylinder is R, and the coefficient of friction $\mu > D/h$, where D is the diameter of the cylinder and h is its height (Fig. A.51).

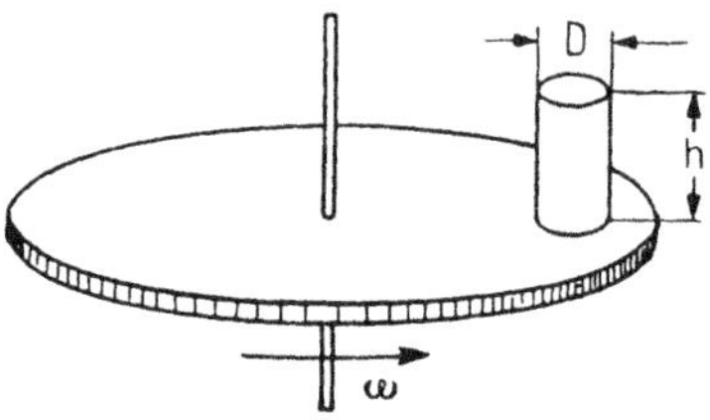

FIG. A.51.

134. Calculate the moment M of the gyroscopic forces acting on the shaft from the propeller if an aeroplane flying at $u = 300$ km hr^{-1} makes a turn of radius $R = 100\,m$. The moment of inertia of the propeller is $J = 7$ kg m^2, and the propeller turns at $N = 1000$ r.p.m.

135. A pendulum that performs 3601·4 vibrations in an hour in Leningrad, performs 3600·0 vibrations in the same time in Moscow. By how much do the accelerations of freely falling bodies differ in these two cities?

136. Four bodies A, B, C and D (Fig. A.52), which can be regarded as material particles, rotating about a centre, always remain along the same straight line at a fixed distance from one another. Forces act between the bodies in accordance with Newton's law of universal gravitation. The masses C and D are equal and negligibly small by comparison with masses A and B, whilst the distance r is very small by comparison with R. What further forces must be exerted from body B on to C and D in order for the distances between all the bodies to remain unchanged?

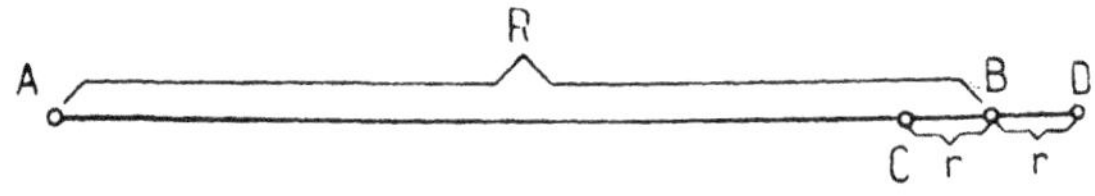

FIG. A.52.

137. A uniform rod is suspended at both ends from two identical strings of length L. The strings are parallel in equilibrium. Find the period T of small vibrations, produced after a rotation of the beam about a vertical axis through its mid-point.

138. A ring of thin wire performs small vibrations about a horizontal axis, like a pendulum (Fig. 90). In one case the axis lies in the plane of the ring (Fig. A.53(a)).

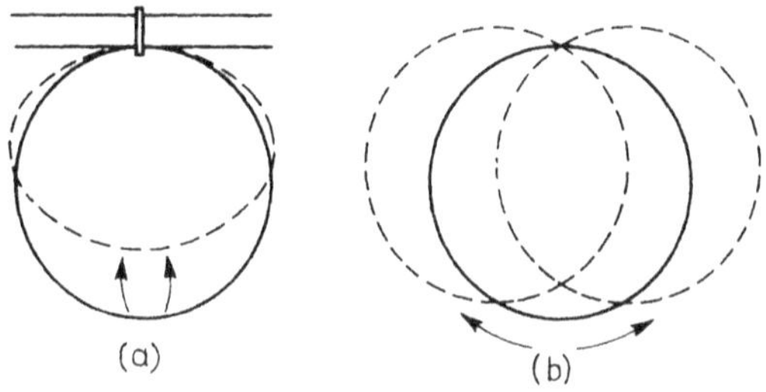

FIG. A.53.

139. A solid uniform disc, radius $r = 10$ cm, vibrates about an axis perpendicular to the plane of the disc and passing through a point of the rim. What length l would a simple pendulum require in order to have the same period of vibration as the disc?

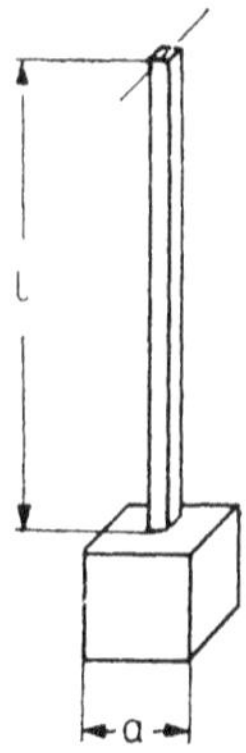

FIG. A.54.

140. A physical pendulum consists of a rod square section suspended at one end, and a load clamped at the other end (Fig. A.54). The load is a cube of side $a = 40$ mm; the rod is of length $l = 400$ mm and side of cross-section $b = 4$ mm; the load and rod are made of the same material. Find the approximate period of vibration T of the pendulum (the rod can be assumed thin in the calculation).

CHAPTER V

141. A lift descends with constant speed. What will be the tension in the cable from which the lift is suspended at the instant when the cable drum suddenly comes to a complete stop? How will the tension vary after this?

142. Calculate the maximum force T of the cable tension and its elongation (see the previous problem), if the coefficient of elasticity of the cable for the length at which the drum stoppage occurred is 1 ton cm^{-1}. The weight of the lift is 3 tons, its speed 10 m sec^{-1}.

143. Could a cable consisting of a thin copper wire in a lead sheath be used for telephone connection to a captive balloon at a height of 300 m. The limit of strength of lead is 2 kg mm^{-2}, its density is 11·4 g cm^{-3}.

144. A steel cable that can support the weight of a lift cabin at rest has a diameter 9 mm. What must be the diameter of the cable if the lift cabin can have an acceleration of $8g$?

145. What uniformly distributed load Q can be supported by a granite block if it is in the form of a regular hexahedron of side $a = 10$ cm and the permissible stress on the compressed face is $p = 45$ kg cm^{-2}?

146. What is the extension of an iron rod hanging from one end due to the action of its own weight? What is the change in its volume?

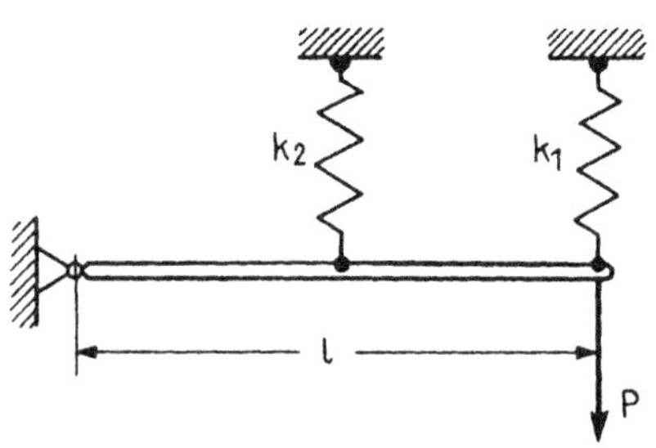

FIG. A.55.

147. A rod of length l is jointed to a wall at one end and is supported by two springs (Fig. A.55); the spring at the end has a coefficient of rigidity k_1, and that at the middle rigidity k_2. Find the extensions of the springs and the stresses in them under the action of a load P applied at the end of the rod. The rod is assumed absolutely rigid and weightless, whilst the springs are not extended in the absence of an external load.

148. A load hangs from three ropes as shown in Fig. A.56. The ropes are all made of the same material, and the two outer ones are identical. Find the ratio of tensions in the rope material, if the deformations on loading are extremely small.

149. A five-times safety factor for strength is chosen for rods. A steel rod with rectangular section (Fig. A.58) has one end fixed into a wall. The ratio of height to width of the rod is 3 : 1. (1) What load P is permissible at the end of the rod, if its length is 15 cm, and width 5 mm? (Breaking tension in the material is 100 kg mm^{-2}.) (2) Find the sag λ on bending at permissible loading. Young's modulus $E = 20{,}000$ kg mm^{-2}.

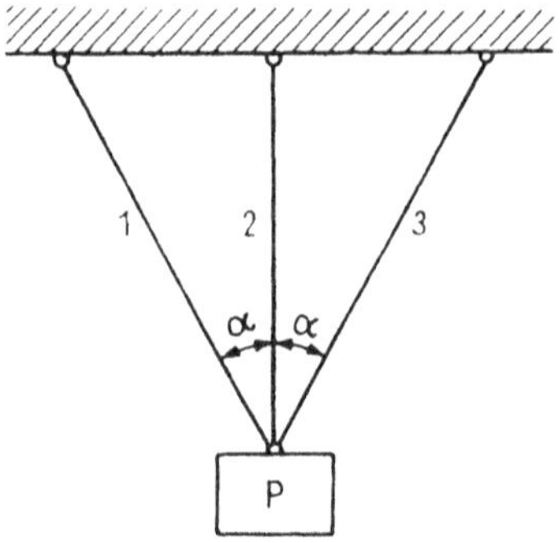

FIG. A.56.

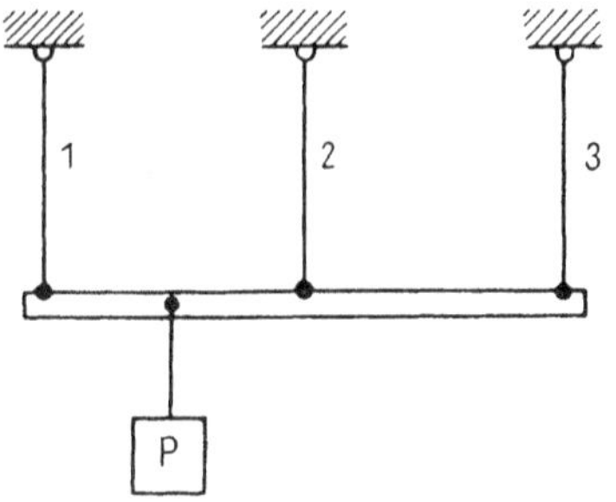

FIG. A.57.

150. The ends of a beam are clamped to supports (Fig. A.59), one support being movable. Find the sag λ of the beam under the action of a force P, applied to the mid-point of the beam. The length of the beam is l, the modulus of elasticity of the material is E and the moment of inertia of the cross-section is J. Regard the beam as weightless and the bending as small.

151. The beam of a balance has a rectangular cross-section with sides $a = 8$ mm (horizontal) and $b = 10$ mm (vertical). The length of the beam is $l = 250$ mm. What is the maximum sag λ of the beam if the balance is designed for a maximum load $P = 500$ g, and Young's modulus for the material is 15,000 kg mm^{-2}.

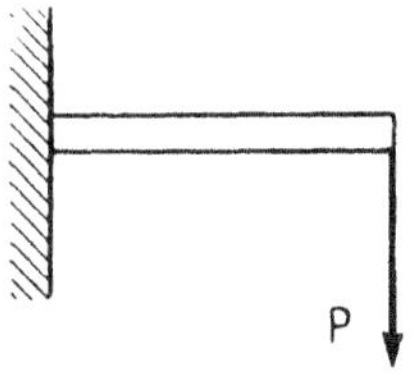

FIG. A.58.

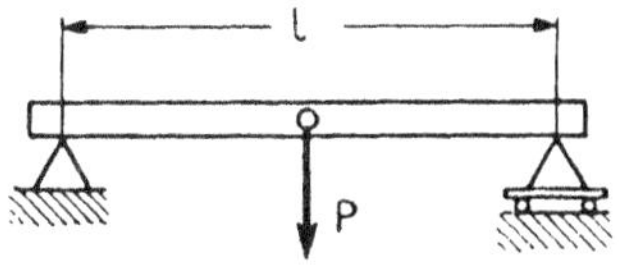

FIG. A.59.

152. The ends of a copper pipe, the outer and inner diameters of which are $D = 20$ mm and $d = 10$ mm, rest on two supports, the distance between which is $l = 400$ mm. The middle of the tube carries a load $P = 90$ kg. Young's modulus for copper is $E = 10^4$ kg mm^{-2}. Find the sag λ of the pipe midway between the supports.

153. The frame of a sensitive galvanometer, rotating between the poles of a magnet, is suspended on a fine platinum wire or thin bronze strip. Find the maximum permissible weight of the frame if the limit of strength of platinum = 30 kg mm^{-2}, and a wire of diameter 4μ is used for the suspension.

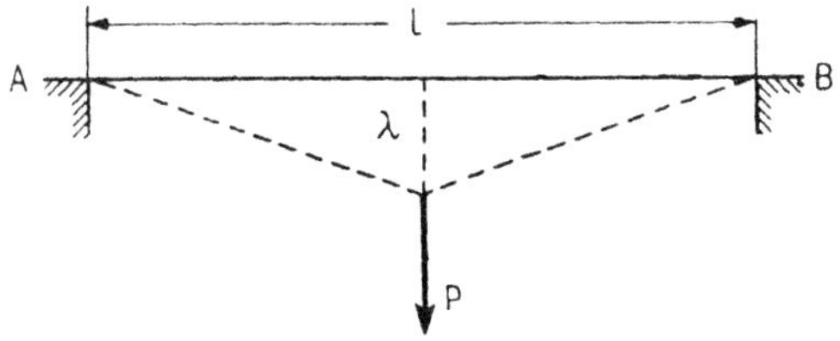

FIG. A.60.

154. A wire stretches between two clamps A and B, a distance l apart (Fig. A.60). A load weight P hangs from the mid-point of the wire and produces bending λ. Find λ as a function of P, given that Young's modulus is E, the diameter of the wire is d and $\lambda/l \ll 1$.

155. A steel wire, diameter $d = 1$ mm, bends round a drum, diameter $D = 2$ m. Find the extra tensions produced in the material of the wire if the modulus of elasticity of steel is $E = 2 \times 10^6$ kg cm^{-2}.

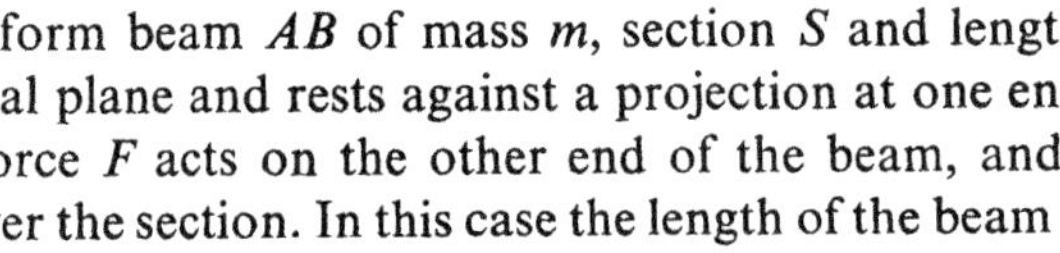

FIG. A.61.

156. A uniform beam AB of mass m, section S and length L is placed on a horizontal plane and rests against a projection at one end (Fig. A.61). A constant force F acts on the other end of the beam, and is uniformly distributed over the section. In this case the length of the beam is well known to diminish by the amount $\Delta L = (1/E)(L/S)\,F$, where E is Young's modulus. The question is, what is the compression of the beam and how is it distributed if the beam does not rest against a support, whilst all the other conditions remain unchanged.

157. It follows from the previous problem that tension exists in a beam in accelerated motion. Will there be tension in a freely falling beam?

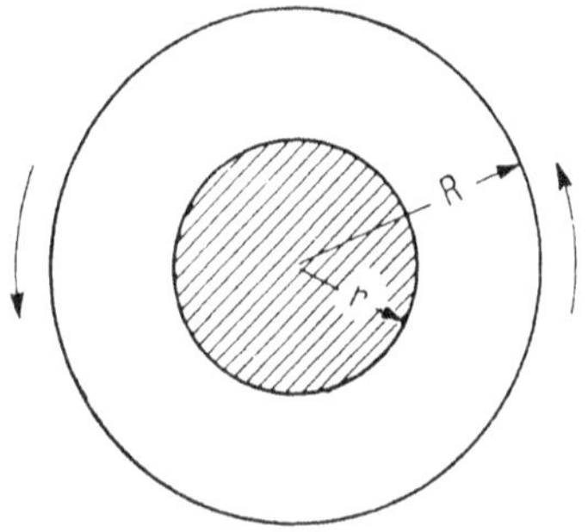

FIG. A.62.

158. A uniform disc of mass M and radius R rotates about its axis with angular acceleration β (Fig. A.62). The forces accelerating the disc are uniformly distributed over its rim. Find the tangential force F acting per unit length of the circumference that forms an imaginary dividing line of radius r in the disc (shown shaded in the sketch).

159. A thin uniform elastic rod of length l, mass M, Young's modulus E, rotates uniformly with angular velocity ω about an axis perpendicular to the rod and passing through one end. Find the distribution of the stress T in the rod and its total elongation ΔL. When calculating the linear deformation and stress, assume that the cross-section is invariable and the elongation small.

160. A hollow metallic sphere, the outer and inner diameters of which are d_1 and d_2, floats on the surface of a liquid. The density of the metal is δ_1, the density of the liquid δ_2. What weight p must be added inside the sphere in order for it to float below the level of the liquid? The compressibility of the sphere can be neglected.

161. What is the lifting force F of 1 m^3 of helium filling a dirigible, if the density of helium relative to air is 0·137, and 1 m^3 of air weighs 1·3 kg?

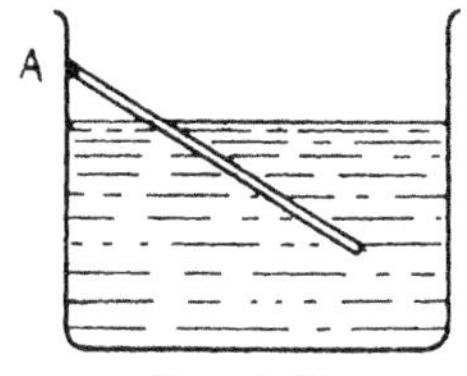

FIG. A.63.

162. At the bottom of a vessel containing liquid (or gas) a body is situated whose specific weight is much greater than the specific weight of the liquid (or gas). Is it possible to cause the body to lift upwards by increasing the pressure on the liquid (or gas)?

163. What is the pressure P of the air in a caisson dropped into water, and intended for underwater work at the bottom of a river of depth h m, if the atmospheric pressure is H mm Hg? The density of mercury is d.

164. A thin rod is clamped at one end to the wall of a vessel, whilst the other end is submerged in water (Fig. A.63). The rod can rotate freely

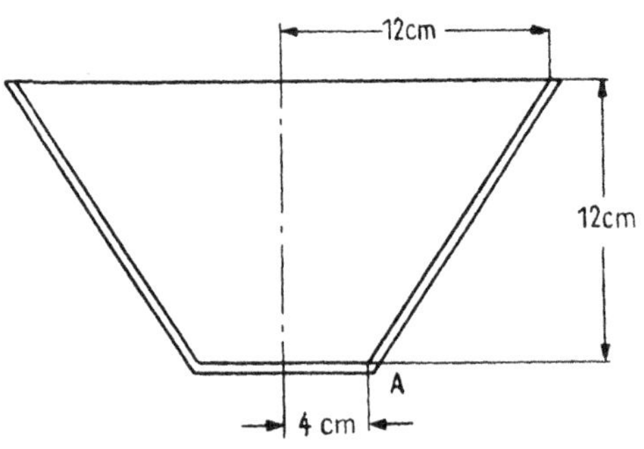

FIG. A.64.

about the horizontal axis of the hinge A, which is above the water level. Find the density δ of the rod material if, at equilibrium, $1/n$ of the rod is not submerged in the water. Neglect capillary forces.

165. Find the pressure force F of the water on the square wall of an aquarium (of side a). What is the height h from the bottom of the aquarium of the point of application of the resultant of the pressure on the wall?

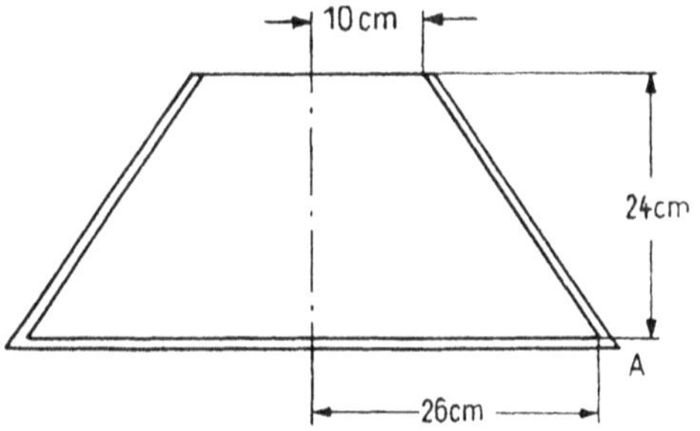

FIG. A.65.

166. A trough has the section shown in Fig. A.64 and is filled to the top with water. Find the pressure P over 1 m length of the side wall and the moment of this force M about the bottom edge A.

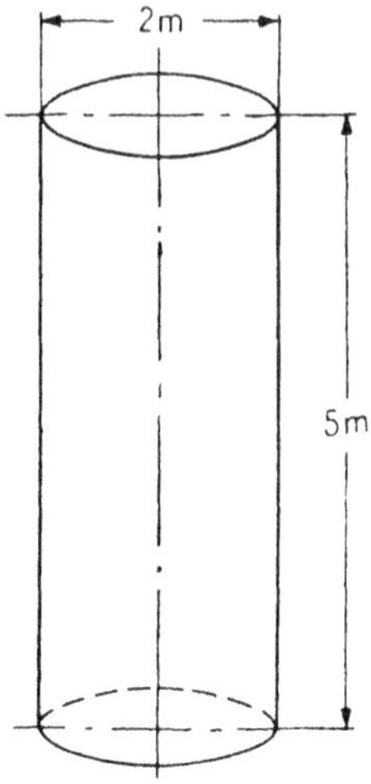

FIG. A.66.

167. Calculate the stress disrupting a vertical seam at a height $H = 1$ m in a cylindrical cistern filled to the top with water. The dimensions of the cistern are shown in Fig. A.66.

168. A vessel containing water stands on a truck. The truck moves in a horizontal direction with acceleration 0·29 g. What is the angle α between the water surface and the horizontal?

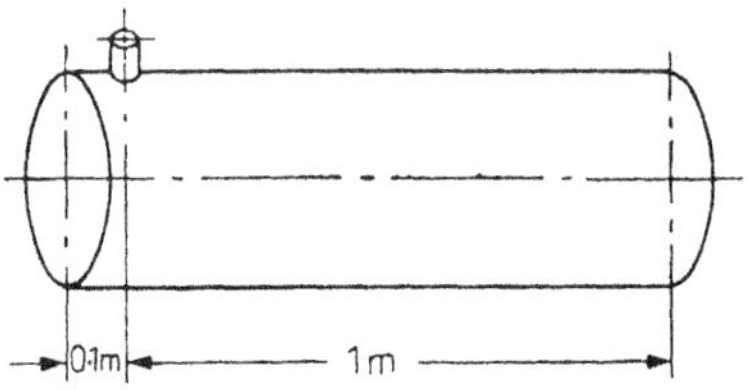

FIG. A.67.

169. A closed cylindrical vessel filled with water has a stopper at the top (Fig. A.67). What is the acceleration a in a horizontal direction along the axis of the vessel that must be communicated to the vessel in order for the stopper to come out, if it can support a pressure of 0·05 atm, and the distance from the opening to one end of the vessel is 1 m, and from the other end 0·1 m.

170. A goods train that includes an enclosed oil tanker moves with a speed v_0 then, as a result of braking, starts to decelerate uniformly and, after traversing a distance s, comes to a stop. Find the pil pressure forces on the back and front walls of the tanker during the period of braking. Assume that the tanker is a rectangular parallelepiped of length l, width a and height h; the oil density is ϱ.

171. A tank is filled with water and oil (density 0·9 g cm^{-3}). What is the initial speed v of flow of water from an opening in the bottom, if the height of the water layer is $h_1 = 1$ m, and of the oil layer $h_2 = 4$ m? (Neglect viscosity.)

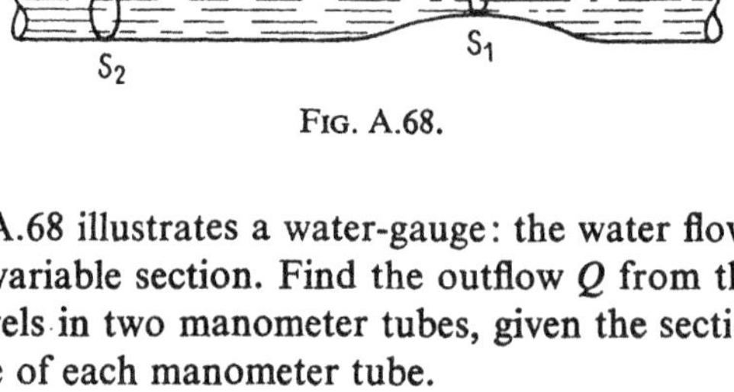

FIG. A.68.

172. Figure A.68 illustrates a water-gauge: the water flows along a horizontal tube of variable section. Find the outflow Q from the difference Δh of the water levels in two manometer tubes, given the sections of the main tube at the base of each manometer tube.

173. Figure A.69 illustrates an arrangement whereby a locomotive can fill up with water whilst travelling. A water-filled channel is arranged along the track. A bent tube is attached to the locomotive and drops into the channel in such a way that its mouth is directed forwards. Calculate the height h to which the water rises if the speed of the train is $v = 36$ km hr^{-1}. Neglect the viscosity of the water.

174. Find the dependence on time of the force F acting on the bottom, area S, of a cylindrical glass into which water is poured from a tea-pot, as illustrated in Fig. A.70. A constant amount Q cm^3 of water pours into the glass per second; ϱ is the density of the water.

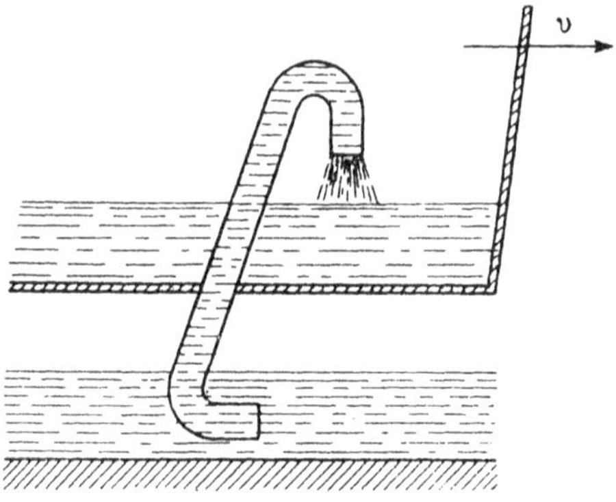

FIG. A.69.

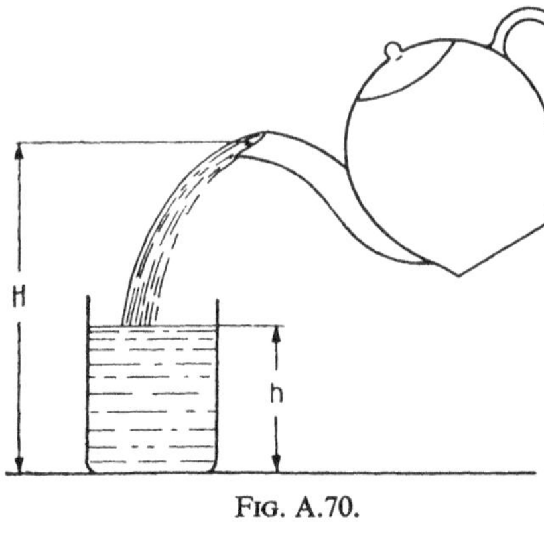

FIG. A.70.

175. A cylindrical vessel filled with water stands on a truck. The height of the water in the vessel is 1 m. On opposite sides of the vessel relative to the motion of the truck there are two taps, each with an opening area of 10 cm^2, one at a height $h_1 = 25$ cm above the bottom of the vessel, the other at a height $h_2 = 50$ cm. What horizontal force F must be applied to the truck in order for it to remain at rest when the taps are open?

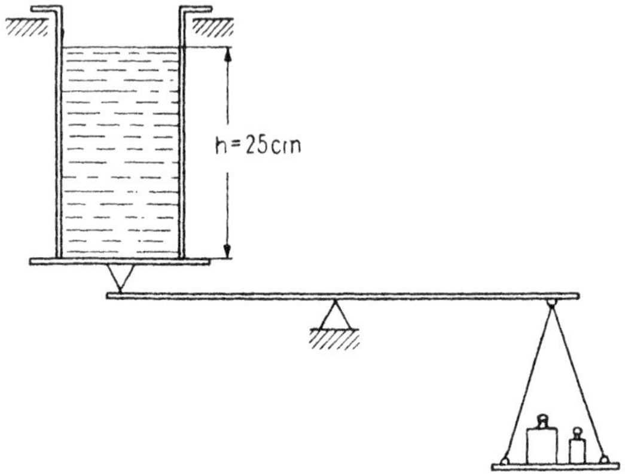

FIG. A.71.

176. Find approximately the maximum power and most advantageous speed of rotation of a water wheel (as illustrated in Fig. A.72), given that the head of water is $h = 5$ m, stream cross-section $S = 0{\cdot}06\ \text{m}^2$, wheel radius $R = 1{\cdot}5$ m. The stream strikes the blades continuously and falls downwards after impact.

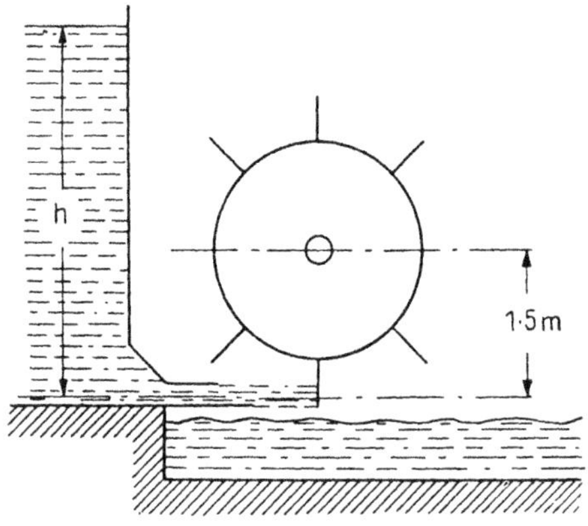

FIG. A.72.

177. A totally enclosed cylinder is filled with water and also contains: a cork, a small piece of lead and a body A whose density is equal to that of water. The cylinder rotates rapidly about its axis. How are the bodies disposed in the cylinder if the axis of rotation is vertical?

178. A cylinder filled with water rotates uniformly at 1 rev sec^{-1} about a vertical axis and carries the water along with it. A smooth horizontal rod, mounted along a diameter of the cylinder and submerged in the water, carries a cube of side 2 cm, made of a material with density $\varrho = 2\ \text{g cm}^{-3}$. The cube is mounted so that it can slide on the rod and is supported by a small spring at a distance of 50 cm from the cylinder axis. Find the spring tension T.

179. In a rotating vessel the pressure on the bottom (see Problem 405) is greater at the wall than at the centre. Why does the water not flow from the wall to the centre when the vessel rotates?

180. Figure A.73 illustrates a familiar experiment on the flow of a viscous fluid along a pipe, and shows the fall in pressure along the pipe. How can the fluid velocity be determined from the data indicated in the figure, if the fluid density is 1 g cm^{-3}.

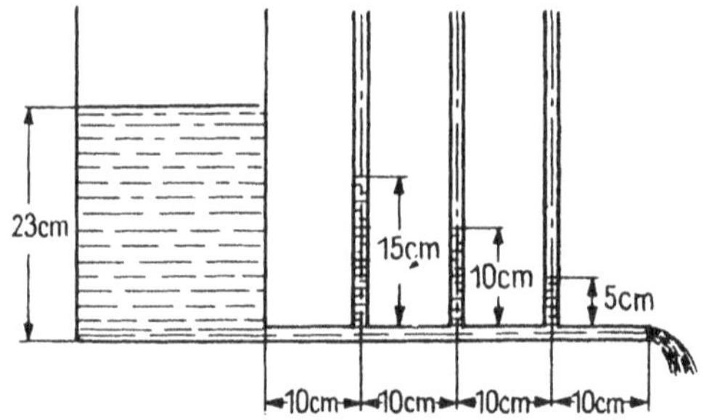

FIG. A.73.

181. A cylindrical vessel that stands on a support has two openings 25 cm apart. The jets from the openings intersect. Find the point of intersection, given that the water level is 26 cm above the upper opening.

182. The density of a gas may be determined as follows. A large glass bulb of capacity V is filled with the gas under test to a pressure H mm Hg and weighed. Let its weight be P. Part of the gas is now removed, so that the pressure falls to h mm Hg. The new weight of the bulb is p. What is the density of the gas at atmospheric pressure?

183. A gas-filled electric lamp contains nitrogen at a pressure of 600 mm Hg. The volume of the lamp is 500 cm^3. How much water will enter the lamp if its tip is broken off under water at standard atmospheric pressure?

184. Find the number n of cycles of the piston of an air pump required to pump a vessel of volume V from a pressure p_1 to a pressure p_2, if the volume corresponding to one cycle of the piston is v. Neglect the dead space.

185. A narrow cylindrical pipe closed at one end contains air which is separated from the outer atmosphere by a column of mercury. If the pipe is turned with the closed end uppermost, the air inside it occupies a length l; whereas when the open end is uppermost, the length occupied is $l' < l$. The length of the mercury cloumn is h. Find the atmospheric pressure.

186. A cylindrical pipette of length l is half submerged in mercury. It is closed by a finger and withdrawn. Part of the mercury flows out. What length of mercury column remains in the tube? The atmospheric pressure is H.

187. A McLeod gauge is required to read pressures up to 0·1 mm Hg. The volume of the manometer sphere must not exceed 150 cm^3, and the capillary length must not exceed 20 cm. What minimum cross-section of capillary is required?

188. In the manometer of the previous problem, how far from the end of the capillary is the marker corresponding to a pressure of 0·00005 mm Hg?

189. Find the density δ of sea water at a depth of 5 km, if the density at the surface of the ocean is $\delta_0 = 1{\cdot}03$ g cm^{-3}, and the compressibility of water within the pressure limits 1–500 atm is equal to $\beta = 47{\cdot}5 \times 10^{-6}$ atm^{-1}

190. The viscosity of water was found by measuring the amount of water flowing from a capillary of length $l = 10$ cm and diameter $d = 1$ mm under the action of the pressure difference produced by a column of water 50 cm high. $V = 217$ cm^3 water flowed out in a time $\tau = 3$ min. Find the viscosity η of water.

191. Find the coefficient σ of surface tension of a liquid if it rises to a height $h = 32{\cdot}6$ mm in a capillary of diameter $D = 1$ mm. The density of the liquid is $\delta = 1$ g cm^{-3}. The edge angle of the meniscus is zero.

192. The surface tension of a liquid can be determined by weighing a drop when it breaks from a capillary and measuring the neck of the drop at the instant of breaking off. It is found that 318 drops weigh 5 g, and $d = 0{\cdot}7$ mm. Find the surface tension.

193. A vertical glass capillary is attached to one arm of a balance and kept in equilibrium by weights on the other arm. What happens to the balance if a vessel containing water is carefully brought up below the capillary so that its end touches the water surface?

194. What is the capillary pressure p in a mercury droplet of diameter $d = 1\ \mu$ at 15°, if the surface tension of mercury at this temperature is $\sigma = 487$ dyne cm^{-1}?

195. There is a crack of width $a = 0{\cdot}02$ mm in the bottom of a vessel. To what height h can the vessel be filled with mercury before any flows out through the crack? The density of mercury is $\varrho = 13{\cdot}6$ g cm^{-1}. The surface tension (at 15°) is $\sigma = 487$ dyne cm^{-1}.

196. Gas bubbles of diameter $d_1 = 0{\cdot}05$ mm separate at the bottom of a pond of depth $h = 2$ m. What will be the diameter d_2 of the bubbles when they rise to the surface? $\sigma = 73$ dyne cm^{-1}.

197. A soap bubble of radius r, filled with air, is contained in a cylinder with a movable piston. The air pressure outside the bubble is at first equal to the atmospheric pressure p_0. On moving the piston slowly the bubble is compressed until its radius is halved. Find the pressure of the outer air in the cylinder at this instant.

198. Two vertical glass plates, partially submerged in liquid, together form a wedge with a very small angle α. Find the height h to which the liquid rises as a function of the distance x from the rib of the wedge.

199. A drop of water of mass $m = 0{\cdot}1$ g is introduced between two plane parallel glass plates, wetted with water, the edge angle being zero. What is the force F of attraction between the plates if they are spaced $d = 10^{-4}$ cm apart. The surface tension of water (at 18°) is $\sigma = 73$ dyne cm^{-1}.

200. Find the direction of the forces acting on two vertical parallel plates partially submerged in liquid in the following cases: (1) both plates are wetted by the liquid, (2) neither of the plates are wetted by the liquid, (3) one plate is wetted and the other not.

201. Find the surface tension σ of the liquid of a loop of rubber band of length l and cross-section S placed on a film of the liquid dilates to a circle of radius R after the film has been punctured inside the loop. Assume that Hooke's law holds for small extensions of the rubber and that Young's modulus for rubber is E.

202. Determine the angle of rotation φ of a disc suspended from an elastic thread when a second similar disc rotates at a distance $h = 1$ cm below it with angular velocity $\omega = 50$ radian sec^{-1}. The radius of each disc is $R = 10$ cm, the modulus of torsion of the thread is $f = 100$ dyne cm rad^{-1}, the coefficient of internal friction of air is $\eta = 1{\cdot}8 \times 10^{-4}$ g sec^{-1} cm^{-1}. Neglect edge effects. Regard the motion of the air between the discs as laminar.

CHAPTER VII

203. The linear coefficient of thermal expansion of steel is $12 \times 10^{-6}\,\mathrm{deg}^{-1}$, Young's modulus $E = 2 \times 10^{12}$ in c.g.s. units. What pressure p must be applied to the flat ends of a steel cylinder for its length to remain unchanged when the temperature rises 100°C?

204. When laying tramway rails they are made equal to one another at the joints. How great are the stresses p appearing in them for temperature variations from $t_2 = -25$°C in winter to $t_2'' = +30$°C in summer, if the laying is done at $t_1 = +15$°C? Young's modulus for iron $E = 2 \times 10^{-6}$ kg cm^{-2}, whilst the linear coefficient of heat expansion $\alpha = 1{\cdot}25 \times 10^{-5}\,\mathrm{deg}^{-1}$.

205. The water flowing from the nozzle of a fire hose has a sufficient speed for the jet to reach a height of 20 m (neglecting air friction). What is the temperature rise of the water if the jet is directed into a closed fixed tank? Assume that the heat developed only goes into heating the water.

206. A thermometer indicates a temperature t_0 in melting ice, and a temperature $-t_H$ in the steam from water boiling at a pressure of H mm Hg. Find an expression for the true temperature t at any intermediate reading t_t of the thermometer, on the assumption that the termometer tube is divided into equal volumes by the scale graduations. The boiling-point of water at a pressure of H mm Hg is T_H.

207. A thermometer in melting ice indicates $t_0 = -3{\cdot}0$°*, whilst the reading is $t_H = 101{\cdot}4$° in the steam from water boiling at a pressure of 760 mm Hg. What temperature does the thermometer register in the vapour of boiling methylated spirits (66·9°)?

208. Two completely identical thermometers are filled at 0° with equal volumes of mercury and toluol. Find the ratio of the length l of a division corresponding to 1° on the mercury thermometer scale to the length l_1 of a division on the toluol thermometer scale. The volumetric expansion coefficient of mercury is α, of toluol α_1, and the linear expansion coefficient of glass is λ.

209. Discuss the behaviour at different temperatures from 0° to 10° of a thermometer filled with water. At what temperatures will the reading of the thermometer be the same? Use the formula

$$V = 1 - 0{\cdot}00006105t + 0{\cdot}000007733t^2,$$

for the volume V of water as a function of the temperature t. Take the volume at 0° as unity.

210. A hydrogen thermometer with a constant volume V_0 undergoes a pressure change from p_0 mm Hg to p_1 mm Hg on heating from $t = 0°$ to t_1. Find the gas temperature t_1 if the volumetric expansion coefficient of the vessel is α and of hydrogen is α_p.

211. A thermometer immersed in water of mass $m = 6{\cdot}7$ g undergoes a temperature rise of $\Delta t = 14{\cdot}6°$ and indicates a temperature $t = 32{\cdot}4°$. What is the temperature $\mathbf{9}°$ of the water prior to the measurement. The water equivalent of the thermometer is $k = 0{\cdot}46$ cal deg^{-1}.

212. Let m g ice at 0° be mixed with m' g water at $t_1°$ in a calorimeter. Find the temperature t of the mixture if the latent heat of melting is λ, and all the ice melts.

213. What amount m of water at $t_0 = 20°$ can be frozen by vaporising $M = 100$ g of sulphuric ether, having a temperature $t_1 = 20°$, a specific heat $q = 0{\cdot}5$ cal g^{-1} deg^{-1}, and a heat of vaporization $r = 90$ cal g^{-1} (assuming that the heat of vaporization is taken entirely from the water)? Assume that the heat of vaporization of the ether is independent of the temperature.

214. The specific heat of zinc is determined by heating a piece of it weighting $m_2 = 235{\cdot}6$ g to $t_2 = 99{\cdot}3°$ and dropping it in a brass calorimeter. The specific heat of brass is $q_1 = 0{\cdot}093$ calg^{-1} deg^{-1}, the mass of the calorimeter and mixer is $m_1 = 100$ g, the mass of water is $m = 209{\cdot}3$ g; the initial temperature of the calorimeter and water is $t_0 = 20{\cdot}5°$. The temperature of the water in the calorimeter rose to $t = 27{\cdot}6°$. Find the specific heat of zinc.

215. Given the necessary precautions, water can be cooled to a temperature $t = -10°$. What mass of ice m is formed from $M = 1$ kg of such water when a particle of ice is thrown into it and freezing thereby produced? Regard the specific heat of the supercooled water as independent of the temperature and equal to the specific heat of ordinary water.

216. When 1 g hydrogen is burned and converted into water, 34,000 cal of heat are produced. How many grams of coal must be burned in order to dissociate 1 g water, if 50 per cent of the heat released by the coal is used? The calorific value of coal is 7000 cal g^{-1}.

217. To determine the volumetric expansion coefficient α of glass, a small glass bulb with elongated ends is weighed, first when empty, then with the amounts of mercury needed to fill it completely at two temperatures, say 0° and t. Let p_0 and p_1 be the weights of mercury at the two temperatures. Find the volumetric expansion coefficient α of glass if the volumetric expansion coefficient of mercury is α_1.

218. A glass bulb with volumetric expansion coefficient α is weighed in liquid at temperatures t and t_1. The weights of the displaced liquids are p

and p_1 respectively. Find the volumetric expansion coefficient α_1 of the liquid in the temperature interval from t to t_1.

219. A picnometer at a temperature t is filled with alcohol, the weight of which is p. (he picnometer plus alcohol is then heated to a temperature t_1 and the excess of alcohol above the previous level is removed by filter paper. The weight of the alcohol is then p_1. Find the mean volumetric expansion coefficient α of alcohol, if the linear expansion coefficient of glass is λ.

220. A barometer has a brass scale. At a temperature $t_1 = 27°$ the height of the mercury column, measured on the scale, is $H_1 = 751{\cdot}3$ mm. Find the height H_0 of the column at $t_0 = 0°$. The coefficient of linear expansion of brass is $\lambda = 0{\cdot}000019$ deg^{-1} and the volumetric expansion coefficient of mercury is $\alpha = 0{\cdot}000182$ deg^{-1}.

221. The linear expansion coefficient λ of German silver is determined with the aid of a lever device (Fig. A.74). The length of the rod A at $t_0 = 0°$ is $l_0 = 23{\cdot}02$ cm. On heating to $t_1 = 99{\cdot}3°$ the arrow of the second lever deflects through the angle $\varphi = 9°30'$. The short arm DE of the second lever is $DE = a = 2{\cdot}5$ cm. The ratio of the arms of the first lever $BC : BD = 1 : 10$. Find λ.

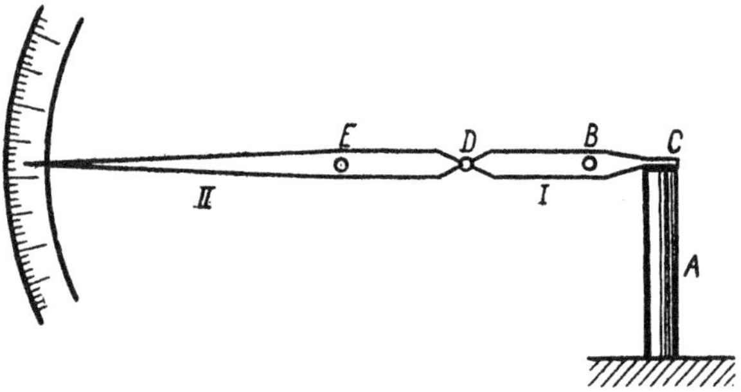

FIG. A.74.

222. The wheel of a locomotive has a radius $r_0 = 1$ m at $t_0 = 0°$. Find the difference in the number of revolutions of the wheel in summer at $t_1 = 25°$ and winter at $t_2 = -25°$ over a journey of length $l = 100$ km. The linear expansion coefficient of the material of the wheel is $\beta = 0{\cdot}000012$ deg^{-1}.

223. A piece of ice floats in a vessel filled with water. Will the water level change when the ice melts, if the final temperature of the water remains 0°?

224. What force P must applied to a steel rod of 1 cm^2 cross-section, in order to stretch it as much as it would be stretched when heated through 1°. The linear expansion coefficient is $\lambda = 12 \times 10^{-6}$ deg^{-1}. Young's modulus $E = 2{\cdot}1 \times 10^6$ kg cm^{-2}.

225. A steel band is driven on to a wagon wheel at $t_1 = 300°$. Find the extension force P on the band at $t_0 = 20°$ of the band cross-section is $S = 20\ \text{cm}^2$. Young's modulus is $2{\cdot}1 \times 10^6\ \text{kg cm}^{-2}$ and the linear expansion coefficient is $\lambda = 12 \times 10^{-6}\ \text{deg}^{-1}$.

226. A badly dried flask at $t = 20°$ contains a mixture of air and water vapour, the partial pressure of which are 0·25 and 0·1 mm Hg respectively. Find the error in the reading of a McLeod gauge attached to the flask for measuring pressure, if the volume of the manometer flask is $V = 50\ \text{cm}^3$, and the capillary radius is $r = 1$ mm. The pressure of water vapor at 20° is 17·5 mm Hg.

227. What volume is occupied by a gram-molecule of ideal gas at a pressure of 3 atm and temperature $T = 400°\text{K}$?

228. An aerostat of volume $V\ \text{m}^3$ was filled with hydrogen at a temperature $t_1 = 15°$. Its temperature rose to $t_2 = 37°$ due to solar radiation at a fixed atmospheric pressure, and the excess gas escaped via an appendix, as a result of which the weight of the aerostat plus gas fell by $Q = 6{\cdot}05$ kg. The specific weight of hydrogen is $\delta_0 = 0{\cdot}000089\ \text{g cm}^{-3}$. Find the volume V of the aerostat.

229. The action of acid on a certain quantity of marble ($CaCO_3$) produces $V = 1320\ \text{cm}^3$ of carbon dioxide (CO_2) at a temperature $t_1 = 22°$ and pressure $p = 1000$ mm Hg. Find the mass Q of the marble entering into the reaction. The density of carbon dioxide at 0° and 760 mm Hg is $\varrho_0 = 0{\cdot}001977\ \text{g cm}^{-3}$.

230. A factory chimney of height $l = 50$ m carries off smoke at a temperature $t_1 = 60°$. Find the static pressure p producing the draught in the chimney. The air temperature is $t_2 = -10°$. The specific weight of air is $d_0 = 1{\cdot}29 \times 10^{-3}\ \text{g cm}^{-3}$.

231. Nitrogen at $T = 300°\text{K}$ is pumped into a thin-walled spherical bulb of weight $P = 1$ kg. Find the maximum amount of nitrogen which can be delivered to the vessel if the permissible stress in the walls is $\sigma = 50\ \text{kg mm}^{-2}$.

232. Two vessels A and B containing air are connected by a capillary with a tap. Vessel A is immersed in a water bath at a temperature $t_1 = 100°$ and vessel B in a refrigerating mixture at $t_2 = -20°$. The vessels were originally disconnected by closing the tap, and the air pressures in A and B were respectively $p_1 = 400$ mm and $p_2 = 150$ mm Hg. Find the pressure established after opening the tap, if the volume of A is $V_1 = 250\ \text{cm}^3$, and of B is $V_2 = 400\ \text{cm}^3$.

233. Express in c.g.s. units the heat Q expended in heating a lead sphere of mass $m = 100$ g through $(t_1 - t_0) = 10°$. The specific heat of lead is $c = 0{\cdot}0307\ \text{cal g}^{-1}\ \text{deg}^{-1}$.

234. Two thermally insulated vessels of volumes $V_1 = 1$ litre and $V_2 = 3$ litres are connected by a pipe with a tap. Prior to opening the tap, the first vessel contains nitrogen at $t_1 = 0°$ and pressure $p_1 = 0{\cdot}5$ atm, and the second vessel argon $t_2 = 100°$ and pressure $p_2 = 1{\cdot}5$ atm. Find the pressure and temperature established in the gas mixture when the tap is opened.

235. What is the change in the level difference $h_1 - h_2$ of water in two communicating capillaries of diameters $d_1 = 0{\cdot}1$ mm and $d_2 = 0{\cdot}3$ mm on heating from 20° to 70°, if the surface tension of water at these respective temperatures are 73 and 64 dyne cm^{-1}.

236. How many molecules of nitrogen does a vessel of 1 litre capacity contain, if the temperature of the nitrogen is 27° and the pressure is 10^{-6} mm Hg?

237. What is the pressure of the gas mixture in a 2·5-litre flask, if it contains 10^{15} molecules nitrogen and $3{\cdot}3 \times 10^{-7}$ gargon? The temperature of the mixture is 150°.

238. Calculate the specific heat at constant pressure of a gas of the following molar composition:

$$20\%\ \text{He} - 30\%\ H_2 - 50\%\ CH_4$$

(The molar composition indicates the relative numbers of moles of the given components to the total number of moles in the mixture.)

239. What is the change in $\gamma = c_p/c_v$ for nitrogen if, under the action of some reagent, a certain fraction of the molecules dissociates into atoms? The ratio of the number of dissociated molecules to the total number of molecules is α.

CHAPTER VIII

240. A steel rod of length $l = 20$ cm and cross-sectional area $S = 3$ cm^3 is heated at one end to $t_1 = 300°$ whilst the other end rests in ice. Assuming that heat transmission occurs exclusively through the rod (without losses from the walls), calculate the mass m of the ice melting in time $\tau = 10$ min. The thermal conductivity of steel is $k = 0{\cdot}16$ cal deg^{-1} sec^{-1} cm^{-1}.

241. A copper percolator is heated on a primus. The water is brought to the boil and releases $m = 2$ g steam per minute. The thickness of the percolator base is $l = 2$ mm, and its area $S = 300$ cm^2. Find the temperature difference $t_2 - t_1$ between the inner and outer surfaces of the base, on the assumption that the base is heated uniformly. The thermal conductivity of copper is $k = 0{\cdot}92$ cal deg^{-1} sec^{-1} cm^{-1}.

242. Solve the previous problem when the base is covered on the inside by a layer of scale $l_1 = 1$ mm thick. The thermal conductivity of the scale is $k_1 = 0{\cdot}003$ cal deg^{-1} sec^{-1} cm^{-1}.

243. Three plates of the same dimensions are combined to form a small pile. The centre plate is lead, and the outer plates silver. The outer surface of one silver plate is held at a constant temperature $t = 100°$. The outer surface of the other silver plate is at $t_3 = 0°$. Find the temperatures t_1 and t_2 at the points of contact of the lead with the silver plates. The thermal conductivity of lead is $k_1 = 30$ cal deg^{-1} hr^{-1} m^{-1}, and of silver $k = 360$ cal deg^{-3} hr^{-1} m^{-1}.

244. Find the amount of heat Q lost per m^2 of wall during a time τ equal to 24 hr when the air temperature in a room is $t_1 = 20°$ and the outdoor temperature is $t_4 = -10°$. The wall thickness is $l = 20$ cm. The thermal conductivity of the wall material is $k = 0{\cdot}003$ cal deg^{-1} sec^{-1} cm^{-1}. The external thermal conductivity at the wall-air boundary is $\alpha = 0{\cdot}0002$ cal deg^{-1} sec^{-1} cm^{-2}. Find also the temperature t_2 of the inner and t_3 of the outer surface of the wall.

245. How much coal has to be burnt in 24 hr in a central heating system if the surface area of the walls and roof of the house is $S = 10{,}000$ m^2, in order to maintain a temperature $t_1 = 18°$ in the rooms, when the outdoor temperature is $t_2 = -22°$? The wall thickness $L = 60$ cm, the internal thermal conductivity of the wall material is $k = 0{\cdot}002$ cal deg^{-1} sec^{-1} cm^{-1}, and the heat leakage per unit area of the roof surface is the same as per unit wall area. The external thermal conductivity at the air–wall boundary is

$\alpha = 0{\cdot}00025$ cal deg^{-1} sec^{-1} cm^{-2}. the calorific value of coal is $r = 7500$ cal g^{-1}

246. Two vessels, filled with liquids at temperatures t_1 and t_2, are joined by a metal rod of length L, cross-section S and thermal conductivity k. The masses and specific heats of the liquids are m_1, m_2 and c_1, c_2 respectively. The vessels and rod are thermally insulated from the surrounding medium. What is the time τ required for the temperature difference to be halved?

247. Liquid with a temperature t_1 is poured into a thin-walled closed metallic vessel. The air temperature outside the vessel is t_3. Find the temperature t_2 of the outer wall, given that the internal thermal conductivity of the metal is k, the external thermal conductivity at the metal–air boundary is α, and at the metal–liquid boundary ∞. The wall thickness is L.

Note. A vessel is regarded as thin-walled when the wall thickness is small compared with its linear dimensions.

248. Find the temperature t_2 in the previous problem in two limiting cases: (1) a very thin metallic vessel, and (2) a vessel made of material of very small thermal conductivity.

249. Liquid with a temperature t_1 is poured into a thin-walled closed metallic vessel with a total surface S. What time τ does it take the liquid to cool to a temperature t_2 if its mass is m, its specific heat c, the air temperature outside the vessel t_3, and the external thermal conductivity at the metal–air boundary α?

250. The water in a pond has a temperature $t_1 = 0°$. The temperature of the surrounding air is $t_2 = -10°$. What thickness x of ice forms after 24 hr, measured from the instant when the water freezes. The thermal conductivity of ice is

$$k = 0{\cdot}0053 \text{ cal deg}^{-1} \text{ sec}^{-1} \text{ cm}^{-1},$$

the latent heat of freezing of water is $\lambda = 80$ cal g^{-1}, the density of ice is $\rho = 0{\cdot}9$ g cm^{-3}.

251. The thermal conductivities of liquids are found by using three copper plates, arranged horizontally, one above the other. The lower plate is washed by a flow of cold water (temperature t_1) the upper plate by warm water (temperature t_3). The space between the lower and the centre is filled with liquid with thermal conductivity k_1, and the space between the centre and upper plates by liquid with thermal conductivity k_2. The distance of the centre plate from the lower is d_1, and from the upper is d_2. Express the thermal conductivity k_2 in terms of k_1, if the steady-state temperature of the centre plate is t_2. In the case when the known liquid is water ($k_1 = 0{\cdot}00143$ cal deg^{-1} sec^{-1} cm^{-1}), and the unknown liquid is benzol, the distances are $d_1 = 1$ mm, $d_2 = 1{\cdot}2$ mm, the temperatures are $t_1 = 80°$, $t_2 = 68{\cdot}6°$, $t_3 = 10°$. Find S_2 for benzol.

252. A rod of cross-section S is held with its ends in solid plates, the distance L between which is maintained constant. The temperature of one plate is then raised, and a constant heat flux Q is established in the rod. What is the pressure p exerted per unit cross-section of the rod if the initial stress in the rod was zero? The thermal conductivity of the rod is k, the linear expansion coefficient is α, Young's modulus is E.

253. Write the van der Waals equation for a gas containing ν moles.

254. The critical temperature of carbon dioxide (CO_2) is 31°C, and the critical pressure 73 atm. Find the critical volume V_{cr} of a gram-molecule of carbon dioxide.

255. Find the constants of the van der Waals equation for nitrogen, if t_{cr} for nitrogen is $-146°$ and $p_{cr} = 33$ atm.

256. A. G. Stoletov (1892) first showed that, in order for a liquid contained in a given volume to be reducible to the critical state, a precise amount of it must be taken. Let us take the following example. A vessel whose volume is $V_1 = 15\ \text{cm}^3$ is to be filled with water at $t_1 = 18°$ in such a way that, when it is heated to the critical temperature (after first pumping out and sealing the vessel), the critical pressure is established in the vessel. On the assumption that water obeys the van der Waals equation of state, find the volume of water must be poured into the vessel, given that the critical temperature of water is $t_{cr} = 374°$, the critical pressure is $p_{cr} = 205{\cdot}5$ atm, the molecular weight is $\mu = 18$, and the density at 18° is $\varrho = 1\ \text{g cm}^{-3}$.

257. Express the critical magnitudes in terms of the constants of the equation of state proposed by Berthelot for describing the behaviour or real gases:

$$\left(p + \frac{a}{Tv^2}\right)(v - b) = RT.$$

258. Find the connection between the pressure, temperature and volume of a gram-molecule at the critical point on the assumption that the substance obeys the Berthelot equation of state (see the previous problem).

259. Regarding the latent heat of vaporisation λ as the work done in overcoming the internal pressure π, find the relationship between π, λ and the density ϱ of a liquid. Assume that the liquid obeys the van der Waals equation.

260. Show that the resultant of all the pressure forces of an ideal gas on the walls of a vertical closed cylindrical vessel containing the gas is equal to the weight of the gas, whatever the length of the cylinder, when and only when the gas density obeys the barometric formula

$$\varrho = \varrho_0 \, e^{-\frac{\mu}{RT} gx},$$

where μ is the molecular weight of the gas, R is the gas constant, T is the temperature (°K) and g is the acceleration due to gravity. The gas temperature is assumed the same throughout the cylinder, and the gas is in equilibrium.

261. Use kinetic theory to show that, when a piston is displaced quasi-statically in a cylinder filled with an ideal monatomic gas, the pressure and volume of the gas are connected by

$$pV^{5/3} = \text{const.}$$

The cylinder walls and piston are impermeable to heat.

Hint. Consider the collision of a molecule on the moving piston and use the fact that the molecule speed is much greater than the piston speed.

262. Solve the previous problem for a diatomic gas. Show that in this case the pressure p and volume V are connected by

$$pV^{7/5} = \text{const.}$$

Hint. Use the theorem on the uniform distribution of the kinetic energy over the various degrees of freedom.

263. Find the most probable (v_m), the mean ($\bar{v}$) and the root mean square ($\sqrt{\bar{v}^2}$) velocities of chlorine molecules at 227°.

264. At what temperature is the root mean square velocity of hydrogen molecules equal to the root mean square velocity of nitrogen molecules at 100°?

265. The mean free path for hydrogen at atmospheric pressure is $\Lambda = 1{\cdot}28 \times 10^{-5}$ cm. Find the kinetic diameter d of a hydrogen molecule.

266. What is the number z of collisions per second of neon molecules at 600°K and 1 mm Hg if the kinetic diameter of a neon molecule is $d = 2{\cdot}04 \times 10^{-8}$ cm?

267. Find the pressure p of hydrogen in a round litre flask, in which the mean free path of a molecule is greater than the dimensions of the vessel. The kinetic diameter of hydrogen is $2{\cdot}2 \times 10^{-8}$ cm, and its temperature 300°K.

268. An ideal gas is heated at constant pressure. What is the variation with temperature of the mean free path Λ and the number z of collisions of its molecules per second?

269. An ideal gas is compressed isothermically. Find Λ and z as functions of the pressure.

270. Find the kinetic energy E of the molecules of 5 litres hydrogen at a pressure of 3 kg cm^{-2}.

271. A mixture of equal weights of hydrogen and helium at 0° is placed in a cylindrical vessel of 1 m^3 volume, closed at the top by a weightless piston which moves without friction. The atmospheric pressure is 740 mm Hg. What is the quantity of heat required to heat the mixture to 200°?

272. Perrin determined Avogadro's number by finding the height distribution of spherical gamboge particles suspended in water. He found that the ratio α of the number of particles in layers spaced at a distance $l = 30\mu$ apart was equal to 2·08. The density of the particles was $\varrho = 1{\cdot}194$ g cm^{-3}. The density of water is $\varrho_0 = 1$ g cm^{-3}. The particle radius was $r = 0{\cdot}212\mu$. Calculate Avogadro's number N on the basis of these data. The water temperature $t = 18°$.

273. The molecular weights of colloidal particles may be determined by studying the distribution of their concentration in the field of the centrifugal force produced by a centrifuge. Find the molecular weight μ of the colloidal particles, given that the ratio of their concentrations at points distances r_2, r_1 from the axis of the centrifuge is α. The density of the particles is ϱ, the density of the solvent ϱ_0. The angular velocity of the centrifuge is ω.

CHAPTER X

274. What must be the speed of a lead bullet if it melts when striking a steel slab? The temperature of the bullet is $t_0 = 27°$, its melting-point is $t_1 = 327°$, its heat of melting $q = 5$ cal g^{-1}, and its specific heat $c = 0{\cdot}03$ cal g^{-1} deg^{-1}.

275. Two gram-molecules of nitrogen at temperature t_0 and pressure p atm are compressed isothermically to a pressure p_1 atm. Find the amount of heat Q released by the gas to the thermostat.

276. In an apparatus designed for measuring the mechanical equivalent of heat, a mixer furnished with scoops rotates in a calorimeter filled with water. The friction of the water on the walls of the calorimeter produces a force which tends to rotate the calorimeter in the direction of rotation of the mixer. The calorimeter rotation is prevented by a load of weight P, attached to a string which wraps round a pulley and fastens to the calorimeter cover on a wheel of radius R. The moment of the load is constant in this device and has an opposite sign to the moment of the friction forces acting on the calorimeter walls from the liquid. When the mixer rotates the load neither rises nor falls. Find the mechanical equivalent E of heat, if the temperature rises by Δt when the mixer has performed N revolutions. The mass of the water in the calorimeter is M, and w is its water equivalent.

277. Find the mechanical equivalent of heat, given that $c_p = 0{\cdot}237$ cal g^{-1} deg^{-1} for air and $\gamma = c_p/c_v = 1{\cdot}41$. The molecular weight of air is $\mu = 28{\cdot}84$.

278. (1) Does an ideal gas get hotter or colder when it expands according to the law $pV^2 =$ const. (2) What is its molar specific heat during this process?

279. 97,000 cal are released when 12 g solid carbon burns to form carbon dioxide (CO_2), and 68,000 cal released when 28 g carbon monoxide (CO) are burned. What amount of heat would be released if 12 g carbon were burned in such a way as to produce pure carbon monoxide?

280. The combustion of hydrogen at 100° to form 1 mole of water vapour releases 58,000 cal. When 1 mole of steam at 100° is condensed, 9700 cal are released. Find the heat of formation of 1 mole of liquid water from hydrogen and oxygen at the same temperature.

281. A heater is switched on in a room for a certain time. The air temperature in the room rises from T_1 to T_2, whilst the pressure remains the same and equal to the outside pressure. Regarding air as an ideal gas, find the amount of heat that goes into increasing the internal energy of the air in the room.

282. What is the amount of heat Q consumend in heating 1 m^3 of air from 0° to 1° at constant volume and an initial pressure $p = 760$ mm Hg? The density of air in normal conditions is $\varrho_0 = 0{\cdot}00129$ g cm^{-3}; $c = 0{\cdot}237$ cal g^{-1} deg^{-1}; $\gamma = c_p/c_v = 1{\cdot}41$.

283. Solve the previous problem on the assumption that the air is heated from 91° to 92°, instead of from 0° to 1°.

284. A cylindrical vessel, closed at the top by a frictionless weightless piston, contains 1 m^3 hydrogen at 0°. The atmospheric pressure is 730 mm Hg. What is the amount Q of heat that must be consumed to heat the hydrogen to 300°?

285. Find the increase ΔU in the internal energy of helium when it expands isobarically from a volume of 5 liters to a volume of 10 liters. The process occurs at a pressure of 2 kg cm^{-2}.

286. A mass $m = 4{\cdot}032$ g hydrogen is mixed with $m_1 = 32$ g oxygen. Their specific heats are $c_p = 3{\cdot}50$ cal g^{-1} deg^{-1} and $c_{1p} = 0{\cdot}218$ cal g^{-1} deg^{-1} respectively. Find the decrease in the internal energy ΔU o fhte mixture on cooling it through $t = 20°$ at constant volume. For both gases, $\gamma = 1{\cdot}40$.

287. A volume V_0 at a temperature $t = 0°$ contains m gram-molecules hydrogen and $\frac{1}{2}m$ gram-molecules oxygen. What is the maximum steam pressure p_t obtained when the mixture is exploded, if the molar specific heat of steam is c, and the gram-molecular heat of formation of water from oxygen and hydrogen is C?

288. A closed vessel contains 1 mole (18 g) water at 0°. How much heat is required to raise the temperature of the system to 100° and at the same time convert all the water to saturated steam? The latent heat of vaporization of water at 100° and constant pressure is 539 cal g^{-1}. Neglect the vapor pressure of saturated steam at 0° and the heat capacity of the walls of the vessel. Neglect also the volume of the water compared with the volume of the saturated steam.

289. The speed of sound in air at 0° is 332 m sec^{-1}. Find the speed of sound in hydrogen at the same temperature. The molecular weight of air is $M = 28{\cdot}8$.

290. Find $\gamma = c_p/c_v$, if the speed of sound in air at 0° and normal pressure $H = 76$ cm Hg is $v = 332$ m sec^{-1} and the density of air is $\varrho = 0{\cdot}001292$ g cm^{-3}.

291. Find the ratio of the speeds of propagation of sound in hydrogen (v) and in carbon dioxide (v_1) at the same temperature. For the first, $\gamma = c_p/c_v = 1{\cdot}4$, for the second $\gamma_1 = c_p/c_v = 1{\cdot}3$. The density of hydrogen is $\varrho = 0{\cdot}0000899$ g cm^{-3}, of carbon dioxide $\varrho_1 = 0{\cdot}001977$ g cm^{-3} (in normal conditions).

292. What is the ratio $\gamma = c_p/c_v$ for argon, if $Q = 254$ cal are required to heat $m = 1$ kg of it through $(t_1 - t_0) = 2°$ at constant pressure $H = 760$ mm Hg, and $Q = 486$ cal heat are released on cooling argon from $t_2 = 100°$ to $t = 0°$ at constant volume $V = 5$l. The initial pressure of the cooling argon is $H_1 = 10$ atm.

293. For argon, $\gamma = c_p/c_v = 1{\cdot}7$. Find the pressure after adiabatic expansion of this gas from a volume $V_1 = 1$ l. to $V_2 = 2$ l. if the initial pressure is $p_1 = 1$ atm.

294. The following method is sometimes used for finding $\gamma = c_p/c_v$. A definite amount of gas, the initial temperature, volume and pressure of which are t_0, V_0 and p_0 respectively, is heated by a platinum wire through which an electrical current passes for a definite time: firstly at constant volume, so that the gas reaches a temperature t_1 and a pressure p_1 (volume V_0), then secondly at constant pressure, so that the temperature becomes t_2 and the volume V_1 (pressure p_0). Show that

$$\gamma = \frac{(p_1 - p_0)V_0}{(V_1 - V_0)p_0}.$$

295. In the Clement–Desormes method for determining c_p/c_v (Fig. A.75), gas is pumped via the tube B into the vessel A so that the pressure p_1 in it is somewhat above atmospheric. The tap C is then suddenly opened. The gas expands adiabatically to atmospheric pressure p_0. After a certain time, when the gas in the vessel again takes room temperature, its pressure becomes p_2. Find an expression for $\gamma = c_p/c_v$ on the basis of these data.

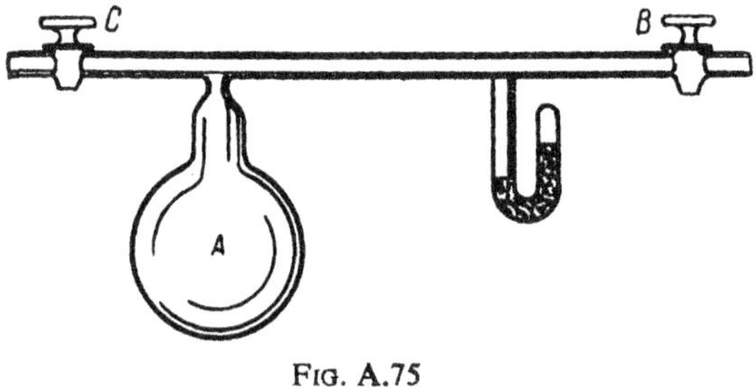

FIG. A.75

296. A device occasionally used for obtaining gases at very high temperatures and pressures consists of a cylinder barrel closed at one end and a piston bullet which enters the barrel at the open end. If the bullet and barrel are provided with a good finish the escape of gas through the gap between them can be made small. Due to the very high temperatures, the strongly compressed gases can be regarded as ideal in these conditions. Estimate the upper limit of the temperature T, the pressure p and the density ϱ of argon, subjected to compression in this device, if a bullet of mass $m = 100$ g enters the barrel, which has a volume $V = 200\ \text{cm}^3$, with initial velocity $v = 250\ \text{m sec}^{-1}$. The initial temperature and pressure are $T_0 = 300°\text{K}$ and $p_0 = 1$ atm respectively.

297. The specific heat of cobalt is $c_1 = 0{\cdot}104\ \text{cal g}^{-1}\ \text{deg}^{-1}$, and of gold $c_2 = 0{\cdot}0312\ \text{cal g}^{-1}\ \text{deg}^{-1}$. Find their atomic specific heats.

298. Find the efficiency of the cycle consisting of two isotherms and two isobars, assuming that the operative substance is an ideal gas.

299. The idea of dynamical heating, devised in 1852 by W. Thomson (Lord Kelvin), is as follows. Fuel is burned in the furnace of a heat engine, which operates a refrigerating machine. The refrigerating machine takes its heat from a natural water reservoir (say from subsoil water) and delivers it to the heating system. Simultaneously the water in the heating system is used to cool the heat engine. Find the theoretical amount of heat (i.e. discounting losses) which the heated location obtains from burning 1 kg coal, using the following data: the calorific value of coal is $q = 800\ \text{cal kg}^{-1}$, the temperature in the steam engine boiler is $t_1 = 210°$; the temperature of the water in the heating system is $t_2 = 60°$; the temperature of the subsoil water is $t_3 = 15°$.

300. Two bulbs with volumes $V_1 = V_2 = V = 1$ litre are connected by a pipe with a tap. The first bulb (volume V_1) contains air at atmospheric pressure, and the second is pumped out to ideal vacuum. Assuming that air obeys the van der Waals equation, and that the bulb walls and pipe are adiabatic, find the change in the gas temperature on opening the tap. The initial temperature is $T = 290°\text{K}$, and for air, $a = 1 \cdot 35 \times 10^6\ \text{atm cm}^6\ \text{mol}^{-2}$.

301. How much heat must be applied to one mole of a van der Waals gas in order for its temperature to remain unchanged when it expands *in vacuo* from a volume V_1 to a volume V_2?

302. Two bodies A and B, heated to different temperatures, are located in a rigid adiabatic envelope and are placed in thermal contact with one another. As a result, heat is transferred from the hotter to the cooler body until their temperatures are the same. Show that the entropy of the system $A + B$ is increased in this process.

303. Temperature and entropy can be taken as the fundamental variables characterizing the state of a body. Draw a graph of the Carnot cycle, taking entropy on the axis of abscissae, and temperature on the axis of ordinates. Use this graph to calculate the efficiency of the cycle.

304. Find the entropy change ΔS of 30 g ice when it becomes vapor, if the initial temperature of the ice is $-40°$ and the temperature of the vapor is 100°. Regard the specific heats of water and ice as constant, and all the processes as occurring at atmospheric pressure. The specific heat of ice is $c = 0{\cdot}5 \text{ cal g}^{-1} \text{ deg}^{-1}$.

305. Find the total entropy change ΔS (water plus iron) when 100 g iron, heated to 300°, ist submerged in water at 15°. The specific heat of iron is $0{\cdot}11 \text{ calg}^{-1} \text{ deg}^{-1}$.

306. Find the entropy rise ΔS of an ideal gas of mass M occupying a volume V_1 when it expands *in vacuo* to a volume V_2 (Gay-Lussac process).

307. Find the entropy change ΔS of 5 g hydrogen, isothermally expanded from a volume of 10 to 25 liters.

308. Two vessels of the same volume contain different ideal gases. The mass of gas in the first vessel is M_1, in the second M_2, whilst the pressure and temperatures of the gases are the same. The vessels are connected to one another, and a diffusion process commences. Find the total change ΔS in the entropy of the system, if the molecular weight of the first gas is μ_1 and of the second μ_2.

309. A heat insulated cylindric vessels is separated into two parts by a piston of negligibly small mass. One side of the piston is occupied by an ideal gas of mass M, molecular weight μ and molar specific heats c_v and c_p, which are temperature-independent, whilst the other side is occupied by a high vacuum. The initial temperature and pressure of the gas are T_0 and p_0. The piston is released and moves freely, so that the gas can occupy the whole of the volume of the cylinder. The piston pressure is then gradually increased, so that the gas volume slowly returns to its initial value. Find the changes in the internal energy and entropy of the gas during the process.

310. Find an expression for the entropy of a mole of van der Waals' gas, the specific heat c_v of which is independent of temperature. Obtain the equation of the adiabat for such a gas.

311. Find the maximum speed attainable by a gas when it escapes adiabatically from a flask, if the absolute temperature of the gas in the flask is T.

312. What is the pressure dependence of the mean velocity of the molecules of an ideal monatomic gas when it contracts or expands adiabatically?

CHAPTER XI

313. The volumetric expansion coefficient α of mercury at 0° and atmospheric pressure is $0{\cdot}00018\ \text{deg}^{-1}$. The compression coefficient is $\varkappa_T = 0{\cdot}0000029\ \text{atm}^{-1}$. Calculate the thermal coefficient of pressure of mercury.

314. What must be the increase in the outside pressure in order to maintain a constant volume of mercury when it is heated from 0° to 10°? (See the previous problem.)

315. Show that the adiabatic and isothermal compressibilities of a physically homogeneous isotropic medium are connected by

$$\frac{1}{V}\left(\frac{\partial V}{\partial p}\right)_{\text{ad}} = \frac{1}{\gamma}\,\frac{1}{V}\left(\frac{\partial V}{\partial p}\right)_T,$$

where $\gamma = c_p/c_v$. Show that this relationship is a consequence of the first law of thermodynamics and the functional relationship between p, V and T (equation of state) only.

216. Find $c_p - c_v$ for a van der Waals gas.

317. The specific heat of mercury at standard atmospheric pressure p_0 and temperature $T = 273°\text{K}$ is $c_p = 0{\cdot}0333\ \text{cal g}^{-1}\ \text{deg}^{-1}$. Find its specific heat at constant volume c_v and the ratio $\gamma = c_p/c_v$, if the coefficient α of thermal expansion and the isothermal coefficient β of compression of mercury are

$$\alpha = \frac{1}{v}\left(\frac{\partial v}{\partial T}\right)_p = 1{\cdot}81 \times 10^{-4}\ \text{deg}^{-1},$$

$$\beta = -\frac{1}{v}\left(\frac{\partial v}{\partial p}\right)_T = 3{\cdot}9 \times 10^{-6}\ \text{atm}^{-1}.$$

The density of mercury is $\varrho = 13{\cdot}6\ \text{g cm}^{-3}$. What part of the difference $c_p - c_v$ goes into the production of external work A and what part into increasing the internal energy of the mercury?

318. It is shown in thermodynamics that the necessary conditions for stability of a physically homogeneous and isotropic substance are: (1) $(\partial p/\partial V)_T < 0$; (2) $c_c > 0$. Use this to show that $c_c > c_p$, and thus $c_p > 0$.

319. The external pressure acting on water is increased whilst at the same time heat is supplied or removed in such a way that the volume of water remains unchanged. Does the water get hotter or colder, if its initial temperature is (1) below 4°, (2) above 4°?

320. Experiment shows that a rubber strap elongates on cooling (if its tension remains constant). Using this fact, show that the strap heats up if it is stretched adiabatically.

321. The enthalpy of a physically homogeneous and isotropic material is the state function defined by $H = U + pV$. By considering an infinitesimal Carnot cycle and applying Carnot's theorem to it, show that the enthalpy H and specific heat c_p satisfy

$$\left(\frac{\partial H}{\partial p}\right)_T = V - T\left(\frac{\partial V}{\partial T}\right)_p; \quad \left(\frac{\partial c_p}{\partial p}\right)_T = -T\left(\frac{\partial^2 V}{\partial T^2}\right)_p.$$

322. Show that a gas which obeys the van der Waals equation with $a = 0$ always heats up in the Joule–Thomson experiment. Find the temperature rise on expansion.

323. Show that a gas which obeys the van der Waals equation with $b = 0$ always cools cown in the Joule–Thomson experiment. Find the temperature drop on expansion.

324. Find the connection between the critical temperature T_k and the inversion temperature T_c of the Joule–Thomson effect for a substance which is subject to the van der Waals equation of state.

325. The values of the van der Waals constants are given below for nitrogen, hydrogen and helium.

	a, atm cm^6 mol^{-2}	b, cm^3 mol^{-1}
Nitrogen	$1{\cdot}35 \times 10^6$	39·6
Hydrogen	0·191	21·8
Helium	0·033	23·4

Use these data to find the inversion points T_i of the differential Joule–Thomson effect for these gases. Find also the temperature change ΔT in the Joule–Thomson experiment if $T = 300°K$ and the pressure drop is $\Delta p = 0'1$ atm.

326. By considering the Carnot cycle for a system consisting of a liquid and its saturated vapour, and applying Carnot's theorem to it, express the derivate of the saturated vapour pressure with respect to temperature (dp/dT) in terms of the specific volumes v_v, v_l of the vapour and liquid and the heat of vaporization, q.

327. Find the pressure of saturated steam at 101°. Regard the steam as an ideal gas.

328. At 0° the water vapour pressure over ice is $p_1 = 4{\cdot}58$ mm Hg. The heat of melting of ice at 0° is $q_1 = 80$ calg^{-1}. The heat of vaporization of water at 0° is $q_2 = 596$ calg^{-1}. Find the water vapour pressure over ice at $t = -1°$.

329. Saturated water vapour at $T = 300$°K is subjected to adiabatic compression and adiabatic expansion. In which of these processes does the vapour become unsaturated, and in which supersaturated?

330. A piece of ice is placed in an adiabatic envelope at 0° and atmospheric pressure. What fraction $\Delta m/m$ of the ice melts, if the ice is subjected to an all-round compression to $p = 100$ atm, and what is the change in temperature of the system? The change in volume of the ice on compression can be neglected. The specific volumes of water and ice are: $v_w = 1$ cm^3 g^{-1}, $v_i = 1{\cdot}11$ cm^3 g^{-1}. The heat of melting of ice is $q = 80$ calg^{-1}. The specific heats of ice and water are connected by $c_i = 0{\cdot}6\, c_w$.

331. The Russian physicist M. P. Avenarius showed that the heat of vaporization is zero at the critical temperature. Verify this by using the Clausius–Clapeyron equation.

332. Find the specific volume of water vapour v_v at 100° and standard pressure, if the boiling point of water is known to be 99·1° at a pressure of 735·5 mm Hg. The talent heat of vaporization at 100° is $q = 539$ calg^{-1}.

333. Find the entropy change ΔS of a substance on heating, if its specific heat c is constant and the volumetric expansion coefficient is zero.

334. If a soap film of thickness $l = 0{\cdot}001$ mm is at the melting-point of ice, find the temperature drop of the film when it is stretched adiabatically in such a way its area is doubled. The density of the soap solution is $\varrho = 1$ gcm^{-3}, the specific heat $c = 1$ calg^{-1} deg^{-1}. The surface tension of the soap solution falls by 0·15 dyne cm^{-1} when the temperature rises 1°.

INDEX

Printed in Dunstable, United Kingdom

91898325R00342